GLOBAL STUDIES ENCYCLOPEDIC DICTIONARY

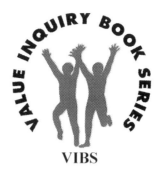

VIBS

Volume 276

Robert Ginsberg
Founding Editor

Leonidas Donskis
Executive Editor

Associate Editors

G. John M. Abbarno	Richard T. Hull
George Allan	Michael Krausz
Gerhold K. Becker	Olli Loukola
Raymond Angelo Belliotti	Mark Letteri
Kenneth A. Bryson	Vincent L. Luizzi
C. Stephen Byrum	Hugh P. McDonald
Robert A. Delfino	Adrianne McEvoy
Rem B. Edwards	J.D. Mininger
Malcolm D. Evans	Danielle Poe
Roland Faber	Peter A. Redpath
Andrew Fitz-Gibbon	Arleen L. F. Salles
Francesc Forn i Argimon	John R. Shook
Daniel B. Gallagher	Eddy Souffrant
William C. Gay	Tuija Takala
Dane R. Gordon	Emil Višňovský
J. Everet Green	Anne Waters
Heta Aleksandra Gylling	James R. Watson
Matti Häyry	John R. Welch
Brian G. Henning	Thomas Woods
Steven V. Hicks	

a volume in
Contemporary Russian Philosophy
CRP
Edited by William C. Gay

GLOBAL STUDIES ENCYCLOPEDIC DICTIONARY

Edited by
Alexander N. Chumakov
Ivan I. Mazour
William C. Gay

Assistant Editor USA, Michael T. Howard
Assistant Editors, Russia, Anastasia V. Mitrofanova
and Vladimir M. Smolkin
Editorial Assistant USA, Jonathan P. Branch

With a Foreword by
Mikhail Gorbachev

Amsterdam - New York, NY 2014

Cover photo: 'Watching Sunset with Bald Eagle, Llano, Texas,' used by permission of Kathryn M. Boepple, (c) Copyright Kathryn M. Boepple 2014"

Cover design: Studio Pollmann

The paper on which this book is printed meets the requirements of "ISO 9706:1994, Information and documentation - Paper for documents - Requirements for permanence".

ISBN: 978-90-420-3854-7
E-Book ISBN: 978-94-012-1097-3
© Editions Rodopi B.V., Amsterdam - New York, NY 2014
Printed in the Netherlands

Contemporary Russian Philosophy
(CRP)

William C. Gay
Editor

Mikhail Yu. Sergeev
Associate Editor

Other Titles in CRP

William C. Gay and Tatiana Alekseeva, eds. *Democracy and the Quest for Justice: Russian and American Perspectives.* 2004. VIBS 148

CONTENTS

FOREWORD		ix
	MIKHAIL GORBACHEV	
PREFACE		xi
INTRODUCTION		
	MICHAEL T. HOWARD	1

A	Acid Mine Drainage – Axial Time	3
B	Bahá'í Faith – Buddhism	30
C	Capitalism – Cyberculture	47
D	Debt Crisis – Dynamic Strategy Theory	122
E	East-West – Evolutionism, Global	142
F	*Factor Four* – Frolov, Ivan Timofeevich	207
G	Gandhism – Gumilev, Lev Nikolaevich	212
H	Hegemonism – Hypothesis	252
I	*Imperatives to Cooperation of North/South* – Issyk-Kul Forum	270
J	Japanese Post-War Economic Miracle – Just War Doctrine	304
K	Khozin, Grigorii Sergeevitsh – Knowledge Economy	312
L	Labor Market, World – *The Limits to Growth*	313
M	Macroshift – Myth	322
N	Nation(alism) – Nuclear Winter	342
O	Opinion, World Public – Ozone Layer Depletion	358
P	Pacifism – Political Science and Global Studies	363
Q	Al Qaeda – Quixote Center	379
R	Racism, Environmental – Russian Philosophical Society	381
S	Sartre, Jean-Paul – Sustainable Development, World Summit	408
T	Teaching – Transparency and Information Technology	438
U	*Umma* – Urbanization	470
V	Values, Universal – Virtualization	476
W	War – World Wars	479
X	Xenogamy, Cultural and Political – Xenophobia	497
Y	Youth	499
Z	Z-Generation – Zinovyev, Aleksandr Aleksandrovich	500

CONTRIBUTORS	501
LIST OF ENTRIES	511
INDEX	517

FOREWORD

In presenting this book to the world reader, I would like to mention two of its distinctive features. First, it deals with a fundamentally vital area, which is of particular importance in a globalizing world. We live in a global world; global challenges and threats are in the forefront of international politics. The study of global problems is gaining increasing prominence in the social sciences, terminology is rapidly evolving and changing, and a new language of global studies is emerging.

Another feature of this publication is its interdisciplinary nature and its international team of contributors, bringing together in fruitful cooperation scholars representing both natural and social sciences. In this volume, 300 prominent scholars from fifty countries worldwide have cooperated to produce a work of over 450 articles that comprise an orchestral harmony.

The editors call this collection of articles an encyclopedic dictionary. It is not what one usually means by a dictionary. As a rule, standard dictionaries incorporate what is already known. In this case, we have a book that considers the future state of the world. They tried to look into the future and were not afraid of presenting innovations in thought on even the most pressing issues of our day. What should be mentioned as well is not only the very broad coverage and deep scholarly analysis, but also—and this is of equal importance—the calm and moderate tone of the authors of this international dictionary, respectful of the diversity of existing viewpoints and positions.

So, it is possible to find common ground and a platform for a calm and constructive discussion even on the most difficult and sensitive issues. Here I should mention one of the lamentable patterns of our time: conflicts are contagious; acting like magnets, they attract the young whose positions and temperaments have not yet solidified. The world community has a pressing task ahead: we need to educate the new generations in a spirit of tolerance, with a broad and comprehensive vision of today's problems.

I have no doubt that this collective work will be of use for people working in a variety of fields and occupations: employees of international organizations, governmental officials, leaders of political parties and movements, religious leaders, politicians and political scientists, the military, and law enforcement officials. I also have no doubt that it will be a resource for young scholars. The broadness and depth of this book are designed not only to educate but also to inspire the minds of the young from all countries of the world to initiate creative inquiry.

I would like it very much, dear reader, if you would enjoy using this book. Although it may not be easy reading, the book is intelligent, rich in content and, I believe, necessary in our complex, turbulent, and fragile world.

M.S. Gorbachev
М. С. Горбачёв

PREFACE

The International Interdisciplinary *Global Studies Encyclopedic Dictionary* has been prepared by a large team of authors from various countries and is published in the Russian and English languages.

By the beginning of the twenty-first century, a new area of interdisciplinary knowledge emerged: global studies. Within its framework, various disciplines research globalization processes and problems engendered by them. A vast scholarly literature has already been published around the world on issues of globalization. This encyclopedic dictionary aims to reflect the very wide range of basic terms and notions related to global studies.

Authors of the contributions in the *Global Studies Encyclopedic Dictionary* primarily include scholars in social, human, natural, and technological sciences and other researchers on globalization problems within ecology, economics, politics, culture, law, and the other fields of knowledge. Beyond established scholars, we also include professionals in economics and business, political and public figures, and governmental officials.

The academic status of this volume is reflected in its Editorial Board and Editors, including, from Russia and over fifty other nations, prominent scholars and politicians, as well as leaders of many of the largest Russian and international organizations, research communities, and state bodies. The volume has been prepared by the Center for Scientific and Applied Programs "Dialog" with the support of the Russian Philosophical Society, the Russian Ecological Academy and with participation of the international non-governmental organization "Concerned Philosophers for Peace," the Paideia project of the Boston University (USA) and the International Public Foundation for Socio-Economic and Political Studies (the Gorbachev Foundation). It contains over 450 contributions from more than 300 authors in fifty countries.

Finally, we want to thank our Assistant Editors, Michael T. Howard in the United States and Anastasia Mitrofanova and Vladimir M. Smolkin in Russia, and our Editorial Assistant Jonathan P. Branch in the United States.

Alexander N. Chumakov, Ivan I. Mazour, and William C. Gay

INTRODUCTION

Since the 1960s, world scholars have sought to better understand what is now called "globalization," recognizing the universal significance and potential dangers associated with, for instance, the ever-quickening pace and accessibility of information exchange, the international mobility of labor markets, the exponential growth of knowledge and service-based industries and accompanying technologies, population fluctuations, and the transformation of traditional value systems worldwide. In their entry on the topic of "Global Studies," editors Ivan Mazour and Alexander Chumakov remark:

> The most important achievement of global studies was the creation of a language for interdisciplinary communication acceptable for different sciences, and the development and upgrading of fundamental key concepts and categories, such as "globalization," "global problem," "ecological crisis," "ecologization of production," "demographic explosion," "global dependency," "world community," "new thinking," and "new humanism."

An earlier volume by the same editors, *Global Studies Encyclopedia* (2003), published in Russian and English editions, won the prestigious Book of the Year award in the Encyclopedia/Dictionary category at the Moscow International Book Exhibition and Fair in 2003. The primary purpose of this new volume, *Global Studies Encyclopedic Dictionary*, is to include many more, yet briefer entries about the key terms and notions in Global Studies. Though readers may encounter vocabularies or approaches with which they not familiar, our editorial goal has involved exposing our readers to the rich variety of voices in the global conversation, rather than enforcing a narrow standard of methodological and philosophical homogeneity.

Michael T. Howard, Assistant Editor

Work Cited

Mazour, I. I., A. N. Chumakov, and William C. Gay, eds. 2003. *Global Studies Encyclopedia*. Moscow: Dialog: Raduga.

ℭℨ A ℬ℘

Acid Mine Drainage is an ecological and economic problem that threatens areas all over the world. Three types of wastewater are typically involved: acid quarry waters formed during construction and exploitation of open-pit mines; acid mine waters formed at driving and exploitation of underground mines; and under-spoil (near-spoil) ones formed in the spoil piles of rock enriched by sulfide mineralization and non-standard off-balance ores. Their common features are: medium acidity (a pH of 1.5–4.5), high mineralization caused by the sulfate compounds of heavy metals, complex salt composition, high reductive-oxidative potential (Eh), and the presence of sulfuric acid.

Until the late 1960s, the underground waters of any deposit area were investigated only for the purpose of defining the watering conditions of ore bodies, exploring possible water inflow to the mines, and taking measurements needed to secure the construction and safe exploitation of a mining complex. At that time, mine waters drained to the surface from mines and quarries were directed, regardless of their acidity and toxicity, to the local drainage network and to the lowered ground areas. In part, they seeped through to the mine workings.

The volume of acid mine drainage formed yearly in individual mining enterprises varies from 1.5 to 5.0 million m^3 and more. More than 30 million m^3 of acid mine drainage is formed yearly in non-ferrous metals mining of the Ural region alone, about 450 million m^3 having been accumulated in the ponds, sumps, and vitriolic lakes.

The ores left in the underground workings become the source of the pollution of water basins on a large scale. Huge volumes of rock mass stripped in the process of exploiting underground mines are continuously and intensively oxidized by atmospheric oxygen assisted by the iron- and sulfur-oxidizing bacteria, and, in some areas, by the endogenous fires that favor the formation of considerable quantities of neogenic sulfates (or neosulfates).

After the so-called wet conservation of a mining complex, when the deposit is closed and the level of underground water is restored, neo-sulfates are intensively dissolved, forming huge volumes of sulfate salt brines and mineralized acid water

In many countries worldwide, the rapid rise of underground water levels results in overflows of acid mine water after the mines are closed. Serious problems concerning both the cost of restoring polluted territories and the scale of pollution thereby arise. When exhausted quarries are flooded, their bowls are turned into lakes with vitriol-laden water, or "vitriolic lakes." If the quarry is buried, then a pool of acid underground water is formed in the porous filling soil.

The problems created by acid drainage include: (1) large water volume and its constant inflow; (2) free sulfuric acid in the water; (3) complex chemical composition and high mineralization; (4) slowly dissolved heavy and poisonous metals exceeding the MPC by 10 to 1,000 times; (5) the need for developing special measures to prevent water from spreading over the surrounding area and to reduce the content of free sulfuric acid and heavy metals.

Global Studies Encyclopedic Dictionary

Sediments resulting from the liming of acid water are the fine tailings accumulated in the special sump ponds that occupy tens of hectares of land containing considerable quantities of copper, zinc, and other valuable, but ecologically harmful, components. In some countries, acid mine drainage containing more than 1 g/l of dissolved copper are processed by liquid electrolytic extraction to obtain cathode copper with an extraction level of about 90 percent, while the other metals are lost. Acid waters that contain less than 1 g/l of copper are neutralized by lime and dumped on the area, or kept and accumulated in vaporizing ponds. At the enterprises where the heap leaching of low-grade ores is used, these waters are utilized as technological solutes, having been preliminarily strengthened with sulfuric acid.

Research shows that the problem of purifying acid mine water has been addressed by specialists examining the extraction of only one metal, rather than all of them, so that so-called piece technology is used. All investigations focus upon a single method (such as extraction, sorption, electro-coagulation, cementation, dialysis, membrane method) for the extraction of the targeted component, leaving the overall problem of acid drainage purification unsolved.

P. Plaul, Z. M. Shulenina

Age of Great Unity: In the *The Book of the Great Unity* (1963), Kang Youwei (1858–1927) builds on the Confucian tradition and envisages an age of great peace and unity as a goal that can be attained through political reform. Constant striving to bring about a better world, which is the central concern of moral philosophy in China, finds its most complete expression in Kang Youwei's invocation of the ideal of the harmony of human life in all spheres. Putting forth universal peace and great unity as the supreme goals of all actions, Kang Youwei worked vigorously to bring about political and social reform. In the age of great peace he states, "The way of the great unity is the high point of fairness, justice, love, and good government." He urges individuals and states to strive to eliminate suffering and draws upon the feeling of sympathy all people share.

My way of saving the people from suffering consists in abolishing the nine spheres of distinctions. (1) Do away with the distinction between states unify the whole world. (2) Do away with class distinction to bring about equality of all people. (3) Do away with racial distinctions so there will be one universal race. (4) Do away with the distinction between physical forms to guarantee the independence of both sexes. (5) Do away with the distinction between families so men may become citizens of Heaven. (6) Do away with the distinction between occupations so that productions may belong to the public. (7) Do away with the spheres of chaos so that universal peace may become the order of the day. (8) Do away with the distinction between species so we may love all sentient beings. (9) Do away with the sphere of suffering so happiness may reach its height."

Kang's view on suffering resembles the Buddhist point of view, but his solutions from distinctly Confucian roots. Kang regarded Confucius as the first and foremost political reformer and sought to revitalize the Confucian ethics. Kang confirms the view of human nature as being inherently good and the basis of morality as the feeling of sympathy for the suffering of others first espoused by Mencius and sustained by the neo-Confucian thinkers.

"In the Age of Great Unity, the world government is engaged in building, reclamation of deserts, and navigation as the primary task. In the Age of Great Peace, all agriculture, industry, and commerce originate with the world government." With the elimination of political hierarchy equality prevails. In the state of Great Unity, there will be no competition among nations or individuals.

"In the age of Great Peace, all people are equal . . . and do not consider position or rank as honor." With no rank, there can be no reliance on power to oppress, or resorting to intrigue or flattery to seek advancement. Individual moral life and communal life are seen as a continuum. By developing ourselves we create an environment in which others may also flourish. Kang recaptures the insight expressed in the Great Learning that the moral development of the individual leads to the good of the world.

"All people work for each other. In the Age of Great Peace, since men's nature is already good and his intelligence is superior, they only rejoice in matters of wisdom and love. New institutions appear every day. Public benefits increase every day. The humane mind gets stronger every day. Knowledge becomes clearer every day. People in the whole world together reach the realm of love, longevity, perfect happiness, and infinite goodness and wisdom."

In Kang Youwei's conception, wisdom consists in possessing the knowledge to accomplish the goal of love." Wisdom is to initiate things, accomplish undertakings, promote utility and benefits, and advance people, while love is to confer benefits extensively on all the people and bring salvation to them, and to benefit things. There is no honor outside of wisdom and love."

Kang Youwei extends sympathy from personal relations to the whole of political domain, combining it with the fundamental concept of Confucian ethics, *ren* (love). *Ren* is the most important concept in Confucius' moral philosophy. Love is the Archimedian lever by which the moral life of human beings is lifted. For Kang Youwei the development of *ren* in all spheres of human action is the aim of moral education and political life. As the range of concern for the well-being of all forms of life expands, it leads directly to extending love to the entire universe.

References: Confucius. *Confucian Analects, The Great Learning, and The Doctrine of the Mean*. Translated by J. Legge. New York: Dover, 1971; Kang Youwei. *The Book of Great Unity*. In *A Source Book in Chinese Philosophy*. Edited by W.-T. Chan. Princeton, NJ: Princeton University Press, 1963.

H. Höchsmann

Aggression: (1) A forceful action or procedure (e.g., unprovoked attack, especially when intended to dominate or master); (2) the practice of making attacks or encroachments, especially: unprovoked violation by one country of the territorial integrity of another; (3) hostile, injurious, or destructive behavior or outlook, especially when caused by frustration.

Thus, the two essential features of aggression are first, the infliction, or tendency to inflict, damage or loss; second, the infliction is unprovoked, making those upon whom it is visited into "innocent victims" as opposed to have

6 *Global Studies Encyclopedic Dictionary*

somehow deserved the infliction. Accordingly, we must consider more deeply those two notions: damage and innocence.

The dictionary definition talks of "forceful" action. But if we take this literally, the definition would be too narrow. For example, the action of gently slipping poison into someone's tea need not be "forceful" but it would do fatal damage to the consumer of that tea. In recent times, there has been considerable fear of "biological" warfare, where the attacker would release dangerous organisms into the atmosphere—not necessarily an especially "forceful" procedure, though this action would surely count as aggression.

Granting, then, that the infliction of damage might not be literally forceful, sudden, or violent, does damage, in and of itself, count as aggression? When do we damage someone? We can easily begin with bodily wounds, disablements, diseases, and death. These injuries can range from very minor, as in scratches, to very major, as with torture-level pain, dismemberment, and death. Would we say that someone who, quite deliberately, and without provocation, made a tiny scratch on another person's body was being "aggressive"? Perhaps not: just odd, in a somewhat unpleasant way. Damage is always a matter of degree, and can be very minor.

Aggression is initiated damage upon persons who have in some sense not done anything to deserve or merit such treatment: "unprovoked," as our dictionary entry has it. This requires us to understand what constitutes the kind of provocation that would justify some level of adverse intervention into another person's life.

A further problem arises when people act as members of groups, or otherwise identify with them—the nation-state being a prime example, but tribal action provides a further major case as does, sometimes, religiously-motivated action. Group occupancy of a particular region can be unclear and its history spotty. Suppose some group claims to be the rightful occupant of land already occupied by another group? Evidence for and against the rival claims may well be lost in the mists and sands of time.

How do individual rights cohere with group rights? Or do they? A historically realistic example concerns the occupancy of America by Europeans. The arriving immigrants soon became aware that there were already people there, but those people, the natives of America, were often hunter-gatherer tribes rather than agricultural ones. Consequently, they occupied and used given areas only briefly, then went elsewhere, then returned, thus "occupying" a very large area but very thinly. Do newly arrived agriculturalists that occupy a small area during the absence of the native tribes aggress on native's territory? Or do the returning tribes now aggress on the agriculturalists, who may have built houses and planted fields in what they thought was vacant land? Clearly negotiation is required for resolving such matters. If war ensues, as it sometimes does and did, it is not entirely obvious who is in the right.

A combination of two methods, of technological exchange and negotiation, is likely to provide the workable peaceful solution to problems of this kind, but it is easy to see that this requires some delicacy and that delicacy was not as often observed in the American case as might have been hoped. The same, we must admit, is generally true of historical conflicts, as for example in the Middle East in the latter half of the twentieth century and continuing.

Acid Mine Drainage – Axial Time

We should take note here of the popular view that people are entitled to such things as minimum incomes, or even to equal incomes or equal initial properties, or perhaps to an equal share of the resources provided by nature rather than human activity. Resolving such questions takes us deep into social philosophy, but if claims of that kind were right, we would certainly need to modify the general thesis that aggression is wrongful. Thomas Aquinas already advanced the claim that people in danger of starvation, say, may take what they need from those who have more, even if the latter achieved their wealth by wholly peaceable means, and in modern times Marxists and others have insisted on that and much more.

Why do people aggress? This important question has received much attention. We may divide the hypotheses into two basic types, these being compatible rather than mutually exclusive:

(1) Invading for gain: Aggression is used as means to further ends. Examples have already been mentioned above: one person or group of people attacks another as a means of survival, though at the cost of the different person or group, or, to enhance the possession of amenities. "Invading for gain," as Thomas Hobbes calls it, is a frequent enough motive for individual theft and for international war. However, which kinds of gains motivate aggression? It is when we consider various answers to this that we see the need for another type of hypothesis: that aggression is a basic instinct, aimed at sheer domination of the other, rather than an independently definable benefit.

(2) An aggressive instinct: Even if humans do have an aggressive instinct, the situation is not necessarily hopeless. There would still be two possible ways to improve matters. First, aggression is not all that simple a phenomenon, psychologically speaking. The will to excel or dominate need not lead to war. For example, it could lead instead to athletic or other competitions. Instead of a World War, we could have a World Cup, a competition with the huge advantage that the losers walk off the field after shaking hands with the victors instead of being buried there, and with the further advantage that the losers can go back and work on their game, hopeful of victory in future. Moreover, sport is much safer for spectators than wars, in which it is all too easy to become an unintended or unavoidable victim even when one is not a direct participant.

Second, as previous analyses have already suggested, aggression cannot possibly be an organism's only instinct. We all need to eat, for example. But aggression, at least among fellow humans, is an extremely inefficient method of keeping oneself alive: cooperative effort at improving our situations is far superior from that point of view. Cooperative effort is better still as a means of enhancing wealth. Because this is so, it is possible that even if human beings have aggressive instincts, they can appreciate, by rational thought, the merits of repressing those instincts, or re-channeling them.

Aggression might be prominent in a personal psychological profile even if it is not a "basic instinct." Students of developmental psychology, and ordinary people in their experience with bringing up children, are well aware that aggression can be developed by some methods of upbringing, and discouraged by alternative methods. For example, the once popular maxim "spare the rod and spoil the child" seems to be nearly the reverse of the truth: teaching children, by example, that violence should be avoided, applied from the earliest age, is much more likely to result in adults who shun violence and use peaceful methods of resolving disputes. A cultural climate can also reinforce aggression. For example, cultivation of the view that one's tribe or nation has been ill used by some other group can be very conducive to aggressive activity. It can be very difficult to reverse the effects of such a climate, as we are becoming increasingly aware in the modern world.

So, while aggression is definable, knowing who is and who is not an aggressor may not be so easy. Ample basis exists for condemning aggression, whether by individuals or groups, but we also have ample justification for the need for further investigation on the containment of conflict, the causes of aggression, and related matters.

References: Freud, S. *Civilization and Its Discontents*. Translated and edited by J. Strachey. New York: W.W. Norton, 1962. Lorenz, K. *On Aggression*. London: Methuen, 1972. Ripstein, A. "Hobbes on World Government and the World Cup." In *Hobbes*. Edited by T. Airaksinen and M. A. Bertman. Brookfield, VT, 1989. Cf. International Law Commission, "Question of Defining Aggression." Accessed 12 October 2013. http://legal.un.org/ilc/texts/7_5.htm.

J. Narveson

Aggression as an International Law Concept (Latin *aggressio*, "attack") refers to the use of armed force by a state against the sovereignty, territorial integrity, or political independence of another state or in any other manner inconsistent with the Charter of the United Nations (UN General Assembly Resolution 3314). Necessary features include the first use or initiative (the first use of armed force by a state), intent, and deliberation. A consensus definition of aggression was adopted by the 29th Session of the UN General Assembly in 1974. Definitions exist for both direct and indirect aggression. Complicity in aggression constitutes a separate crime.

"Direct aggression" is aggression committed by a state's regular armed forces. Direct aggression, regardless of a declaration of war, includes the following: (1) invasion or attack by the armed forces of a state on the territory of another state; (2) any military occupation of a state's territory, however temporary, resulting from such invasion or attack; (3) any annexation by the use of force of the territory of another state or part thereof; (4) bombardment by the armed forces of a state against the territory of another state, or the use of any weapons by a state against the territory of another state; (5) blockade of the ports or coasts of a state by the armed forces of another state; (6) attack by the armed forces of a state on the land, sea or air forces, or marine and air fleets of another state; (7) use of armed forces of one state stationed within the territory of another state with the agreement of the receiving state, in contravention of the conditions provided for in the agreement or any extension of their presence in such territory beyond the termination of the agreement.

Armed bands, groups, irregulars, or mercenaries who are formally not a part of a state's regular armed forces, or who conceal such affiliations while acting in behalf or under mandate of a state, commit "indirect aggression."

"Complicity" in aggression refers to the action of a state allowing its territory, which it has placed at the disposal of another state, to be used by that other state for perpetrating an act of aggression against a third state.

The notion of aggression is not limited to the actions listed above. The UN Security Council has the right to determine whether an act of aggression has been committed, and has the right to take into consideration all the circumstances of each particular case of aggression. No consideration, whether political, economic, military, or otherwise may serve as a justification for aggression. Any aggression, whether direct, indirect, or complicitous, is qualified as a violation of international law and gives rise to international responsibility. No territorial acquisition or special advantage resulting from aggression is or shall be recognized as lawful.

International responses to the crime of aggression include coercive actions directed toward suppression of aggression, the restoration of peace, various measures aiming at the removal of the consequences of an act of aggression, and the prevention of renewed aggression. International law provides for political penalties for aggression, subjecting the aggressor state to temporary restrictions of its sovereignty and imposing financial responsibility in the form of repayment of damages resulting from the act of aggression. People guilty of planning, preparing, unleashing, or committing aggression are criminally responsible. States committing aggression with the use of weapons of mass destruction bear special responsibility. Upon a UN Security Council decision, non-military sanctions or, if necessary, coercive actions may be applied against an aggressor by the United Nations member states. A state against which an aggression has been committed has the right of individual or collective self-defense. Military actions of a state responding to an act of aggression in the course of self-defense are considered non-aggressive and legitimate. Armed struggle for freedom and independence and the participation of irregular armed groups and volunteers in assisting national liberation movements are not considered to be aggression.

S. V. Moshkin

Agricultural Adaptive Strategies refer to strategies of intensifying agricultural production with the use of high technologies, the differentiated utilization of natural, biological, technogenic, labor, and other resources and the creation of crop types capable of exploiting natural and anthropogenic environmental resources during photosynthesis with the greatest efficacy while resisting abiotic and biotic stressors.

Based on the ability of green plants to use the almost inexhaustible and ecologically safe resources of the sun and atmosphere in conditions of mostly chemically induced productivity, agriculture has turned into one of the most conspicuous threats to nature in terms of resource and energy depletion. Still, it fails to satisfy the increasing demand of the Earth's population for food and production of raw materials. A sharp reduction in the number of crops, the wide utilization of genetically homogeneous types and hybrids, and the use of pesticides have resulted in the appearance of more aggressive and virulent strains of pathogens and an increasing harmfulness of many kinds of insects and weeds. Rates and scales of water and wind erosion of soil, salinity, and

10   *Global Studies Encyclopedic Dictionary*

swamping in conditions of technogenic-intensive agriculture have reached a catastrophic level in many countries of the world. If chemical-technogenic intensification of plant cultivation continues, the severity of the "pesticides boomerang" will constantly increase, giving plant-host parasites the advantage in the "evolution dance" of genotypic variability.

The reduction in rates of productivity and croppage of major agricultural crops (e.g., rice, wheat, and corn), the reduction of arable land area up to 0.16 hectares per capita, and the constant rise in the cost of exhaustible resources and energy form a real threat of a global food crisis in the twenty-first century, most likely reaching the magnitude and severity of crisis situations in other spheres.

The transition to an adaptive strategy of nature management assumes the human ability to form an environment compatible with biosphere. Adaptive strategy also includes designing agricultural ecological systems and agricultural landscapes that unite high efficiency, ecological stability, and the ability to self-regulate and improve our interactions with the environment.

The adaptive system of agricultural management is aimed at the preservation of habitat and the improvement of the quality of human life, undercutting damaging agricultural practices to establish a new basis for the use of ecologically safe, inexhaustible, and reproducible resources. The adaptive nature of the new strategy requires a multivariant, dynamic, and high-technological approach and, consequently, an ability to integrate and use technological achievements of applied and fundamental science.

Human civilization cannot function according to economic laws alone, abandoning the other components of humanity. Since agriculture is closely connected to specific features of life and religion in every nation, it influences the quality of not only the material, but the spiritual human environment. Food supply and preservation of the environment become closely interconnected, and the future of human civilization depends to a lesser degree on economic than on ecological factors. Due to changes in paradigms of nature management, which now favor an orientation toward the utilization of inexhaustible and reproducible resources, the strategy of adaptive intensification of agriculture appears economically justified, socially acceptable, and viable in the long term.

The complete self-renewal of resources in highly productive agricultural ecological systems cannot always be ensured. However, a reduction of the costs of exhaustible resources per each additional unit of general and used biomass provides an initial measure of success. Another distinctive feature of the new strategy consists in focusing on the expansion of both the productive and environmental functions of agricultural ecological systems and agricultural landscapes. Since the amount and quality of the crop largely depend on environmental factors, including weather, adaptive strategy pays special attention to: comprehensive analysis and risk reduction at the stages of the agricultural-ecological zoning of the territory; the definition of the type structure of crops and formation of crop rotations; the selection of adaptive types and technologies; and the design of agricultural ecological systems and agricultural landscapes.

One of the most important features of adaptive intensification strategy is a possible increase in forecasting and pre-adaptive abilities of agriculture. This approach must thereby take account of possible global and local climate fluctuations, weather conditions, demographic situations, market conditions, food demand, and other factors to anticipate and reduce risk.

A. A. Zhuchenko

Agricultural Afforestation is a system of forestry measures aimed at enhancing soil, hydrological, and climatic conditions of individual territories to increase the efficiency of agricultural land. Agricultural afforestation is one of the major natural types of melioration that contributes to sustainable agricultural management and food security.

At the end of the nineteenth and the beginning of the twentieth century, countries with a transitional type of economy could not avoid losses when implementing agricultural reforms. Agricultural production followed the extensive model whereby sown areas were expanded. Expansion of cultivated areas was achieved mainly through stubbing, plowing up steppe virgin lands and swarded hillside areas. The shrinkage of forested areas, wood burning, the mineralization of manure, abscission, and breakdown of soil humus has reduced the energy resources of ecosystems by 20–25 percent.

The accelerating pace of agricultural land use results in depleting soil fertility, preventing any possibility of self-recovery in the natural cycles of the biosphere. The biggest losses of natural organic substance in the soils occur as they are getting cultivated, especially by moldboard plowing. Some scientists believe that the cultivation of fertile loam over a seventy-year period results in 50 percent loss of organic matter in the surface layer. Though meadow soils accumulate organic matter, they do so by a very slow process. Also, an excessive increase of sawn acreage at the expense of forests produces a hot, dry wind in steppes and forest-steppes, which over-dries large fields, impeding their resistance to erosion. Natural disasters recur more frequently. In the twentieth century, droughts occurred on the average of every three to four years, compared with only once per ten years during the nineteenth century. Before that, droughts plagued most regions even more rarely, once every twenty years.

In the majority of countries with a transitional type of economy, development of land for agricultural purposes is excessive. For example, in the Ukraine agricultural acreage constitutes 72.2 percent of the territory (41.8 million hectares), the plowed areas occupy 57.1 percent (32.6 million hectares of the arable land); the plowed land is 79.8 percent of the acreage, while meadows and pastures occupy only 12.9 percent. By comparison, the plowed territory of the United Kingdom, France, and the Federal Republic of Germany constitutes between 28.6 and 31.8 percent, the sawn acreage is 37.3 to 55.6 percent, and meadows occupy 18 to 25 percent. After the return to raw-crop farming in the 1960s, eroded arable land territory increased by 2 million hectares, or by 25.6 percent. Currently, more than 14.9 million hectares of farming land, or 35.2 percent, are affected by water and wind erosion. The annual increment of eroded land in the Ukraine is between eighty and ninety thousand hectares. The total loss of humus as a result of erosion and other destructive processes approaches forty-two million tons.

Losses because of unbalanced application of organic matter constitute eighteen million tons, while erosion per se accounts for twenty-four million tons—more than 56 percent of the total loss. Such a tendency in the land ten-

Global Studies Encyclopedic Dictionary

ure resulted in a 9 percent reduction of the humus content in arable land from 3.5 (in 1961) to 3.2 percent (in 1981). During the last century this reduction has reached 25 to 30 percent in some regions. Meanwhile, in 1996, the forest wind belt occupied 440 thousand hectares (with a forested area of 1.5 percent) as compared to 491 thousand hectares in 1961. Forests protect only a third of farming lands. It takes nature from 1.5 to 7 thousand years to form a 10 to 29 cm sized fertile ball; hence the losses suffered by soils as a result of destructive processes are irreplaceable.

Land cultivation in the Ukraine exceeds ecologically accepted standards and is sensitive to abnormal climatic fluctuations. When forest protection is reduced, large fields are prone to withering sun radiation and wind aggravation, contributing with other agrotechnical factors to soil erosion.

V. D. Baitala

Agricultural Decollectivization can be defined as a transfer of property rights from state and collective farms (Russian: *sovkhozi* and *kolkhozi*) to individual or corporate ownership. With the breakup of communist regimes, all Central and Eastern European countries have embarked on some form of decollectivization, and many developing countries have attempted to switch from common to privately defined land resources. Decollectivization may be partial (the transformation of collective forms into successor enterprises that preserve existing farm boundaries) or radical (the dissolution of existing management structures and their replacement by individual farms). Radical decollectivization may be accomplished by restitution (the return of land or assets to pre-collectivized owners and their heirs), distribution (to members, workers or the wider rural population), or auctions. The most common form of decollectivization in Central Europe has been restitution. The decollectivization debate has thus included arguments about the rights to private and common property and how these should be governed.

Proponents of partial decollectivization have stressed the need for preserving existing farm boundaries to maintain economies of scale and the role of collective resources in maintaining local cohesion. Opponents have typically seen partial decollectivization as "just changing the sign on the door," in a move that merely preserves existing rural elites. These critics maintain that more radical decollectivization, based on claims for historical justice and prospects for improving agricultural efficiency, is required. The historical justice argument for restitution is based on the premise that land and capital were unfairly expropriated and that the main beneficiaries of reform should be those who lost the right to land under collectivization. Restitution presents a set of problems, as witnessed recently in the Central and Eastern European countries. The pre-Communist land structure in many countries was extremely fragmented, especially after the land reforms of the early twentieth century, and concerns that fragmented ownership creates a new class of peasant farmers who cannot adequately support themselves have been raised. Heirs of restituted land may not be engaged in agriculture, have sector specific skills, or even reside in the country. The rights of heirs versus existing collective farm members and workers have been fiercely contested. Economically, ex-

Acid Mine Drainage – Axial Time

cessive fragmentation has led to a growth in subsistence production (non-marketed output) with low returns and efficiency.

The economic argument for decollectivization rests on the assumption that private individuals will more effectively manage land than collective bodies. This debate has a long history and centers on claims that decollectivization may promote higher production to the extent that the new farms created will farm more efficiently. Collectivized farms will be inherently less productive as the effective supervision of workers may be costly and practically difficult because of the spatially dispersed nature of agricultural production. In contrast, family farming "internalizes" incentives and the predominance of family farms in North America and Western Europe is often used to justify the claim that they are a superior form. Efficiency studies on farming in Central Europe, however, appear to show that, in general, collectivized farms can match the efficiency of individual farms, albeit through a large reduction in the labor employed and greater managerial autonomy. Advances in productivity come at the expense of collective ideals, sometimes referred to as the "degenerate or die" conundrum.

Regardless of arguments about what form of farm structures should emerge, agricultural employment is declining throughout the world. This reduction in labor has given rise to social and political tensions in many countries. Whatever form of farm structure emerges, attention must be given to the stimulation of the rural non-farm economy.

M. Gorton

AIDS (Acquired Immune Deficiency Syndrome) is a pathological state in which host defenses are weakened as a result of immune system lesion caused by the human immunodeficiency virus (HIV). AIDS was first identified in the late 1970s and cases have been registered in all countries of the world, producing demographically significant consequences in epidemic proportions.

During the 1980s, HIV/AIDS appeared to be more dangerous primarily to, perhaps even limited to, some risk groups (gay males who engaged in anal sexual intercourse). This turned out to be false. Anyone of any age or sexual orientation can contract AIDS. Twenty years ago, optimists believed that, considering the modern level of medical science, a vaccine against AIDS would be found quickly. These hopes were not justified. By the 1990s, the disease (at least in some countries) had transgressed the boundaries of all risk groups. AIDS has become epidemic in more than twenty African countries. In some nations, a quarter of the sexually active population is infected. According to some estimates, by 1999, about 33.6 million people worldwide were HIV positive (most of them have already died or will die in a few years, as 16.3 million people have died since the beginning of the epidemic). As recently as the 1990s, experts believed that HIV/AIDS in the developing countries was responsible for only 2 percent of deaths. Since that time, AIDS has become the primary cause of death in Africa and the fourth one worldwide. Nevertheless, AIDS deaths have not been significant when compared to the global balance of births and deaths. In 1999, 2.6 million people died from AIDS, 51.4 million people died from other causes, and 131 million people were born.

14 *Global Studies Encyclopedic Dictionary*

Some estimates project that in the countries where HIV/AIDS is widespread (such as South Africa, Botswana and Zimbabwe) it could reduce population growth to zero in the near future, after which population in these countries will start to decrease. Destabilization of the AIDS-affected societies may have spillover effects far beyond the boundaries of these countries. The economic consequences of an epidemic would exacerbate social conflicts and increase political instability. Consequently, new streams of refugees may emerge, carrying this lethal virus to distant regions.

World Health Organization data currently offer the following facts about HIV/AIDS incidence worldwide:

(1) HIV continues to be a major global public health issue, having claimed more than 25 million lives over the past three decades.

(2) There were approximately 35.3 [32.2–38.8] million people living with HIV in 2012.

(3) Sub-Saharan Africa is the most affected region, with nearly 1 in every 20 adults living with HIV. Sixty nine per cent of all people living with HIV are living in this region.

(4) HIV infection is usually diagnosed through blood tests detecting the presence or absence of HIV antibodies.

(5) There is no cure for HIV infection. However, effective treatment with antiretroviral drugs can control the virus so that people with HIV can enjoy healthy and productive lives.

(6) In 2012, more than 9.7 million people living with HIV were receiving antiretroviral therapy (ART) in low- and middle-income countries.

Reference: World Health Organization. Accessed 07 November 2013. http://www. http://www.who.int/mediacentre/factsheets/fs360/en/

A. D. Korolev, L. V. Sipina

Alarmism (Italian *all'arme*, "to arms!"; French *alarme*, "alarm," "feeling of threat or danger") The concept of "alarmism" was introduced into science and philosophy in the late 1960s and early 1970s, as a response to pessimistic theories of social development, futurology, and prognostication. Alarmism was widespread in the public consciousness of the advanced Western countries and was deliberately exacerbated by a section of the scientific establishment to encourage research on preventing potential crises.

The catastrophes of the twentieth century—two world wars, genocide, the disintegration of empires, and social revolutions—paved the way for alarmism in its current form; which alternates between periods of alarm and periods of calmness (such as war and peace, crisis, and recession). At the extreme, alarmism is focused exclusively on the continuing negative consequences of human activity, which are held to nullify all positive accomplishments. The most alarming anticipations focused on ecology. People became aware that technological civilization was developing at the expense of the environment.

Several other negative environmental consequences of the scientific and technological revolution (including the depletion of resources and irreversible physical and chemical changes in the land, water, and atmosphere) strongly alarmed the scientific community. These could result in planetary climate transformation: searing of atmosphere and soil, general water shortage, melting of polar and high mountain glaciers causing a rise and cooling of oceans, changes of water conditions in all river basins, the melting of permafrost, and, finally, the flooding of littoral lowlands with the loss of coastland up to 100 km inland. The latter would lead to the collapse of the largest cities and migration of people far inland. Solving these problems of planetary scale would require a much higher degree of organization on the part of world community than was expected at the height of alarmist moods or even currently. Alarmism was further exacerbated by the general systemic crisis that was accompanied by the aggravation of the global problems, such as the population explosion in the Third World countries, pandemic hunger, and infectious diseases.

I. I. Kravchenko

Alienation is the state or experience of being isolated from a group or an activity to which one should belong or in which one should be involved. A sociological concept developed by several classical and contemporary theorists, alienation is a condition characterized by low integration with the social group, poor shared values, a high degree of isolation from other individuals or between an individual and the group(s) to which the individual belongs, such as between an individual and its family, or in the work environment.

Melvin Seeman identifies six types of alienation. (1) *Powerlessness* refers to the state of helplessness related to the social and political forces, as when we realize that we have no power over the course of our own actions. (2) *Meaninglessness*: when someone does not understand their own situation and thereby cannot envisage the results of their behavior, they are unable to control what they are able to foresee. (3) Anomie: a state of *normlessness*, wherein there is a general breakdown of order and law in the society. This state arises in the individual in the form of anxiety and worry about personal security. (4) *Cultural-estrangement*: manifested when individuals such as intellectuals regard social goals as being unworthy of being pursued because they do not adhere to societal values. (5) *Social-isolation*: most pure and concrete manifestation of alienation, the individual is actually alone among the others and cannot find the appropriate way to live with and among others. (6) *Estrangement*, as used by Karl Marx to convey his notion of alienation.

Marx identified three major forces that foster alienation in capitalist societies. (1) *The presence of private property*, which estranges the worker from its product because this system makes the object produced by workers belong to someone else. (2) In the system of market economy the product of human labor follows its own laws or the laws of the market and capital and therefore a worker's product does not only stand "opposed" to them, rather, it stands "independent of the producer." *The transformation of human labor into a commodity* arises when workers and their labor are regarded as another form of capital; to make ends meet, workers are forced to sell their labor in return for wages, and the necessary capital which allows them to exist facilitates the further production of capital. The capital works through workers as through any other kind of commodity, increasing their dehumanization. (3) *The system of division of labor* rendered by the factory mode of production entails the total dehumanization of workers, who are turned into an accessory of the

16 *Global Studies Encyclopedic Dictionary*

machine. Personal contribution to the end product becomes insignificant and could not ever offer workers the satisfaction of "creation."

Max Weber took Marx's task further, identifying the forces of alienation in modern Western society not only from and economic standpoint, but with an account of social and political forces. Some principles of rationalization that underlie the complex interplay between capitalism, democracy, and bureaucracy can be described. The bureaucracy is the best agent of rationalization because of its impersonal rules, precision, calculability and accountability; it represents the most formidable threat of alienation in the modern world. What Marx identified in the case of the factory worker, Weber recognized on the level of the entire middle class, which is affected by the bureaucracy: administrators, teachers, scientists, and so on. This group cannot appropriate what they produce either, because of the complete separation of the means of private life from the means of administration. A state of powerlessness thus arises because the bureaucrat has no power over the whole organization and those who are subject to the bureaucratic machinery are equally powerless against it. Once established, the system works autonomous and independently. If Weber's diagnosis of capitalism is correct, this account would be only more so for socialism.

Unlike his predecessors, Karl Mannheim further develops other aspects of G. W. F. Hegel's theory of alienation, examining the phenomenon in the cultural sphere. The results of humankind's creative work become, in time, objects of culture progressively separate from the power of their creator, eventually coming be governed by their own laws. The culture gains autonomy and develops into a completely alienated realm for humans, reaching beyond and "against the individual." From then on culture dictates the actions of individuals, destroying inner creativity and forcing people to follow its external rules. This situation entails a feeling of helplessness and meaninglessness for modern humans, who have no understanding of these social forces and thus no control over their lives. Thus, people live in fear and anxiety.

These three supra-structures, the economic, the administrative, and the cultural systems irrefutably carry the seeds of possible alienation mechanisms and thrive on the dehumanization of humans. Recently, the reality of the economic globalization seems to imply the dismissal of politics and thus a low capacity to act on things and to act on society's choices, so that one is forced to obey to the impersonal and mechanical rules of the marketplace. The expansion of Western models of social organization/administration adds a new threat of alienation in the modern world in the form of ever-greater bureaucratization. Since all societies have cultural systems, it follows that in every society there will be some degree of alienation.

B. Popoveniuc

Alterglobalism is a worldwide movement aimed at changing the main parameters of globalism. Initially a part of the anti-globalization movement, alterglobalism now stands apart from common protest against globalization, or anti-globalism. First articulated in 2002, at social forums in Porto Alegre and Florence, alterglobalism does not simply criticize globalization's negative consequences for humanity, but also proposes another globalization strategy, which would be humanistic in content, might be carried out for multiple participants, and could result in the creation of global civil society. Alterglobalism joins forces that do not accept globalization's ideology, forms and consequences, but recognize its objective characteristics and express hope for globalization with a human face.

A. G. Kosichenko

Alternative Movements: In sociology, alternative movement refers to movements that seek limited social change (e.g., Mothers against Drunk Driving); these may be in the social, political, cultural, or religious spheres. Many alternative movements emerged in the West during the mid-1970s. These movements include traditional democratic movements with a prefix "eco" (ecosocialists, ecoanarchists, and ecofeminists), New Left political initiatives (such as communitarian movement), and various cultural initiatives. Having begun in the United States and Western Europe, alternative movements have gradually spread worldwide.

Participants and ideologists of alternative movements are united not by a common name but by the ideas and methods of political struggle they share. These movements are called "alternative" because they offer values, a world outlook, political ideals, and a political culture that are alternative to the dominant ones.

A. V. Mitrofanova

Anthropocentrism (Greek *anthropos*, "human"; and Latin *centrum*, "focus," "center") Anthropocentrism is a worldview and scientific approach recognizing human beings as the center and ultimate goal of the universe. All phenomena and relationships are seen from the viewpoint of their relevance to humans and their interests (above all in relation to other species of animals and plants). Anthropocentrism may contribute to an anti-ecological orientation in human activities as it considers other manifestations of existence as inferior and permits in relation to them any kind of destructive actions.

A. D. Korolev

Anthropogenic Crises are historical crises caused by the excessive load of human activity on the environment. Most of these crises in history were of local character, but some acquired regional or continental scale.

The precondition for such a crisis was usually the internal imbalance between the technological potential of the society and the quality of the regulation mechanisms produced by culture. Internal imbalance provoked a burst of ecological or geopolitical aggression, accompanied by some mental processes and states. As a result, the society, which was unable to foresee the deferred consequences of its activity, undermined the natural and organizational bases of its existence.

Some historical cases covered an extensive region, but the society managed to find a solution, requiring the reorganization of spiritual culture, economic technologies, and all vital activities. In many respects revolutionary reorganizations grew irreversible and acquired global value.

18 *Global Studies Encyclopedic Dictionary*

Approximately six crucial revolutions became turning points of human history and pre-history. Each was preceded by an endo-exogenous crisis, which threatened the existence of the historically most advanced societies.

The first of these revolutions was the "Paleolithic Revolution," which occurred more than 1.5 million years ago and initiated the appearance of standardized instruments, the regular use of fire, and, probably, the transition of most hominoids from a collective to hunting lifestyle. The formation of protocultural regulations in the Lower-Paleolithic age restricted internal aggression by applying it to "strangers." This event was preceded by "the first existential crisis" in human pre-history. Having introduced the regular use of instruments, hominoids broke several natural balances, in particular, the balance between the natural armament of animals (such as fangs, horns, and hoofs) and the instinctual prohibition of intra-species killings (a population-centric instinct). The efficiency of artificial means of attack quickly surpassed corporal means of defense and mental inhibition, thus threatening the further existence of hominids. This evolutionary challenge resulted in the first protoculture and protomorale or non-instinctive mechanisms of collective regulation.

The second revolution has been named the "Upper-Paleolithic Revolution," or "cultural revolution of Cro-Magnons." Between thirty and thirty-five thousand years ago this transition completed the displacement of Neanderthals. This revolution also produced an increase in the use of stone as raw material and instruments made of bones and horns (when people were given relative independence from natural sources of flint). There was a notable improvement in sign systems of communication (which might have included articulated speech) and the first appearance of two-dimensional images (such as rock paintings). Paleoanthropoids, who created an advanced Mustje culture and for about 150 thousand years dominated their contemporaries of the neoanthropoid type (Cro-Magnons), made use of these advances. Yet Mustje culture experienced a severe crisis at that time. Two hypotheses have been developed to explain the essence of this crisis. One is based on the fact that the significant variability of the material culture of Neanderthals is combined with the absence of traces of "spiritual industry." Freedom of choice of physical actions together with a shortage of spiritual regulators (undeveloped animistic thinking characteristic for Upper-Paleolithic cultures) generated a neurotic syndrome, which revealed itself in asocial conduct with bursts of unguided aggressive energy. Another hypothesis connects the crisis of late Mustje culture to ecology. Neanderthals hit upon an idea to burn out vegetation, increasing thus the efficiency of landscapes, though it resulted in a pernicious reduction of biodiversity.

The third, or "Neolithic Revolution," occurred between 10,000 and 8,000 BCE, ushering a transition from inefficient means of appropriation, such as hunting and gathering, to a productive economy of agriculture and the use of domestic animals. This transition was accompanied by a radical change of thinking and the replacement of normative genocide and cannibalism with rudimentary forms of collective work involving a symbiosis of agricultural and "aggressive" tribes. Deep, complex reorganization was the answer to the Upper-Paleolithic crisis, aggravated by the unprecedented development of hunting technologies and resulting in the destruction of populations and whole species of animals and the toughening of competition between tribes. During the Upper-Paleolithic crisis former demographic growth was replaced by a sharp reduction of the population and only with the development of agricultural devices did the population again started to grow fast.

The fourth revolution, the "City Revolution," took place between 5,000 and 3,000 BCE, initiating the formation of large human agglomerations, the construction of irrigational channels, the appearance of writing and the first legal documents regulating coexistence in conditions of high concentration, and cooperation of large collectives. This revolution followed the spread of bronze instruments, another demographic burst, and an aggravation of competition for fertile lands.

The "Revolution of Axial Times" occurred in the middle of the last thousand years BCE, when advanced but still poorly interconnected societies produced thinkers, politicians, and warlords of a new type, including Zoroaster, Judaic prophets, Socrates, Buddha, Confucius, Cyrus, Asoka, Chuang-tze, and others, who transformed the image of human culture. Authoritative mythological thinking was for the first time superseded by critical thinking, forming the concepts of good and evil and the idea of a person as a sovereign bearer of a moral choice. The higher authority of individual self-control was created as an alternative to the phobia of the gods that dominated before. The purposes and methods of conducting war changed: the number of victims ceased to serve as a criterion for fighting skill and a subject for boasting, and primitive violence and terror partly gave way to methods of reconnaissance and "political demagogy."

The sixth revolution was the "Industrial Revolution," which introduced scientifically informed "know-how" and was accompanied by the development and spread of the ideals of humanism, equality, democracy, international and individual laws, and the valuation of the phenomena of war and peace. The Industrial Revolution was preceded by an extended crisis in agricultural culture in Western and Eastern Europe between the eleventh and seventeenth century, involving uncontrolled extensive growth, deforestation, the destruction of ecosystems, and mass-killings. The development of agricultural technologies has produced the next evolutionary deadlock. Industrial development, having increased human efficacy, energized demographic growth and ecological and geopolitical ambitions.

Finally, during the "Information Revolution" of the late twentieth century and beyond, information and knowledge took on a dominating role in the economy, culture, and society. This revolution has also created the perception that the planetary civilization is close to the next crisis. Within a hundred-year period, weaponry became three to six times more powerful. Intelligence has acquired such an operational power that means of restraint formed by previous historical experience have ceased to meet new demands.

A. P. Nazaretyan

Anthropogenic Environmental Changes are changes to the environment that occur as a result of human activity. Primitive people—collectors, hunters, and fishers, and large herbivorous animals and predators, did little damage to the environment, as they used only resources of natural ecosystems. When the share of net biological production available to people was exceeded by population, or the share was diminished because of natural fluctuations, primitive nations either moved to new sites or the population was reduced, restoring damaged ecosystems. A gatherer needed about 500 hectares of land, and the hunter needed twice as much. Therefore, the population remained relatively stable. Development of hunting and fishing technologies, and the knowledge of how to use fire, did not essentially change the environment.

The situation changed after the development of agriculture. Natural ecosystems were replaced by monoculture agricultural and livestock systems. A farmer needed just one hectare of land, and this hectare output was sufficient for an entire family. First, forests or steppes were burned, then, using the biomass of the natural ecosystem as fertilizer, cultured plants were grown, which were almost completely consumed by humans and then almost completely consumed. These practices introduced anthropogenic changes of the environment, involving: (1) destruction of natural ecosystems and the resulting leakage of carbon, nitrogen, phosphorus, and other substances into environment; (2) change of the reflectivity of the terrestrial surface due to the change of vegetation; (3) lowering of continental water turnover as a result of less intensive transpiration of moisture by cultured plants, the efficiency of which is always lower than the efficiency of natural ecosystems.

Over time, this process became planetary. Within 150 years, the United States destroyed almost all natural ecosystems on its territories, compared to the 2,000 years over which this process occurred in Europe. The Industrial Revolution accelerated environmental destruction to establish enterprises and an economic infrastructure, introducing a new mechanism of anthropogenic change that has polluted nearly all environments. Prior to the twentieth century, anthropogenic changes of the environment developed at a local level, but in the twentieth century, they became global. The economic activities of humankind directly influence the concentration of gases in atmosphere, and substances in fresh water and soil, quickly reducing biodiversity. A total reorganization of the environment is a result of this process.

K. S. Losev

Antiglobalism is a political and social movement that began in the late twentieth century, to promote the integration of economies, nations, and cultures on the basis of democracy, social justice, and respect for national and cultural independence and uniqueness. Antiglobalists do not generally use this name to identify themselves, since the movement has not limited its efforts to opposing global integration, but proclaims the necessity of changing the predominant liberal capitalist form of globalization.

Antiglobalism is an international movement for globalization "from below," in accordance with the interests of citizens. It unites diverse social classes and groups, from the unemployed to upper middle class representatives and parliamentarians. Every social stratum and group, except for the world ruling class, participates in this movement. Recently it became popular in Russia as well. Antiglobalism originated from social protests during the international summits; since 1999, it has become especially visible.

Acid Mine Drainage – Axial Time

This spontaneous movement without a single center is a new, worldwide form of social organization maintaining informal contacts via Internet. It includes the whole range of coalitions of various nongovernmental movements and organizations (trade unions, women's, youth, ecological, etc.), and left-wing political parties (such as traditional communists, and "Trotskyists").

The movement has already conducted several world-scale actions. The most significant were the protest demonstrations and alternative social forums organized during various international conferences and summits of the World Trade Organization, International Monetary Fund, and the World Bank in Seattle, Washington, Quebec, Prague, Brussels, and Genoa. The positive antiglobalist program was discussed at the World Social Forums (*Fórum Social Mundial*) in Porto Alegre, Brazil in 2001 and 2002. More than 70,000 participants representing 4,900 organizations and movements from 123 countries were present at the 2002 forum, whose main slogans were "The world is not merchandise" and "Another world is possible."

The antiglobalist movement arose, above all, due to the emergence of global trade and global markets, with their severe regulations, moral break-downs, and reduction of human beings to the level economic animals. The movement seeks to address the social, economic, and political atmosphere in which the majority of humankind is deprived of democratic rights, freedom, and opportunities to participate in decision-making and live a dignified life. Because of these global developments, the gap between the rich and the poor has widened. The world division of labor undermines equal rights to modern education and knowledge and to access to the heritage of world civilization, rights enjoyed primarily by a handful of intellectuals from the developed countries.

Antiglobalists oppose the path of world development imposed "from the top," since it harshly intensifies injustice and suppresses the individual. They suggest their own, alternative and more humane way of world development. By means of dialogue, joint inquiry, theoretical investigations, and the practical activity of every country, they think a more socially equal, technically and economically advanced, and individual-friendly world could be created. The theoretical foundations of the alternative movement are emerging. Several Nobel Prize laureates and such famous scholars as Immanuel Wallerstein, Samir Amin, and others take part in their formulation.

The antiglobalist movement cooperates closely with older social movements, primarily with trade unions and left-wing parties. Their relations are based on the principles of tolerance, pluralism, openness, mobilization, and cooperation. Now, an international committee uniting hundreds of associations and movements, which is also open for new ones, has been formed. Each organization can initiate seminars, conferences, and demonstrations on its own, which can be sponsored by the participating organizations.

Thus, the antiglobalist movement could be regarded as a movement against the complex mechanism of global power that accumulates the most important power resources, the unique political and military machinery (allowing it to control weapons of mass destruction), the mass media, and the central international political institutions. This power mechanism is setting up the rules of a game uniting the agents of globalization into a single (though contradictory) force. This global nomenclature predetermines world politics

Global Studies Encyclopedic Dictionary

and economy and calls this system a democracy. The extremely powerful international capital opposes employees who are no longer protected by trade unions and governments.

The rise of antiglobalism is also connected with the need to counteract the negative aspects of globalization. The movement thereby opposes Western imperialist propaganda, which imposing its values on other nations and countries. The struggle to protect the originality and uniqueness of cultures and to grant the opportunity to choose their own way of development is reflected in the antiglobalist movement.

A. V. Buzgalin, M. Y. Pavlov

Arendt, Hannah (1906–1975) is one of the most important political and philosophical thinkers of the twentieth century. She spent her formative years in Königsberg/Kaliningrad, where she became acquainted with Immanuel Kant's philosophy. At Marburg University she came to know Martin Heidegger and adopted what is known as the phenomenological method. She also studied Theology, Philosophy, and Classics at Freiburg and Heidelberg. She was a friend of Karl Jaspers throughout her life. Jewish, she fled to France in 1933, and in 1941, she immigrated to the United States. Her comprehensive study *The Origins of Totalitarianism* (1951), made her well known. It is a comparative study politically and historically analyzing Nazism and Stalinism. Her report on the Eichmann trial in Jerusalem, entitled *A Report on the Banality of Evil* (1963) has also attracted great attention.

In her main work, *The Human Condition* (1959), Arendt exposes pluralism as a fundamental quality of action, for action is always connected with speech. Following Aristotle's characterization of a human being as "a living being capable of speech," she addresses the realm of pluralistic action through the division of human activities in labor, work, and action. Former philosophy only divided between work and contemplation. Her theory of human activities results from profound analysis of the ancient polis described by Plato and Aristotle. Labor means all activities needed for the process of everyday life, consumption, and natural regeneration. In this regard, labor is the realm of fertility.

Work is the quality of *Homo faber* (man the creator), who constructs a house as a start into our modern technical world. The process of making is determined by the categories of means and end. In this sense, the attitude of the worker means utilitarianism.

While labor and work can be done without words, action always combines different persons who are forced to declare by speech who they are. By speech and action a person enters into the web of human relationships. Setting new beginnings by every action, a person performs his or her own life story in this web. It is evident that in this concept all human affairs are frail and unpredictable. Hannah Arendt accentuates that her philosophy of the human condition especially criticizes Karl Marx.

Justice is part of human affairs. That is why justice cannot be achieved using the scheme of means and end. In the web of human affairs only promises and treaties can produce "certain islands of predictability"—or "isolated islands of certainty in an ocean of uncertainty." On this point Hannah Arendt refers to the Roman legal system as a main element of political life.

Acid Mine Drainage – Axial Time   23

Consequently, Arendt distinguishes between political violence and political power. Violence is mute and is exercised according to the categories of means and end because modern people like to substitute action by work, whereas political power, which tries to establish the stability of justice, is maintained by mutual promise and contract. The culture of sincere communication is a condition of such a just political power.

As Hannah Arendt's political theory is based on phenomenological anthropology, it can be regarded as an example of a global political concept. Her teaching is a lesson on how humanity can survive on Earth.

References: Arendt, Hannah. *The Origins of Totalitarianism.* New York: Harcourt Brace,1951. *The Human Condition.* Garden City, NY: Doubleday Anchor Book, 1959. *A Report on the Banality of Evil.* London: Faber, 1963.

W. Busch

Asian Development Bank (ADB) is an international bank, whose task is stimulation of economic growth and acceleration of economic development in the ADB developing member-countries. The ADB formation was formed in 1966, when fifteen governments ratified it. The Bank began its operations on 19 December 1966. By the late 1980s, forty-seven countries were members, thirty of them located in Asia. The United States became a member of the ADB in accordance with the Asian Development Bank law (1966). ADB membership is available for members and associated members of the UN Committee on Economical and Social Problems of the Asia-Pacific Region, for the other countries of the region, and for developed countries from other regions-members of the United Nations or one of its special agencies. The fifteen developed ADB member-countries from other regions include the United States, Canada, Great Britain, and fifteen West European countries.

The banking operations of the ADB have two main sources of funding: conventional capital sources and special sources. The conventional capital sources include the shares in capital paid by the ADB member countries, borrowings on the world capital market, and incomes on investments. The special funds include contributions of the developed member countries, incomes from special funds, loans and investments, and sums transferred from conventional capital sources according to a special decision of the ADB Board of Directors. The special funds of the ADB comprise a window of "preferential loans" under low interest rates to satisfy the needs of smaller and poorer members. In 1987, the Bank established the Asian Development Fund with the purpose of consolidating special funds on a continual and organized basis. However, the separate fund for financing technical assistance operations still exists.

V. D. Mekhryakov

Atheism is an attitude of skepticism toward claims of the existence of any sort of God or gods. Atheism is as old as theism and comes in almost as many shapes and sizes. Ancient India, China, and Greece all developed distinctive forms of atheist thought, each of which has made fundamental contributions to their respective civilizations. In India, strands of atheism go back about 2,600 years to the Carvaka School, the early Samkhya theorists, the Jains, and the Ājīvikas. Siddhartha Gautama (563–483 BCE), known to the world as the Buddha, was explicitly atheist. In China, Kongfuzi (551–479 BCE) and his

24 *Global Studies Encyclopedic Dictionary*

followers, Wang Chong (27–97 CE), and the school now called philosophical Taoism were all skeptical of the existence of God or gods, or at least unconcerned about the question. In Greece, Diagoras of Melos (fifth century BCE) was the earliest known atheist. Several of the Pre-Socratic philosophers were broadly atheistic, as were the Epicureans who followed them.

After the rise of Christianity in the West, atheism was little more than a memory until the Enlightenment. The cultural reawakening of the Renaissance and widespread disgust at the carnage caused by the wars of religion stimulated renewed interest in atheism. But when Thomas H. Huxley (1825–1895) coined the word agnostic in 1869, it looked as if atheism would become redundant. Agnosticism was popular because it was seen as a less dogmatic attitude than atheism. Antony Flew (1923–2010), who was the first to make the distinction between negative and positive atheism, cleared up this perception. Flew's other main contributions have been to demonstrate that the burden of proof regarding the existence of God or gods lies with the theist, who is making the greater claim, and with his Falsification Challenge, which shows that much theistic thinking is so circular it is incapable of being falsified.

Negative atheism, the weaker variant of atheism, says that faith in God or gods is not justified. This approach stresses the improbability, rather than the impossibility, of God. One classic expression of negative atheism comes from Charles Bradlaugh's *A Plea for Atheism* (1864): "The Atheist does not say 'there is no God,' but he says, 'I know not what you mean by God . . . I do not deny God, because I cannot deny that of which . . . by its affirmer, is so imperfect that he is unable to define it to me.'" In his classic study of atheism (1990), Michael Martin concludes that negative atheism is justified because the sentence "God exists" is neither true nor false, but meaningless.

Positive atheism is the stronger variant of atheism, which says that it is justified not to believe in God or gods. This approach argues for the impossibility of God, not simply God's improbability. Martin argues that positive atheism is justified if all the available evidence supporting belief in God can be shown to be inadequate and that no acceptable beneficial reasons can be stated to believe that God exists. Martin demonstrates that these conditions exist and that positive atheism is a philosophically sound position.

Another powerful argument is the moral argument for atheism, which shows that some of the dictates of scripture involve acts of such immorality that it is inconceivable a loving, perfect God could condone such activity.

Atheism involves the questioning of the belief in God or Gods on three grounds: because it is very probably false that such a God exists, because the concept of God is meaningless or incoherent, and because the concept of God being proposed is so empty of real substance that it is little more than atheism in disguise. The first approach is most telling against the crudest anthropomorphic notions of God. The second approach is most telling against the mainstream God of the Judeo-Christian tradition, and the third approach is most telling against some of the abstractions put forward by liberal theologians such as Paul Tillich or the theistically inclined physicists.

References: Bradlaugh, C. *A Plea for Atheism*. London: A. and H. Bradlaugh Bonne, 1895. Martin, M. *Atheism*. Philadelphia: Temple University Press, 1990.

W. Cooke

Atmosphere (Greek *atmos*, "vapor" and *sphaira*, "sphere") is the gaseous envelope of the Earth and other celestial bodies such as planets and stars. The Earth's atmosphere is a mix of a multitude of gases dominated by N_2 (78.08 percent) and O_2 (20.95 percent). Extremely important are also CO_2 (0.03 percent), O_3 (2.5×10^{-6} percent), and H_2O (0.01^{-1} percent). Methane, aerosols, and a quantity of ions are present in addition to the free electrons at great heights.

Nitrogen, carbon dioxide, and water vapor come to the atmosphere from the deep-laid layers of the Earth because of the degassing of the upper mantle and crust regions, mainly in the course of volcanic activity. The predominant part of the oxygen component is formed in the course of plant photosynthesis according to the reaction $CO_2 + H_2O$ + light energy (hn) = glucose + O_2. Ozone appears due to molecular oxygen (O_2) associating with the atomic one (O). The free electrons of the ionosphere are generated due to the influence of the solar x-ray, ultraviolet, and corpuscular radiation.

Chemically, nitrogen is extremely passive. Therefore, in spite of its overwhelming quantitative domination over all the other components, it is actually of no importance in the atmosphere. The key position is occupied by oxygen, which not only provides energy to living organisms, but also protects them from deadly short-wave solar radiation. Along with its "offspring" ozone, oxygen is a guard of life on Earth and, at the same time, a product of its activity. Carbon dioxide, another very important element of the atmosphere, has equally diversified functions. On the one hand, it generates oxygen, at the same time providing the building material for autotrophic plants. On the other hand, it is responsible for the so-called greenhouse effect, which makes the surface temperature of Earth favorable for organic forms. Without CO_2, the average (global) temperature of the Earth's surface would be unbearable for the higher life forms, as low as -18C instead of 15C as it is now. The mechanism of the greenhouse effect is simple: CO_2, CH_4, and O_3 along with water vapor let pass only the short-wave (visible) part of solar radiation, but hold back the long-wave thermal radiation re-radiated by the Earth's surface. Calculations show that temperature grows (drops) by about 3 °C as the CO_2 content in the atmosphere increases (decreases) two times.

Meteorological observations became regular as late as the second half of the nineteenth century. Nevertheless, both historical sources and paleoclimatic studies allow us to reconstruct, to a great degree of reliability, not only the relatively recent past of the atmosphere, but also its ancient history. The results of investigations concerning the early stages of the history of the Earth's atmosphere (from the Archean period to the Cenozoic) can be formulated as follows: (1) The great masses of liquid water (oceans) were formed at the earliest stage of Earth's history. (2) In all the previous geological epochs, the climate was warmer than the present-day one by 3–11 °C. (3) The climate of the ancient past was marked by uniformity and stability—the temperature in the polar and equatorial regions differed, apparently, by not more than 10 °C (now it amounts to between 60–70 °C). (4) The content of oxygen in the original atmosphere was negligibly small and began to grow noticeably as late as two billion years ago (though contrarily the CO_2 content dropped simultaneously more than 100 times). (5) The existing latitudinal and seasonal contrast of temperature began to develop 30–40 million years ago.

Among the natural factors affecting the climate dynamics, the most significant are volcanic activity, the variation of small gaseous components and cloudage, the changing relief of the Earth's surface, the variation of pressure over the oceans, and the secular course of solar activity. In the twentieth century, the climate began to be affected by anthropogenic factors. Thus, the burning of fossil fuel and application of modern agricultural technologies led to the multiple growth of accumulation rate of carbon dioxide and methane, the main greenhouse gases, in the atmosphere. Sharply growing also was the atmospheric emission of sulfur, aerosols of industrial origin, and Freon, which actively destroy ozone.

Natural climate changes in the twentieth century are vague. Long-term trends (Phanerozoic Era) appear to attest to the prediction that a further cooling period is coming. This is supported by the prognostic estimates of solar and volcanic activity. Yet if one takes into account the temperature trends of the last 300–400 thousand years, especially those of the last century, a trend toward further warming prevails. The extent of anthropogenic influence on the climate is not yet sufficiently clear. The rapid growth of the CO_2 and CH_4 concentrations resulting from industrial activity is undeniably contributing to the greenhouse effect. But the relationship between the depletion of the ozone layer and industrial waste of Freon remains, in many ways, open. Nevertheless, evidence shows that the uncontrolled growth of energy consumption worldwide is promoting the intensification atmospheric disruption from the state of its already rather unstable equilibrium. In the conditions of increasing dependence of modern civilization on the environment, the problems of monitoring and regulating the global climate assume a special importance.

The considerable rarity of the atmosphere at great heights has produced variations of gas composition and temperatures are much higher there than in the near-surface layers. Thus, the measurements of mesospheric and ionospheric parameters with weather rockets and ionic probes show that in the last forty to fifty years the temperature at the heights of 70–300 km has changed by 15–25 °C, tens of times higher than its maximum variation in the near-surface layers of the air. Such supersensitivity of mesosphere and ionosphere opens new prospects for the control and mid-term forecasting of global climate trends.

G. V. Givishvili

Augustine of Hippo, Perspective on Peace: In the *City of God*, Augustine puts forward the thesis that peace is the ultimate goal of all actions. Every being to seeks its own peace; if we love in the right way, that is "loving what ought to be loved"—for that is what virtue consists in—there will be no conflict among people. Augustine emphasizes the individual and religious grounds for peace. In regarding peace as bringing happiness, which is the ultimate good that all human actions aim at, he builds on the eudaemonian ethics of the Roman world, derived from Aristotle's view of happiness (*eudaemonia*) in the *Nicomachean Ethics*.

Augustine characterizes peace as both a physical and a mental state. "The peace of body and soul is a duly ordered life and death of living creatures." Peace is also an individual and a social goal. Identifying peace with "ultimate good will," Augustine states: "The peace between men is an ordered agreement of mind with mind." From the peace between individuals Augustine believes that we can achieve "the peace of the whole universe, the tranquility of order." By order Augustine means the "arrangement of things equal and unequal in a pattern which assigns to each its proper portion."

Augustine argues that in wanting to secure personal welfare even an "unsociable and savage" creature seeks peace. For any being to preserve its own life there has to be some degree of peace in the sense of the ordered relation among its constituent parts. In his discussion of evil Augustine draws our attention to the presence of the element of the good in all things. Augustine's view of evil is that it is a deficiency of good and not a necessary aspect of existence. Augustine does not deny the existence of evil but asserts that there is always some good: "There exists then a nature in which there is no evil, in which no evil can exist; but there cannot exist a nature in which there is no good." He does not think of humans or nature as fallen or radically corrupt but as moving toward perfection.

Augustine affirms peace as the goal of all political actions. Augustine asserts, "Peace is the instinctive aim of all creatures, and is even the ultimate purpose of war," "since even the wicked when they go to war do so to defend the peace of their own people." Given the choice between peace and war even the most aggressive persons and states would choose peace if they could impose their own terms of peace. The source of conflict is seen to reside not in the ends that we seek but in the means: the person who "seeks to impose his own dominion on fellow man" seeks "peace of injustice." If the goals were divergent there would be no immediate way to resolve the conflicts.

To counteract this propensity to impose individual will to the detriment of the well-being of others, Augustine specifies two rules: "to do no harm to anyone, and to help everyone whenever possible." Augustine believes that when the natural activities of those who seek their own well-being are guided by these rules peace will follow. We are immediately aware of rules such as "Do not do to others what you would not have done to you." Augustine refers to this injunction as natural law, which is "written on our minds" or, "written in our hearts." While he upholds acting in conformity with the moral law he also stresses the freedom of the individual.

For Augustine, virtue consists in acting freely from the love of righteousness. All moral actions lead to the highest of all of our loves—the love of God. The love of God is simply the love of the soul for the highest good attainable untainted by "the pleasures of the body." For Augustine "the two chief percepts" of Christian ethics are "the love of God" and "the love of neighbor." Augustine takes the love of God and our neighbors as coinciding with self-love. In choosing as the object of our love the highest being from which the greatest good can come, we come to know how to love ourselves in

28   *Global Studies Encyclopedic Dictionary*

the right way. From the immensity of the love of God flows the love for all people, which is the source of all goodness. Augustine takes the love of God as generating universal love for all. For Augustine, if we love as we ought to love, there is no conflict between self-love and concern for others. Hence the celebrated saying, "Love and do what you will."

References: Aristotle. *Nicomachean Ethics*. Translated by W. D. Ross. London: Oxford University Press, 1955. Augustine. *City of God*. Translated by H. Bettenson. Harmondsworth: Penguin, 1984.

H. Höchsmann

Authoritarianism (Latin *auctoritas*, "authority" "influence") Authoritarianism refers to a system of government typical of anti-democratic political regimes, often combined with personal dictatorship, characterized by absolute (blind) obedience to authority. Among historical forms of authoritarianism are Asian despotism, tyrannical and absolutist governments of the past, police and fascist regimes, and different kinds of totalitarianism.

V. M. Smolkin

Autotrophy of Humankind (Greek *auto*, "self" and *trophe*, "food" or "nutrition"; literally "self-feeding") Autotrophy of humankind is a concept related to the study of the noosphere suggested by Vladimir I. Vernadsky. The only autotrophic organisms on the planet are green plants. They form the bulk of living substance and accomplish their nutrition through the assimilation of inorganic elements using solar energy. All other organisms, including human beings, feed on other living organisms ranging from plants to the higher orders of carnivores. Human beings reside on the top of the ecological pyramid, being omnivorous and employing as industrial resources inanimate matter.

Human consumption of animate and inanimate natural resources is increasing in geometrical progression. An increasing lack of mineral, or non-renewable, resources is envisaged in the near future, as is a lack of renewable resources essential for survival, such as oxygen, fresh water, bioresources, and climatic stability. By some assessments, the anthropogenic load exceeds by more than ten times the tolerable limits of the biosphere. Human beings, in their industrial activities, do not obey the laws of the ecological pyramid; they somewhat reverse its direction. We thereby face a dilemma: human beings should either obey the laws of the ecological pyramid and take their frugal (in point of quantity) place in the upper part of its cone as a heterotrophic being, or turn to the autotrophical way of nutrition and thus relieve the biosphere of its surplus load. Modern scientific development allows progression in this direction.

As early as 1927, Vernadsky expressed his hope for the possibility of a transition to autotrophy, pinning his hopes mainly on the first serious achievements in the area of the artificial synthesis of inorganic compounds. He supposed that eventually it would be possible to achieve the synthesis of foodstuffs, so that people would no longer be dependent on other living organisms for nutrition. However, this problem has not been solved, at least to the point of providing for all of our nutritional needs

The biosphere can be relieved from the surplus load of human activity only if people are independent from the mineral fuel that produces energy for human purposes. Energy, which is related to higher chemical and thermal pollution, should supplant food as a source of concern.

The most paradoxical aspect is that this energy producing activity takes place on a planet that is flooded with powerful energetic streams coming from outer space. This energy is not of the density required for human consumption, but it is possible to transform it from a dispersed state into a concentrated one. Our small planet annually catches and transforms into different useful kinds of motion some 1021 J of energy. Human beings do not need this much energy. Besides, technological methods of concentrating space streams can be made more versatile and efficient than those employed in the chlorophyll grains of a green leaf. Even the operation of these grains, as experiments have demonstrated, can be enhanced by a factor of eight to ten.

Modern science supplies everything necessary to solve the problem of alternative (renewable) energy. Humankind will have to turn to renewable sources of energy. Releasing the energy from the substance of the planet, human beings became the highest entropy factor, fatal for every living thing, because life is the eternal fight with the entropy. Hence, human beings must become autotrophic in obtaining food and energy or they will continue to ruin the biosphere and thus doom themselves to self-destruction. Autotrophy presents a viable way of advancing the noosphere and implementing a strategy of sustainable development.

E. V. Girusov

Axial Time (German *Achsenzeit***)** is a short historical period, in the course of which the fundamental characteristics of social life change. Karl Jaspers introduced the notion of axial time to designate decisive turns in the flow of events with regard to a specific period. He applied the notion to characterize world history between 800 and 200 BCE when radical changes occurred nearly simultaneously in three different regions of the planet: Europe, India and China. Jaspers spoke about the explosion of the human spirit when the notion of the commonality emerges. He termed these historical shifts as "axial," a turn from one qualitative stage of social development to another. The notion of "axial time" is in many aspects abstract and speculative. It should not be taken as a mathematically precise means for determining how many such turning points have already taken place in history or how many may occur in the future. Rather, this notion permits us to demonstrate how irregular historical changes randomly contribute to a social breakthrough. In recent years, as the interest in global history has grown, the term has become increasingly validated and has acquired new meanings. The broader usage of the term now allows us to designate other turning points in world history as "axial times," including that of globalization.

A. N. Chumakov

ೞ B ಐ

Bahá'í Faith: A global religion founded by Mirza Hussein Ali, a Persian prophet who took the religious name of Baha'u'llah, which means "Glory of God." Bahá'u'lláh was born in Iran in 1817. As a young man he joined a religious sect of Bábism whose followers were expecting a new revelation that had to be delivered by the coming messenger of God. The sect was initiated in Iran in 1844, and soon was suppressed by the government. Its leader, Báb (1819–1850) (whose name means "gate" in Arabic) was executed in 1850. After Báb's martyrdom Bahá'u'lláh came to the forefront of the movement. In 1863, he proclaimed himself the promised messenger.

Bahá'u'lláh, founder of the new religion that came to be known as the Bahá'í Faith, preached for the next twenty-nine years of his life spent in exile. After Bahá'u'lláh's death, in accordance with his will, his son `Abdu'l-Bahá' ("Servant of Glory") was appointed as the head of the Bahá'í community. Later the leadership was passed on to the grandson of `Abdu'l-Bahá,' Shoghí Effendí, and, in 1963, to the Universal House of Justice, whose first members were elected by the representatives of the Bahá'í community.

The principles of the Bahá'í religion reflect its main purpose, namely, the achievement of global unity of humankind. According to Bahá'u'lláh, such a unity cannot be reached without a spiritual revival and human unification under the guidance of one faith. This is a necessary but not sufficient condition for the success of globalization, however. In the sphere of politics it is imperative to create a world federation and an international tribunal that would represent the interests of all nations and maintain universal peace. In the social domain there is a need for a balanced economic development among different countries, protection of equal human rights regardless of religion, race, ethnicity, social status, and gender. In the sphere of culture one needs an obligatory universal education, acceptance of a common script, and harmonious development of science and religion. Finally, on the existential level every individual must strive independently for truth and overcome those inclinations that lead to conflict, especially all kinds of fanaticism and intolerance.

The administrative structure of the Bahá'í religion is based on the principle of democratic centralism. Members of the local community, twenty-one years and older, once a year elect a Local Spiritual Assembly that consists of nine members and governs the affairs of its locality. Delegates from Local Spiritual Assemblies every year elect nine members of the National Spiritual Assembly. Every five years, the members of National Spiritual Assemblies of all countries elect the Universal House of Justice, located in Haifa, Israel, which is the supreme governing body of Bahá'í. Its decisions are made on the basis of consensus of its nine members or, if such a consensus cannot be reached, by a simple majority.

Currently, more than five million Bahá'ís reside in some 100,000 localities across 233 countries and dependent territories. Its members represent

2112 racial, ethnic, and tribal groups. The Faith is recognized as the second-most geographically widespread religion after Christianity.

Reference: Bahá'í International Community. Accessed 12 October 2013. http://www.bahai.org/faq/#FE70B466-AE6D-512A-73C653BB40B0B413.

M. Yu. Sergeev

Baikal Lake is the oldest lake on Earth (twenty-five million years) and one of the largest lakes in the world. Its water-surface area is compatible with the area of such countries as Belgium, Denmark, and Holland. Its length is 636 km, its width varies from 27–80 km, and the length of the coastline is about 2,000 km. Situated in the South of Eastern Siberia, Russian natives call it "the sacred sea"; Buryat-Mongols call it "the universal ocean."

Baikal is a geographical center for the origin of biological species and currently contains more than 2,600 different species. Of these, 84 percent are endemic to the lake, found only in this area. For example, a gauzy viviparous fish called golomyanka lives only in Baikal. Baikal significantly surpasses not only continental reservoirs, but also such seas as the Azov Sea, the White Sea, and the Baltic Sea in terms of the number of organisms it supports. Baikal's ecological system is a unique combination of various species of flora, fauna, and geological objects, which have an immense importance for the preservation of the Earth's gene pool and for understanding its biological and geological evolution. The thickness of the bottom sediments in several places in Baikal reaches ten kilometers. Information about climate changes and the geological history of the continent during the last twenty-five to thirty million years is "encoded" in the bottom sediments of the lake.

Baikal occupies a special place not only in the history of nature, but also in the history of peoples of the world. It is the cradle of many tribes and nations of antiquity. In the authoritative opinion of Alexey P. Okladnikov, during the Glacial Epoch, the ancestors of North American Indians came from the Baikal area.

UNESCO's World Heritage Committee in the course of its twelfth session (held in Merida, Mexico, 2–7 December 1996) recognized Baikal as an example of a remarkable freshwater ecological system. The inclusion of Baikal into the World Heritage List has broadened the possibilities for greater renown of its remarkable value at regional, national, and world levels. On 2 April 1999, the State Duma of the Russian Federation adopted the federal law "On the protection of Lake Baikal," which was approved by the President of the Russian Federation Boris N. Yeltsin on 1 May 1999. According to this law, a special regime of economic and other activities is stipulated that shall be observed in accordance with the principle of equilibrium in solving of social and economic problems and objectives of protecting the unique ecological system of Lake Baikal. The law establishes the legal foundation for the transition of the Baikal region to sustainable development.

The Baikal region occupies the territory of southeast of Siberia and the northern part of the Mongolian Republic. Its total area is 1 million km^2, which is compatible with the territory of France, Germany, and Italy taken together. The Russian part of the territory of the Baikal region is distributed among

32 *Global Studies Encyclopedic Dictionary*

subjects of the Russian Federation, with 73 percent in the Buryat Republic, 21 percent in Chita oblast, and 6 percent in Irkutsk oblast.

V. V. Mantatov, L. V. Mantatova

Barefoot Revolution (1988) is the fifteenth report to the Club of Rome and was drawn up under the guidance of Bertrand Schneider, a well-known expert in the problems of the developing countries. Proceeding from the practical results of several small-scale projects, carried out by various research groups in Asia, Africa, and Latin America, the authors attempted to analyze one of the most complex contemporary problems: possible trends in the development of Third World countries in the context of the relationship between the rich "North" and the poor "South." This relationship, as outlined in the report, should facilitate is the balanced transfer of technologies into less-developed countries. When attempting such transfer, we must assess, in every individual case, both the extent of a country's development and all possible negative cultural, social, ecological implications of the transfer. As a result, the initiative should shift from the North to the South; the priority in politics should be satisfying the needs of rural populations with more emphasis being placed on local production and consumption of food products, and not on their import and export.

Reference: Schneider, Bertrand. *The Barefoot Revolution*. London: IT, 1988.

A. N. Chumakov

Beyond the Age of Waste (1976) is the fourth report to the Club of Rome, completed under the direction of English physicist and Nobel Prize winner Dennis Gabor and Italian expert in management, Umberto Colombo. This report acknowledged the importance of sociopolitical factors, which must necessarily be taken into account when solving global problems. At this point, representatives of the Club revealed more optimism concerning social development and noted that the achieved level of science and technology allows humankind to cope with the problems it faces. However, as the authors of the report noted, aspects of social character are the main obstacles in the way of implementing necessary measures. They opened new opportunities for further objective analysis, but did not offer concrete ways of reorganizing the society, demanding special analysis to be carried out in subsequent reports.

Reference: Gabor, D. et al. *Beyond the Age of Waste*. 2nd ed. New York: Pergamon Press, 1981.

A. N. Chumakov

Beyond the Limits to Growth (1987) is the sixteenth report to the Club of Rome. This was not a traditional report commissioned by the Club, but one targeted to handle a specific problem. Eduard Pestel, a member of the Club's Executive Council, distinguished German scholar, and political figure, wrote the book in the form of his reflections on the evolution of theoretical and practical activities of the Club over the fifteen years since its first and most sensational report *The Limits to Growth* (Meadows, et al., 1972). He wanted to commemorate that date and Aurelio Peccei, the founder and first president of the Club of Rome.

Bahá'í Faith – Buddhism 33

Evaluating heated discussions that ensued around the world after the publication of *The Limits to Growth*, Pestel stressed their role in drawing the attention of millions of people to "world problems," making them face the danger and the necessity of a radical review of the values overshadowed by material interests. He also stresses the uselessness of continued debates, even concerning more perfect world models, since they have already generated considerable effects and cannot remain as effective. Instead, we should use the accumulated experience that enables us to search for new practical ways of reaching reliable development policy. Concerning guidelines for such a development, Pestel offers conclusions from *Microelectronics and Society* (Günter, 1982), assigning the main role to new technologies. Pestel believes all countries are determined to strengthen their technological status, since a society's capacity to effectively handle problems largely depends on available technology.

Additionally, the report defines the level of understanding of various global problems and makes generalizations concerning scientific and technological progress and its social consequences.

References: Günter, F. *Microelectronics and Society*. New York: Pergamon, 1982. Pestel, E. *Beyond the Limits to Growth*. New York: Universe Books, 1987.

A. N. Chumakov

Biocentrism (Greek *bios*, "life"; and Latin *centrum*, "focus") is the view that regards every organism as having a unique biological value, as a result of the development of the bios. Biocentrism per se is an ancient doctrine associated with the ideas of natural philosophy considering nature in its wholeness. The concept of "biocentrism" became widely used in cultural and axiological contexts at the end of the twentieth century.

From the position of biocentrism, human beings are not the end of planetary evolution but all varieties of living beings in the delicate equilibrium with the environment, or biosphere. A human being is only a part of this great variety. Contemporary biology understands human beings as rooted in nature (connected with nature by thousands of threads of genetic and cultural coevolution).

Biocentrism is opposed to the anthropocentric view that places humanity at the center of creation, which has brought about ecological catastrophe. According to biocentrism, anthropocentrism splits nature into two categories: human beings and the environment. Therefore, anthropocentrism reflected in the consciousness of an individual or of many people composing classes, states, nations, or races, inevitably leads to some sort of discrimination because human needs are controversial, unpredictable, and sometimes insupportably huge. To overcome the possible consequences of anthropocentrism, constitutions of each country must include provisions about the presumption of human responsibility to the other components of the bios considered as subjects having the right to live. Human needs (of both an individual and the whole population) should be satisfied only if they do not damage interests of the variety of life on Earth as a whole. Supporters of biocentrism favor inclusion of this provision in the constitutions of all countries, thus solving such issues from juridical and moral positions.

The preservation of the bios and the dynamic equilibrium of the biosphere is an indispensable condition for the existence of human civilization.

34 *Global Studies Encyclopedic Dictionary*

Biocentrism suggests that human beings should consider themselves not only citizens of their countries but also, first, citizens of the bios. They should respect not only the constitutions of their countries, but also the unwritten Constitution of the Bios. Biocentrism proposes that we live for the sake of the preservation of planetary equilibrium at all levels-from a resource-balanced biosphere able to preserve all varieties of species of plants, animals, and microorganisms to the balance between an organism and its ecosystem.

Philosophical ideas of biocentrism are supported and developed by the International Biopolitical Organization and the Committee for Biological Education (CBE) of the International Union for Biological Sciences (IUBS).

M. V. Gusev

Biodiversity is derived from the combination of "biological" and "diversity." Henry Walter Bates (1892) coined the term in *The Naturalist on the River Amazons*, where he described his impressions of meeting nearly 700 different species of butterflies during an hour walk.

Diversity is a concept dealing with the scale of variability and distinctions between quantities and object groups. Biological diversity deals with diversity in the living world. The term "biodiversity" is usually used to describe living organisms' number, variety, and variability. This term involves a multitude of different biological exponents and serves as a synonym for the concept of "life on Earth."

In the scientific context, the notion of diversity may be linked to such fundamental concepts as genes, species, and ecosystems, which correspond to the three fundamental, hierarchically connected levels of the organization of life on the planet. The phenomenon of biological diversity is determined by the fundamental characteristic of biological macromolecules, especially the ability of nucleic acids to spontaneously change their structure, leading to genomic alterations and hereditary variability. On this biochemical base, diversity emerges as a result of three independent processes: spontaneous genetic variations (mutations), effects of natural selection in mixed populations, and geographic and reproductive isolation. These processes lead to the further taxonomic and ecological differentiation on the following levels of biological ecosystems: the levels of species, cenosis, and ecosystem.

Life diversity has been a subject of investigation since ancient times. The first systems of nature, known from Aristotle's works, present an analysis of biological diversity. Carl von Linne first described a scientific methodological basis for the description of biodiversity in his *System of Nature* (1735). Charles Darwin's *The Origin of Species By Means of Natural Selection* (1859) radically changed the conception of nature. This study ended long searches by naturalists and specialists in systematization who had been trying to find a reasonable explanation for so many different and similar characteristics among the observed organisms. Darwin was able to give a logical and convincing explanation of the origin of these alterations, that is, natural selection. He was the first to grasp the link between the natural selection and hereditary alterations of the population to expose variability and explain its root.

Modern evolutionary theory, or neo-Darwinism, contains several statements not entirely original to Darwin's theory. The main mechanism is still natural selection, although our ideas the corpuscular nature of genes allows us to picture more completely the origin of variability as a result of mutations and variability preservation in a hidden condition in diploid organisms, in addition to the re-shuffling of genes in the process of genetic recombination, providing a constant source of new gene combinations available to natural selection. Evolution proceeds within two stages: variability appears and the purposefulness of variability emerges under the influence of natural selection.

The alterations that emerged under the influence of natural selection may have various consequences depending on living conditions. A process occurs when the environmental conditions determining natural selection are very homogeneous for the entire species or population of a natural habitat. Species adaptation to the environment grows steadily, and with the environmental change the entire species changes. Noticeable alterations may occur within a long period. Hence, the genetic structure of the separate line of consequent generations changes gradually and equally from generation to generation. This process is called philetic evolution. Another type evolution occurs when various populations of one species somehow become isolated from each other, so that natural selection affects them individually; different populations acquire different alterations. Thus, isolated populations continue changing until one original species splits into two or three different ones. The described process is known as the process of species formation.

The formation of species is significant: it leads to the subdivision of one genetic population into a row of subgroups. Each subgroup represents an independent evolutionary line with its own resources for philetic alterations.

At the same time, divergence is continuous, and does not break off after a group reaches the state of a species, but proceeds, resulting in the emergence of higher taxonomic categories.

The studies conducted by geneticists in the first half of the twentieth century form the basis for contemporary conceptions of the problem of biological diversity. These studies revealed how genetic diversity can originate in an outwardly homogeneous population.

Biodiversity has become one of the most widespread notions in scientific literature, the nature protection movement, and international affairs. Scientific explorations proved that a sufficient level of natural diversity on our planet is a crucial condition for the normal functioning of the ecosystem and of the entire biosphere. Presently, biological diversity is regarded as a basic component in the system of organisms. In some countries the concept of biological diversity forms the basis of state ecological policy if the state strives to preserve its biological resources to provide stable economic development.

The concept of biodiversity was brought into wide scientific use in 1972, at the Stockholm environmental conference of the United Nations, where ecologists convinced the political leaders of the world community countries that the protection of living nature must become a priority in any human activity on Earth. The International Union of Biological Sciences, when implementing scientific elaboration of the program called "Biological Diversity," established a special work team at the General Assembly in Canada in 1982. The problems of the preservation of the diversity of life were brought to the forefront in many countries. Academician M. S. Gilyarov took

36   *Global Studies Encyclopedic Dictionary*

an active part in the scientific and organizational efforts and became one of the founders of this large international project.

Adoption of the international Convention on Biological Diversity at the UN Conference on the Environment, held in Rio de Janeiro, 1992, was of great significance. Currently, the Convention on Biological Diversity has been signed by the representatives of 180 states, including Russia (1995), which undertook responsibility for protecting the living nature of one-seventh of the land of the planet.

References: Bates, H. W. *The Naturalist on the River Amazons.* London: J. Murray, 1892; Darwin, Charles. *The Origin of Species by Means of Natural Selection.* New York: Modern Library, 1936; Linne, Carl von. *Systema Naturae.* Reprinted from 10th ed. New York: Stechert-Hafner Service Agency, 1964.

D. A. Krivolutsky, N. N. Drozdov

Bioethics (Greek: *bios*, "life"; and *ethos*, "habit" or "temper") is an area of knowledge and practice that studies and solves ethical and legal problems in biomedical science, in practical health protection, and in the development and application of biotechnologies. The term was coined in 1969, by Van Rensselaer Potter, an American physician who interpreted bioethics in the spirit of ethical naturalism as a concept in whose framework the most important moral requirement is the conservation of life on earth. Later, this term, and the term "biomedical ethics" whose meaning is similar, was mainly used to designate the current condition of medical ethics.

The difference between traditional medical ethics and bioethics is that the latter is not the corporate ethics of a professional community: the paternalistic approach in relations between a physician and a patient, characteristic of traditional medical ethics, is being replaced by the principle of respect toward the independence of a patient (or subject in biomedical research).

This principle is a specification of Kant's idea that each person is an end in itself and should not be considered a means of solving any problems at all, even the problems of commonwealth, as applied to biomedical practice. Consequently, physicians' and patients' rights and obligations are being rearranged. For example, the principle of informed consent requires any medical (diagnostic, preventive, curative, or exploratory) intervention to be based on a voluntary and deliberate consent of its subject.

Another reason bioethics appeared is the spread of new medical technologies that made it possible to act deeply and intensively on the body and mentality of a human being. People have to face new ethical problems that, in most cases, have no analogues in the previous experience of humankind. These include: (1) organ transplantation, artificial reproduction techniques, and reproductive human cloning technologies, (2) means of life-supporting treatment that facilitate the prolongation of life processes in the body of a person who lost consciousness, whether for months or even years, and (3) possibilities and prospects of diagnostic, therapeutic, and even eugenic interventions in the human genome. These problems also impact the identity of a human being, the definition of the beginning and the end of an individual human life, the acceptability of different criteria of a person's death including the criteria of complete or partial brain death, the property right to genetic information, and the question of the protection of its confidentiality.

In solving all these and many other bioethical problems, the most important role is played by cultural and historical traditions that are different in different countries, regions of the world, and religions. As a result, the search for specific solutions sometimes requires more than the experience, knowledge, and competence of professionals only. The solutions acceptable from the social and cultural point of view can be those developed in the process of a broad public discussion that make it possible to identify and to reconcile the values and interests of different social groups.

Bioethics is an interdisciplinary area of knowledge that attracts the attention of physicians, biologists, philosophers, lawyers, theologians, culturologists, sociologists, and so on. The practical efficacy of this sphere is to a significant degree related to the fact that in the process of its development, institutional structures and mechanisms of moral and ethical control in biomedicine, which proved their effectiveness, were offered and refined. These include ethics committees, established in scientific research institutions, which run experiments involving people or animals. Any research project that involves such experiments must obtain permission of the corresponding ethics committee. Apart from experts, such committees include people who are not professionals in biomedicine and thus can be independent of the interests of the professional community.

One of the noticeable trends in the development of bioethical problems is the ever-increasing use, in addition to ethical regulation, of legal control of the processes for the creation and use of up-to-date biomedical technologies. Such control is being performed not only on a national level, but also on an international basis. Problems of bioethics occupy an ever more notable place in the activities of international organizations, not only medical ones, such as the World Health Organization or the World Medical Association, but also the United Nations, UNESCO, and the Council of Europe that develop normative documents regulating the practice of health protection and biomedical research. Documents such as the UNESCO Universal Declaration of Human Genome and Human Rights (1997), the Convention of the Council of Europe for the Protection of Human Rights and Dignity of the Human Being with Regard to the Application of Biology and Medicine (1997), and the Declaration Prohibiting Human Cloning currently being developed by the United Nations testify to the globalization of bioethics and its problems.

Another sign of bioethics globalization is the intensification of interest in problems of the use of new biomedical technologies to solve problems of health protection in poor regions of the world and the just distribution of both benefits and costs and risks originating from the scientific and technical progress in biomedicine between different countries. The subject raised more often lately is that global bioethics is supposed to pay attention, not only to individuals, but also to social values, namely, to identifying, formulating, and protecting the interests of vulnerable social groups.

B. G. Yudin

Biogeochemical Cycle is the process of continuous exchange of biogenes (necessary life substances), carried out by biota (the totality of all natural organisms) and the environment. Biota consists of: (1) organisms-producers, which, using inorganic substances of the environment and solar energy, synthesize organics (vegetative organisms-photosynthetics, chemosynthetics comprise an insignificant share in biota); and (2) organisms-consumers, which decompose (mineralize) organics back to initial substances, returning them to the environment through complex alimentary chains.

A limited number of biogenes are available to modern organisms. For example, the basic biogene carbon is consumed as carbonic gas, dissolved in the atmosphere or in the waters of the world's oceans. The resources of carbon both in the atmosphere and in the ocean are limited: converting to carbon, there is about 700 billion tons of carbonic gas in the atmosphere; approximately the same amount can be found in the world's oceans. If producers produce annually about 100 billion tons of organic substances, converting to carbon consumed by consumers, the biogeochemical cycle should be completed with fine precision. If the difference between synthesis and decomposing will amount to 1 percent or less, life that consists of photosynthetics and consumers, living at their expense, may disappear fast.

Initial cycle cells are assemblages of organisms with the characteristic size of up to tens of meters, including producers and consumers. Let us take a forest tree as an example. This system is subject to the law of large numbers as the producers here are leaves or fir-needles, which work independently from each other and even compete for light, and the consumers are microorganisms and hyphae of fungi, and insects, which are independent from each other. The fluctuation in such a system is equaled to one divided by the square root of several independent elements. Nikolay V. Timofeev-Resovsky called these assemblages "elementary units of the biological cycle." The biogeochemical cycle is the mechanism of regulation of the environment: when the concentration of biogenes is disturbed, the cycle becomes incomplete, and this or that biogene either is withdrawn out of the environment as an excess, or is added to it, because biota should maintain the conditions in which their concentration is optimal.

K. S. Losev

Biogeography is a science exploring living organisms and their associations in relation to their geographic distribution and allocation over the surface of the Earth, including land, fresh water, seas, and oceans. Associations and organisms are the objects of biogeography and of biology and ecology. It is one of the fundamental geographic disciplines. Knowledge of its basics is necessary to understand a wide range of issues such as the use and protection of nature, environment monitoring, medical geography, and human ecology.

As a geographic science, biogeography investigates primarily the objects' allocation in space, their interaction with each other and with the environment, the most important regularities of the planet and its regions of vegetable covering, and animal population structure and dynamics. Biogeography uses many geographic methods, among which the comparative geographic and the cartographic methods play the most important role. A deep knowledge of vegetable and animal organisms' biological characteristics, physiology, and ecology is also imperative, and an ability to employ the information about the interaction of organisms and associations with each other and with the environment is essential.

Much of the distribution of organisms and associations is defined not only by its biological peculiarities, but by the geological history of planetary and the regional development. Past biological distribution and natural conditions of different geological epochs form the subject of such sciences as paleontology, historical geology, and paleogeography. Data provided by these sciences is widely used in biogeography to explain specific biological distributions.

Biogeography acquired a special importance when the majority of nations proclaimed the transition to the conception of steady development at the UN conference on environment and development (Rio de Janeiro, 1992). This strategy is based on balanced economic development favorable to the environment and preservation of natural resources, to answer the needs of contemporary and future human generations. Biogeographic methods are necessary to use when carrying out, for example, the Paneuropean Strategy of Biological and Landscape Diversity, signed by the Heads of the Nature Protection Departments of fifty-five European countries in 1995. The main goal at this point is working out optimal methods and modes in production, taking into account the natural peculiarities of regions, reducing the anthropogenic load on the ecosystems and maintaining biological diversity.

The inhabitants of various regions of land and areas of water differ, but we may reveal these differences in many ways, depending on the concepts, approaches, and methods utilized. Floristic and faunistic investigations constitute the opening stage of regional biogeographic study. A range of concepts and terms are used at this point, such as flora, the totality of plant taxons growing on a territory, fauna, the totality of animal taxons, and microflora, the latter concept applied to the microorganisms. The flora and the fauna of a region form in the aggregate its biota.

These terms may be applied not only to the whole totality of plants, fungi, microorganisms, and animals, but also to the large subdivisions of the vegetable and animal world. Thus, we may speak of the spore plants' flora, ferns' flora, fungi' flora (mycoflora). Accordingly, we speak of the mammals' fauna (theriofauna), birds' fauna (ornithofauna), or insects' fauna (enthomofauna).

All these concepts are capacious enough to allow the analysis of species composition and the other taxons' composition and diversity, their origin, and spatial and temporal interrelations.

Historical development of the local flora and fauna is spontaneously brought about through the processes of species formation. Differences between the floras and the faunas are explained mostly by the geological history of each region. The analysis of floras and faunas over space and time explains the historical biogeography, telling us in which ways, at what time, and from which taxonomic groups a contemporary or fossilized fauna and flora were formed.

Alterations of flora and fauna are caused by physical and geographic environment changes and by inner mechanisms inherent to all living organisms. These include philetic evolution and temporal phenotype and genotype changes, by which every species, genus, family, or any other taxon exists on

40 *Global Studies Encyclopedic Dictionary*

the Earth surface for a limited period, transforming gradually into a new species, genus, or family. One of the most important geographic (chorological) characteristics used in the analysis of flora and fauna is natural habitat.

Another method of analysis represented in ecological biogeography involves exploring the distribution regularities of plant, animal, and microbe associations. Neighboring plants, animals, and microorganisms establish various relations with each other and with the environment and make associations (cenoses) different from each other in species composition, species numbers and in structure.

In almost every cenosis, we may select plant association (phitocenosis), animal population, and microorganisms association. The elementary unit of land differentiation in ecological biogeography is biogeocenosis—a homogeneous plot of the earth surface with a definite composition of animated and static components, united through metabolism and the flow of energy in one complex. The boundaries of biogeocenosis are marked after vegetation, which is one of the most important and easily selected (physiognomic) components of biogeocenosis.

The term "biome" refers to a complex selected on the basis of vegetative physiognomic signs and representing a biogeocenosis totality of a zone, such as tundra, taiga, and mixed woods. Similar biomes of different continents have different biota. For example, the conditions in broadleaf woods are similar in the European part of Russia and the eastern part of North America. However, in spite of essential resemblance of appearance and structure of associations, they consist of different plant and animal species. The spatial structure and appearance of associations is formed under the influence of the similarity of ecological conditions, while the flora and fauna (and taxons composing them) are shaped by their own developmental regularities. The biota of the North American broadleaf woods has richer species composition including rare species that were defined primarily by conditions during the previous geological epochs.

N. N. Drozdov, D. A. Krivolutsky

Biosphere (Greek *bios*, "life" and *sphaire*, "sphere") is one of the covers (spheres) of the Earth composed mostly through past and present activity of living entities. In 1875, Austrian geologist Eduard Suess introduced the concept of Earth covers. The water cover of the Earth was named the hydrosphere, the firm cover was named the lithosphere, and the area of the Earth's crust having life was named the biosphere. Earlier, the gaseous cover of the Earth was termed "atmosphere." The term "biosphere" became popular half a century later when Vladimir I. Vernadsky coined the term for the Earth's specific cover whose physical, chemical, and energy parameters are defined by the past and present activity of living organisms. The concept of the biosphere is the basis of the modern paradigm of the relations between humankind and nature: human beings are a creation and a part of the biosphere; if they destroy it they will undermine their own essential conditions of life.

The biosphere is the cover of the Earth where life is diffuse; "living substance" is present there that determines the chemical composition and energy processes in the atmosphere, the hydrosphere, and in the top layer of the lithosphere and topsoil. The biosphere is a united dynamic system on the Earth's surface formed and regulated by life.

Vernadsky's (1926) concept of the biosphere is commonly accepted not only because it is scientifically well grounded, but also because it has become vitally important for humankind. This concept allows us to realize that in our epoch of large-scale environmental transformations, all the resources of the biosphere are limited and exhaustible. Therefore, the world community is responsible for environmental preservation and for securing the survival of future generations.

The biosphere includes: (1) a lower part of the air cover (atmosphere), or troposphere, where life can exist, up to a height of 10–15 km; (2) all the water cover (hydrosphere) where life permeates down to the greatest depths of the World's oceans, exceeding 11 km beneath the surface; and (3) the upper part of the firm cover (lithosphere), composed of residual soil usually extending 30–60 m and sometimes more. Residual soil is a complex of geological sediments produced by decomposition (such as products of oxidation, aquation, and hydrolysis) and the leaching of various rocks. Beyond the bounds of the residual soil, life can be found only sporadically. For instance, some microorganisms were found in oil-bearing waters at a depth of 4,500 m. If we include in the biosphere the atmospheric strata, where the transfer of inactive germs is possible, then its vertical ambit amounts to twenty-five to forty kilometers, although special rocket-installed traps have revealed the presence of microorganisms at heights up to 85 km.

Life processes influence not only the region of active life but also the upper layers of the stratosphere, where mineralogical and elemental composition was formed by the biospheres of the geological past. According to Vernadsky, its capacity amounts to between five and six kilometers. The main agents producing the stratosphere are organisms, water, and wind converting and mixing sedimentary rocks after rising above the water level.

Within the limits of the biosphere regions exist where active life is impossible. For example, in the upper layers of the troposphere and in the coldest, hottest, and driest regions of Earth organisms can only be found in a quiescent state. These regions of the biosphere are known as the parabiosphere.

Often the term "ecosphere" is used to designate the layer of the planet that is the most abundant with life. The ecosphere is a biospheric layer where biogenic migration of organic matter prevails over their transfer by physical factors. The ecosphere's capacity amounts to only some tens of meters and its total area comes to approximately one-third of the Earth's surface.

Apart from the presence of living substance, the biosphere has three other typical characteristics: (1) it contains a considerable amount of liquid water, (2) it is irradiated by the energy flux of powerful sun rays, and (3) interfaces among substances exist in the gas, liquid, and solid phases. Therefore, a continuous circulation of matter and energy occurs where living organisms play the most active part.

The biosphere accumulates and redistributes huge fluxes of matter and energy. This process is possible because of chemical characteristics of cyclic or, according to Vernadsky's geochemical classification, "organogenic" ele-

42 *Global Studies Encyclopedic Dictionary*

ments named for their ability to take part in the many chemically reversible processes, while the geochemical history of all these elements can be described in terms of circular processes or cycles. Vernadsky named this biotic circulation "the order of the biosphere." The geochemical activity of living substance is of greatest importance here.

Reference: Vernadsky, V. I. *The Biosphere.* New York: Copernicus, (1926) 1998.

N. N. Drozdov, D. A. Krivolutsky

Biosphere Resources are a body of factors, sustaining the existence and evolution of the biosphere as a stable system, supporting its entire species, including human beings, with everything necessary for their existence.

Biosphere resources are necessary to sustain life on Earth. They include: (1) the air comprised of gases in the ratios necessary to sustain life, (2) fresh water, clean enough to sustain life, (3) soil fertility, (4) solar and other energy, (5) climate stability to ensure ecological balance on the planet, (6) an ozone screen in the upper atmosphere sufficiently dense to sustain life, (7) biodiversity, and (8) stability of such crucial parameters as temperature, pressure, humidity. The distinctive feature of biosphere resources is that they are either created or stipulated by life-sustaining activities of the organisms themselves. At the super-species level the teleology of biospheric processes is attained by a complex metabolic correlation among individual organisms, species, and the abiotic environment. Scale and rates of biogenic metabolic processes by far exceed geological processes. Since life on Earth began, all abiotic substance on the planet's surface has been processed by live organisms several thousand times and has been biogenically transformed by live substance.

All universal properties of the biosphere, including the bluish-green color of the planet, have been created by live organisms and signify the presence of life. Even the influence of space on the Earth's surface undergoes serious transformation. For example, sun rays and hard space radiation, having passed through the ozone screen created by living forms and through the atmosphere, become a vital condition for sustaining life, though in outer space these factors destroy any forms of life.

Biosphere resources fall into the category of renewable resources (according to the established geological classification). But the scale of human activity, which has sharply risen within the last 100 years, has put all renewable resources on the brink of depletion. For example, burning all the mineral fuel contained in the Earth's crust would require ten times more oxygen than its content in the atmosphere. The ratio shows to what extent modern energy industry, 90 percent of which runs on fuel, is detrimental to the biosphere.

E. V. Girusov

Biotechnology (Greek *bios*, "life," and *techne*, "art" "crafting skill," and *logos*, "word" "studies") Biotechnology is an interdisciplinary area that originated within the biological, chemical, and technical sciences and employs organisms and biological processes in manufacturing. The development of biotechnology is connected with the solution of such global problems as hunger eradication, energy shortages, and the scarcity of mineral resources, health improvement, and environmental protection. The development of biotechnologies was especially intense in the second half of the twentieth century in connection with discoveries in the field of molecular biology. Biotechnology is one of the most important components of the technological revolution that in many respects determines the tendency toward globalization of the world's industry and economy.

Hungarian biochemist Karl Ereki coined the term "biotechnology" in 1919. It designates processes connected with the activity of organisms and tissue cells producing useful products. As early as the sixth century BCE, a beer recipe was written on cuneiform tablets. In the nineteenth century, Pasteur's discoveries determined the role of microorganisms in the main processes involved in empirically discovered biotechnologies traditionally used in various national cultures, such as wine-making, bread baking, cheese making, and flax "maceration."

The development of industrial biotechnologies has been greatly influenced by Fleming's discovery of penicillin in the 1940s. The discovery of the method of genes amplification and the creation of recombinant DNA in the 1970s led, in the 1980s, to qualitatively new lines of biotechnology, including genetic engineering, and the artificial creation of bacteria.

An important line of biotechnology is the production of enzyme systems immobilized on organic structural supporting materials. Such systems catalyze several reactions and, since the 1980s, serve as the sensors able to determine the concentration of different substances. Such sensors are used in increasing frequency in chemistry and ecology.

From the point of view of ecology, the capability of some bacteria to produce methane was known in the eighteenth century. Modern biotechnology is able to cultivate some special cultures of bacteria that can utilize organic waste converting it into fuel gas. The prospect for the usage of the Botriacoccus Isohrysis algae biomass as an organic fuel, by the calorie content not yielding to petroleum, is being considered. The modified cells of spinach in the presence of the enzyme hydrogenase are able to produce hydrogen from water, thus marking an important step in hydrogen power production.

In Japan, animal waste is converted into biogas and biomass. In India, about one million gas plants for biogas production are in operation, and in China, about seven million small biogas plants are operating with a volume of 10–15 l., which is sufficient to provide a family of five with the fuel gas.

B. G. Rezhabek

Bipolar World is a geopolitical world order of the second half of the twentieth century, when the world was split militarily and politically into two opposing camps. One of the camps was headed by the United States, backed by West-European states, and the other (largely formed of East-European states) was led by the Soviet Union. Meanwhile, the many Third World countries (neutral and non-aligned countries of Asia, Africa, and Latin America) were not perceived as a pole of power.

The bipolar world order was based on the balance of power that took shape after World War II and resulted from the division of the world into two systems (socialist and capitalist) along with the disintegration of the colonial system. A tense struggle for world leadership and supremacy developed. The

44 *Global Studies Encyclopedic Dictionary*

bipolar world was based on the existence of nuclear weapons, the arms race, and the militarization of the economy, raising mutual "nuclear vulnerability." The balance of fear enabled "peaceful" development of the world community.

The bipolar world forced other states involved in conflict situations to accept potential intervention by the superpowers. Some states maneuvered between the two superpowers. The possibility of the intervention of the Soviet Union and the United States restrained the aggressiveness of non-nuclear states. Possession of nuclear weapons became a catalyst in the "race for power" and was complemented by a conventional arms race.

The ideal of the bipolar world entailed the preservation of rough parity between the superpowers and maintenance of considerable disparity between them and the rest of the world. Disintegration of the colonial system and the rise of new independent states and centers of power undermined the influence of the superpowers and, hence, the world of bipolarity.

Y. M. Pavlov

The Bologna Process: The Bologna Process is a trend toward globalization in European higher education aimed at the formation of a single European educational area. The Bologna process has been documented in the form of the Bologna Declaration (1999) named after the town of Bologna, Italy, where the first European university was founded. Forty countries, including Russia (2003) participate in the Bologna process.

The Bologna process entails a specific type of education meeting the demands of a democratic, highly technological, informational market society characterized by an orientation toward personality, freedom of choice with regard to an individualized path of education, and the informational nature of educational activities. Among the main aims of the Bologna process are: a central role of universities in the development of European cultural values; the need to adjust education to social demands; mobility of citizens to facilitate their access to educational centers; unification of degrees, qualifications, and competences of graduates; respect for various cultures, languages, and national systems of students; enhancement of university autonomy and academic freedom; and the continuity of education.

The Bologna process equalizes differences in levels of higher education in Western Europe and Russia according to the formula "bachelor to master" and opens the possibility for Russian higher education diplomas to be recognized in Europe.

M. D. Shchelkunov

Brain Drain refers to the international migration of highly skilled workers, often thought to result from the globalization of the world economy and expansion of the activities of transnational corporations. Globalization leads to the transformation of the migration flows engendered by capital flows and fuels the migration of highly skilled professionals who move from one country to another as employees of transnational corporations. As transnational corporations extend their operations to the developing countries, they send skilled employees thus allowing researchers to speak about a remarkably new stage in the migration processes occurring in developed and developing countries. The emigration of highly skilled specialists from developing countries is

Bahá'í Faith – Buddhism　　　　　45

compensated by the immigration of highly skilled specialists from developed countries, thus producing mutually beneficial brain exchange.

Although brain exchange is of great significance for the global economy, one should still consider negative effects produced by the continuing brain drain. Thus, according to the UN, over the last thirty years, the financial losses of the developing countries due to the brain drain exceeded sixty billion United States dollars. The irreversible losses of the developing countries due to the brain drain virtually mean a multibillion gain by the developed countries, eventually leading to the increase of economic inequality in the world, and undermining the concept and practice of sustainable development.

I. V. Ivakhnyuk

Buddhism is a religion indigenous to the Indian subcontinent that encompasses a variety of traditions, beliefs and practices largely based on teachings attributed to Siddhartha Gautama, commonly known as the Buddha, meaning "the awakened one." The Buddha lived and taught in the eastern part of the Indian subcontinent sometime between the sixth and fourth century BCE.

Spreading over Central Asia, reaching as far as China and Japan, by 1000 CE, Buddhism had embraced all the Indian Ocean basin nations and a large part of East and South-East Asia. In each region, Buddhism interacted with local folk beliefs and cults, giving birth to several trends, schools, and sects, which, in turn, interacted, adopting new ideas, forms, and peculiarities. Due to its flexibility and transformative capability, Buddhism attracted people from various social classes and spread widely where the Buddhist doctrine responded to the interests and aspirations of the local population.

Since Buddhist doctrine transmits universally important values, Buddhism was adopted easily, especially after its message had been translated into the native languages. Buddhism regarded individual life as a fleeting moment of immersion in the abyss of suffering. An individual could escape from this suffering after refusing to commit typical rational but wrong everyday life acts, especially those affecting not only that individual, but their environment as well. Buddhism with its concept of inactivity could enable an individual to approach the state of dilution into the ultimate universal bliss or endow upon them a position within the Buddhist hierarchy of heaven.

A prince from a successful Gang basin dynasty, Siddhartha abandoned the world at the age of twenty-nine, experienced intricate spiritual trials, headed the largest philosophical and religious school, and, after his death at the age of eighty, left several disciples and an enormous philosophical and didactic heritage. In the beginning, the doctrine was transmitted orally; later it took the shape of a written canon containing a code of fundamental statements of the new religion that subsequently became a worldwide one. Soon after Buddha's death the first Buddhist Council was held where his disciples systematized the doctrine and arranged Buddha's heritage into three basic groups: parables, rules of conduct, and statements of doctrine.

Buddhist doctrine was finally recorded in writing in 80 BCE, in Ceylon. It constituted the so-called Pali canon that, together with the "Milinda's questions," formed the ideological and practical base of the ancient Buddhist orthodox school of Theravada. Theravada gave birth to one of the largest contemporary Buddhist schools of Chinayana ("the smaller chariot") proclaiming the narrow way of salvation.

Buddhist doctrine is based on the belief that suffering is the essence of life. Every suffering, though, has a reason. Knowing the reason enables us to put an end to suffering. It can be ended in a special conscious way. Every individual can adopt these four Buddhist truths, if they follow the noble octal way of salvation. This way includes the following regulations: right understanding, right aspiration, right speech, righteous behavior, righteous living mode, righteous eagerness in spiritual exercises, right and controlled by reason perception of life, and right self-absorption leading to the state of nirvana.

An individual is the focus of the entire Buddhist worldview. Each individual's karma (voluntary and conscious positive and negative acts predestinating the individual's path in the series of reincarnation) is determined by his or her conscious actions. That is why the Buddhist doctrine pays so much attention to the clearness of terminology and audibility and lucidity of formulation.

The idea of the individual soul is important for Buddhist doctrine. Here the difference between Eastern and Western ways of expressing ideas is clearly demonstrated. Buddha kept silent in response to a hermit's question of whether "I" exists. Then he explained to his disciple that any verbal response to such a question would entail an opinion and would adjoin his doctrine to any of the Brahmans' schools, while keeping silent, he maintained his independent approach.

Currently, a hundred million people adhere to the Buddhist doctrine. The Buddhist concept of universal harmony and peace is valuable and attractive for the entire world community. At the same time, the concept of karma and the connected conception of sansara circle (which is the eternal movement involving all the living individuals and forcing them to concentrate all efforts and voluntary actions on salvation) sometimes contradict modern political, moral, and ethical principles, the idea of human rights, and the values of post-industrial society. However, the humanistic ideas of Buddhism and its art remain an imperishable cultural heritage of humankind.

P. M. Kozhin

౦౩ C ౮౦

Capitalism: The word "capitalism" owes much of its prominence to its major enemy, Karl Marx, whose definition of it remains quite satisfactory: that socioeconomic system in which capital, the "means of production," is privately owned. A subsidiary distinction is that between the sums of money that are invested in those means of production, and the productive factors themselves that are purchased with it, the term "capital" being used to stand for either. Productive forces, in turn, might be further distinguished, into (1) investment in physical assets such as productive machinery, which Marx called "fixed" capital, and (2) the labor-time purchased (or the money spent on it) to put the physical assets into operation. Note, however, that the presence of capital in either of these latter senses is not sufficient to call the society in which it occurs a "capitalist" one, for the deciding factor is not whether the society has capital—all societies have at least a little—but, how it is controlled.

The opposite of capitalism is socialism, the system in which the society's capital is, as we may put it, owned by society at large. Note that ownership of capital would not be absent in a socialist society: rather, it would be in the hands of "society." However, that immediately raises major questions of what that would mean and how it would be effected. In practice, it would have to be a matter of control over capital by persons representing, or at least claiming to represent, the interests of people at large.

Capitalism and socialism can be matters of degree. Some of the capital of society could be in private hands, and other parts directed by public agencies. "Mixed" economies are dominant in today's world, where all of the "advanced" countries allow substantial private ownership of capital, while at the same time operating many functions such as education, health, and welfare services. Some do not recognize the latter as "socialist" on the ground that the "production" that Marx refers to is only "material" production, as of food, clothing, houses, automobiles, and other consumer goods. However, it is not clear why this distinction should be of fundamental importance, since all of the listed functions—education, health, welfare—can also be privately produced, and the issue whether they ought or ought not to be appears to be just the same issue in either case.

Capitalism is closely allied with the notion of the "free market"; its defenders prefer the term "free enterprise," which is appropriate when we recognize that if capital is private, that means that whether a given capital expenditure is made is decided by some individual or group of individuals, acting on their own rather than under the direction of political agencies. This raises an important question: What are the motives of those who do decide to invest capital in one way or another? The standard answer is, private profit: the person making this investment does so in the hope of "making" money, that is to say, of being able to sell the products of the enterprise for more money than it costs to make them, and to be in control of those further moneys. Note, however, that profit making is not a logically necessary element of the definition capitalism. Non-profit enterprises have been established in many countries—persons invest money in some activity, not for their own

48   *Global Studies Encyclopedic Dictionary*

profit, but for some charitable purpose. It is possible for enterprises to be jointly owned by their own employees—the "cooperative" or "co-op"; here the profit is not private to someone or few persons who do not themselves work in the enterprise, but is shared by all in some way. Sometimes the word "capitalism" is restricted to the system in which business enterprises are not cooperative in that sense, but that restriction seems somewhat arbitrary.

We should note here that capitalism involves a commitment to specific ethical principles. Arguably, they involve a commitment to just one ethical principle: the right to private property (income), in a sense of that term broad enough to include the "ownership" of one's self. Agents in a free-market society proceed by making offers to people who have what those agents want; they do not proceed by using force or by fraud. In saying this, we immediately encounter a problem familiar in the expounding of any ideological or normative thesis, for obviously it sometimes happens that people do increase their income by theft, or by overt violence, and certainly by telling lies. However, what constitutes a free-market system is not the deviations, but the situation where the principle of private rights is respected. In a market system that is functioning strictly as such, with no deviations, all transactions between persons are voluntary: no one is compelled to make whatever decisions he or she may make about what to do with that individual's time, personal capacities, income, or property.

Proponents of capitalism argue that if that condition is realized, then it is reasonable to suppose that each person will do its best with life. Or rather, that a person will do the best, in its view of what is best. Critics may argue, by contrast, that an individual's view of what is good may not be the best view, that a person requires the intervention of social agencies to do well. To this, supporters will reply that those "social agencies" are just more people, and it is hard to see why they should be any better at running one's life than the individual. They may go further and point out that if these other agencies are actually given power over that individual, then what is to prevent those agencies from exploiting their client, who after all is deprived of control?

In an obvious way, capitalism has won and socialism has lost the ideological debate. At the same time, every "advanced" country has an extensive public sector, and in most cases, those sectors represent half or more of the country's Gross Domestic Product. To justify this in the face of the arguments supporting capitalism, it must be held that the special kind of goods that education, health, and welfare consist in are somehow different from the "consumer" sector with its cars, homes, foods, restaurants, and so on.

Today, a small minority of thinkers—the "libertarians" and more specifically the "anarcho-capitalists" or "market anarchists"—hold that this is all a mistake, and that society would do still better if government left even those goods to the workings of the market (or, in the case of the market anarchists, ceased to exist altogether!). Whether a society that puts half the decisions about what to do with people's money into the hands of politically selected public agencies should really be called "capitalist" is also an interesting question. If not, we should say that instead of capitalism's having "won," the situation is currently pretty much of a draw, with the "capitalist welfare state" the virtually universal result.

J. Narveson

Capital Flight refers to the drain and transfer of financial flows from peripheral or semi-peripheral countries to the central states of the capitalist system. The strongest movement of financial resources from the developing states toward the developed ones has been taking place since the 1980s. Capital flight is caused mostly by the total indebtedness of some countries to others, such that the amount to be repaid exponentially exceeds credits obtained. Additionally, since peripheral and semi-peripheral countries are characterized by less favorable conditions for capital growth, capital moves toward more developed states. All these factors are related to the functioning of offshore centers, through which basic capital is transferred from various countries. In peripheral and semi-peripheral countries, capital tends to leave due to political instability, high level of corruption and low labor productivity. According to estimations of analysts, about $1 billion was annually exported from Bulgaria throughout 1990s, and Bulgarian citizens own about 7 billion dollars outside Bulgaria, which substantially exceeds the total amount of all the loans and foreign aid to the country provided for the same period. Even bank deposits of poor countries are kept in banks of developed nations.

In the late 1990s, about 70 percent of the world's available capital was transferred to the United States as investments (shares, debentures, assets of absorbed companies). The United States also receives capital fleeing Western Europe. After the collapse of the state socialist system in Eastern Europe and the former Soviet Union, hundreds of billions of dollars moved to West European and North American countries in the form of deposits in their banks, or in securities.

V. Prodanov

Capital Punishment, or the death penalty, is a legally enforced deprivation of life based on a court decision; a lawful infliction of the extreme penalty on a person convicted of a grave offense. The morality of this practice is the subject of public debate, in which philosophical and ethical arguments play an essential role.

Two main viewpoints on the nature and goals of the death penalty have emerged. According to one view, the death penalty is prevents grave offenses by eliminating criminals from society and by deterring potential criminals from committing crimes based on the real threat of death. According to the second viewpoint, advocated by Kant, among many others, the death penalty constitutes retribution for grave crimes committed, which must follow irrespective of any considerations of expediency. Any utilitarian considerations are irrelevant. In this view, the death penalty is recognized as an absolute imperative and is conditioned only by the nature of the crime.

The first theoretical arguments in favor of abolishing capital punishment were given by Cesare Beccaria, Italian lawyer and educator (1764), who held that it is not so much the severity of punishment that deters crime, as its inevitability. Voltaire opposed the death penalty. G. W. F. Hegel, John Stuart Mill,

50   *Global Studies Encyclopedic Dictionary*

and Montesquieu, while supporting the death penalty in general, believed it should be restricted. In Russia, Leo Tolstoy and Feodor Dostoevsky advocated the abolition of death penalty.

All the principal arguments for and against capital punishment can be divided into the pragmatic, moral-pragmatic, and moral ones. One of the strongest arguments of Beccaria and his supporters was that many observations had failed to establish an empirical correlation between the death penalty and the number of grave offenses. Thus, the view of the death penalty as the means of preventing grave crimes is rendered baseless. We can also argue that the death penalty is an ineffective means of struggling with evil, for it creates a semblance of struggle, distracting attention from taking adequate measures really needed to prevent grave crimes. Finally, the possibility of wrongful conviction stands as a sufficient reason, from the standpoint of justice, for the abolition of the death penalty.

Apart from these largely pragmatic arguments, the opponents of the death penalty often put forward a moral thesis of the inviolability of human life. Nobody, including the state, has any right to infringe on life, even on a criminal's life. As for just retribution, it cannot be based on the principle of equal retribution (*lex talionis*), which is typical for ancient societies. The state cannot to any degree act like a murderer, reciprocating with the same actions.

Both the reality and procedure of executing capital punishment are horrible, involving a planned, cold-blooded process of killing. As the criminal has to be punished anyway, the death penalty is often substituted with a life sentence as an alternative measure of punishment. That punishment may turn out to be even more severe than the death penalty. There have been many cases when criminals preferred the death penalty to a life sentence.

Reference: Beccaria, Cesare, and Voltaire. *An Essay on Crimes and Punishment*. Inscribed by I. H. Tiffany. Philadelphia: Printed by William Young, 1793.

B. O. Nikolaichev

Capital, Transnational: Transnational capital is capital that does not possess a national identity. It is earmarked and bunched in tax shields and financial stashes. From the beginning of the 1990s, the migration of capital increased drastically. On the one hand, there were objective causes of this: a great number of global and local financial crises occurred in conjunction with technological advancements that allowed the immediate payment of transactions. On the other hand, a substantial growth of "gray" and "black" capital transpired as the result of business criminalization, tax evasion, illegal arms and drugs trade. Legalization of this capital is carried out in offshore or in countries where it is legislated that banks need not disclose information about their clients. With the help of various financial schemes this capital is sheltered under the cover of these jurisdictions. Transnational corporations using different tax planning schemes also bunch their capital offshore and in these countries.

Nominally, capital that has left its parent country and resides in economically favored zones belongs to companies, usually offshore, located in other countries. However the governments of these countries (according to their laws and regulations) do not require the taxation of this capital or restrict its movement. At this stage, this capital changes status and becomes transnational. At the next stage, this transnational capital is invested in commercial activities. Countries with the highest returns and lowest risks are chosen for such investments (presently these are usually developing regions of South East Asia and South America). Most of the transnational capital returns to the parent countries in the form of foreign investments.

At the investment stage, transnational capital materializes and becomes joint stock, bank, venture, floating, and loan capital. This capital can be invested in criminal activity, terrorist organizations, political organizations and parties, and consumer purchases. The profit from these investments, again with the help of different schemes, appears in tax shields and financial stashes, thus ending the cycle. Transnational capital is often used for short-term investments in stock markets for speculative purposes. The transnational transference of this capital is especially notable, as it can be used in many countries during over a short time.

A. V. Ishkhanov

Center and Periphery are the basic notions in dependency theory: the center is the community of highly developed capitalistic countries of North America, West Europe, Japan, Australia, New Zealand and the periphery is the community of developing countries of Latin America, Asia and Africa. Countries of the highly developed center are distinguished by an economy based on modern science, by intensive, resource-, labor- and energy-saving technologies, by high indices of socio-economical and cultural development (per capita gross domestic product, life expectance, popularity of expensive and prestige durable goods, development of scientific and cultural infrastructure, literacy and education of people), by the existence of powerful states built upon political democracy and by liberalism as the leading ideology.

Countries of the underdeveloped periphery are characterized by an economy based on outdated resource-, labor- and energy-consuming, "dirty" technologies and agricultural or raw-material economic development, by a low index of socio-economical and cultural development, by political and public life oscillating between authoritarianism with the high level of used governmental and by non-governmental violence and unstable democracy.

Countries of the center, through transnational corporations, international economic institutions and international credit and financial system, exploit manpower and natural resources of countries of the periphery, and interfere into their political and social life.

The periphery is dependent upon the center, and its development is just a reproduction of backwardness and poverty.

V. I. Dobren'kov, A. B. Rakhmanov

Chernobyl Catastrophe was an nuclear accident that occurred at the Chernobyl nuclear power plant (CNPP) on 26 April 1986, when the fourth reactor broke down and a great amount of radioactivity was released into the environment. The territories of Byelorussia, the Russian Federation (the European part), and the Ukraine suffered most of all. The first radioactive pollution after the Chernobyl disaster was temporary, and caused by the short-lived radionuclide release, which later lost its ecological magnitude. Among the short-lived

52   *Global Studies Encyclopedic Dictionary*

radionuclides the iodine isotopes and the tellurium were the most considerable. A year after the disaster, the radiation status of the majority of the European territory of the Soviet Union was primarily defined by the long-lived cesium isotopes: cesium-134 and cesium-137, and, after three years, by cesium-137. However, in the nearest zone (mainly in the alienation zone) strontium-90, plutonium-239, plutonium-240, plutonium-238, americium-241, curium-244, cerium-144, and rutenium-106 appeared to be—and remain—the most formidable radionuclides.

The disaster after-effects had an essential impact on the alteration of radioactive pollution areas in not only the territories of the former Soviet Union and Russian territories, but also the whole Northern Hemisphere.

Y. A. Israel

China, Globalization Problems in: In 1972, China took part in the meeting on environmental problems held by United Nations in Stockholm, Sweden. It was the country's first acquaintance with the problem of globalization. It was also the time of the "proletarian cultural revolution" in China, so the Chinese government could not devote full attention to the problem, but recognition of its importance was closely related to the beginning of the reform-and-openness policy at the end of 1970s. Openness meant that the Chinese economy would take an active part in world economy, import foreign investments and advanced technologies, and sell its production in the world market. This circumstance stimulated study of globalization problems for two reasons. First, openness in external contacts resulted in joining the Chinese economy to common global market. This is the very basis of economic globalization. Second, the openness policy removed ideological bans, allowing new ideas to begin to penetrate into China from abroad, including those of globalization.

Since the 1990s, the problem of globalization gradually became a focus of attention of Chinese society. In 1992, a UN Forum on environmental problems and development took place in Rio de Janeiro, Brazil. The important document "Agenda 21," a UN economic agenda for the twenty-first century, was adopted there. After the Forum, the Chinese government actively concerned itself with the problem. In 1994, it published *China's Agenda 21— White Paper on China's Population, Environment, and Development in the 21st Century*, which acknowledged that the population of China is very large, resources are scarce, production is outdated, and the country is in the process of rapid industrialization and urbanization.

To ensure sustainable development, China should on no account follow the old way of Western countries, where everything is reduced to quantitative economic growth based on consumption. The problem of scarce resources is a global problem. Therefore, all countries of the world should jointly manage the challenges facing humankind. In the opinion of Chinese government: (1) economic development should be coordinated with environmental protection (conservation); (2) environmental protection is a common problem of humankind, however the developed countries bear greater responsibility; (3) international cooperation should be based on respect for state sovereignty; (4) environmental protection cannot be separated from peace and stability all over the world; and (5) while solving environmental problems one should not lose

Capitalism – Cyberculture 53

sight of practical interests of various states and long-term interests of the world as a whole. At the same time, it was pointed out that even countries as undeveloped as China can eliminate poverty, ensure the stability of society, accumulate funds and technology, improve environment and provide sustainable development only on basis of rapid economic growth.

The Chinese government has a great respect for economic globalization, but it is against the global-scale unification of politics and economics. It approves establishment of common world economy, but at the same time, supports political multiplicity and cultural diversity. Some scientists do not agree with such an approach. They believe that the creation of common policy and culture is inevitable under the conditions of common world economy formation. While fighting against political and cultural hegemonism of the United States on a global scale, it is necessary to seek the co-existence of global democracy with various national cultures.

Since 1990s, the works of Chinese scientists on the problem of globalization have grown rapidly. The study of this problem in China has the following distinctive features:

(1) It features a wide range of investigated problems. Not only philosophers, but also scientists are engaged in the study of this problem. Economists, political scientists, historians, literary students, artists and even government officials have show great interest in it. As a result, a wide coverage of problems has taken place, of which the most important are: the concept of globalization; economic globalization; Marxism and globalization; Dan Syao Pin theory and globalization; national, regional, and governmental globalization; globalization and China; globalization's effect on the economies of various states; globalization and foreign affairs; finance and globalization; and globalization and culture problems.

When discussing such an extensive subject as globalization's effect on the economies of different countries, Chinese scientists pay attention to the following specific problems: the ability of the developing countries to participate in economic globalization; what the attitude of developing countries to economic globalization should be; the choice of policies of developing countries in the economic globalization process; the uneven development of economic globalization and the developing countries; warnings to developing countries regarding the direction of economic globalization; social injustice in the twenty-first century; economic globalization and regional cooperation; economic globalization and European integration; the place and role of the United States in the process of economic globalization; the German model of globalization; globalization and state welfare; globalization and foreign-policy strategy of the United States in the twenty-first century; globalization and the policy of democracy in the countries of Latin America; Russia and economic globalization; the benefits and drawbacks of economic globalization and the response of the developing countries; the determination of the extent of participation in economic globalization; globalization and Japan; globalization in Asia; and globalization and national safety.

(2) A strong influence of Marxist philosophy is apparent. China is presently the only large state in the world acting under the flag of socialism. The present-day Marxism does not have such a critical and competitive spirit as

before, but still it is the main ideology within Chinese society. Marxism has had great influence on study of globalization in China. Some scientists apply the Marx-Engels' "theory of world history" to explain tendencies of modern globalization. They point at the fact that *German Ideology* (1846) and *The Communist Manifesto* (1848) contain foresight about the unavoidable global expansion of capitalism and the establishment of economic, political, and cultural uniformity of various countries, which will mean the beginning of true "world history."

The globalization we face today proves Marx and Engels's ideas. Chinese scientists applies Marxist method of class analysis to tendencies of present-day globalization. They believe that globalization rests on a global-scale international division of labor. The United States and other developed countries of the West, based on their capital and technology, act as capitalists in the world economic system, while China and other developing countries mainly rest upon their cheap work force to fabricate products for Western countries, acting as world proletariat. Present-day international relations have gained a class nature.

Yet another group of scientists connects the globalization problem with the historical destiny of capitalism and the future of socialism and capitalism. In their opinion, the appearance of globalization proves that capitalism threatens the very existence of the humankind. Its development has come to its historical end and the replacement of capitalism by communism becomes the inevitable result of historical development.

(3) The main distinctive feature of the study of globalization in China is its fundamental practical rationalism. The majority of Chinese scientists are not interested in the problems of the nature of globalization, its moral estimation, and the future of socialism and capitalism. They prefer studying very specific technical problems. Chinese scientists consider globalization mainly as an objective and historically inevitable tendency, which is directed by the developed capitalist countries. It cannot be resisted and should be accepted. They study the following questions from different viewpoints: How should China respond to globalization? How can China avoid or weaken the adverse effect of globalization? How can China use favorable factors and to promote development of the country?

Owing to the strong influence of practical rationalism, there exists no anti-globalist movement in China, as distinct from developed countries in the West. The economic success of China to a considerable degree may be explained by the fact that China managed to seize the opportunity and to use efficiently the chance provided by globalization.

By studying globalization, Chinese scientists accentuate an active role of Chinese traditional philosophy. Some scientists point to the fact that two thousand years ago, an idea of world unity was set up in Confucian canon *Liyun*. Under world unity, the private property and selfish (individualistic) ideas will disappear and all people will become brothers. This idea is identical with the ideal of globalization.

A. Qinian

China, Liberalism in Contemporary: For nearly half a century, the term liberalism almost disappeared in China's political lexicon, only mentioned when it was criticized as a bourgeois ideology or symbol of egoism. At the turn of the twenty-first century, however, there has been a revival of liberal political philosophy, echoing the situation a century ago when Western liberalism first came to China.

(1) The Revival of Liberalism: A conference entitled "Cultural China: Thoughts and Trends in Transition" was held at Princeton University in the United States in the mid-1990s. The debate was soon echoed in China: a group of professors and researchers with liberal viewpoints appealed to the new liberalism clearly and stoutly. A major symbol of the revival was the publication of a series of articles in memory of Isaiah Berlin, who had just died, in *South Daily News Weekend*, the popular liberal newspaper in Canton. The tone of those articles positively evaluated the significance of Berlin's liberalism to China's reality and future. Three dead Chinese scholars (economist Gu Zhun, historian Chen Yinke, and free writer Wang Xiaobo) were also commemorated for their liberal ideas.

Liberals have always been criticized by new Leftists, whose position is closer to the orthodox doctrine. Through these disputes voiced in academic journals and on Internet sites, liberals expound their basic ideas and views on several important relationships and issues: individualism and collectivism, free market economy and Stalinist socialism, globalization and nationalism, human rights and national sovereignty, liberty and justice, democracy and dictatorship.

(2) Individualism Reconsidered: Liberals argue that it is time to re-evaluate individualism; the collectivist argument is an ambiguous expression that often conceals some dangerous or totalitarian intentions. Liberals stress that the individual is the basic unit of a society, taking priority both in the genetic and ontological sense. Although the social influence of the individual is evident, the difference in personality and individuality of people in the same circumstance proves the existence of individual subjectivity and uniqueness. In that sense, freedom is first and foremost individualistic, and collective and national freedom is secondary, because it presupposes individual liberty. Professor Xu Youyu of the Philosophy Institute of CASS points out: "the core of liberalism is the affirmation of individual value and esteem, respect and protection of personal rights and interests. A profound understanding of liberalism would necessarily break from the traditional conception that oppresses individuality and takes the nation and collective as the only value and it also requires a clear distinction between individualism and egoism or selfishness" (quoted in He Le, 2011, p. 136).

New liberals emphasize that private property is a prerequisite for guaranteeing personal freedom; so they advocate amending the constitution and promoting the rule of law for private ownership. They also hold that the biggest threat to individual liberty is often from government; so, there must be the separation of powers and checks and balances among them. Contrary to the orthodox doctrine that condemns private ownership and demands an omnipotent government, liberals stress that if most of a society's property is concentrated in the government's hands, it would provide a solid base for despot-

56 *Global Studies Encyclopedic Dictionary*

ism; therefore, civil society must be separated from the government, and likewise capital from political power.

(3) Human Rights and Sovereignty: Since China suffered so many human rights disasters throughout the twentieth century, the new liberals claim that the issue should not be taken as purely domestic, ignoring international standards and supervision. On the contrary, narrow nationalist propaganda and ideology play an important role in domestic politics and international relations. *The China that Can Say No* (Zhang Zangzang et al., 1996) aroused so much anti-Western fever in China that the new liberals have warned many times that such a dangerous attitude would cost too much for an open China. They cite Alexis de Tocqueville's argument that there are at least two kinds of patriotism, one is narrow, instinctive, and irrational, the other is universal, sane and rational; they stress that the Chinese people need rational patriotism.

Liberals emphasize that human rights include not only the right to survive, but also the right to equal treatment, freedom of movement, conscience or belief, expression, assembly, and other political actions. Also, a nation's sovereignty cannot always be superior to its citizens' human rights; you cannot imagine a nation executing its sovereignty and enjoying its national pride with most of its citizens' basic rights not being protected.

(4) Re-Evaluation of System Choice: Liberals hold that communist revolutions deviated from the mainstream of world civilization—the direction toward freedom and democracy, and that Deng Xiaoping's reform and open-door policy was a regress in that direction. However, it should not stop half way: for future progress, a liquidation of the past is necessary. So, they require that all the past revolutions or revolutionary movements, from the Russian Bolshevik Revolution, the May Fourth Movement to Mao's Cultural Revolution, must be re-evaluated from a liberal democratic perspective. Leftists, on the other hand, hold that liquidation is "to stubbornly maintain the Cold War thinking," and they refuse to deny the revolutionary history of the twentieth century since the 1911 revolution, because those revolutions were of historical necessity. Such an excuse for past revolutions is unacceptable to liberals; they insist that moral responsibility for the Cold War cannot be divided equally between the two sides. As a matter of fact, what the Soviet side did was wrong in economic policy, political orientation, and human rights records; China unfortunately stood on the wrong side.

(5) Freedom, Equality, and Justice: New leftists criticize Liberals on the grounds that that they take freedom as an excuse to suppress equality. New leftists contend that a few intellectuals are concerned only about their own freedom of speech, not the suffering of workers, and they attribute current injustice and corruption, including the draining of state property to private persons, to the market economy and internationalization. Therefore, they want to regress to tight market control by the government and try to refute liberals who maintain that legitimate privatization will resolve the current social strains.

It is not fair, liberals argue, to let the market take all the blame and be the scapegoat of all injustice and corruption, because the real reason for those social problems is the centralized, unchecked, and omnipotent political power interwoven into the market. Such a power structure and tradition would produce huge corruption and social injustice even if there were no market.

Capitalism – Cyberculture 57

Therefore, the way to get rid of injustice is not to restrict or abolish the market, but to improve and consummate it, to protect private property, develop democracy, and implement social justice.

(6) Democracy: Direct or Indirect? New leftists repeat the Maoist slogan of direct, or grand, democracy, which is the populist view that everyone participates in the policy-making process. Liberals, by contrast, argue that it is a high-sounding democracy without feasible procedures, since its scope is too large to fulfill and it implies some dangerous tendencies. Without due process and civil liberties, direct democracy presupposes a political power that is unrestricted and expanding limitlessly.

Indirect democracy is liberal democracy whose aim is to protect citizens' rights to liberty; it is a mixed political structure, such as representative or parliamentary democracy, that has multi-layers and multi-checks in its decision-making process; also, it is a mixture of constitutionalism, democracy, and republic, taking advantage of different polities to avoid the inherent limitations of pure polity. Comparing the two forms of democracy, some liberals say, it is acceptable that, when indirect democracy is developing, some direct forms of democracy can be undertaken, but it is dangerous that complete or pure direct democracy be constructed from the beginning of the institutional set-up with indirect democracy being totally put aside.

Some liberal professors in the Central Party College suggest that the Chinese Communist Party itself must be democratized, selecting their own leaders. Liberals have put forward some proposals of gradually enlarging the scope of direct election of government leaders, urging current top leaders of the communist party to declare a definite timetable for political reform. Chinese liberals stress that China should not be excluded from the general course of the economic and political progress of the world. Arguing furiously with new leftists, liberals contend that by staying in the idealistic and authoritarian model China's reform will reach a dead end of a monopoly of power and corruption.

References: Z. Zangzang, Z. Xiaobo, S. Qiang, T. Zhengyu, Q. Bian, and G. Qingsheng. *Zhōngguó kěyǐ shuō bù* (The China that Can Say No) Beijing: Zhonghua gongshang lianhe chubanshe, 1996; He Le. "Liberalism and Its Impact on China's Reform." In *The People's Republic of China Today*. Edited by Zhiqun Zhu, p. 136. Hackensack, NJ: World Scientific, 2011.

S. Gu

Chinese Enterprises, Transitional Activity of: China's entry into the World Trade Organization marks a new stage in carrying out the policy of openness to the world and participation in international economic competition. Encouragement of transnational activity of Chinese enterprises is a new strategic policy of the Chinese government. A stable and rapidly developing Chinese economy lets Chinese enterprises actively operate in the international economic arena.

The strengthening globalization tendency is characterized by the growth of foreign investment. There were only 7,000 transnational corporations (TNC) in the mid-1970s, and by 2000, this number has grown up to 63,000 with about 700,000 foreign branches. These TNCs produce one third of the world's gross national product. Trade within and among transnational corporations amounts to 60 percent of international trade. Direct investments made

58   *Global Studies Encyclopedic Dictionary*

by multinational corporations exceed 90 percent of global direct investments. The 500 largest international TNCs introduce 70 percent of new technologies. According to data of the UN Conference on Trade and Development, in 1990, there were only twenty-four TNCs among the 100 largest international economic entities, while in 2000, this number increased up to twenty-nine.

Foreign investment activity of Chinese enterprises began at the end of the 1980s. By July 2002, there had been 6,758 enterprises with Chinese capital (not mentioning financial investments) in almost 160 countries and regions. Chinese investments, stipulated in investment agreements, amount to over $8.7 billion USD. Foreign investment activity by Chinese enterprises started with establishment of trade organizations providing export and import activities. Gradually, the sphere of their interests expanded to include agriculture, construction, transport, development, and manufacturing of new types of products, development of natural resources, communications, finance and insurance, public catering, and tourism. In terms of investment forms, most of the trade organizations are enterprises with purely Chinese capital; the industrial organizations contain foreign capital as well; organizations dealing with the development of natural resources are mostly co-operatives. Though Chinese enterprises have just begun their activity in the foreign markets, some corporations have already attracted attention due to their achievements. The world market has witnessed the emergence of the first Chinese TNCs. Since 1998, the corporation "Higher" has established about thirteen factories in the United States, Italy, and other countries. These enterprises work in the area of the development of new kinds of products, manufacturing, and sales. In 2001, Higher was fifth on the list of home appliance sellers in the United States market, as the total sales of home appliances amounted to $150 million USD.

In spite of the successful establishment of foreign enterprises, this transnational activity is still at the initial stage. Foreign investments are rare; the total sum of foreign investments amounts to 0.15 percent of the present sum of the world direct foreign investments. Direct foreign investments of Chinese enterprises amount to less than 1 percent of the fixed capital investment in China, whereas the average world index exceeds 5 percent. The ratio between investments into foreign countries and from foreign countries accounts to 1:0.011. This index is much lower than the corresponding figure in the developed countries, which accounts to 1:1.1, and concedes even to the index of developing countries that accounts to 1:0.13. Average foreign investments of Chinese enterprises amount to over $1.3 million USD, whereas the average foreign investments of the developed countries amount to six million United States dollars. Chinese foreign investments are excessively concentrated (about 60 percent) in South-East Asia. The structure of Chinese foreign investments is insufficiently rational regarding the sectors of the economy. More than 60 percent of enterprises and investments are in the trade sphere, whereas production takes only 23.2 percent. Besides, management and economic efficiency are far from ideal.

The policy of openness to the world means both attracting foreign investments and entering foreign markets. The only way to ensure full openness is mutual investments. Activity in both domestic and foreign markets allows one to use domestic and foreign resources, domestic and foreign markets. China needs competitive and powerful TNCs that will help to raise the international competitiveness and international prestige of the country. Twenty years of reforms show that many Chinese enterprises are capable of entering the world market.

Several macroeconomic factors, however, constrain full-scale foreign investments: imperfect policy and legislation; the absence of legislation regulating economic activities abroad. The state does not clearly define its structural policy and preferences for foreign investments. Selection and approval are too complicated; state currency control is excessively strict. Restrictions are imposed on the currency sources, on bank credits for foreign investments, on reinvestments of the profit gained from foreign investments. State policy on foreign investment encouragement is directed toward only a few enterprises and sectors. Besides, no preferential terms for foreign investors have been established. Recently, Chinese enterprises began to apply new methods of investment. Earlier they created new companies; the new investment methods include the issuing of shares for international securities markets, shares exchange with foreign companies, and purchase of foreign companies.

L. Yetsin

Chinese Philosophy and Globalization: Compared to Western philosophy, Chinese philosophy has distinctive features specific to Chinese culture. Science appeared in the West from rational thinking, while Chinese philosophy has developed from a love of wisdom (principally for worldly wisdom). The focus of attention in Chinese philosophy is a spiritual (inner) life of an individual, which cannot be subjected to rigorous scientific study.

Confucianism, Buddhism, and Taoism are the substance of Chinese philosophy. Confucius did not believe in gods, nor did he deny their existence. Buddhism as a religion has turned China into specific cultural phenomenon and has lost its religious signs (for instance, belief in gods or in supernatural powers). Just as in Buddhism, Taoism in China is not a religion in the strict sense of the word. Buddhism and Taoism are not religious teachings: they are human wisdoms. Chinese philosophy is the love for human wisdom and not the love for the wisdom of a god (or gods). Reason and belief never conflicted in Chinese philosophy; they coexisted peacefully.

As compared to Western philosophy, concepts in Chinese philosophy are never strictly defined. It comes out rather as a set of senses, defined depending on specific situation. Such a philosophy, without clear concepts or strictly logical reasoning, cannot be a systemic one. Chinese philosophy never sought after a system (wisdom has no system). Philosophy is a love for wisdom and not the deference to reason or knowledge.

Every nation has its own idea, which is especially distinct in its national philosophy. Chinese philosophy expresses the essence of Chinese idea. In the age of globalization the traditional Chinese philosophy may significantly enrich resources of world philosophy. China, the very ancient country with rich cultural tradition, should take part in the international globalization process and in the dialog of cultures. Chinese philosophy should not isolate itself; it

60   *Global Studies Encyclopedic Dictionary*

should carry on a fruitful dialogue with other cultures, in particular, with Western philosophy.

Z. Baichun

Christianity (Ancient Greek *Christos*, "Christ," from the Hebrew *Mašíaḥ*, "the anointed one" and Latin suffix *ian*, used to form adjectives, and *itas*, used to form nouns, "a state of being") Christianity is a monotheistic religion based on the life and teachings of Jesus Christ as presented in the New Testament. Christianity is the world's largest religion, with approximately 2.2 billion adherents, known as Christians.

Christianity emerged in the first century CE in Palestine, which was a part of the Roman Empire, against the background of a severe economic, political, and psychological crisis, accompanied by social and ethnic inequality and oppression. Initially, it spread among the oppressed (slaves, libertines, craft workers, and paupers), and was characterized by radical anti-Roman attitudes and uprisings against the imperial domination. In the second and third century, as the radicalism and reconciliation with the Roman authorities progressed, the participation of Roman senators in Christian communities grew. As long as this tendency was increasing, the authorities reduced cruel persecution of Christians and supported the new religion. Eventually, in the fourth century, Christianity became the official religion of the Roman Empire.

Ideologically, ritually, and institutionally, Christianity can be shown to have influences stemming from Roman polytheism, Egyptian cults of Osiris and Isis, Iranian Mithraism, and Greek cults of Kibena and Attis. However, the Judaic ideas of monotheism, its messianism, and eschatology played a special role in the development of Christianity. It emerged as a Judeo-Christian sect, not simply adopting any existing religious or cultural material, but creating a new religious complex: theology, culture, and traditions. In addition, Stoicism, neo-Pythagoreanism, Platonism, and neo-Platonism, and Aristotelianism provided metaphysical, ontological, ethical, logical, and psychological categories and terms for the emerging Christian thought and allegoric, dialogic, and symbolic ways of interpretation the educated Christians knew very well.

Christian doctrine, tenets, ritual, and organizational structure were created for several centuries through the efforts of the early Christian theologians and to the initial Ecumenical Councils, were subject to heated debates and confrontations. The Creed of Nicene and Chalcedon, adopted by the first Ecumenical Council in Nicene in 325, was of principal importance. (The second universal council was held in Constantinople in 381.)

Christian tenets include: creationism, providentialism, the threefoldness of God expressed in His three manifestations—God the Father, God the Son, and God the Holy Spirit, the incarnation, the resurrection of Christ, the atonement, and the second coming of Christ. Rituals include sacraments, ceremonies, and holidays.

The economic, socio-political, geopolitical, and cultural conditions prevalent at the time when Christianity was emerging had an impact on the specifics of interconfessional developments, particularities of the creed, ritual, and organizational structures. Differences between the Western and the Eastern Church increased and finally resulted in a Schism between Eastern Orthodoxy and Roman Catholicism (1054). In the sixteenth century, Reformation and Protestantism emerged. These three are now the central branches of Christianity. Other recognized sects of Christianity include: a group of non-Chalcedonic churches has been established (they seceded before or after the fourth Ecumenical Council in Chalcedon in 451 and do not accept the teaching about Jesus as it is formulated in the Nicene Creed); the Armenian Apostolic Church; the Coptic Church; the Ethiopian Church; the Syriac (Jacobit, Syriac-Jacobit) Church; the Malankar Syriac Church; the Eritrean Church; the Assyrian Church of the East; and the Malabar Church (the Christians of Apostle Thomas), as well as many smaller Christian communities.

I. N. Yablokov

Christianity, Orthodox (Greek *orthos*, "straight," and *doxa*, "opinion") The Orthodox Christian Church is one of the two major divisions of Christianity (the other being the Roman Catholic Church, or Latin) that resulted from the medieval division of Chalcedonian Christianity, also known as the East-West Schism (1054). It is the second largest Christian Church in the world, with an estimated 225–300 million adherents.

Historically, the differences between Orthodoxy and Roman Catholicism are numerous and profound and since Vatican II, the differences have widened (Azkoul, 1994), though Pope Francis may seek to reunite the two Churches. The differences fall into these categories (1) *Faith and Reason*: The Latins place a high value on human reason, building on the results of philosophy and science. Orthodoxy does not change tenets of faith based on any findings of science. (2) *Development of Doctrine*: Orthodoxy does not, like the Latins, believe that the theology should change over time. (3) *Belief about God*: The Roman Catholic Church believes that the existence of God can be proven by reason, while Orthodoxy holds that the knowledge of God is planted in nature; unless we have the gift of faith, we cannot know God. Also, Orthodoxy distinguishes between the essence of God (what God Is) and God's uncreated energies (the means by which God acts). (4) *Belief about Christ*: Orthodoxy teaches that Christ voluntarily gave his life as the "ransom for many," while the Latins teach that Christ's sacrifice was just satisfaction for the sin of Adam, whose debt had been transferred to all humanity. (5) *View of the Church* (ecclesiology): The Roman Catholic Church views the Pope as the Bishop of all the Church, while Orthodoxy views all Bishops as equal. (6) *Holy Canons*: Unlike the Roman Catholic Church, the Orthodox Church does not think of canons as laws regulating human relationships or securing human rights, but as means training in virtue meant to produce holiness. The Roman Catholic Church continually revises Canon Law. The Orthodox Church, while it may add canons, never changes the original canons. (7) *Mysteries*: Both the Orthodox and the Roman Catholics recognize at least seven Sacraments or Mysteries: The Eucharist, Baptism, Chrismation (Confirmation), Ordination, Penance, Marriage and Unction, but the belief in the nature of the sacraments differs; e.g., the Orthodox do not believe in transubstantiation. (8) *Nature of Humankind*: Both Churches believe that humankind was originally created good. Orthodoxy believe that Adam's sin introduced

62 *Global Studies Encyclopedic Dictionary*

death into the world, and with it, the passions (the Cardinal Sins), and it places heavy emphasis on the need to master the passions. (9) *Mother of God*: Both Churches believe that Mary was the virgin Mother of God, but the Orthodox reject the doctrine of the Immaculate Conception of Mary. (10) *Icons*. Both Churches employ icons, but the Roman Catholic Church uses both flat and 3D (statues) depictions, while the Orthodox Church holds theological reasons for using only flat depictions. (11) *Purgatory*. The Roman Catholic Church believes in purgatory, where souls undergo a cleansing before entering the Kingdom of Heaven, while the Orthodox Church believes that souls ender Hades (a place of the dead) to await rejoining of the body with the soul upon the second coming. There are also a number of other, minor differences.

The Church's structure is composed of several self-governing ecclesial bodies, each geographically (and often nationally) distinct but unified in theology and worship. Each self-governing body (autocephalous jurisdiction), often but not always encompassing a nation, is shepherded by a Holy Synod whose duty, among other things, is to preserve and teach the apostolic and patristic traditions and related church practices. Like the Catholic Church, Anglican Communion, Assyrian Church of the East, Oriental Orthodoxy and some other churches, Orthodox bishops trace their lineage back to the apostles through the process of apostolic succession.

Some Orthodox Churches do not recognize the autocephaly of others; for example, the Russian Orthodox Church (the ROC) does not recognize the autonomy of the Ukrainian Orthodox Church, while the American Orthodox Church does. The ROC recognizes fifteen autocephalous and four autonomous churches. The Chinese Orthodox Church is also formally autonomous (since 1957) but so far it has no primate, lacks temples, and members. The four most ancient autocephalous churches (the Patriarchates of Constantinople, Jerusalem, Alexandria, and Antioch) are especially respected. The Patriarch of Constantinople is titled "ecumenical," but does not possess the power of the pope and is considered "the first among the equals." The following churches are also autocephalous: the Russian, the Georgian, the Serbian, the Romanian (autocephaly proclaimed in 1885), the Bulgarian, the Cyprian, the Hellenic (Greek), the Albanian, the Polish (autocephaly granted by the Constantinople Patriarchate in 1924–1925, recognized by the ROC in 1948), the Czechoslovak (autocephaly granted by the ROC in 1951), and the American (autocephaly granted by the ROC in 1970). Autonomous churches include: the Finnish (since 1923, under the aegis of the Constantinople Patriarchate), the Church of Sinai, the Japanese (since 1970), and the Cretan.

All the churches share a liturgical relationship with each other, and dioceses and other administrative entities of one may be located on the "canonical territory" of the other. However, this does not prevent the churches from being different. For example, the Jerusalem, the Russian, the Serbian, and the Georgian churches use the Julian calendar (the old style), while all the rest use the Gregorian one (the new style). Many people think this "unity in diversity" is an advantage of Orthodoxy making it flexible and easily adjustable to every cultural environment. At the same time, each nation's desire to have "its own" autocephalous church can be seen as an obstacle to the unity of the Orthodox world.

Many churches have no liturgical relationship with those listed above:

Capitalism – Cyberculture 63

(1) The "Ancient Eastern Churches" include pre-Chalcedonian or monophysite churches, namely, the Armenian, the Coptic, the Jacobyte (Syriac), the Malankar Syriac (in Southern India), the Malabar, the Ethiopian, and the Eritrean (or, those who did not recognize the decisions of the Chalcedonian Council of 451), and pre-Efesian or Nestorian churches (Assyrian or Chaldean) (or, those who did not recognize the decisions of the Efesian Council of 431). The Nestorians think that God entered a man Jesus after his birth. The monophysites believe that the divine nature of Christ fully displaced his human nature. These churches consider themselves Orthodox, but the others do not recognize them as such.

(2) The Uniate churches are those keeping Orthodox rites but recognizing papal authority and Catholic dogmas. About twenty-one such churches exist. Some Uniate churches (e.g., in Ukraine) are very influential because they play an important role in the process of national identity building.

(3) Churches that did not recognize the reforms held by the "official" Orthodox churches fall in this category. These are various groups of Russian Old Believers (*staroobriadtsi*) that seceded in the seventeenth century (for example, the Russian Ancient Orthodox Church with its center in the town of Novozybkovo); the Russian Orthodox Church Abroad (seceded after the socialist revolution in 1917); the Genuine Orthodox or Catacomb Church (seceded in 1917, functioned underground in the Soviet Union); "Old Style" churches, such as the Church of Genuine Orthodox Christians of Greece (seceded from the Greek Orthodox Church having refused to observe the Gregorian calendar), and the Old Style Orthodox Church of Bulgaria.

Reference: Azkoul, F. M. "What Are the Differences between Orthodoxy and Roman Catholicism?," *The Orthodox Christian Witness* 27:48, and 28:6 and 8 (1994).

A. V. Mitrofanova

City, Creative: The "creative city" is a conception of a city where work, home, and pastime are integrated into a single, environmentally friendly living space. It reflects a new paradigm for the development of world civilization, which is conditioned by the requirements of the informational society and proceeds from the assumption that creative abilities are peculiar to every citizen. Social orientation with a creative human, or *Homo creatoris*, in its center comes to take the place of machine and technique worship and the cult of consumerism.

In the industrial age, urban space consisted of three clearly distinct areas: place of employment, pleasure resort, and leisure place. With the advent of the new economics and hi-tech production the emphasis was put on a new system of organizing interaction human, or technology (the idea of "technopolis" first appeared in 1960s in the United States). In 1970s, the conception of technopolis was implemented at the state level in Japan, involving the deliberate usage of scientific and cultural potential in the interests of prosperity of the Japanese nation.

The idea of technoparks, which rest upon innovation, also served to organize work. The interaction between humans and nature was reflected in the concept of ecopolis, which allows the combination of the advantages of urban and rural habitat. However, with the disintegration of the places of employment, rest and leisure is abolished only in the concept of a creative city. The

64   *Global Studies Encyclopedic Dictionary*

environment is considered as a single whole, and work is regarded as a creative act, hence the concept of fixed time of work, rest and leisure disappears. The creative city becomes the field of public space and the place of exchange of ideas and innovations, being an alternative to the still dominating tendency of privatization of public space. Authority delegated to lower levels substitutes for centralization, encouraging a variety of lifestyles, traditions, multiculturalism, tolerance, and self-management, so that pride and self-respect are cultivated. The creation of a multitude of city centers with their distinctive features, which serve as public spaces to generate creative ideas, is promoted.

The technology of transforming a depressive city into a prosperous and creative one was developed in Great Britain within the framework of CTI project (Creative Town Initiative) applied in Huddersfield, near Leeds. The idea of a creative city is being implemented in such cities as Helsinki, Finland; Dubna, Russia; and Ilyichevsk and Ovideopol in the Ukraine.

S. Udovik

City Development, Urban and Ecological Aspects of: In the course of human history, the growth and development of cities influenced the life of society. Urban lifestyle implies development of industry, motor transport, waste processing. All these are anthropogenic kinds of activity, which create urbanized space around them.

Mass urbanization is mainly a phenomenon of the twentieth century. In 1900, no more than 14 percent of the world population lived in towns. A city is an artificial habitat, created by human beings for their comfort and aims to free themselves from biological dependency on the natural environment. The city phenomenon enabled human beings to rise above the natural surroundings enough to forget about their affiliation with the natural world.

The Industrial revolution caused the appearance of big cities. Industrial development in cities and the corresponding growth of urban populations are considered to be the basic premises of mass urbanization. First, it was timber that served as the main energy source, then coal took its place, and later on, oil provided energy and became the basis of urbanization. Cheap oil provided fuel for vehicles and industry, and enabled technological processes to concentrate. The growth of urban, not rural regions was promoted by the economic policy of the government as well, because it made the foodstuffs cheaper in the cities. It was through the devastation of rural regions that many cities became enriched. All this served as a fuse to the urban explosion.

During the growth and development of cities, anthropogenic influence was laid on the natural components of urban and suburban, areas. Ultimately, the rapid growth of population makes the changes in nature more evident; changes take place in landscape, topsoil, flora and fauna, atmospheric air, and reservoirs, in the end leading to the violation of the balance of the biosphere.

If one considers the character of the interaction between urban population and natural surroundings, we can see the specific periods of system development. When a town appeared, when the influence on nature was not big in its scope, the environment remained in essence unchanged. As economic objects and activities increasingly influenced the natural complex, the natural environment was degraded.

Capitalism – Cyberculture

Today, economic activity makes quantitative and qualitative changes to the circulation of substances. In recent years, anxiety has been growing in recent years about the ill effects of chemical substances on human health and habitat. Chemical elements and their different mixes are amassed in the natural environment, artificially taken into nature and unusual to it. These materials, spreading into the atmosphere, include dust, gases, smoke, and solid particles; these materials then settle on the soil surface. In the last few years the ill effect of many substances was noted, such as pesticides, vinyl chloride and polychlorided biphenyl, dioxins, lead, mercury, other heavy metals, and chlorofluorocarbons. The presence of chemical pollution violates natural processes of the self-purification of the natural environment.

Some substances are deposited into nature in especially large quantities: sodium, chlorine, iron, vanadium, fluorine, copper, and barium are wasted in quantities of hundreds of thousands and tens of millions tons annually.

Deforestation and bog reclamation has greatly influenced the circulation of basic biophils, which are present in living substance and soil humus as the main components. These kinds of anthropogenic influence cause soil destruction and erosion acceleration. Nowadays, the anthropogenic flood of solid matter and atmogenic waste reach ten billion tons a year, a value that is close to that of eventual global denudation, which is 23–25 billion tons per year.

Urbanization elements from cities reach towns, conquering the material sides of human life, and forming a specific social environment, and spiritual life of society. The multidimensional process of urbanization has an effect on every side of human life.

Under the influence of the human-caused impacts, the quality of the environment worsens, threatening human health. Population health is threatened not only by the pollution of atmospheric air, soil, drinking water, and remoteness from natural areas, and noise and other factors. The net or, as medics say, combined effect of these different factors in their totality further augment this threat through different environments. In modern cities, there is a specific profile of population pathology: occupational diseases, neurosis, cardiovascular system diseases, traumatism, and respiratory tract diseases. With new chemical substances taking their place in the way of life the number of diseases such as intoxication and allergies has also increased. It is also known that a person gets tired more quickly in a city, and an urbanized environment favors the appearance of mental and psychosomatic diseases.

The atmospheric influence on soil-and-vegetative cover is connected to the fall of acid precipitates, washing calcium, humus, and microelements out of soils, leading to the violation of the photosynthesis process, which leads to a slowing down of plant growth and even to their death.

The high sensitivity of trees (birch and oak especially) to air pollution has been known for a long time. The net effect of both factors leads to a noticeable decrease of soil fertility and the extinction of woods.

Precipitation from the acidic atmospheric air accelerates the processes destructive to cultural monuments, causing the chemical destruction of human-generated objects and engineering constructions. The main chemical agents in the rains are dissolved sulfuric and nitric acids formed during sulfur and nitrogen oxidation reactions.

66 *Global Studies Encyclopedic Dictionary*

The atmosphere is characterized by an extremely high dynamism, caused by the rapid movement of air masses in horizontal and vertical directions and the diversity of the physical and chemical reactions occurring within it.

City populations are threatened where the exhaust from car engines pollutes atmospheric air by exceeding permissible levels of the concentration of nitrogen dioxide, carbon oxide, sulfurous anhydride, and solid particles. The most dangerous of the substances wasted by vehicle engines is lead, which can cause detrimental effects to every organic system. In children, high amounts of lead exposure can cause hyperirritability, stupor ataxia, coma, convulsions, or even death. With long-term exposure, lead can be amassed in the brain where it causes the IQ to decrease. All industrial countries are now switching to lead-free paints and gasoline.

Trash burning is also a source of air pollution in cities. This process ought to go under thorough control, because trash burning is the main source of dioxins, which are extremely toxic substances.

It is vital to find multi-factor approaches to industrial development that would allow us to conserve the ability of the environment to support our normal life. It is important to reach stable, ecologically safe industrial development. Reduction of air pollution and improvement of water and soil quality can be reached with the help of new ecologically pure technologies. Federal government and local autonomous bodies must control the process of nature management by a branch system of monitoring and controlling the state of the natural environment and checking that every managerial decision matches standards of ecological permissibility.

The process of urbanization develops at an extremely high speed, and though urbanization is a progressive process as a whole, it demands management, because the uncontrolled development of the urbanized areas violates the balance of the natural environment and has negative consequences.

A. G. Ishkov, O. V. Yakovenko

Civil Disobedience: Civil disobedience is an attempt to bring about a change in the law or in government policy through the violation of a law that is believed to be immoral, unconstitutional, or irreligious. In all probability, violations of law believed to be unjust have occurred, as long as there have been differences between king, or ruler, and subject, or citizen. In civil disobedience, there is a perceived conflict between two obligations. Particular individuals believe that the demands of morality or religion, often expressed as the requirements of conscience, require that they violate a legal command when there is a perceived conflict between the two.

But while civil disobedience involves efforts by the disobedient to disavow personal complicity with an unjust law, it requires that the disobedient agent see themselves as otherwise citizens, or members, of the political order. Therefore, the Hebrew midwives' disobedience might be interpreted as an act of nonviolent insurrection, inasmuch as the Hebrews were an oppressed people in Egyptian bondage. By contrast, in the tradition of civil disobedience, the purpose of violating a law, as a form of protest, has been to appeal to the conscience of the community in having injustice recognized and the law repealed. Thus, civil disobedience, as a form of political action, presupposes something more, namely, the ability to recognize the difference between the legal order, on one hand, and civil society—usually directed by religion or morality—on the other hand.

This objective in civil disobedience is almost always associated with the efforts of the disobedient to disavow personal complicity with an unjust system supported by obedience to the unjust law. Other objectives may be closely related: some may violate a law to obtain the opportunity for an appeal and a challenge to the constitutionality of the law; others may seek to bring government officials to the negotiating table. Others, following Mohandas Gandhi's example in South Africa, may seek to flood the jails and courtrooms enough to hamper the oppressive processes of the state. Thus, in Sophocles' play *Antigone* (441 BCE) we find a clear representation of civil disobedience in Antigone's refusal to obey Creon's decree due to her interpretation of her moral duty to bury her brother in conformity with the requirements of the gods. But, as playwright, Sophocles also appeals to his audience to consult their consciences about Antigone's plight, and this represents a third feature of civil disobedience: an appeal to public conscience, or the moral sensibility of the community, to consider whether a law or command is just.

The conscientious refusal to obey, as demonstrated in the speech and behavior of the disobedient, and the open, public appeal to conscience are among the features that distinguish civil disobedience from the self-interested behavior of the criminal or the subversive behavior of the revolutionary. These limited and civil aspects of civil disobedience can be captured by two further defining conditions: the public and deliberate nature of the violation of law, and the willingness to accept punishment for the regime's judgment on the disobedience. The civil disobedients' argument is not with the government as a whole, but with a particular part deemed to be immoral.

Beyond these five conditions, widely recognized as necessary for civil disobedience, there is little agreement on additional defining characteristics. But, if civil disobedience is to succeed as a conscientious protest against an unjust law, then it is plausible to require the following as a sixth condition. There must be a sufficiently clear and logical link between the law or policy that is the object of protest and the law that is violated. For the most part, accounts of civil disobedience do not distinguish between "direct" disobedience, involving violation of the law believed to be unjust, and "indirect" disobedience where the law violated is not the target of the protest. Whereas Gandhi committed direct civil disobedience in the famous Salt Satyagraha by making salt from seawater, in direct contravention of law in British India, Thoreau's refusal to pay taxes in Massachusetts was an instance of indirect civil disobedience. Henry David Thoreau was directly opposed to slavery, and to injustices against Native Americans and the aggression of the Mexican-American War; the payment of taxes supported oppressive government policies.

A final common characteristic, nonviolence, is probably most controversial. Some commentators believe that nonviolence is implied by the notion of civil disobedience, but others such as Thoreau held that some such acts could be violent, as for example, John Brown's raid at Harper's Ferry. For the most part, those regarding violence as legitimate, when necessary, relate this claim to their more general justification for civil disobedience. It is morally permis-

68 *Global Studies Encyclopedic Dictionary*

sible to act violently, on the view of some, when violence is the only way to avert a very great evil. For Thoreau, because governments act with naked force and coercion rather than on the basis of principle, a government committing injustice cannot demand a nonviolent response from protesters.

Often, in the absence of some final and uncontested substantive test of the justice of a law, it is necessary to rely on tests of the sincerity and intensity of activist' commitment to respond in a way they believe to be incumbent on them. For this reason, the pattern developed by Gandhi (and advocated as well by Martin Luther King Jr., Vaclav Havel, and many others) is telling. Gandhi required self-examination and self-purification, he refused to inflict suffering on others, he accepted blows without retaliation, and he accepted arrest and punishment. There was nothing in his bearing that communicated interest in personal gain, nor could he have acted as he did without the deepest commitment to justice. The example of Gandhi compels our respect; in his earnest desire for truth and social justice, he affirms the heights to which we all must aspire.

When Gandhi was convicted for sedition for publishing anti-government views in *Young India*, he was allowed to address the court before being sentenced. He told the judge that if he truly believed in the justice of the system of law that he was entrusted to enforce, then he must impose on him the maximum penalty allowed by law. Gandhi's case illustrates why civil disobedience is of such importance for the integrity of a social order. When they stand by their deeds, the civilly disobedient require us to re-affirm the justice of our laws. If we can be confident of the justice of our cause, then we must impose the punishment that the disobedient act calls out. But we must have the courage to offer a moral defense of the law; we cannot hide any inconsistencies between our creeds and what the law requires in practice. Nor should we want to.

R. P. Churchill

Civil Society represents social relations and institutions functioning independently, while remaining capable of influencing, political power, and is composed of a community of independent individuals and independent social actors. Civil society is self-sufficient, involving relations among people as market participants, owners, partners, competitors, neighbors, members of public associations and movements, churches, friendly associations, and clans. The institutions of civil society do not possess power; yet this society spontaneously generates a legal system.

The establishment of this type of society requires specific institutional preconditions, including at least minimal democratic rights and freedoms, making the individual autonomy and self-organization of citizens with common interests and aims both possible and legal. Civil society presupposes elements of market relations that are more or less autonomous and have relatively independent internal connections and rules. Civil society consists in horizontal social connections and independent institutions and associations established by free and responsible individuals for the protection of private interests, thus presupposing a democratic society free from authoritarian rule, forcible methods of governing, and totalitarianism and violence. Civil society

Capitalism – Cyberculture 69

implies respect for the law and morality and the prevalence of economic and private spheres over the state and political sphere.

The main components of civil society are political parties and institutions such as human rights advocacy associations, a free press, a parliament, and an independent court. These institutions oppose the state to a degree and aim at controlling state activity and subordinating it in the public interest. For example, according to the Declaration of Independence, citizens of the United States have the right to overthrow their government "when a long Train of Abuses and Usurpations, pursuing invariably the same Object, evinces a Design to reduce them under absolute Despotism, it is their Right, it is their Duty, to throw off such Government, and to provide new Guards for their future Security."

The main principles of civil society include: (1) economic freedom—a multiplicity of property forms, market relations, competition; (2) recognition and protection of human and civil rights; (3) democratic power providing the equality of all citizens in the eyes of the law and the legal security of an individual; (4) a constitutional state based on a principle of division and interaction of authorities; (5) political and ideological pluralism supposing legal opposition; (6) freedom of thought and speech, and the independence of the mass media; (7) non-interference of the state in people's private life; (8) the mutual duties and responsibilities of citizens; and (9) the effective social policy of the state.

The theoretical concept of civil society was formed on the basis of Western liberalism stating that a reasonably organized society is capable of creating a perfect order based on common sense, science, and justice. A state based on these principles moves toward personal freedom, political and economic activism of all citizens, and social security.

Civil society implies that the structure of social relations makes a citizen the main element of public life. Civil society is not just a set of legal and political mechanisms and institutions, but a state of mind and morality through which a person becomes an active participant in the decision-making process able to influence society and the state.

B. I. Zelenko

Civilian Police, International: The international civilian police are an independent element of the UN international peacekeeping missions. Its main task is supervising or monitoring local civilian police divisions in order to provide effective law enforcement, maintain public order, and guarantee respect for fundamental human rights and freedoms. The international civilian police may also perform the following functions: (1) Monitor the activities of local police and other civil and legal (law enforcement) authorities. (2) Assist local police in patrolling. (3) Observe crime scenes. (4) Investigate crimes. (5) Supervise and monitor flow of refugees and forced migrants. (6) Control meetings, demonstrations, and other public activities. (7) Observe penitentiary facilities and custodial conditions. (8) Assist humanitarian relief organizations and their political activities. (9) Assist in holding elections.

The UN has established fifteen peacekeeping missions worldwide, and 7,530 police officers from seventy-two countries take part in six of them. Those who comprise the force are selected from nations worldwide according to their

70   *Global Studies Encyclopedic Dictionary*

ability to perform professional duties under emergency conditions. Their growing role in conducting peacekeeping missions is based not on their numbers or "might," but on their high quality of assistance to conflict resolution. The quality of civilian police activities depends directly on the force's command of peacekeeping technologies thus facilitating the operation's goals, which include: (1) assistance in providing humanitarian relief; (2) human rights advocacy and other activities of the mission; (3) monitoring agreements; (4) human rights protection and the activities of local law enforcement authorities; (5) advising local police to promote humanity and effectiveness; (6) gathering information about various situations and incidents; (7) training local law enforcement personnel in modern methods of policing.

The command system of the UN civilian police consists of two levels. The first is strategic, implemented on the basis of the UN's principal organs responsible for conducting peacekeeping operations, including the General Assembly, the Security Council, the Trusteeship Council, the Economic and Social Council, the International Court of Justice, the Secretariat (headed by the Secretary General), and the Department of Peacekeeping Operations. The second is operational and oriented toward functional tasks. The operational level includes three basic echelons: central headquarters, regional headquarters, and police stations. Each of them employs specific management techniques. This level's structures are very flexible, and the choice of a specific management system is determined by the following factors: (1) the geography of the host country; (2) the military and political situation in the region; (3) financial and economic factors; (4) the civilian police mandate. Regional commanders, who are assisted by political and legal advisors, head regional headquarters, which also consist of several departments: (1) operations and personnel; (2) human rights supervision; (3) local police reform.

V. I. Kudashov

Civilization: (Latin *civis*, "citizen") (1) The sum of all positive achievements of humankind. (2) A progressive development of the world. (3) A normative vision of an advanced social order (mostly Western).

Various interpretations of the notion "world civilization" correlate with various meanings of this idea. Discussions concerning the status of a world civilization are central for contemporary social science. The two primary perspectives include: the position that world civilization is a reality and the view that world civilization is an ideal normative construction.

Our understanding of world civilization is linked to the contemporary global processes permeating every sphere of human activity. One extreme in the wide range of opinions concerning globalization involves an optimistic praise for the elimination of differences, of Euro-Atlantic hegemony, or a "triumph of one civilization over the others" (Braudel, [1962] 1993). At the other extreme is the negative perception of globalization, which has led to the archaization of non-Western societies.

A correct definition of world civilization must account for its juxtaposition with local civilizations. The existence of world and local civilizations and the correlation between them is a problem complicated by several methodological difficulties related to the polysemy of the term itself, which de-

Capitalism – Cyberculture

notes the entire hierarchy of socio-cultural entities. First, the term "civilization" can be used to designate specific ethno-social organisms (such as Mayan, Babylonian, and Sumerian civilizations), or relatively ethnically homogeneous entities. Second, the term can be applied to wider sociocultural entities sharing a largely common culture (such as Hellenic, European, Latin American, and Russian civilization). Third, within the formation approach, the word "civilization" stands for a group of sociocultural entities belonging to the same historical type (such as slave-owning or feudal civilization). Finally, the notion can be used to describe entire social and cultural achievements of humankind, or "world civilization."

World civilization, being historically limited, is linked to social progress, capturing all human social, material, and spiritual attainments irrespective of their specific regional, ethnic, cultural, and political features. Civilization is characterized by the mechanisms of social succession and continuity, ensuring preservation and transfer of the heritage common for all humankind. World civilization is independent of peculiarities of the existing sociocultural entities and local civilizations. Local civilizations have established features and attributes, including traditional culture, language, habitat, economic, and spiritual unity. But any specific form of civilization usually expresses historically conditioned values and phenomena left behind in the course of its evolution.

The notion of world civilization accentuates cultural and social achievements of this entity, their consistent growth, enrichment, and dissemination. The material, cultural, and social achievements of world civilization are consequently compared with universal values.

The issue of the "universal" becomes especially significant when a society in transition faces the choice between the old, obsolescent social order, and the new one. When gaining a foothold, a new civilization must formulate an attitude toward the previously accumulated universal heritage in order to find its place in world history.

The tendency toward globalization, generated by new technological, cultural, and political conditions, causes the leveling of local civilizational specifics. Often, the inability to painlessly assimilate alien, forcefully imposed values and achievements gives rise to political and ideological movements that attempt to defend historical and civilizational specifics (such as Slavophilism, Euroasianism, Afrocentrism, and religious fundamentalism), thus hampering unification and preventing constructive interaction and mutual understanding.

World civilization is a result of the labor of many generations, eras, countries, and continents, involving achievements and knowledge that have endured the test of time and that are reflected in the memory and social ideas of humanity. Thus, intercivilizational dialogue, tolerance, and peaceful resolution of intercivilization conflicts seem to be the best foundation for the future of nations.

Reference: Braudel, F. *A History of Civilizations*. Translated by Richard Mayne. New York: Penguin Books, 1993.

M. M. Mchedlova

72   *Global Studies Encyclopedic Dictionary*

Civilization Development, Types of: Types of civilization development refers to general features of historical evolution, characteristic of some original civilizations, their common reproduction, and development of features of social life. Two types have been identified—traditional and technogenic—characterizing the variety of civilizations that replaced primitive state and archaic communities. Each is represented by a variety of civilizations.

The majority of the twenty-one civilizations described by Arnold J. Toynbee belonged to the traditional type (including Ancient Egypt and Babylon, Ancient India and China, Ancient Greek civilization, and the medieval societies of the West and the East). This development type preceded the technogenic type, which appeared in Europe between the fourteenth and sixteenth century. Technogenic societies are frequently called "Western," because of their region of origin, opposing them to the traditional "Eastern." At present, though, the technogenic type of civilization development dominates the planet.

The formation of technogenic civilization was preceded by important mutations of two traditional national cultures: the culture of the ancient Greek polis and the culture of the European Christian Middle Ages. The synthesis of their achievements during the Renaissance and the further development of ideas of a new worldview during the Reformation and the Enlightenment generated a core system of values that underlies the technogenic civilization. Technological progress became its fundamental development process. As a result, social communications, attitudes, and institutions change radically over the course of each generation.

Traditional cultures never aimed at transformation of the world and maintenance of an individual's authority over nature. The latter attitude, however, dominates in technogenic cultures, involving not only natural, but also social objects, which become subjects of social technologies.

A core value of technogenic civilization involves an understanding of nature as the inorganic world, represented by a specific naturally ordered field of objects acting as materials and resources for human activity. Those resources were presumed boundless, to be taken from nature in increasing numbers. The traditional understanding of nature opposed it, considering nature as a living organism and an individual as a tiny part of this larger whole.

In technogenic culture the active, sovereign individual is of prime value. In traditional cultures a personality was defined first through a person's involvement in strict and firm (usually birth-given) family-clan, caste, and class relations. Technogenic civilization, by contrast, affirms the ideal of loose individuality, and of the self-sufficient person who can be included in various social arrangements, having equal rights with others.

Priorities of individual freedom and human rights, which traditional cultures did not know, are connected to this understanding.

The values of technogenic civilization assign a special place to the value of innovations and progress, whereas in traditional societies, innovations were always limited by tradition and were masked as tradition. In traditional cultures the conception of cyclic development and cyclic time, where past has a priority over the future, dominated (it was considered that heroes and wise persons of the past left precepts, doctrines, examples of acts, which define tradition and serve as learning patterns). In the culture of technogenic societies time is understood as directed from the past to the future; development is associated with progress and the value of the past yields to the value of the future (the "Golden Age" is not in the past, it is in the future).

The success of reformatory activity, resulting in positive results for individual and social progress, is directed, in technogenic culture, by the knowledge of natural laws. Scientific rationality in this type of culture takes a dominant place in the system of human knowledge and actively influences its every form.

This system of values gives birth to many other features of technogenic culture. The specified values act as a genetic code of technogenic civilization, according to which it replicates and develops.

Upon their formation, technogenic societies attack traditional civilizations, forcing them to change. Sometimes these changes occur as a result of military seizure or colonization, but more often they are the result of modernization, which traditional societies must carry out under pressure from technogenic civilization. This was the way Japan, for example, was reformed when it chose the path of technogenic development after the Maydi reforms. This was the way of Russia, which went through several epochs of modernization, based on transplantation of the Western experience.

The technogenic type of development unifies public life to a much greater degree than the traditional one. The science and education of technological progress and the expanding markets generate new modes of thinking and life, transforming traditional cultures. The process of globalization appears as the result of the expansion of technogenic civilization, which is implanted in various regions of the world.

Technogenic civilization gave many achievements to humankind. Scientific-technological progress and economic growth resulted in a new quality of life, ensured a growing consumption level, enriched health services, and increased the average life-span.

At the same time, technogenic civilization generated the global crises of the second half of the twentieth century, the worsening of which may result in the self-destruction of humankind. The resolution of these crises will demand the radical change of some of the basic values of technogenic civilization, the first of which is the attitude of a person to nature and the ideals of domination oriented at forceful transformation of objects.

Post-industrial society can be considered as a transient stage on the way to a new type of civilization development, taking into account that: (1) conditions for the solution of ecological and other global problems will be actively created; (2) individuals are empowered to play a major role in developing this new type of civilization; (3) the emerging civilization is characterized as a society transitioning to dominance of non-materialistic values, shifting from exponentially growing goods-power consumption to information consumption. Transition to a new type of civilization development is one of the possible scenarios of the future. However, humankind can create catastrophic scenarios, connected to a tendency toward maximum prolongation of the technogenic type of development.

V. S. Stiopin

74   *Global Studies Encyclopedic Dictionary*

Civilizations, Diversity of: Unlike a popular idea of an inevitable clash of civilizations we suggest that the existing diversity of cultures and civilizations is a way of their coexistence. The focus is on unity, on unity in diversity. Such an approach would be the most fruitful for understanding the phenomenon of globalization in its humanistic interpretation.

Insisting on the priority of unity, one should certainly not forget about diversity of the world. Unity without diversity makes the world a military barrack. In this case, diversity seems to be a barrier in the way to universalization of the world. The hegemonic globalists count on the erosion of cultural originality, on its subordination to a standard of "civilized life." The collapse of all great empires and their "great ideologies" has demonstrated that any attempt to impose a "standard" world order and lifestyle is ultimately doomed to failure and defeat. One should not forget that opposing political and cultural diversity causes, as a rule, a defensive response in the form of conservative traditionalism and fundamentalism. That is one of the "traps" of globalization, and its dangers should not be underestimated.

V. I. Tolstykh

Civilization, Evolution of: Scientific and technical progress is amazingly vindictive; the more innovations we introduce, the more consequences we have to consider. The progressively growing quantity of unknown causes fear and discomfort. We frequently demonstrate fear of the future, which is characteristic mostly for the developed regions, not for the poor developing countries.

Meanwhile, a historically unprecedented increase in the volume of practically applied knowledge has happened due to the evident advantages this knowledge gives. Information makes networks that connect people. In Goethe's times and even long before parcels, letters, and decrees were delivered by various means of transportation. Information networks using communication channels later replaced even telephone systems. These closed circuits have a decentralizing function. Our participation becomes irrelevant to our location. Differences between cities and countries dissolve similarly to differences of cultures, due to mobility and integration of closed information networks. Political actions become as independent as the means of communication. Public opinion does not occupy a specific space, and those aspiring to have political influence should act through mass media.

Modern civilization increases the gaming space for our activities and behavior concerning our choice of education and profession, mobility, and leisure. Thus, we should be able to determine our way of life ourselves. Morality is a set of rules for individual and public self-determination; the more independently we live, the more we need morality.

The dynamics of knowledge production is limited by the borders of time; new knowledge should extend and be consumed in the process of the communication of scientists, researchers, and teachers. Our unprecedented wealth depends on scientific and technological progress, but the more wealth we have, the less prone we are to bear the costs for obtaining this welfare.

This explains, why sensitivity to, say, the ecological consequences of progress increases together with the number of positive results of this progress. Therefore, successes of "natural sciences" simultaneously stimulate development of "cultural sciences" that do not multiply our technical skills, but guide us on our way between our origins and the future.

H. Lübbe

Civilization, Global: Global civilization s the totality of local civilizations, countries, peoples, and ethnic groups manifesting the diversity of humankind, and the level of their sociocultural, technological, and economic development. The preconditions of global civilization were laid during the neolythic revolution in the sixth to seventh millennia BCE, when the transition to the economy of reproduction took place and was accomplished in the third millennium BCE when the market economy, international trade, the state, and classes emerged. Global civilization was developing through recurrent interchange of world civilizations (neolythic, early class civilization, ancient, medieval, early industrial, industrial, and, since the beginning of the twenty-first century, post-industrial) and generations of local civilizations. The new phase of global civilization began at the end of the ecological, technological, geoeconomic, geopolitical, and sociocultural space. The main contradictions of global civilization in the twenty-first century are: polarization of levels of technological and economic development of rich and poor countries and local civilizations, the emergence of international terrorism as the clash of civilizations and the impact of globalization on civilizational and cultural diversity. These contradictions may be resolved on the basis of dialog and partnership among civilizations.

Y. V. Yakovets

Civilizations, Local: Local Civilizations are cultures and societies developed by particular nations or ethnic groups united by common history, technological and economic space, and political relations. Local civilizations emerged in the third to fourth millenniums BCE in a narrow region north of the equator. The first generation of local civilizations was represented by civilizations of Ancient Egypt, Mesopotamia, India, and China. The second generation included the Mediterranean civilizations (ancient Greece and ancient Rome), and those of Persia, India, China. Native American civilizations in the South, Central and North America developed independently. As to the third generation (Middle Ages, early industrial and industrial world civilizations), the world was dominated by the expansion of Western European civilization, which spread over most of the planet (Western Europe, North America, Latin America, Oceania), including the European-immigrant dominated colonies in Africa and India. Western European civilization also exerted influence on the Eurasian civilization. The fourth generation of local civilization emerged in the late twentieth century. It includes Western (Western European, Eastern European, North American, Latin American), Oriental (Japanese, Chinese, Indian, Buddhist, Muslim) and mixed (Eurasian, African and Oceanic) civilizations. Over thousands of years different civilizations have interacted with each other, and this interaction produced different forms—from military clashes and wars to dialogue, collaboration and partnership.

Y. V. Yakovets

Climate (Greek *klima*, "region" "slope"; thus slope of the Earth from equator to pole) Climate refers to a measure of the average pattern of variation

76 *Global Studies Encyclopedic Dictionary*

in temperature, humidity, atmospheric pressure, wind, precipitation, atmospheric particle count and other meteorological variables in a given region over long periods of time. Climate is different than weather, in that weather only describes the short-term conditions of these variables in a given region. It is one of the most important geographic characteristics. The Earth's climate is conditioned by the existence of the Earth's air covering, or atmosphere, which, together with the hydrosphere (World Ocean), the cryosphere (glaciers and permafrost), the surface of continents, and the biosphere, composes the global climate system. The climate extracts energy from solar radiation, the angle of which is different on different latitudes and in different seasons due to the spherical shape of the Earth and the angle of the rotation axis of the planet. The interaction of solar radiation, the extremely mobile atmosphere, and different types of surface on the planet creates peculiar air conditions that are regarded as weather of various types. A complex of meteorological parameters is used to characterize the climate (including atmospheric pressure, air temperature, absolute and relative humidity, clouds, wind speed and direction, and precipitation). Average amounts for a period of some years represent a climate norm, and deviations are regarded as anomalies (positive and negative). We can monitor the scales of climate changes by taking into account the amount and stability of anomalies.

The most important factor of weather formation is the circulation of the atmosphere, producing anticyclones and cyclones, dry and fair weather, and cloudy, rainy days. The main role in the activation of circulation of the atmosphere is played by contrasts of underlayers, detectable on the surfaces of oceans and land, hot deserts, and glaciers in the southern and northern extremities of the planet. Oceanic flows transport great amounts of warmth from the tropics to higher latitudes (Gulf Stream) and cold in the opposite direction (Peru stream), creating stable climate anomalies. Even minor deviations in thermal conditions and dynamics of major streams strongly influence changes of global climate. Their effects can be contrary in different regions. During recent years a strong dependence of global climate on periodical changes in temperature of the El-Nino stream (in tropical part of the Pacific Ocean) has been observed.

Weather that exists within a given region for many years makes its climate. Alexander Humboldt was among the first to present this definition of climate at the beginning of the nineteenth century. Aleksandr I. Voeikov first described global climates in 1884. Milutin Milanković presented mathematical calculations of climate changes on the Earth in 1930. He expressed an idea about astronomically predetermined periodical fluctuations (tempos) of the Earth's climate. Periods of climate cycles that are 100 thousand years long reflect fluctuations of the eccentricity of the Earth's orbit; 40-thousand-year periods reflect changes of the planet's axis angle, 20-thousand-year periods reflect the effect of the precession of equinoxes, forty-year annual and quasi-biannual cycles reflect changes of characteristics of solar activity determined by the number of sun spots. Emergence of cyclones is impeded by the influence of other climate-formation factors (including tectonic activity, changes of the ratio of land and ocean, and human industrial activity).

Capitalism – Cyberculture 77

Long-term climate changes are defined in accordance with geological and paleogeographic data, short-term or historical are defined on the basis of analysis of archeological evidence and annals, and modern changes are defined in accordance with instrumental meteorological monitoring data. In recent times scientists have used drilling of glaciers in Greenland and Antarctica. The isotopic analysis of core materials helps to identify very accurately the temperature, amount of precipitations, and gas composition of the atmosphere for the last 200 to 300 thousand years.

Analysis of the glacier core from the Antarctic station "Vostok" showed a close connection between the amount of carbon dioxide in the atmosphere and its temperature. Scientists have documented that CO_2 and other greenhouse gases in the atmosphere have increased to at least a five times higher concentration over the past 200 years. The largest amount of greenhouse gases in the atmosphere appeared during the last decade of the twentieth century. This process is the cause of a recent global warming that has no precedents during the last millennium. The rising of the temperature leads to glaciers melting in the polar regions and in the areas of the permafrost, occupying about 24 percent of the land surface in the Northern hemisphere. Ice melting will result in a significant rise of the world ocean's level, and methane, the most powerful greenhouse gas, will ooze out of the permafrost. According to some scenarios, the average rise of temperature in different regions of the Earth by the middle of the twenty-first century will be between 1–4 °C and the weather on the planet will change.

V. A. Markin

Club of Budapest is a public association uniting people in the arts who share the goal of forming a sustainable and humane world. Hungarian scholar In 1993, Ervin Laslo, one of the founders. proposed an organization analogous to the Club of Rome. The Club of Budapest unites outstanding politicians, public and religious figures, artists, writers, and scholars as honorary members. The fact that the world now undergoes global changes is the central idea of the Club of Budapest: old values are replaced with new ones and we have to understand that we can change the world only by having changes ourselves first. The Club of Budapest considers its mission to be the cultivation of the creative energy of individuals, industries, and states so that the whole of humanity can coexist peacefully and cooperate for the purpose of the common good. This idea is part of the "Manifesto about the Spirit of the Planetary Consciousness" adopted by the Club of Budapest in 1996. This document postulates the necessity of finding a new way of thinking as a precondition for responsible ways of living and acting; it includes a call for planetary consciousness. Planetary consciousness is defined as knowing and feeling the interdependence of the principle of solidarity of humanity and the conscious adaptation of the ethics and ethos needed for this consciousness. Development of planetary consciousness implies that the world governed from above is being replaced with the self-governing world. To achieve these goals, some international projects have been worked out validating the necessity of forming planetary consciousness for individuals, communities, and social institutions.

L. F. Matronina

78 *Global Studies Encyclopedic Dictionary*

Club of Rome is a nongovernmental, informal organization concerned with global studies. The origins of this organization date back to 1968, when, on the personal initiative of distinguished public figure Aurelio Peccei (subsequently a founder and first president of the Club of Rome), approximately thirty European natural and social scientists gathered in Rome and formed a "Standing Committee." A decision was made that Club membership should be restricted to one hundred people representing a cross-section of modern progressive humankind. As a result, outstanding public, political, and state figures and representatives of business and financial circles, including those from developing countries, became members of the Club of Rome. At present, about thirty National Associations assist the Club of Rome in different countries of the world, including Russia. These associations act on the basis of the Charter of Associations adopted in Warsaw in 1987, following the general line of the Club of Rome.

The principal goal of the Club of Rome is to study global issues, to work out solutions for universal problems, and to draw the attention of the world public to them. During the first two years the initiators of the Club looked for international supporters until they came to the conclusion that neither the scientific community nor wider population fully realized the danger faced by humankind. Concern was expressed with respect to what was happening nearby, although some authoritative statements and warning voices were also heard. These statements gradually mounted but were unorganized and spontaneous and, proceeding from individual specialists, failed to find any support at all. The founders of the Club of Rome soon realized that appeals and calls alone could not change the situation. A different mode of action was necessary that would make people look differently at the modern world and themselves within it. Aurelio Peccei wrote that in that period the majority of people were ready to welcome the Club of Rome, but on the condition, however, that it would in no way interfere with their day-to-day affairs and infringe upon their interests. He also noted that it seemed the Club's words were forgotten even before they were heard. People seemed so light-hearted not only because those appeals came from an unknown organization, which did not identify itself with any political party or ideology and had no unified system of values or consolidated opinion, but also because of people's indifference toward their own future. They did not see, did not realize, and, what was most disconcerting for the members of the Club of Rome, did not want to realize the imminent dangers.

A sensation was needed. Only something absolutely extraordinary and shocking, like a bomb explosion, could make people come to think over their fate and to see the imminent dangers. In search of such an opportunity, attention was drawn to the work of Jay Wright Forrester, who attempted to simulate the dynamics of world processes with the help of mathematical models and computer technologies. This new and unusual enterprise was in line with the principle goals of the Club of Rome.

The Club of Rome has made significant contributions to the history of global studies because of its being first among many futurological organizations that sprang up in the wake of the aggravation of global problems. It succeeded in engaging distinguished scholars and practitioners, and in accom-

Capitalism – Cyberculture
79

plishing the task of principle importance: it showed to all humankind the danger of current tendencies in the development of world civilization. The members of The Club of Rome succeeded in capturing the attention so persistently sought after; non-ordinary conclusions and prognoses of their reports aroused a significant response on the part of the world public and scientific and political circles and became a factor of significant influence in shaping mass consciousness on the planet. The practical recommendations of many reports are often used nowadays in making social and economic forecasts concerning individual countries and regions. Various sciences are widely applying the methodological principles developed and first applied by that organization.

The Club's activities are evaluated both negatively and positively; however, balanced evaluations and careful consideration of the results prevail. According to German scientist E. Gartner, the Club earned great sympathy with many critically minded scholars and commentators by virtue of its revealing things otherwise systematically underestimated and concealed in the Western world to preserve the image of prosperous "post-industrial" society. Many high evaluations can be found in the works of Russian specialists; for example, Nikita N. Moiseev wrote that the Club of Rome studies assert the objective necessity to search for new ways of development for our civilization and the necessity of a new understanding of the processes of world development. These studies also demonstrated the advanced level of modern scientific and social thought and reflected the humanistic position of many Western people and their attitude toward the most acute issues of modernity. If some of the negative tendencies and grave prognoses have not materialized, this is beyond any doubt a merit of the Club of Rome.

References: Chumakov, Alexander N. *Filosofiya globalnykh problem* (Philosophy of Global Problems). Moscow: Znanie, 1994. King, Alexander. "The Club of Rome: Reaffirmation of a Mission," *Interdisciplinary Science Reviews* 11:1 (1986), 13–18. Gartner, E. "On the 'Second Phase' of the Work of the Club of Rome," *Scientific World* 20:4 (1976). Peccei, Aurelio. *The Human Quality*. New York: Pergamon, 1977.

A. N. Chumakov

Club of Rome, Reports to the: Reports to the Club of Rome document research projects made for and under the aegis of the Club of Rome, which analyze the most important world problems, developing scientific methods of modern global studies, offering practical recommendations, and presenting alternative scenarios of world development. Reports have been prepared by independent working groups composed of respected scholars and highly qualified specialists who are only given the theme and guaranteed research funding by the Club. In the absence of its own funds, the Club attracts money from various foundations and sponsors. At the same time, the Club does not influence the process of work, nor its outcome and conclusions; the authors are fully independent in implementing their creative plans. The final outcome of each project is discussed, and approval given at an annual conference of the Club with the participation of invited scientists, public and political figures, and the press, after which the reports are published.

The Club of Rome has commissioned over twenty reports, and prepared one organizational report, *First Global Revolution* (1991) by Alexander King,

80   *Global Studies Encyclopedic Dictionary*

President, and Bertrand Schneider, General Secretary. Other reports have addressed issues including difficult conditions facing Latin America.

A. N. Chumakov

Cluster: The theory and practice of cluster formation and development was a response to the challenge of globalization. Globalization homogenizes economic areas contributing to a unipolar world. In the 1990s, the clusterizing economic activity in opposition to globalization grew stronger through the localization of regional activity, which is based on trades and uses individual regional features.

Michael Porter elaborated the basic concepts of cluster competitive capacity theory in 1990. The cluster principal differs from various preceding strategies of regional development, especially "growth centers," which were popular in the 1970s, and marks a transition to positive feedback instead of negative one. Instead of relying upon administrative decisions to create "growth centers," direct governmental regulation, governmental support of socially significant, but noncompetitive businesses and the use of regional economic "leveling" triggers positive development convolution. This model uses the concept of economic "effectiveness," inspires initiative from below, is grounded on the organization of "autonomous starting mechanisms" in problem regions (especially on "start-up" or venture enterprises), and follows the Public Private Partnership criterion (PPP). The classic example of a cluster is the Silicon Valley in the United States, founded in 1990, with the efforts of local authorities and Stanford University. At present, nearly 20,000 venture start-up enterprises have been established within this cluster, with 2.5 million employees.

Though clusters are regional, many have become known worldwide: Detroit's automobile cluster, Hollywood's cinema cluster, Switzerland's bank and watch cluster, the financial cluster in the City region in London, the fashion clusters in Milan and Paris, and the high-tech and aerospace clusters in Seattle. Silicon Fen near Cambridge, UK, and the high-tech cluster in Bangalore, India are the newest thriving clusters. The vast potential of cluster approach is demonstrated in Finland, near the Polar circle, where the IT-telecom cluster near Oulu University has become a world leader in the industry. The Nokia Company was its driving force. The cluster comprises more than 120 companies. Its favorable environment contributed to the creation of another cluster, Medipolis, in the medicine and biotechnology field, which includes more than fifty companies. The cluster approach together with PPP criterion allows Finland to lead in world competition.

The following conditions are required for cluster formation: (1) creation of a relevant climate of innovation on the macrolevel (e.g., country) and regional level; (2) identification and use of regional infrastructure advantages; (3) use of historical heritage and cultural peculiarities and traditions; (4) inspiration of competition on the local level by regional authorities. Local authorities support the formation of start-up enterprises, consultation and educational services to businesspersons, areas for beginners in business, and negotiation areas between business and authority as needed. The development of business culture is necessary. Educational centers, research and development technologies, and implementation are required. Local authorities stand as a customer to order the cluster output and services on the competitive basis.

Resource concentration, including the attraction of skilled workers, affects the resolution of specific regional problems. Forming social capital based on partnership and trust between business and authorities, the knowledge of economics and establishment of social justice are required.

With the high educational level and strong scientific and technical potential of the population, only the predominating administrative approaches and the priority of state enterprises prevents the implementation of the tools of cluster technology in Russia and Ukraine. Therefore, Russia and Ukraine rank relatively low in the world countries competitive capacity list. "Science towns" in Russia should be promoted as clusters and the idea of creating ferrous metallurgy and metalloid clusters in Northern Russia. The "Podolye Pervy" cluster is under formation in Khmelnitsky City, Ukraine, along with a handicrafts cluster in the Carpathian region.

S. Udovik

Coevolution is the co-development of interacting elements within a single system, developing and preserving the integrity thereof. Initially, coevolution was introduced in biology to define evolutionary mutual adaptation of species.

The possible scope of coevolution is very large: not only can species co-evolve, but other biological systems, too: communities of organisms, ecocenoses and ethnoses, cultures, states, and economical systems (such as exchange companies). Evolution is naturally understood as development in the broad sense of the word, rather than that of biological reductionism.

The most interesting, non-degenerated coevolution types presuppose a sort of approach between two interrelated evolving systems without converging to a single common image. Instead, a mutual adaptation is achieved involving a change in one system that initiates the change in the other one that does not result in consequences undesirable for the first system. With these interactions a (relative) symmetry and equivalence of coevolving systems is required that presupposes each to have a set of potential possible changes or "responses" to other system transformations.

Coevolution to the social sphere occurs as the antithesis of conflict development (the most typical scheme is when one of two coexisting systems suppresses the other one in the course of competition for some resources). Coevolution preserves diversity in an ecosystem where evolving elements coexist, determining a uniform coevolutionary preference against conflict development. Social coevolution initiated the tendency to understand this concept as one of the basic concepts in a wide range of theoretical constructs, regarding development on the highest generalization level, such as within universal evolutionism. However, no known convincing examples of coevolution exist beyond biological and social fields.

The idea of the coevolution of nature and society, biosphere and man in universal evolutionism has been proposed as a line of development that can solve ecological problems when followed. This idea can be productive under two conditions. First, humans must correct their developmental strategy, reducing their disturbance in biosphere to a level that does not exceed its regulatory and compensatory capabilities. Second, as a result of such coevolution, the biosphere must develop so as to increase response capacity on human

82   *Global Studies Encyclopedic Dictionary*

driven disturbance. The first condition is necessary to overcome ecological crisis; the corresponding processes actually occur, the question is whether their amount is satisfactory, so that humans have time to change their development strategy before facing disaster.

The predominating role in the evolution of biosphere belongs to the biota. Biota evolution is performed via speciation. The disappearance and appearance of species almost always causes a surge of changes in ecosystems, in which a species establishes its ecological niche. The speed of this process has been estimated. Paleontologically, the average species' lifetime is approx. three million years. According to modern understanding, about 10,000 years are required to naturally form a new biological species. Human social evolution occurs with all genetic constants, preserved and realized through the interrelated development of social structures, public conscience, production systems, science and technology, and material and intellectual culture. While analyzing nature and society, the main concern of coevolution lies in human influence on biosphere, which alters qualitative characteristics, type, and structure mainly due to scientific and technological advances and technical evolution. The latter is actualized through innovation, which, for some, is similar to speciation in biota.

Material production and its management, like biota, have a "spontaneous structure." Innovation, or new elements appearance in manufacturing and management techniques, and the denial to use any of the elements, as a rule, provokes a surge of other innovations in the relevant "technological niche." But the rate of technical revolution as compared to biological evolution increases exponentially.

With such a difference in biological and technical rates (three decimal exponents) it is incorrect to speak about the coevolution of nature and society. Due to time restrictions, the biosphere cannot react to human innovations by forming new species, which would be adapted to the consequences of innovation, in a sufficient mode or scale. For humans, such a biotic reaction to anthropological influence would be very desirable (many problems might be resolved, if, for instance, bacteria appeared which decomposes polyethylene, turns aluminum waste to bauxites and nephelines, or resists soil acidulation.).

Human-invoked biotic speciation in order to "enhance" its coevolutionary capacities (such as through the controlled influence of the genetic apparatus of natural species), even if possible, does not make the notion of nature and society coevolution more correct or constructive. Implementing this possibility would mean the end of natural biosphere evolution, turning biota into a man-controlled system. However, if applied to social systems, coevolution represents the only possible means to obtain sustainable development. Thus, a coevolutionary approach is opposed to the statement of inevitable civilizational conflict.

Reference: Gorshkov, V. V. et al. "Biotic Control of the Environment," *Russian Journal of Ecology* 30.2 (1999), 87–96.

V. I. Danilov-Danilyan

Cold War: Originally the term was employed in the fourteenth century to describe the long-lasting struggle between Muslims and Christians for politi-

Capitalism – Cyberculture

cal dominance in the Iberian Peninsula. After World War II, the term "Cold War" denoted the ideological conflict and hostility between the Soviet bloc, led by the Soviet Union, and the West under the leadership of the United States. Although protracted, this fight was dubbed "cold" mainly on the account that its participants conducted it in Europe without recourse to "hot warfare," or actual fighting. Mutual nuclear deterrence made this situation possible. Both the superpowers accumulated enough nuclear warheads to destroy the world many times over, and both the White House and the Kremlin were ready to use these weapons of mass destruction to maintain their spheres of influence in Europe, "freezing" the Cold War division of the continent until 1989. On other continents, "hot" hostilities between these two camps tended to erupt periodically, but usually were fought by Third World proxies. Invariably, Soviet-supported guerillas attacked official governments with sympathies to the United States and vice versa. The only examples of conflicts that involved United States and Soviet troops include the Vietnam War (1965–1975) and the Afghanistan War (1979–1989), respectively. The Korean War (1950–1953) was the only instance when UN troops were deployed. United States soldiers also constituted part of these troops, the vast majority of whom were drawn from among the South Korean population.

The legitimizing premise of the Cold War was solely ideological, involving a conflict between two Western ideologies that vied for dominance over the entire world. The Kremlin aimed at spreading communism all over the globe with enforced classlessness, collectivism, centrally planned economies, and the vision of the final disappearance (withering) of the nation- state. Western critics debunked this theory pointing out that under the guise of communism the Soviet Union continued the imperial project commenced during the times of the Russian Empire. To communism the West opposed democracy, individualism, market-oriented (capitalist) economy, and collaboration of nation-states. Soviet propaganda denounced this alternative as carried out at the cost of the exploited worker and in the imperial interest of the United States. The West's final victory indicated that economically the Soviet system was less viable, and less attuned to the needs of the average man. Both the systems mostly benefited the elites, but in the West a much larger and diversified elite emerged than in the tiny inner cores of the communist parties in the Soviet bloc. Moreover, the volume of generated wealth per capita and the quality of living enjoyed by an average person in the West was comparable to that accorded to communist elites in the Soviet bloc.

From the vantage of geopolitics the Cold War was a continuation of World War II by different means, because after 1945, no peace conference (promised by the Allies in the Potsdam Protocol) was organized to legalize the effects of the World War II. In the light of international law, such legalization occurred only in 1990, with the signing of the Two-plus-Four Treaty (involving the two Germanys and four Allies) by the wartime Allies (France, the Soviet Union, the United Kingdom, and the United States).

The Cold War division of Europe was grounded in the splitting of Germany into West and East Germany although the Potsdam Protocol stipulated that after the termination of the Allied occupation Germany should be reconstituted as a single state. Another foundation of the Iron Curtain that cut the conti-

84 *Global Studies Encyclopedic Dictionary*

nent into Soviet-dominated Eastern Europe and United States-protected Western Europe was the unprecedented territorial shifting of Poland 300 kilometers westward. In agreement with the secret Ribbentrop-Molotov Pact (1939) the Kremlin retained one-third of prewar Poland's territory east of the River Bug (Curzon line), while, on the other hand, in 1945, it "indemnified" Warsaw with east German territories east of the Oder-Neisse line (with the exclusion of the northern half of East Prussia made into the Soviet Kaliningrad Oblast).

Poland was the most staunchly anti-Soviet and anti-communist state in Central Europe, but this arrangement made it into Moscow's unwilling captive. In the absence of binding international regulations, only the Soviet Union was powerful enough to ensure Polish rule on Poland's share of the formerly German territories. Similarly East Germany was not a viable state without the Soviet propping.

After World War II, the Western-Soviet alliance rapidly unraveled. The symbolical commencement of the Cold War is the Soviet blockade of West Berlin (1948–1949). From the vantage of ideology, however, we can observe that the Cold War had actually begun in 1917, with the ascent of the Bolsheviks to power, and their proclamation of worldwide challenge to capitalism and democracy. Some saw the end of the Cold War during the second half of the 1950s, in the wake of de-Stalinization. Their hopes were dashed by the Cuban missile crisis (1962) when the Kremlin attempted to install nuclear missiles on Cuba mere 150 kilometers away from the United States. Another period of relaxation in the Cold War tension was called détente (the French word for "slackening"). It lasted from the 1972 summit meeting at Moscow between Richard Nixon and the Soviet First Secretary Leonid Brezhnev to the Soviet intervention in Afghanistan (1979). Almost all the Polish adult population organized in the anti-communist Trades Union Solidarity; under the Kremlin's pressure Warsaw imposed martial law in Poland (1981). This triggered another bout of the Cold War in Europe, which lasted until Mikhail Gorbachev's rise to power in 1985.

Faced with the daunting economic problems, Gorbachev sought to disengage the thinly spread Soviet forces and resources from abroad in an effort to restructure the Soviet Union to make it a viable state. A dedicated reformer, Gorbachev introduced the policies of glasnost and perestroika to the USSR. The West applauded his efforts, but the attempted dual economic and political reform necessitated abandoning the Kremlin's control over the Soviet bloc and precipitated the break-up of the Soviet Union (1991). Unexpectedly, the West "won" the Cold War, symbolized by the fall of the Berlin Wall, though political pundits had predicted that would last much longer.

With the Cold War over, it was possible to legalize the decisions of the Potsdam Protocol in the light of international law. Germany reunited, and Poland's western border was fully recognized. Both these momentous events became the cornerstone of lasting peace and of the post-Cold War order in Europe. With the terror of nuclear deterrence over, "hot" local wars erupted following the breakup of Yugoslavia and in some of the post-Soviet states.

T. Kamusella

Colonialism (Latin *colonia*, "settlement") Colonialism describes policies and actions of a strong state that establishes territorial control over an underdeveloped country or nation, most often accompanied by exploitation of the local population.

Western European colonialism began as a vehicle for world economic development because colonies provided capital accumulation in the metropolitan states and new markets for their goods. A consequence of this unprecedented growth of trade was the emergence of the world market and the movement of the center of economic activity from the Mediterranean Sea to the Atlantic Ocean. Many European ports (e.g., Lisbon, Seville, and London) became centers of trade. Antwerp, the largest international center of trade and financial operations, became the richest city in Europe. This development of trade contributed to the emergence of banks in the sixteenth century. Transnational firms and international trade, in turn, stimulated the development of capital and the transcendence of national borders.

The world colonial system was shaped during the final stages of the territorial division of the world in the last third of the nineteenth to the beginning of the twentieth century, establishing a system of domination by powerful nations over colonies and dependent nations, primarily in Africa and South East Asia. Great Britain, Spain, France, Russia, Germany, the US, Belgium, Italy, and Japan were the main colonial powers. Their joint colonial territories in the beginning of the twentieth century covered more than one half of the Earth (72 million km^2) where one third of the planetary population lived (560 million people). Great Britain, the largest colonial power of the world, dominated India, Africa, America, Australia and New Zealand. The United States dominated the Caribbean, the Pacific, Central America, and the Far East. France controlled Africa and Indochina, Germany possessed colonies in Africa and Japan held territories in Manchuria, Korea, China, and South East Asia. Russia promoted its interests and influence in the Far and Middle East and in Central Asia. After the Second World War the world colonial system collapsed as a result of national liberation movements in the colonies.

A. N. Chumakov

Communication The world and society are increasingly perceived as *sub specie communicationis*. Many speak about the "linguistic turn" in which, during the twentieth century, increasing focus has been placed on the relationship between philosophy and language (Rorty, 1967). Karl-Otto Apel (1972) spoke of the "communication a priori." Interdisciplinary tendencies have opened new possibilities in comprehending the modern diverse world.

Five axioms of communication are put forth by Gregory Bateson (1972): (1) One cannot not communicate. Every form of behavior is a form of communication,. Because behavior has no counterpart, it is impossible not to communicate. (2) Every communication has a content and relationship aspect such that the latter classifies the former and is therefore a metacommunication: meaning that all communication includes, apart from the plain meaning of words, more information. (3) Each communication process depends on punctuation of the communication partners: both the sender and the receiver of information structure the communication flow differently and therefore interpret their own behavior during communicating as merely a reaction to the other's behavior. (4) each person communicates simultaneously in both digital and analogue forms (verbal and behavioral), and (5) The structure of communication processes can be either symmetrical (based on equal power between communicators) or complementary (an interaction based on differences in power).

Quality communication includes cognitive, axiological, emotional, and functional dimensions that are mutually complementary. To be able to act in a sensible way, it is necessary to react to the dynamics of the readjusting world and society through education, and comprehension and acknowledgment of values. Use of information technologies has the potential to give rise to negative phenomena (such as loss of interest toward personal contacts). At the same time, modern information and communication technologies can be conducive to the further development of human potential.

References: Rorty, R. *The Linguistic Turn.* Chicago: University of Chicago Press, 1967. Apel, K.-O. "The A Priori of Communication and the Foundation of the Humanities," *Continental Philosophy Review* 5:1 (1972), 3–37. Bateson, G. *Steps to an Ecology of Mind.* San Francisco: Chandler, 1972.

I. Semradova

Communication, Global: Global communication is a form of intercultural dialogue in which all cultures are drawn into the global communication space. This space imposes a dialogue based on cultural similarities, rather than dissimilarities, which radically influences the content of intercultural dialogue.

The amount of information and the speed of processing are growing, affecting even the nature of interpersonal communication. The amount of information is literally overwhelming. A share of meaningless or superfluous information (commercials and announcements, brief news, and film-loops) is growing at the same time. Human beings gradually get accustomed to consuming information without considering any meaning. A situation arises in which consciousness is empty of any meaning, and is simultaneously flooded by a tremendous amount of information. Information bytes are overwhelming the lives of many people, densely packing every minute and causing biological clocks to run faster. Currently the system of communication is imposing its own laws and regulations on intercultural dialogue. Cultures are immersing, so to speak, into a different external milieu, which permeates intercultural dialogue.

This process manifests itself in the integrative tendencies in the language of intercultural communication. One of the consequences is the subordination of all languages to the one that is most capable of proliferation, owing to political, scientific, technological, and other causes. The world has started to either speak the language of the countries that dominate it or be subordinated to the language of a technical super-culture (e.g., computer culture).

Prevalent in the integrated communication space are common stereotypes, evaluations, parameters of required behavior, and its popular, or simplest, elements. While creating many conveniences, this trend robs intercultural dialogue of any meaning. We can understand any person at any point on Earth, but at the level of coincidence or even identity of meanings. Dialogue does not occur as a mutually enriching factor in the course of learning from each other, but as a pseudo-dialogue where parties to communication do not listen to their counterparts, hearing only themselves. A sharp increase in the speed of the disintegration of old values and the compression of time occurs in this process. Consequently, synchronization of culture is disrupted, because new formations are emerging so fast that they fail to adapt in due time to the traditional system. People do not have time "to absorb" new values through correlation with the preceding ones, and they begin to consume them. A rapid overburdening of language occurs, with words and expressions which can be understood by younger people, but which is beyond the comprehension of the middle generation.

Today, mass culture is manifested as "pop" culture, bearing all the characteristics of the lower culture, though not at the local level, but at the level of the emerging integrative culture, as manifested in the notion of "show." Show is an integrative (mass) formation where individual creativity is subordinated to participation or simulated participation. Participation as such becomes a form of communication without the transfer of any meaning. Inside a show, a showpiece is inseparable from the listening audiences and technical facilities of reproduction. The pop music performer and listener is one whole, and we cannot imagine them separately. Show has penetrated the lives of people so that any event, even the most tragic one, can become an object of entertainment, unless the observer is personally involved in it. The whole world currently has become a global show performed according to the laws of the genre.

V. V. Mironov

Communism: (Latin *communis*, "common") Communism can refer to (1) a utopian vision of the future utopian society; (2) an ideology of some political parties and movements; or (3) a specific way of life.

(1) Communism as a utopia looks to locate society at the same time in the past (the primeval communism) and in an ideal future, when society would unite the best characteristics of the past (no classes, no states, no trade and monetary system, no legal institution of family) with the achievements of modernity (such as human freedom, economic wealth, refined culture). The global scope of the communist utopia is of principal importance: the communist system is based on human universality.

Although usually communism is connected with the names of Karl Marx and Friedrich Engels, their predecessors, the utopian socialist writers of the nineteenth century, are commonly called "communists" as well. Especially interesting are writings by Charles Fourier, who emphasized the globality of the new social order he called "harmony." The central difference between utopian communism of the past and the "scientific communism" of Marx and Engels is not the characteristics of the utopia itself but the fact that Marx worked out a theory of proletarian revolution as the means to achieve the state of communism.

In Marxism, communism is the second (after socialism) phase in the development of a communist society. A socialist society still contains some elements of capitalism (such as class differentiation, commodity economy, and the state and its enforcement institutions). In a communist society, there are no classes, no state, no trade and monetary system; this society is based on a

88 *Global Studies Encyclopedic Dictionary*

principle "from everyone according to their capabilities, to everyone according to their needs." However, we cannot find, in the writings of Marx, Engels, or Lenin, more specific characteristics of the future society. Nevertheless, some Marxists, such as August Bebel, still theorized on this topic, and were quite interested in reconstructing daily life (one of the reiterant traits of communism is mechanization and socialization of housework).

Soviet social theorists worked out an elaborate theory of a protracted transition from socialism to communism. No bright utopian visions were allowed; the only official attempt of this kind was undertaken in the Communist Party of the Soviet Union (CPSU) Program adopted by the twenty-second CPSU Congress (1961) that proclaimed that socialism-building in the Soviet Union was in general complete and that the transition to communism would take place during the lifetime of the present generation. This Program offered no new theoretical approaches to communism but it indicated some of its concrete characteristics, such as free public utilities, free public transportation, and free catering. However, the realization of this Program was stopped after Nikita Khrushchev lost his office. At that time, a new concept of "fully developed socialism" was introduced and the transition to communism again moved to the uncertain future. Nevertheless, a vast literature emerged describing the future communist society, though in the genre of "science fiction" (e.g., Ivan Efremov's *The Nebula of Andromeda* [1957]).

(2) Communism as an Ideology: During the twentieth century, communist ideology was the foundation for the worldwide activity of communist parties. Caring not too much about the specific characteristics of the communist utopia, these parties concentrated on the issues of taking over power and moving the forces of revolutionary struggle. Initially, they were prone to armed assumption of power by proletarians, but when, beyond the Bolshevik revolution, communist revolutions in Europe in the early part of the twentieth century failed, most of the communist parties rejected the perspective of an armed uprising and adopted legitimate forms of struggle (when this was possible). After the Second World War communist parties legitimately won power in Eastern Europe. Up to the end of the 1950s, most of the world communist parties considered the Soviet system an exemplary communist society.

In the 1960s, after Stalin's regime had been condemned in the Soviet Union, many communists outside the Soviet Union lost their illusions concerning Soviet communism. The new ideologists, considering themselves creative followers of Marx (such as Herbert Marcuse), argued that the working class had lost its revolutionary drive and that revolution was presently represented by the youth and various marginal groups who were not a part of capitalist society. The Communist movement of the 1960s was heavily influenced by the theories of Mao Tse Tung and was supported by some other ideologists from the developing countries (such as Franz Fanon). They insisted that revolutionary struggle now takes place not between the bourgeoisie and the proletariat within each country, but globally between the developed world ("the world city") and the developing world ("the world village"). The main revolutionary force was considered not proletariat but the poorest peasants. Communist movement began to erode. On the one hand, pro-Soviet communist parties financially supported by the Soviet Union were still alive, and, on the other hand, new, non-conventional approaches to revolution emerged (including, for example, liberation theology which is a fusion of Marxism and Catholicism). After the world socialist system ceased to exist, most of the conventional communist parties disappeared, while the non-traditional versions are not always associated with communism.

(3) Communism as a Way of Life: Communism, understood as a way of life, dates back to Nikolai Chernyshevsky, a Russian socialist of the nineteenth century. In his 1905 novel, *What Is to Be Done?* Chernyshevsky insisted that, being aware of the basic characteristics of the future, people in a communist society can and should transfer some of these characteristics into the present. After the October Revolution of 1917, an idea gained influence in the Soviet Union that even before "full" communism is built, the Soviet people can and should transfer some elements of the future communist society into their everyday life. Terms like "communist education," "communist household," and "communist morality" were introduced.

Communism as a way of life spread beyond the Soviet Union. In the 1960s, various communes and other social projects emerged in Europe that could be considered elements of the future communist society implanted into the present. Non-traditional versions of communism, such as liberation theology, also introduced communist elements into daily life. For example, liberation theology implied so-called basic ecclesiastic communes (*communidades eclesiales de base*, or CEBs) performing, among other things, some of the social functions that governments in Latin America refused to do.

A. V. Mitrofanova

Competitiveness: Economic competition is studied across different categories/levels, including (1) goods and services; (2) manufacturer or corporate, (3) industrial, and (4) among countries. There is tight internal and external interdependence among these levels; e.g., competitiveness of countries and industries ultimately depends on the ability of manufacturers to competitively produce goods.

In the globalizing world, some factors in the foreground of world competitiveness are not price-driven, the most important of which are quality, novelty, technological advancement, and intellect content. Therefore, most countries provide commodity competitiveness using innovations and the development of highly technological commodities. Creating these is impossible without scientific and technological development potential, assessed by innovation cost indicators. These represent a country's innovation ability and, in addition to research and development costs, takes into account design and marketing costs, the number of employees in the scientific sphere, the number of patents received within the country and abroad, the cost of protecting intellectual property, and the development of the educational sphere.

V. A. Galichin

Composite Development Indices (Integral Development Indices) are a combination of economic, social, demographic, and ecological indicators, which help determine the overall development of a country or region.

For a long time, the notion of development in economics was only associated with the increase of goods and services output determined by gross

Global Studies Encyclopedic Dictionary

domestic product (GDP) and GDP per capita indicators. In the 1970s, GDP and GDP per capita as socio-economic development indicators were replaced by composite development indices. Among them was the Index of Physical Quality of Life (PQLI), proposed by Morris David Morris in 1979, that employed three indicators, namely: infant mortality rate, life expectancy at age one, and adult literacy. The drawback of the PQLI was that that it only approximates actual welfare within a state without taking into consideration the average per-capita income. In the 1980s, other composite indices were presented. The index of demographic transition (1984) considered three demographic indicators, the index of human suffering (1987) examined ten socio-economic indicators, and the index of Quality of Life in Urban Areas (1990) calculated using ten socio-economic and ecological indicators. However none of the proposed indices could fully supplant GDP as an indicator of development. Because indicators used for the indices calculation lacked validity, indices could be ambiguously interpreted and had poor statistic database. These drawbacks were to some extent overcome in UNDP Human Development Index.

I. A. Aleshkovsky

Computer Ethics: With the introduction of computers, the personal and social life of every person on this planet has been profoundly altered. Technology had always influenced people's lives, but on a very visible scale: tools and materials were within the power of the craftsperson or user. Computers are universal tools that can be shaped and molded to perform nearly any task. Computers can do anything we can describe in terms of input, process and output. We find them everywhere: first on the desk, but more and more in daily life appliances: the refrigerator, the microwave, the car, electronic devices, and telecommunication apparatuses, but also in the workplace, in education, in the health care, in political and military life, in the world of business and banking, and in the media.

Computers are black boxes that contain programs (software) and data that are almost invisible unless one has an apparatus that can read the bits and bytes as information. Information is data that are processed by programs. One finds it as illegible bits and bytes on machine level. But as human beings we can give meaning to this information so that information becomes knowledge, or significant information. As Bacon wrote, knowledge is power; meaning that not especially the apparatuses themselves, but the information contained in them can have a thorough impact on people and society, at good or at bad. This situation poses serious ethical dilemmas, exacerbating old problems and forcing us to readjust ordinary moral standards in new fields.

A further complex problem emerges when computers transpose information into commandos for machines and apparatuses without any human interference. The impact of it can make operations (for good or bad purposes) much faster and far-reaching in scope, than what we can foresee, and, possibly, stop when specific limits are exceeded. How do we protect ourselves against dangers that are caused by computer apparatuses? What is still the role of people? Are humans only the last appendage of the more and more autonomous machine? Or are we not only the inventors, but also does the ultimate controllers of this technology? Is our role that of the ultimate measure

Capitalism – Cyberculture 91

by which the use of computers is justified and the computer only a means to our ends or is it exactly the inverse? We cannot deny that technology changes our behavior. Automated apparatuses do so in a profound, though not always perceptible manner.

This new technology demands new considerations for the designers of computer technology, such as:

First we must consider questions concerning the input. Is it ethical to put every bit of information on a computer? Can everyone do this? Who or what permits them to do that? Is this fair? Here the question arises concerning the spamming, hacking, and cracking of protected computer systems, which leaves behind cookies and spy ware on a person's computer to trace and re-trieve personal information or discover a person's whereabouts, (and hence the concern for the respect for privacy), the question of whether we can put anything (secret information, sexually explicit pictures, racial and religious hatred) on internet, the question of installing parameters that adjust machines and apparatuses in functioning and the question of a fair and equitable acces-sibility for users.

Second we confront questions concerning the processing of data: Here arise questions concerning accuracy in processing data to transform it into useful information, questions concerning intellectual property, encryption technology and data protection, the question of monopolies on software and intellectual property and the question of free software.

Third questions concerning the output arise: who can obtain the infor-mation? To what use may it be employed? Should it be checked? Here arises questions pertaining to the illegitimate downloading and copying software or data bases, the question of protecting children against offending contents, questions concerning automatic functioning of machines and apparatuses without human interference, the question of fair access, the confidentiality of some information and even the concept of stealing is at stake. The traditional concept of theft has to be broadened to that of using or appropriating one's intellectual property without proper authorization.

Ethical values in the debate include justice and equality, respect for pri-vacy and property, the freedom of the individual, and the security of the soci-ety. Many countries have approved laws that regulate the access to infor-mation stored in computers, or which regulate the use of apparatuses that in-fluence one's privacy. However, new technology develops much faster than new (legal) standards can be written. Merging different functions in one de-vice (for example, the new generation of handheld cell phones with the possi-bility of sending and receiving messages and pictures worldwide) changes traditional concepts and requires a new thinking concerning standards and values. Furthermore there is still the important question of the short lifecycle of the computer products. A lot of these products contain materials that dam-age the environment or they contain rare material that provokes local feuds among the communities of origin. For this reason ethical reflection is neces-sary on the part of designers, users, and lawmakers to foresee and anticipate the impact of computer technology at different levels and for different stake-

92 *Global Studies Encyclopedic Dictionary*

holders, or those who are affected and have legitimate rights in relation to the activities of computer-induced apparatuses.

H. Lodewyckx

Concerned Philosophers for Peace (CPP): CPP was initiated as a response to the increased militarism of the Reagan Administration, especially in relation to its deployment of Euromissiles and policies on nuclear weapons. Subsequently, the organization progressed from a critique focused on nuclear war fighting strategies to the promotion of co-operative endeavors, first, with Soviet and, now, Russian philosophers. Increasingly, the organization has offered criticism of military actions of the United States. Ongoing efforts have also been made to link with other movements connected with the quest for peace, especially within feminism. Since its inception in 1981, CPP has become the largest, most active organization of professional philosophers in North America oriented to the critique of militarism and the search for a just and lasting peace. It has published a newsletter since 1981, and numerous monographs and anthologies of essays on issues of war and peace. Through its meetings, newsletter, special book series, and various international affiliations, Concerned Philosophers for Peace has augmented Peace Studies as a recognized area of philosophical specialization.

W. C. Gay

Conflict, Political: Political conflict is a theoretical and practical struggle of subjects of a political system, which mobilizes groups struggling for authority with the purpose of modifying, transforming, or keeping the social order, political statuses, and institutes. Political conflict is usually considered to be an extreme form of political action to those or other institutes representing entrenched social groups, representing the opposition of various political forces on specific questions. At the same time, it is the integral, immanent feature of the world of a political system, marking its very beginning. Disagreements, competing points of view, collisions, appeals to authority and demands for sanction represent what we refer to as political conflict. This is not only a category of political science, but also language of the theory of politics, a mentality.

Political conflicts arise from conflicts of interests, rivalries, and social struggle among professional, ethno-confessional, and other groups and trade disputes and the redistribution and recognition of political authority and dynamics of leadership (such as the deputy, the president, minister, the judge, and the leader of political party) in institutes and structures of this authority. Among sources of political conflict include the hierarchical structure of the political sphere, unequal distribution of the political rights and freedom, forms and levels of participation in a political life, differences in political ideologies and the policies of established social groups. In times of political conflict, groups compete for specific (indivisible) resources such as authority and powers. Politics thus differ from economics, where all things have an exchange value, since, for instance, political freedom and independence cannot be understood in terms of fair exchange. This fact determines the distinct bitterness of political conflict and the urgency expressed by those engaged therein.

Political conflict endeavors to mobilize the greatest number of supporters from each of the conflicting groups. Political conflict should also be distinguished from legal or juridical disputes, which represent struggle on the grounds of the law or between subjects of the law and does not assume an appeal to the uninitiated. On the contrary, a political conflict seeks to mobilize the uninitiated, and the scales of mobilization can influence the outcome of any given conflict.

A. V. Glukhova

Conflict, Social: (Latin *conflictus*, "collision," "struggle") Social conflict is defined as a clash of interests, goals, and demands of individuals or groups in the process of social interaction. In sociology and political theory, different methodological approaches have been articulated concerning the interpretation of this term. One is a behavioral approach to conflict as a social process, in which an individual or group seeks to achieve their goals and realize their interests by eliminating or subduing another individual or a group pursuing similar goals and interests. Conflict can also arise among groups pursuing different goals, but using the same means of achieving them.

In conflict, there is always an awareness of the opponent's presence and ambitions; the goals of both parties are clearly identified. Conflicts are based on antagonism, non-recognition of the alien system of values, and negative bias against the opponent. American sociologist Lewis A. Coser believes that conflict is a behavior that entails a struggle between rival parties for scarce resources and includes attempts to neutralize opponents, do harm to them, or eliminate them. In defining conflict as a struggle, one stresses the intensity of the interaction between the subjects, which may assume the form of an open confrontation. The structural approach defines social conflict in terms of social contradictions linked with differences in the lifestyles of people belonging to different social groups, the disparity of their life opportunities, which in turn influence in a way their worldview, or the ideal image of their future. Structuralists define social conflict either as a relationship between elements, which can be characterized as objective (latent) or subjective (evident) opposites (Ralf Dahrendorf). Awareness of contradictions is not necessarily required for identifying the relations as conflicting. At the initial stage of a conflict, the parties can be both aware and unaware of the opposition. In the course of conflict the quasi-groups rallying around an alleged identity of interests based on common social positions organize themselves under favorable conditions into groups of interests with well-defined goals and programs. Conflict as a social interaction has a wide scope of forms of manifestation ranging from parliamentary debates or civil war to peace talks or a strike.

The phenomenon of social conflict has been studied within different methodological paradigms: dialectic theory of conflict (Marx, Dahrendorf), functionalism (Georgs Simmel, Coser), general theory of conflict (Kenneth E. Boulding), theory of structural violence (Johan Galtung), and psychological theory of international conflict (C. R. Mitchell).

A change of paradigm in the study of conflict has always been closely linked with social changes. Conflicts perform an important integrative function in a society as a form of interaction between social groups that provides for an exchange of benefits and functions, correction of interests, and relations with the help of a feedback system.

94   *Global Studies Encyclopedic Dictionary*

A conflicting system in its moderate form with controllable and institutionalized conflicts is more stable in the long run than a system based on subduing the interests of different groups. A conflicting system ensures a variability of forms of development, providing opportunities for finding alternative solutions for the most important problems. This function makes the existence of different interest groups legitimate, promoting the formation of new groups, opening the way for structural improvements, and changing orientations of the society as a whole. Interaction between equal conflicting parties ensures flexibility of the social organization and its higher adaptive capability with respect to the impact of the external environment. In this regard, conflict interaction is often preferred to other forms of contacts, including cooperation, since the latter may consequently stand in the way of social changes.

A. V. Dmitriev, V. K. Egorov, A. V. Glukhova

Consciousness, Global: Globalist consciousness is an acknowledgement of the degree and scale of the threats to human civilization that, due to the mass media, are rapidly becoming the property of nearly the whole population of the planet.

The formation of global civilization that started during the epoch of the great geographic discoveries and colonial conquests and continued through the foundation of the world market, and then of the world economy and the emergence of the world policy accompanying these processes (the struggle between the leading European capitalistic states, first, for the sharing of the world, and then for its repartition, culminating in the two World Wars), ended by the second half of the nineteenth century with several global problems of our times that threaten further existence of humanity on Earth. First, among these problems manifesting the modern world character, the danger of a new World War involving nuclear (and later, new chemical, biological, climatic, and even geological) weapons appeared. Then, the planetary consequences of the menace caused by human industrial activity were realized; they were inflicted, first, by the developed countries' industries whose wastes, dumped into the atmosphere, threatened dramatic climate changes all over the planet and, as a consequence, the extinction of life. The pollution of the world's oceans and the destruction of forests create conditions for global "oxygen starvation," insofar as it destroys natural sources for atmospheric oxygen reduction (while industrial substitutes for wood and bioplankton do not exist). Destruction of the ozone layer, brought about primarily through the use of Freon in industrial and household refrigeration, means, in perspective, the annihilation of life as a result of intense space radiation.

At the same time, along with these menaces a global danger of famine for the rapidly growing population of the Earth was realized. It relates mostly to underdeveloped countries comprising the majority of the planet's population, but in the long run it concerns developed countries as well.

The problem of the exhaustion of material resources for industry that cannot be restored grew more urgent. The Chernobyl accident became a highly perceptible demonstration of the reality of a global threat: it turned out that a "peaceful atom" is fraught with danger no less than a "military atom," especially due to the rapid growth of atomic power engineering. This accident attracted public attention to the problem of human-caused catastrophe as a whole.

Political reforms, having put an end to the Cold War and balancing on the brink of a "hot" war, which had seemed to be able to soothe the acute situation in the world, split up not long ago into two irreconcilably hostile camps, increased, on the contrary, the menace of a grand human-caused catastrophe in all the countries of the former socialist camp owing to obsoleteness of industrial (mostly power and chemical) enterprises, expiration of many of the naval, industrial, and exploratory atomic reactors, bacteriological, chemical and ordinary arms reserves, and the lack of funds and qualified human resources able to prevent dangerous situations and develop effective technologies in order to control, dismantle, and deactivate all the above-mentioned sources of danger. Acknowledgement of the degree and scale of these threats by the educated part of the population due to the mass media rapidly becomes the property of the more masses of the population of the planet. This was the process of the birth of globalist consciousness, with the peace movement under the slogan "Against Atomic Death" as the first stage, followed by the so-called ecological alarmism, whose program was shaped in publications of the Club of Rome reports. Such alarmism forms the prevailing motive of the today's globalist consciousness that is being widely and quickly promoted not only by the mass media, but also through the growing mobility of the population of the planet (emigration and immigration, tourism and various international contacts). Thus, globalist consciousness deviated from the subject of ecological alarmism. Today, almost everything that happens in different regions of the planet and attracts public attention-politics (especially its scandals), sport competitions, so-called public life events, beauty contests, and criminal news, becomes or is able to become the property of the whole Earth's population and evoke mass and not always predictable and adequate responses.

In this aspect, mass globalist consciousness is very mixed, fragmentary, and changeable. At the same time, when actual grave events, able to evoke remote and large-scale consequences, happen, this mass consciousness proves to be capable of immediate consolidation, splitting up and originating mass actions. As an example we may take the public reaction to the 11 September 2001 terrorist acts in the United States, after which the problem of international terrorism and any events somehow or other connected to it (such as military operations in Afghanistan and information about terrorists' possible access to the nuclear weapons and other weapons of mass destruction) were brought to the forefront. The issues of the death penalty, euthanasia, abortion legislation, cloning, and genetic engineering as a whole become subjects of popular discussion and causes of mass actions.

On the other hand, both globalist consciousness and individual consciousness may be as well applicable to regions, cultures, and situations. That is why national, racial, and religious solidarity or dissension, not to mention political leaders,' music and sport stars' "promotion," forms the sphere where the laws of mass psychology work. This sphere of mass consciousness is manipulated by the modern local, regional, or global mass media, as long as authorities, the Mafia, and other owners use them. Thus, on the one hand, acting

96 *Global Studies Encyclopedic Dictionary*

as the most important means of the efficient global consciousness' formation, the mass media, on the other hand, make the "nervous system" of such a human entity that is called the "crowd." At this point, any crowd, from a street one to the "global" one, contains the characteristics of a schizoid personality: it easily follows the mood that may often sharply and suddenly switch. As a whole, it is inclined to irrational behavior and originates in individuals disposed to various kinds of radicalism and dangerous or unexpected actions. As a result, "globalist consciousness," and mass consciousness, shows the signs of sick-paranoid or schizoid-consciousness. In such a capacity it has become the subject of special studies nowadays.

At the same time, in some contemporary publications the term "globalist consciousness" is used to designate the mentality of people professionally interested in global problems. In this case, the agents of this consciousness would be the Club of Rome and the corresponding United Nations or government committees' members, and influential social organizations whose interests exceed the limits of local troubles, financial and transnational industrial companies' leaders and managers whose interests cover an essential part of the Earth territory and population, and everyone who is affirming that the whole planet should be the area of their interest and care.

A. F. Zotov

Consolation: In the West, the role of consolation for philosophy is apparent in antiquity, with the Epicureans and the Stoics who dealt with the pain linked to death, and later with Boetius. Augustine realized in Christian philosophy an important innovation, linking the presence of the ideal and the consolation of human pain. The Christian contribution to the concept of consolation articulates this philosophical turning point in two respects, the incarnation of God in humankind and the historic dimension of humans, two founding elements of an eschatological doctrine of salvation.

Other areas of the study of consolation include the development of psychology. Modern psychology conserves in its premises the traces of a philosophical undertaking geared toward palliating the deficiencies and pains of the human soul. In philosophy, since Montaigne and Rousseau, existentialists, for whom individual existence and sufferings remains the primary issue, have often produced novels or short stories. In the recent period, characterized as post-modern, where theoretically the great classical schemes have lost their aura or totally collapsed, we witness the return of philosophy as a consolation through new practices such as the philosophical consultation, the *"café philosophique"* conceived as a collective philosophical dialogue, or the editing of philosophical writings intended for the public.

O. Brenifier

Constitutional Ideal, Global: The global constitutional ideal is a search of legal and constitutional ideals in both historical legal thought and contemporary juridical theory and practice. The notion of a legal and constitutional ideal was mentioned in Immanuel Kant's "rational law theory" ([1785] 1998), in Max Weber's "formal and rational ideal type of law" ([1903–1917] 1990) and in the recommendations made by Alexis de Tocqueville in his *De la démocratie en Amérique* (1888). Some constructive considerations concerning the building of a constitutional ideal can be found in the works by many "natural law" and "social contract" theory representatives, utopian socialists, and Marxists.

This idea was emphasized and applied to constitutional law by Carl Schmitt (1928) when he wrote about "the ideal constitutional concept of a civic legal state." It became one of the sources of the future global constitutional ideal despite that Schmitt's level of conceptualization had been limited to analyzing only the most developed constitutions of that time and that the severe class confrontation in the world divided between two systems had made an earnest global conceptualization hardly possible. This semi-forgotten idea again attracted scholarly attention in the end of the 1990s, following the emergence of the new, planetary, economic, political, scientific, technological, and other phenomena.

These problems can only be solved by common efforts of a planetary scope that would enormously increase individual positive capacities. It has become evident that there are objective common, global interests directly touching each human being's life interests independent on their individual characteristics. These are the right to life, to a healthy environment, to appropriate living standard, to social security, to access to culture, education, and health, to privacy, and some other rights and freedoms. Such common interests are inevitably becoming a part of the constitutional ideal and are reflected in the constitutions of many states although they are carried into practice by different means.

At the level of separate countries, a legal aspect of this trend has been reflected especially in the constitutions adopted after World War II, and it becomes increasingly emphasized globally in the activity of the United Nations and its Security Council, in the development of the Helsinki process, in international agreements and programs coordinating international activity in the spheres of human rights, education, health, culture, communications, information, environmental protection, disarmament, nuclear energy development, and fighting crime and international terrorism.

Universal and regional international organizations, intergovernmental and non-governmental organizations and institutions are taking part in these developments of increasing international significance. Human survival and progress depend in many aspects on this regulatory activity.

A new human consciousness begins to form—the one of persons' feeling themselves not just as a part of a separate ethnic community, but as "world citizens" who move across the national and state borders that have become too narrow for their self-realization and self-knowledge, and who are able to overcome the crisis of the passing era and to build the civilization of the third millennium. These new human beings will be the objects of the global constitutional ideal and the main criterion of its progress. The new constitutional ideal exists on the universal level of democracy and humanism reflecting completely different, harmonious relations among humans, industry, science, technology, and nature, without which human survival in the third millennium is impossible. At the same time, one should clearly distinguish between the global constitutional ideal (reflecting the essential, substantial, moral, and ethical aspects and building a kind of conceptual ground for

98   *Global Studies Encyclopedic Dictionary*

the new type of constitution) and the ideal model of constitution (an optimally structured, formed, and organized legal document).

The universal roots of the global constitutional ideal are securely grounded in the massive all-civilizational foundation of modern society, which is much more deep and fertile than those artificially ideologized surface layers that were engendered by economic differences eventually disappearing after having encountered real life.

Following Alexander Solzhenitsyn, we can acknowledge that the optimal constitution is not the one that achieved the maximal prevalence of the majority over a minority using the state mechanism and its violent methods, but the one that helps to reach an agreement between the majority and all the possible minorities using democratic procedures and considering mutual interests.

Accordingly, a constitution builder's task is perhaps to find, with the assistance of legal, moral, ethical and other norms, stimuli and mechanisms, compromising solutions formed on the basis of universal interests and values via democratic legal procedures and considering all parties' interests. At the same time, social consensus, being the purpose of the ideal constitution, is the more stable and constructive the less cohesive methods are used to reach it, which means that the path to realization of the global constitutional ideal is perhaps through diminishing violence in international and domestic life in order to achieve non-violence.

The turning point humanity experiences nowadays demands changing our very vision of characteristics and criteria of the progress, our life principles, and introducing a new scale of values and priorities for the future civilization. That is why non-violence moves to the foreground.

Practical movement toward the global constitutional ideal (a model, essential conceptual ground that nearly each modern constitution strives toward) allows for a free and broad range of various individual characteristics reflecting national, psychological, regional, historical, traditional, cultural, and other peculiarities of states and peoples, is lengthy in its progress in many directions and has impact on the development of various old and new institutions. This ongoing process demonstrates a comprehensive trend toward formation of "universal law" and its more private component—the global constitutional ideal. The general process of humanization of law makes human being its foundation and makes individual interest its criterion of central social progress.

Globalization and similarity of the problems faced by humans, nations, states, humanity as a whole, and convergence of living conditions engender similarity of many legal, including constitutional, solutions, create common economic, cultural, informational spaces, and justify legal unification making possible the achievement of the common good, that is, the material ground for building "universal" law in spite of increasing diversity of specific conditions in various countries. All these factors influence both international and constitutional law and all concrete legislative fields composing the foundation for globalization of general trends of contemporary law development.

Being very close to global problems of modernity, international law is usually the first responder to them. Given that the threat of nuclear war, ecological disasters, unpredictable genetic transformations, and energy and food problems has made humankind mortal for the first time in our history, some-

Capitalism – Cyberculture   99

thing like the whole human community's "self-preservation instinct" begins to develop. As a result the necessity emerges to provide, first, the world social organism with a comprehensive tool for this self-preservation and for effective self-regulation and self-development.

That is why, especially after World War II, the basic international law principles (or, mutually agreed provisions admitted and respected by the international entities and vested in universal and other international acts) began to evolve as "self-preserving" principles directly related to promoting international peace and security. These principles include such important provisions as non-use of force or threat of force; territorial integrity of states; inviolability of borders; peaceful resolution of international disputes and disarmament.

Against the background of changing political thinking and the end of bloc confrontation, the common international law foundation directed toward the international community's effective development has become stronger. These are the principles of sovereign states' equality; non-interference; equality and self-determination of nations and peoples; cooperation; respect for human rights; fair implementation of international obligations.

Via constitutional law, democratic principles, norms, and standards are transferred into specific spheres of the law. Often, in the beginning, these principles can only be found in the legislation of the most advanced states and then, sometimes with their assistance, they penetrate the other countries and regions. There is a constructive mutual penetration of the advanced legal ideas, principles, institutions, mechanisms and norms of international and constitutional law and specific law, in which the norms of international and constitutional law, being more general and close to universal values, play the central part. In this process, some basic global ideal penetrating the development of each sphere of law will crystallize. Scholars recognize a trend toward approximation of law systems and unification of modern constitutions that also demonstrates an increasing tendency toward the emergence of the universal law for our civilization.

This movement toward the global constitutional ideal will be most likely carried into practice via regional socio-culturally and economically homogeneous centers that are formed within integrated groupings throughout the world that should finally unite. One of the groupings closest to this ideal is the contemporary European Union. Its basic documents and principles can become a kind of experimental polygon for working out and bringing into being the global constitutional ideal. Within the EU, "the European citizen" is now emerging as an intermediary stage on the way to "the world citizen."

References: Weber, M. The Methodology of the Social Sciences. Edited by E. A. Shils and H. A. Finch. Translated by Shils. New York: Free Press, (1903–1917) 1997. Kant, I. Groundwork for the Metaphysics of Morals. New York: Cambridge University Press,(1785) 1998. de Tocqueville, A. *De la démocratie en Amérique* (Democracy in America). Paris: C. Lévy, 1888. Schmitt, C. *Verfassungslehre* (Constitutional Doctrine). Munich: Leipzig, Duncker & Humblot, 1928.

S. Y. Kashkin

Constitutional State: One in which exercise of power is constrained by law, which is enshrined in a constitution that recognizes as its principal features a division of powers, an independent judiciary, the legitimacy of government, legal protection of citizens against violation of their rights by the state authority, and compensation for damages inflicted by a public body.

In the modern global world, the concept of a constitutional state represents a kind of universal ideal for building a mutual relationship between state, society, and persons. The essence of a constitutional state is consistent democracy, popular sovereignty as a source of power, and subordination of state to civil society. In a constitutional state, the rule of law guarantees the protection of citizens against arbitrary acts of the state and its bodies (from the German *Obrigkeitsstaat*, "a state based on the arbitrary use of power"). A constitutional state's most important principle is the supremacy of law, by which is meant that both citizens and the government are subject to standing laws.

The constitutional state is not an end in itself, but a socially and historically conditioned form of expressing, implementing, and protecting freedom and equality. A specific level of development of a given society determines the nature of this freedom, its quality and quantity. That level is determined by the degree of equality among people.

In a constitutional state, the state machinery, its bodies, institutions, and officials serve the entire society, not just one of its factions; the state is responsible before citizens; individuals, their security, rights and freedoms, dignity, and honor are the highest value; the observance and protection of human rights and freedoms are the main obligation of a state.

However, the meaning of a constitutional state should not be reduced to the protection of individuals against the state. It both restricts and encourages state activity, to guarantee human rights and freedoms, justice and legal protection. When adopting a legal act, a constitutional state accepts concrete obligations to its citizens and defines the legal measures of responsibility that its official representatives bear for actions committed on behalf of the state.

O. E. Kutafin

Consumer Advocacy is a movement for the protection of consumers' rights in their relations with sellers; an association of consumers in unions either as a public venture or under the aegis of the state.

In the West, between 1998 and 1999, the improvement of consumers' rights consisted in the ability to solve specific protection problems in cases of: manufacturer's bankruptcy, sale of products of inferior quality by manufacturers, and obtrusive advertising in the field of telecommunications.

Between 2000 and 2001, a new trend in the development of international commercial law appeared in the West (including consumers' rights) related to the emergence of a new social type dubbed "consumer 2002." This part is acted by a well-to-do person of middle age surrounded by lots of things acquired over many years and who is trying to get rid of them. This social type is characterized by a refusal to acquire jewelry or make showy purchases and memorable gifts. Preference is given to short-lived things, compact items, and even such intangible acquisitions as superficial knowledge in some field. In the material domain the attention of "consumer 2002" is primarily directed toward acquisition of products coming from electrical household appliance, furniture, and food industries: computers and office equipment, stylish furniture and groceries. The new type of consumer presents a challenge to the consumer-merchant. People of new social type prefer quality to quantity and do not increase acquiring property with income growth. This consumer type does not promote growth of demand for consumer goods produced by multinational companies. These consumers reject the merchant's conception of an ever growing spending in property acquisitions and choose to spend on non-material products, such as education, health care, and other social services.

The appearance of the "consumer 2002" has brought about changes in the hierarchy of consumer priorities and preferences. In economically advanced societies the purpose of the movement in defense of consumers' rights, contrary to materially-minded and utilitarian standards peculiar to the so-called consumer society, is shifting to values that define the quality of life: ecological well-being, health care (not only medicine but also a healthy way of life), education and culture, access to information, opportunities for creative development and self-actualization. Current practice testifies to the fact that catering to needs in general and those of each consumer in particular has become one of the major problems and final objectives in any business undertaking both in the manufacture of commodities and in the service sector.

In connection with such changes, for instance, in the judicial practice of Great Britain, normative regulation of terms in contracts with consumers in the European Union of 1994 was subjected to scrupulous analysis.

The arduous pace of life in modern post-industrial society, the density of population, bureaucratization, alienation, the modern state with its military, technological, and propagandistic potentials force the "consumer 2002" to seek a refuge for protection from the pressures of the external world. Therefore, solitude, or privacy becomes a necessity, a kind of an intangible consumer product. In the United States of the 1970s, the conception of the right to privacy gradually became established in court practice and led to the recognition of the inviolability of private life as one of the fundamental values safeguarded by the Constitution. This conception provided the legal basis for legislative restrictions on new forms of state control over personality, such as collection and use of personal data or telephone eaves dropping and electronic surveillance.

Privacy is a peculiar non-material consumer product performing two functions: (1) a mechanism of social psychological adaptation of personality and the right for independent assessment and decision-making and (2) the selection of a line of behavior. The slogan of the modern consumer is, "There can be no individuality without privacy."

In the United States, the institution of privacy protection exists simultaneously in three fields: common law, where it is applied in relations between the subjects of private law; constitutional law governing the relations between an individual and the state; and in legislation, by assigning a specific authority to an individual who secures the legal right to control the inviolability of some aspects of their private life within the limits set by a specific legislative act.

In Russia, the movement in defense of consumers' rights consists mainly mass media popularization, informational/educational publications, and television (e.g., "Expertise," on RTR TV). The law "On Defense of Consumers'

102 *Global Studies Encyclopedic Dictionary*

Rights" underwent three revisions in 2000. Frequent revisions of this law and publication of commentaries thereon is evidence of the high attention to consumers' rights on the part of the state represented by the Federal Anti-Monopoly Committee. Consumer law in Russia is segregated as a separate legal field.

O. V. Dyogteva

Contamination with Heavy Metals is one of the most widespread forms of contamination of the natural environment, atmosphere, superficial and subterranean water, natural biogeocenoses, and ecosystems, or the population habitat that ensures reproduction of most of renewable types of life-critical resources. The heavy metals are those with density higher than that of iron (7874 kg/m^3): Pb, Cu, Zn, Ni, Cd, Co, Sb, Sn, Bi, Hg, and some others. In natural geospheres (litho-, hydro-, atmo-, and biosphere), they are normally characterized by a relatively low content, and they participate in the circulation of natural substances as necessary trace elements ensuring critical reactions in biological metabolism systems. In elevated concentrations, many of them become extremely strong toxicants that cannot be utilized by ecosystems and negatively impact biota and contacting population.

Depending on their negative impact, heavy metals and other toxic chemical elements are divided into several danger classes. Their degrees of danger can also vary according to their host environments (soil, water, air, food). Heavy metal contamination is normally related to natural or human mobilization of lithospheric substances containing different concentrations of heavy metals that become involved in natural circulation of solid, liquid, and gaseous phases, followed by migration or accumulation in different depositing environments (air, water, soil, biological objects), producing a negative impact on contacting biological systems and populations. Nature management practice shows that heavy metal contamination is usually complex.

A particular case of contamination with heavy metals induced by human activity is related to the development of human society and the intensification of the use of natural resources. According to existing assessments, only eighteen chemical elements (including some heavy metals) were used for economic purposes in ancient times, in the eighteenth century there were twenty-eight, in the nineteenth century, sixty-two, and at present, we use all elements known on Earth including neptunium, plutonium, and several radioactive isotopes that do not exist in natural conditions.

The toxic waste of motor vehicles creates a high ecological burden in cities and areas adjacent to motorways with intensive traffic. According to some assessments, during the sixty years of active use of leaded gasoline, the aggregate release of lead in the Northern Hemisphere countries was approximately 2.6 million tons.

The maximum allowable levels of heavy metal contamination, from the point of view of human health safety, are determined on the basis of the special hygienic indices MAC (Maximum Allowable Concentration), ASIL (Approximate Safe Impact Level), and AAQ (Approximate Allowable Quantity); the latter two are temporary values that substitute for MAC until it has been determined in a given locality. Specific heavy metals attain MAC differently in different environments; thus we have MACs for atmospheric air, water, natural and artificial water bodies, and soil. Hygienic MACs are determined for several heavy metals in food, and biological MACs have been set that determine the limit level of content of several components in a human organism. In the practice of nature management research, we must consider the possibility of the complex impact of contamination with different heavy metals. For this purpose, integral values should be calculated on the basis of the summation of the concentration characteristics of heavy metals. The index of aggregate soil contamination with chemical elements (e.g., Zn), expressed as the total of concentrations of the entire aggregate of contaminating metals in a local background, is widely known. Experiments permit the determination of threshold limits of the aggregate contamination with different metals endangering different populations.

V. I. Morozov

Corporations, Transnational (TNC) Transnational corporation refers to a multinational corporation whose industrial and market activity takes place beyond the borders of a nation-state; a large company with assets in several countries; or several companies of various national origins, dominating one or several spheres of economy, or capable of influencing a specific industrial sector within and even beyond the country. TNCs are based on international division of labor; on increasing internationalization and globalization of the world; on the new connections and interrelations in world trade. TNCs are a major factor of economic globalization.

Transnational corporations signify intensification and internationalization of financial and industrial relations. Global industrial and trade networks formed by TNCs and their affiliates are the source and the mediator of globalization. They create the material base for international industry that is the core of the international economy.

A TNC has the following characteristic features: (1) active participation in world economic development and the international division of labor; (2) relative independence of capital flows from intranational processes; (3) influences the world economy by operating in many countries.

Enterprises (companies) become transnational corporations in the course of the struggle for an increase in profit. TNCs demonstrate that the domestic level of capital concentration production has not simply overgrown national borders, but has become a resource base for international production.

Traditionally, capital export is one of main forms of the internationalization of capital. An exporting economic actor turns into a set of corporations, operating in various countries. Internationalization of the added value production becomes an essential feature of the internationalization of a transnational corporation's capital.

In the 1990s, the number of TNC branches increased from 100–500 thousand units. A national company operating in foreign markets, as a rule, has no foreign branches and investments, but sells products manufactured in the country of its origin. A TNC employs workers from many countries; it is characterized by the internationalization of production, while a national company is characterized only by internationalization of sales.

104   *Global Studies Encyclopedic Dictionary*

TNC branches become not only a part of the world economy, but its center. Therefore, they can be used as an instrument of economic policy serving the interests of the countries of their origin. As a result, nation-states gradually lose control over a part (the most advanced) of their economies and, accordingly, over their economic policies (first, the budget policy). They become unable to make independent economic decisions, because the most important of them go through TNCs and the headquarters of transnational banks.

TNC activity traditionally embraces the largest and the most modern enterprises operating mostly in the international market and having an impressive share in the production (services) of the related branches. Powerful TNCs coordinate production and communication of enterprises located all over the world; they manage international cooperation and specialization on a global level. The industrial universality of TNCs allows them to carry out an industrial-trade policy providing highly effective industrial and trade planning, dynamic investment policy, and research at the national, continental, and international level for all divisions (branches) of the parent corporation as a whole entity. Manipulating the policy of transfer prices, TNC satellite companies, operating in various countries, skillfully bypass national legislations. As a result, the tendency to decrease the profit rate is neutralized.

This leads to global changes in the structure of the international division of labor. Trade now mostly takes place between the enterprises of one TNC located in different countries (over 40 percent of world trade). US, Canadian, West European, Japanese, Latin American, and Asian industries are managed by several TNCs, extracting profits from the organization of specialized and diversified production at the global level. Production that does not take into account state borders assumes there should be a sales network belonging to one TNC or several interconnected TNCs operating in many countries. Since separate enterprises are tightly connected with the parent corporation through industrial-technological and financial processes, the latter carries out a uniform global (or continental) strategy, applying it to all enterprises both on its "own" and on the "foreign" territory.

R. I. Khasbulatov

Corruption: (Latin *corrupcio*, "break," "spoil," "damage") Corruption refers to the venality of state officials, functionaries, and employees of proprietary and other public organizations. The content of corruption is determined by the total of actions or inactions expressed in illegal receipt of funds, propriety, services or benefits by a person authorized to perform state or other socially significant duties and granting these advantages. The problem of corrupt practice inheres to a variable extent in all countries. The venality of officials is conditioned by many factors, including: the standard of living, the development of legal culture, the social meaning and level of responsibility associated with official status, and the level of judiciary and media independence.

Manifested in diverse ways, corruption can be classified as either ethical deviations or violations of law. Among ethical deviations are official actions that, while not directly prohibited by law, are generally assessed as unjustifiable. Ethical deviations include visiting banquets arranged by corporations that seek to influence legislation, informal communication with individuals of scandalous repute, habitation in a house or an apartment, the price of which does not correspond to official's pay, and the use of others' expensive cars. Public moral principles have little in common with the term and rarely coincide with governmental and non-governmental service ethics. A commercial employee is often permitted much more latitude than a state official has in a similar situation.

Corruption as violation of law is classified as follows: civil infractions, disciplinary corruption delinquency, administrative corruption delinquency, and criminal corruption. Civil infractions include accepting presents by officials, employees or their relatives from individuals who seek official preference for their own gains. The distinctive feature of this case is that the present is given in exchange for legal activity. Disciplinary corruption delinquency refers to the use of official status to gain advantages, which presuppose disciplinary punishment. Often, the line between disciplinary corruption delinquency and criminal corruption is so relative that it is not always obvious. Thus, the issue of subordinate guilt is resolved depending on a superior's discretion. Administrative corruption delinquency involves the taking of other's property by an official or an employee by appropriation or misapplication. The extent of culpability is crucial in this case. Criminal corruption is a penal act, which directly undermines the authority and legitimate interests of the service and is expressed in unlawful acceptance of any advantages, property or rights for property, services or benefits; or the accordance of such advantage, property, services or benefits.

I. M. Matskevich

Cosmism (Greek *cosmos*, "order,"'"ordered space") is a psychological and philosophical phenomenon reflected in different cultural spheres-religion, philosophy, art, literature, and everyday consciousness; it is a deep human and cosmic experience, expressed in a rational way by philosophical outlooks and principles corresponding to various stages of human cultural history.

Cosmism is understood as: (1) a specific "world outlook" or "world perception," connected with the realization of global unity; (2) a spread of a "cosmic point of view" to humankind, its future and past being, on the one hand, a philosophical expression of the idea of the unity of human beings and the universe, on the other hand, of space explorations; (3) a philosophical idea of the "active evolution" of humans and the world, which is directed by reason, and the space expansion of humankind is just a part of this program; (4) barely tangible meanings of human inner cosmos (connected, naturally, with the physical cosmos); (5) "the last word of philosophy of technology," assuming that cosmism justifies the most anti-ecological and antihuman features of modern civilization. (This last point of view, known as "anticosmism," deals not with cosmism itself, but with its adherents who are preoccupied with technologies.) In some interpretations, cosmism is a synonym not for "cosmic," but for common earthly situations projected into space.

Sometimes cosmism is esteemed as a world outlook specific only to Russian philosophy at the beginning of the twentieth century. If earlier thinkers concerned with the scientific and technological exploitation of the physical cosmos were considered cosmists, now there is a tendency toward an ex-

106 *Global Studies Encyclopedic Dictionary*

tremely broad interpretation of this phenomenon: many outstanding Russian philosophers, at the end of the nineteenth century and the beginning of the twentieth century, are considered to be cosmists, including those who did not call themselves "cosmists" at that time. The term "Russian cosmism" has become so popular partly because of the ideas of Russian philosophy of that period, but mainly as the result of social and cultural factors of the present time. The term, which emerged in the 1960s and 1970s, soon after space exploration began, marks the revival of Russian religious and philosophical ideas. There are different trends in cosmism, including Russian cosmism: natural scientific, religious and philosophical, poetic and artistic.

The phenomenon of cosmism is not limited to scientific and philosophical ideas—it has a complex, multi-level structure. The first and the deepest level is the area of the collective unconscious in the human psyche, where cosmism is manifested in the form of one of the most universal archetypes: "a cosmic human." The next level is the projection of the "cosmic human" archetype to different social and cultural phenomena. Cosmic unity is regarded either as God's creation or as a unity of the natural world including human beings. Analyzing rationalized cultural meanings of cosmism, one should always distinguish the idea of cosmism (that human beings and the cosmos are related and humans are an integral part of cosmos) and the principles of cosmism in which this idea is expressed. The first principle is seen through world culture from the time of myths to the present day. Tsiolkovsky formulated it in following words: "Our fate depends upon the destiny of the Universe."

For many cosmists, the cosmos is a supernatural, transcendental existence, embracing the physical cosmos as something fallen, low, and "created" (but at the same time full of symbols of the transcendental cosmos). Human beings are regarded as creatures that belong to both of these worlds. Cosmism is understood first as a transcendentation of the cosmos of paramount symbols and supreme values. The principles of cosmism can also be interpreted within the context of the "scientific outlook" (Vernadsky) as a natural existence, and human beings as a part of the natural integrity. The interaction of human beings and the cosmos is regulated in accordance with natural and social laws; it is studied, projected, and realized by means of science and technology. The existence of the transcendental cosmos within the frameworks of this interpretation is usually either denied or not considered.

Five different types of cosmism can be (1) Theoanthropocosmism, a Christian-oriented version of cosmism. This approach includes such different systems as the common deed philosophy of Nikolai Fedorov and the all-unity philosophy of Vladimir Soloviev. This religious and philosophical cosmism focuses on the idea of God-human, involving the spiritual and moral perfection and inner transfiguration of humans. (2) Anthropocosmism, developed within the framework of science by Vernadsky, Chizhevsky, Kholodny, and others, has two main axiological ideas: the technologically oriented and the humanistically oriented. The first used to dominate space exploration projects, and the second version is a third type of cosmism, esoteric anthropocosmism, developed within the context of the "secret doctrine" of Helena Blavatsky and the philosophical concepts of Nicolas and Helena Roerich. It is based on the idea of the cosmic influence on human beings. This version of

Capitalism – Cyberculture 107

cosmism is widespread within mass consciousness. (4) The cosmic philosophy of Konstantin Tsiolkovsky, occupies its own niche in cosmism. Axiological orientations of the cosmic philosophy influenced many modern concepts of space exploration, both Russian and foreign. Nearly all of them are based on some idea from Tsiolkovsky, adapting his ideas to social and cultural realities of the cosmic era. These ideas are present in every modern discussion about the prospects of space activities and their influence on society. (5) Different eclectic versions of cosmism that unite, sometimes randomly, ideas and values of the other versions mentioned above.

Cosmism is closely related to globalism. Many modern global problems were predicted long before their presentation by the cosmist thinkers. Cosmism and globalism are connected so tightly that anticosmism is gaining more and more power now. Anticosmism views itself as a humanistic concern about the negative consequences of the scientific and technical revolution, including space exploration (e.g., space militarism and anti-environmentalism). As a rule, adherents of anticosmism underestimate the necessity and inevitability of peaceful space exploration to solve global problems.

V. V. Kazyutinsky

Crime, International: International Crime is a general concept designating socially dangerous actions or operations often committed on the territories of two or more states in which they are prohibited on penalty of criminal responsibility. Hence, international crime can also be transnational crime. We should distinguish as well "world crime" (an aggregate of all the crimes committed in the world over a specific period of time) and "crimes against humanity" (committed in one country, such as genocide or the destruction of cultural artifacts). International crime primarily includes illegal manufacturing and sale of banned products or products withdrawn from current circulation, rendering of prohibited services, and trade of illegally produced goods.

Criminals' principal objectives are to derive super-profits. To achieve these goals, they engage in those kinds of activities that allow obtaining tremendous income quickly enough and out of state control.

At the same time, international crime is well armed. The criminals do not hesitate to utilize such tools as hired assassinations, explosions, and arsons of offices or apartments to give short shrift to disagreeable partners, intractable competitors, "too lawful" officials, or clerks. Nevertheless, we have to admit that the "enforcing impact" is not the end in itself, but just a way to achieve the above-mentioned goals.

The following types of international crime represent both traditional and contemporary trends: production and distribution of illegal drugs; trade in illegal arms; human trafficking; theft and export of cultural treasures and works of art (including special collections); falsification of works of art and sale as the originals; illegal sale of flora and fauna; sale of human organs for transplantation; Internet distribution of pornography (especially child pornography); computer network cracking and fraud; software piracy; and automobile theft and smuggling.

Transnational crime is currently an exceptionally serious social danger. The better any criminal activity is organized, the more dangerous it is for the world community or an individual country. International crime is characterized by the highest level of organization, a natural and compulsory term of its existence. Most often, we speak about a strongly manipulated structure with a complicated hierarchy, frame of branches, and clear rules and regulations for participants at all levels, supplied with all the modern facilities and tools to realize its goals and counteract the machinery of law. The main factor increasing the global danger of transnational crime is the extraordinarily high gravity of its consequences.

The Mafia generates more than $500 billion annually from illegal drug sales. This constitutes the "reward" for the millions of ruined lives, mainly youth. Experts witness that drug distribution puts the young generation in a range of countries on the brink of survival.

The illegal weapons market offers wide opportunities for the most dangerous extremist powers to unleash local wars, organize terrorist acts, and initiate bloody ethnic and political conflicts.

Every year, more than a million people become objects of smuggling. Illegal (hence, uncontrolled) circulation of chemical and nuclear materials, able to become weapons of mass destruction in the hands of some groups and to drag the world into nuclear disaster, especially menace the world community. Computer piracy causes annually billions of losses to the business world.

Terrorism occupies special place in the rank of international criminal infringements. It could be a part of the general crime, a well-organized economic one, for example, but it often has a distinct political reason. Extremist organizations may act as executives or fighting squads in so-called liberation movements (for instance, the Irish Republican army, established in 1919, as a military wing of the Shin Fein party, exercises terrorist methods of fighting for the integration of Ireland until recently). Fighting squads consisting of foreign mercenaries function in Russia (territory of Chechnya and adjoining regions).

A complex of subjective and objective factors increases the intensity of transnational criminal processes, strengthens the positions of criminal syndicates and makes them stable. World economy and policy globalization, simplification of international connections, the "lucidity" of many of the interstate borders, the global finance system and world market formation, development of modern computer and information technologies, the increase in migration flow—all these attributes of the contemporary stage of development of world civilization have not only a positive humanitarian aspect, but also the prerequisites to be used with opposite purpose—to expand international crime.

We also have to keep in mind, that the owners and managers of the international criminal business skillfully and subtly make use of the weak spots in the process of globalization. The criminal syndicates and their branches are guided in many countries by experienced and skilled professionals, authoritative and severe individuals holding enormous funding support and facilities, and possessing an extensive system of contacts including governmental contacts. Their own security is ensured by means of generous bribes to corrupted officials, and of responsible members of the law-enforcement system; by the ability to quickly and flexibly change the strategy, to smartly use the weakness or inertness of the authorities or gaps in legislation.

Capitalism – Cyberculture 109

Presently, the world community distinctly recognizes the tremendous threat of international crime and especially of its well-organized forms. Several UN documents and regional political and scientific forums prove it. The events of 11 September 2001, in the United States revealed that only integration of all the forces and coordination of the actions of civilized countries are able to withstand terrorism and international crime. We speak about one of the largest problems of international politics, in which the reasonable solution could only be acquired as a result of intense permanent exertions. It requires the elaboration and coordination of antiterrorist and criminal legislation to be implemented and activated, the exchange of information and arrangement to be conducted, and appropriate systematical recommendations to be prepared. There should be improved mechanisms of interaction and inter-support, allowing to expose and to neutralize the criminal leaders and the officials rendering direct or indirect assistance to them. Special attention should be paid to elaboration and reliability intensification of the defense system of witnesses, victims, and law-defense authorities actively fighting international crime. The detriment of the international criminal finance warrants effectively fighting it, and therefore it is also of essential significance.

Some international crimes are specified in the regulations of international military tribunals; e.g., the Nuremberg tribunal and Tokyo tribunals, established after the World War II. Resolutions of the UN General Assembly of 1946 and 1947 confirmed its global significance. The UN defines crimes against peace and security of mankind and military crimes: planning, preparation, unleashing or prosecution of aggressive war; the public call to unleashing an aggressive war; the development, manufacturing, accumulation, acquisition or distribution of weapons of mass destruction; the use of prohibited means and methods of prosecution of war; genocide; ecocide; mercenaries; and the attack on people or on the institutions under international protection. Responsibility for the crimes comes irrespective of whether they are provided for by the legislation of the country where the crimes were committed. Others are provided for by international agreements (conventions). These crimes are provided for by national legislation of the countries that joined respective conventions and include terrorism, taking hostages, piracy and others. Such crimes are able to encroach on the interests of international law and order.

In accordance with the resolution of the UN General Assembly of 20 December 1993, a conference of the heads of states and governments and other higher officials was held in Naples, Italy, in 1994. The conference adopted a global plan of actions against organized transnational crime. Responsibility for conventional crimes comes in accordance with the national legislations of the states that joined the conventions.

M. M. Babaev, A. V. Kladkov

Culture (Latin *cultura*, "cultivation" "care") Culture comprises the norms and patterns of behavior that are consecrated by tradition and obligatory for representatives of an ethnos. Culture provides the form of translation of social experience when each new generation masters not only the subjects of the cultural world, the skills and techniques of their technological attitude toward nature, but also cultural values and patterns of behavior. Through regulations

110 *Global Studies Encyclopedic Dictionary*

of social experience, culture forms stable art and cognitive canons, concepts of what is beautiful and ugly, good and evil, and attitudes toward nature and society, what is real, and what is obligatory.

Reproduction of culture as the whole is carried out at three levels: (1) the conservation of culture and its basic grounds, which hide behind a verbal and symbolic cover; (2) the renovation of culture, institutions, or knowledge; (3) translation of culture-specifically, the material world of culture as the world of an individual's socialization. All three levels, characterizing culture in its wide spectrum of forms (science, technology, art, religion, philosophy, politics, and economics), at the same time reveal the structure, mode of activity, and integrity of culture.

In the global sense, cultural doctrine, to some extent, is about specific differences of cultures and the basic types of world culture. Thus we must admit that criteria for creation of cultural typologies differ greatly. Usually a criterion is a basis or major feature according to which separate cultures are united into a specific type. The problem is many different criteria exist, and representatives of different sciences introduce them, producing wide diversity in definitions: there are "catacomb cultures," "primitive cultures," and at the same time "pagan," "Christian," and "Muslim" cultures. The most general criterion that helps to define culture is its existence in geographic space and historical time. The situation in culturology allows us to see both the general differences in human life activity in specific geographic space from biologic forms of life and the qualitative peculiarity of historically concrete forms of this activity at various stages of social development—within the framework of specific epochs and civilization formations.

The typology of culture also allows us to characterize special features of the consciousness and behavior of people in concrete spheres of public life, to consider culture "from within." As culture is a complete system, historical and geographic space may give life to a symbiosis of surprisingly various and unique cultures, which makes the cultural image of peoples, countries, and continents unique as well. This cultural variety creates preconditions for the correlation of various human societies in global space and time, which cannot be just a speculative comparison, since world cultures are a product of long historical development during which cultures themselves have transformed, modified, and changed their internal structure, and being influenced by each other. These observations allow us to speak about both general and individual features of cultural-historical development.

But first culture is a powerful mechanism of anthropological influence; a method of adaptation of an individual to the cultural needs of a society and at the same time a method of individual application of total ethnic and national experience and self-actualization of personality in the cultural space of the ethnos. In this sense a cultures roots can be found in the depths of the generic level, and its vitality is a consequence of its uniqueness and originality with all its traditions, skills and national images taken together. The cultural world is the uniform space uniting a person and the nature around him or her, which influences the social psychology of an ethnos, forms national character, and determines the orientation of its practical activities.

Thus the consideration of culture as an independent integrity, replicated through long historical period, is justified. Cultural ontology includes the mechanism of transfer of social experience through each generation, mastering the object world, skills, and methods of the technological attitude to nature and the cultural values and patterns of behavior. Thus specificity of culture is materialized both in type of social organization and in personality type. Technological and economic levels of cultural development emerge that are carried by its social representatives (groups, classes, organizations). Even at this level the integrating role of culture is revealed in the formation of uniform systems of art, technological and cognitive values, and patterns of behavior.

In the global development of humankind, the locality of unique cultures is observed. Meanwhile, diverse faces of cultures show the uniform image of humankind. The sequence of beliefs, masterpieces of art, philosophical ideas, discoveries, and customs are freely distributed along a chronological axis without opposition of archaic and modern, advanced and reactionary, developed and retrograde, without being divided by socio-economic formations and sociopolitical distinctions. In this case culture is examined from within as a complete formation, and the core of it is an individual, a valuable unit, important for the formation. Culture is at the intersection of the ambivalent aspirations of society: to keep historical originality and the ethical and aesthetic potential and to join in the global and trans-temporal unity of humankind.

G. V. Drach

Culture, Cognitive: Cognitive culture is the set of the historically specific means and methods, norms and prejudices, and traditions and ideals that familiarize the individual and collective subject with knowledge, social standards and cultural values. Cognitive culture accumulates experience connected not only with a level and degree of development of theoretical, scientific activity, but also with the conditions for the formation and functioning of daily and ordinary cognition, fixing the level of development of the person and the society and of their creative, mental abilities as they apply to the real world. Cognitive culture accumulates both abstract knowledge and its appropriate "dimension": activities pertaining to manufacture, storage, translation, and materialization.

The promotion of cognitive culture as a philosophical and culturological concept has been demonstrated by the wide use of the categories of ethical, aesthetic, political, and legal culture in science and education. As a theoretical construct, "cognitive culture" is similar to the concepts such as the "culture of thinking," "cognitive traditions," "ideals and norms of scientific cognition," and "culture of competition."

The main elements of cognitive culture are the traditions of cognition. Under the conditions of globalism, the situation is significant for traditional societies, compelled to consider the fact that in Western countries and even in science there is the general refusal of traditions, of methodologies, and a transition to "anarchical epistemology" (Paul Feyerabend). Tradition's role in cultural life has become ambiguous and inconsistent. Tradition not only unifies the culture, but also consists in the particular rather than the general and universal, and thus plays a role in disuniting and opposing people and civilizations. At the same time the idea of continuity in culture is realized as the

112 *Global Studies Encyclopedic Dictionary*

result of tradition. Occupying a special position in the structure of cognitive culture, tradition brings the influence of the ethnic and the sacral, though the main component of the latter in cognitive culture is faith.

Thus, cognitive culture focuses philosophical attention not only on the elements of gnosiology, which deals mainly with the parameters of the intersubjective knowledge of a material world, but also on the features of formation of personal and religious knowledge.

Specificity of cognitive culture is determined by mentality, the intellectual and spiritual makeup of an ethnos, which constitutes the lowest layer of the consciousness. Perhaps in this respect it is meaningful to speak about national and international cognitive culture in terms of globalism, in that these basic structures of perception are becoming more planetary in scope. Individual, personal and domestic elements of cognitive culture, in its theoretical, specific and scientific manifestations form an immense field of research. Their temporal and spatial parameters preserve the originality of cognitive cultures in the face of global homogenization.

A powerful element within cognitive culture is culture of thinking, which commands, to some extent, the subordination of cognition to formal and logic rules, or logical thinking. As some of the integral components of cognitive culture involve mistakes, error, lie, deceit, prejudices, the critical stance provided by logical and structural thinking offsets these pitfalls.

M. L. Bilalov

Culture Globalization is the formation of a universal, worldwide culture resulting from the inherent global trends within culture, or from its involvement in globalization. The most radical and consistent opponents of culture globalization suggest that: (1) There is a clear distinction between culture and civilization; globalization is a primarily or exclusively civilization-related process. (2) Spirit plays the central part in the origination and existence of cultures; for civilization, the leading role is played by the material component. (3) There is no spiritual progress. (4) Structural homogenization of a culture is not possible. (5) Culture is, first and foremost, a hierarchical construction.

What is meant is not so much denying the inevitability, advantages, or feasibility of culture globalization, but asserting the impossibility of a globalization of the spiritual. Integration of the spiritual, and homogenization in the sphere of culture are considered impossible; their practical implementation is understood as leading to the end of history. History ends (1) due to the spiritual unity achieved (a religious utopia, according to which this level can be designated as unity, nirvana, or social harmony), or (2) as a result of unity of all people bringing them into a state of spiritual "equilibrium," leveling their needs and desires (a profane option, providing the state of mass culture self-sufficiency) into what looks equally utopian.

K. Z. Akopyan

Culture, Local: The term "local culture" commonly refers to the characteristics of experience of everyday life in specific, identifiable localities. It reflects ordinary people's feelings of appropriateness, comfort, and correctness—attributes that define personal preferences and changing tastes is a traditional state of culture (Encyclopedia Britannica).

Capitalism – Cyberculture   113

Local culture appears to be a complete symbolic system of cultural values testifying to the maturity of the individual and humanity. The maturity acts as one of the principles of local culture, which could be, as well, expressed on the level of the products of human spiritual creativity in music, literature, architecture, and philosophy. In each case we obtain a complete composition.

As far as the local culture is based on ethnic customs and values, it cultivates a vigorous inner skeleton, which guarantees "immunity" against another culture and disallows any hostile elements and influences (and this is another feature of the local culture). As a consequence, the opposition of "home-alien" dominating in the given cultures arises, identifying "home" or inner-cultural, as genuine and "alien" as a rejection of home, so then hostile or false. While admitting other cultures, any local culture always considers the self to be the highest achievement of the whole human culture.

Thus, the local cultures' dialogue takes the form of adjustment of one culture to another within the frames of an intercultural semiotic realm by means of the cultural values' interpretation, while matching the two cultures' stereotypes. Here, the examination process of the original lack of convergence in cultural spheres contributes to both cultures enriching themselves with new values. Language becomes the basic conversational tool, and professional language skill provides the capability to understand another culture. Knowing a foreign language I indispensably interpret or adapt the values of a foreign culture to my own. While comparing both cultures I necessarily realize the value and peculiarity of my own culture. Hence, if we make attempts to interpret the dialogue of the local cultures in a semiotic way, we discover that it is being materialized, according to Lotman's felicitous expression, inside the semio-sphere as a specific semiotic realm embracing not only the assembly of various languages, but also a sociocultural field of their functioning.

As the bourgeoisie and bourgeois science emerged, local culture became subject to the pressure of scientific and technological progress. In the beginning, culture tried to exclude the science regarding it as an out-of-culture phenomenon. Then gradually the role of science and the significance of technique grew to the level where the culture itself had to be re-interpreted.

Reference: *Encyclopedia Britannica Online*, s.v. "local culture." Accessed 9 November 2013. http://www.britannica.com/EBchecked/topic/766184/local-culture.

V. V. Mironov

Culture, Mass: (Latin *massa*, "lump," "piece," and *cultura*, "cultivation") Mass culture can refer to: (1) a totality of sociocultural phenomena that became apparent on different historical stages as a result of the processes of the social differentiation of culture; (2) a social phenomenon that meets specific criteria and characteristics (gravitation toward universality, orientation on a middle linguistic semiotic norm, primacy of compensating-entertaining and psychotherapeutic functions, reliance on new communication technologies, and the commercial character of functioning); (3) a type (style) of art culture (in meaning close to the sense of popular culture-pop-culture, or commercial culture and art), which significantly differs from art in its "classical" and "folk" versions and is connected with the withdrawal of art from the sphere of the "vitally serious" to the sphere of fictional, entertaining, and with a loss of

114 *Global Studies Encyclopedic Dictionary*

a "sense of a style" or "measure"; (4) a special layer of modern culture whose development is connected with strengthening of processes of integration and intensification of mass media development, which contributes to the spread of standardized cultural forms, norms, and models, and to the formation of transnational culture; (5) a specific system of characters, a language of culture for the information society, which assures the action of mechanisms of manipulation and unification and entails a formation of a specific organization of everyday consciousness and its virtualization.

The origins of this layer of culture can be found in different cultural and historical periods—in the church culture of the Middle Ages, in fine arts (especially in engravings of the fifteenth and sixteenth century). A compensating-psychological function was one of the initial social functions of mass culture, as the world of religious values was based on achieving harmony of human life based on a balance of "the world of wishes and the world of possibilities."

By the end of the twentieth century, the following directions of modern mass culture had been formed: the industry of "childhood subculture"; a school for the masses; a system of national (state) ideology and propaganda; social mythology of the masses; political movements of the masses; mass media; the industry of entertaining and health-improving leisure and of physical rehabilitation of a person and correction of his or her physical image; professional sports; a system of organizing and stimulating consumer demand.

Frequently, the heterogeneity of mass culture is presented as a variety of cultural forms. However, this variety is just a façade; behind lightness, insignificance, and mosaic-type patchiness, is concealed uniformity and clichés meeting the low level of cultural requirements of people involved in the process of deculturation. This contributes to formation of a specific type of passive individual whose number increases in all developed countries. Commercialization of culture replicates a mass human being almost incapable of self-organization.

During periods of social transition, the adaptive and recreational functions of mass culture revive to contribute into the maintenance of general structures of values and meaning. However, they are mainly based on game, entertainment, and spectacle. It is not accidental that the golden age of Russian mass culture coincides with the periods of transition (the 1920s–1930s; the second half of the 1990s), when new tendencies crystallize and abrupt change of cultural styles and forms occurs. In such periods, mass culture that is always present as an independent subculture begins to devour all the heterogeneous "non-established" cultural material, displacing the other subcultures. The transformations of the 1990s in Russia brought culture closer to reality but, at the same time, they moved human existence toward pseudo-culture having stripped culture of its educational functions.

Fascinated by the high life encoded in commercials, a human being is unable to consume a supposed richness of the uniformed cultural life, with the mirages and virtual realities of mass culture. Unification and the stereotyping of behavior, tastes, and ideals "override" the pluralism of cultural forms. Liberating humans from making decisions, together with designing and mythologizing reality through mass media, forms a "television" and "computer" generation that, according to Marshall McLuhan, is a maker of the present and its victim.

Relations between "high" and mass ("low") cultures can undergo serious changes. Replication of masterpieces removes an element of sacrality from high culture; art works become more accessible thanks to new channels for their distribution and storage. Mass culture does not weaken high reflexivity of elitist art, but obscures it. Mass culture adapts high art for an unprepared consumer not caring about the preservation of the deep laid meaning of an artwork. New forms of interaction between mass and elite cultures allow for the suggestion that their opposition is a simplification of a complex situation of modern culture. Although mass culture is still often considered in terms of cultural and informational imperialism, we cannot help understanding that integration and differentiation in culture contributes to activation of mechanisms of its self-organization. Hence, we can talk about the convergence of elite and mass culture through mass media and advertising.

The carrier of mass culture is basically youth. Mass culture attracts the young by lack of a hierarchy of cultural conventions. This demonstrates the breaking of succession of cultural forms and models, the destruction of the correlation between traditions and innovation, and the degradation and depersonalization of culture.

Each aspect of mass culture is a part, an inseparable element, of an integral multi-level system of culture as a whole—an independent form of culture. Connections between the part and the whole become most apparent in design and advertising. Similar to mass culture they possess commercial value, technological effectiveness, position of status, and fashion; they participate in the formation of a consumer consciousness, stereotypes of behavior, leveling of values, and simplification of meanings. Advertising is an especially powerful means of socialization; it influences upon sociocultural differentiation, upon culture as a whole, because it performs its informational and entertaining function being between mass media and art. Being the meeting point of the world of goods and the world of values, it frequently transforms cultural meanings, mythologizes consciousness by eroding the verbal system of communications. Besides, advertising acts as a distinctive mechanism of self-development of mass culture; it helps mass culture to spread not only nationally, but also globally.

In the age of globalization, advertising transforms mass culture into an industry. Re-evaluation of various mass culture theories in Russia is caused by a rejection of the ideological approach to evaluation of cultural phenomena, and recognition of cultural diversity. Mass culture is regarded as one of subcultures, and its positive and negative sides are theoretically analyzed.

O. N. Astafieva

Cultural Heritage: (1) common cultural property, passed from generation to generation, ensuring the stability of social and cultural formations of different levels, and (2) the aggregate of spiritual and material activity objects, which present historical and ethical value. Stability of civilization is possible when there is a specific cultural area, featured by spatial boundaries, distinctive cultural phenomena, and material and spiritual values, defining specifics and the essence of the civilization: language, religion, art, traditions and so on.

116   *Global Studies Encyclopedic Dictionary*

Cultural heritage is passed via these objective forms and specific deep psychological mechanisms connected to specific behavioral and psychical stereotypes, spiritual identification, and collective human images.

M. M. Mchedlova

Cultural Identity and Globalization: At the end of the twentieth century, the cultural factor has occupied an unprecedented interest within international contemporary relations. This interest dates back to early globalization and to confirmed predictions that there would be many conflicts between the West and the others civilizations, namely Islamic ones.

If globalization has political, social, economic, and cultural dimensions, we find that the cultural dimension is the first and most important for researchers of third world countries, because this is the dimension that has a relation with the identity and aims at nationalism. This cultural dimension, by talking about the world citizen who affiliates with world identity, seeks a merger that leaves aside other separating identities.

Although globalization aims at unifying human societies and putting them into one small village, we see that the philosophy of cultural conflicts has tried to create new factors for international conflicts after the collapse of the Soviet Union and the end of the Cold War.

Although globalization has economic roots and political consequences, it has shown cultural superiority in the global environment; so many researchers see that the effect of culture on globalization and the effect of globalization on culture is a matter of concern.

Some Arab researchers see that the results of globalization are most dangerous for the developing countries. The dangers occur when a culture is merged with a new economic and trading process. When a culture comes to be spread all over the world, and at a large scale, globalization presents itself as a global culture that leads to eradicating and deforming the identities of other nations and changing the national personality, in addition to the effect of global information and media communications, which constitutes threats to the diversity of culture.

Many think that globalization is not only a means for commodity movement or closing distances and removing borders, but also believe it to aim for having one common and comprehensive culture as well, through which the economy and information can be supported. In this way, priority would be given to culture, regarding it as the decisive factor; so economic factors would be secondary.

Many views, opinions, and proposals have been given about the effect of globalization on identity. One large a group of people (in Islamic and Arab countries and some of third world countries) have seen that globalization is a plan or a strategy aiming at invading other parts of the world and threatening domestic cultures, while many others have concluded that globalization does not threaten identities or the eradication of cultures. Instead, globalization can be adapted since humans live with various identities.

So, conflict the between globalization and culture is not necessary. Various identities can co-exist; positive interactions and creativity are compatible with globalization. Nevertheless, some still fear that identity will collapse if globalization functions as a cultural invasion.

Within the framework of talking about the effect of globalization on identity, some note the role of the media in helping globalization to impose western culture, eliminating domestic and national identities. Researchers have different points of view on the ability of the media to influence culture. Some contend that mass media information has formed identities at the domestic, regional, and continental levels or at least reconstituted them. Others say that the mass media has played a less influential role.

We conclude that information technology has closed distances and removed territorial barriers that were basic factors of identity and affiliation for groups living in a specific land, making it easy to separate place from identity and rapidly crossing political and culture borders that had previously separated peoples. New global societies have emerged, weakening domestic identities. Some add that western advertisements and media contents widened the global expansion of consumer culture and created links among people, regardless the place or area.

Still, others see that mass media information has had marvelous effects in gathering people separated by geography, language, religion, race, and literacy. Mass media information also lessens the diversity among people and populations who live inside countries' borders. For these analysts, media information is the most powerful and decisive force. In this way, communication technology is the most effective power. Identities are formed and constituted by this technology.

Contemporary researchers of culture colonialism agree that technology affected the formation of national and international identities as it made it easy for people to have culture awareness that crystallized national, and external, identities. Contemporary researchers say that it is not clear that traversing nationality (cornering information economies) leads to cultural harmony that rises above domestic and national differences.

This analysis proposes that the relation between media information (and information media) and culture identities is very complicated. Media information alone cannot achieve united solidarity or forge a national identity that transcends other nationalities or forms of similarity or resemblance. Perplexity and deformation arises from merging awareness with identity. The problem is that we put the media information factor before identity itself. However, if we put media information, identity, and place together in one cultural framework, this would make media information part of society building. Media information may have a role, but it would not be easy for it to have a power that can generate common relations and connections.

Generally speaking, researchers agree that culture is relative and an open to change. In both the third world and western world, analysts view globalization as a strategy that aims at invading the other parts of the world, introducing a single behavioral and moral pattern or mode through the information media, financed by multinational companies, in a way that threatens the cultural identity of local populations. Other researchers see that the national and cultural identities are not necessarily affected by electronic information,

118   *Global Studies Encyclopedic Dictionary*

so that realizing globalization depends on how strongly people can struggle, surrender, or weaken before the cultural invasion of the information media.

S. Nasser

Cultural Identity and Standardization: The most common explanation of globalization views the world as directed toward unity and unifying measures through the technological, commercial cultural synchronism that have emerged in the West. In this regard, globalization is connected with modernization, with the result of mixing all peoples of the world into one international society. This aspect of globalization has been addressed within several fields. According to international economics, this trend arises from global production and financing in conjunction with the heightened concentration of international relations. Sociology concentrates on the global increase of social intensity and the establishment of international society. Cultural studies stress how world communications and the unifying of cultural measures globally (such as Coca-Cola® imperialism and the global expansion of McDonald's® restaurants) impose cultural and intellectual standardization.

This process can be summarized in giving the political infrastructure an institutional shape that is closely attached to the imperialistic and hierarchical formation of international bodies of organization such as the United Nations and the different Aid Associations. Some claim that originality offers alternative visions for the world situation, while seeing the East as wrong and somehow illogical. In recent years many writings on globalization have appeared. Most of these writings concentrate on the imperialistic aspects. They expose the domination of central cultures and the spread of American values, consumer goods, and types of life. A fear in Europe in the late 1950s and 1960s emerged among the intellectual elite related to danger imposed by American culture, epitomized in the domination of Coca-Cola® culture. This theme has been developed further in studies in cultural sociology.

With globalization some important aspects appear such as:

(1) Globalization enhanced and supported regional inclinations and national tendencies. The EU, for example, was established as a reaction to the American-Japanese economic challenge.

(2) Globalization through different aspects goes side-by-side with modernization that at its core involves alienation.

(3) The idea of international cultural synchronism lacks many things because it ignores opposite events such as the effect of foreign cultures on the West. The new western cultures ignore the fact their cultures include in their deep roots oriental influences. Europe has been receiving cultural trends from the Islamic East up to the fourteenth century. The domination of the West appeared only subsequently.

(4) A prevailing measure in international policy concerns how the West adopt democracy inside its countries while exercising imperialism outside its borders. America is exercising imperialism on all levels in Iraq. However, while using such expressions as democracy, human rights, and freedom, it mixes these expressions with various religious slogans. Globalization does

Capitalism – Cyberculture 119

not need the same type of imperialism as it had in the 1950s and 1960s. It is clear now that imperialism has returned under the form of unipolarism.

(5) Cultural interbreeding or xenogamy is a process by which meanings and types mingle together, allowing that they originated in a variety of historical sources. Xenogamy or interbreeding is a mixture of Asian, African, European, and American influences.

Z. Atta

Cultural Universals (Anthropological Universal or Human Universals) are the ultimate foundations of societies. The variety of cultural phenomena at all levels makes a complete system on the basis of these ultimate foundations of each historically defined culture. Outlook universals (cultural categories) are categories accumulating historically gained social experience; using the system of these categories, a person evaluates, comprehends, and experiences the world. They should not be confused with philosophical categories, which are the result of reflections on cultural universals. Outlook universals may function and develop beyond philosophical reflection. They are inherent even in cultures having no advanced forms of philosophical knowledge (such as Ancient Egypt, or Babylon).

Cultural universals can be divided into two large and interconnected categories. One consists of categories reflecting the most general, attributive characteristics of the objects involved in human activity. They act as the basic structures of human consciousness and have universal character, because any object (natural or social), including objects of thinking, may become subject to activity. They are reflected in such categories as "space," time," "contingency," and "necessity."

Besides these categories, culture in its historical development engenders another type of categories, reflecting the role of a person as a subject of activity, the structure of human communications, their personal attitude toward other people and society, toward the goals and values of social life. This second bloc of cultural universals includes such categories as "person," "society," "good," "evil," "justice," and "freedom."

A two-sided correlation exists between the two types of cultural universals that reflect connections between subject and object and between subject and subject. Therefore, cultural universals emerge, evolve, and function as a holistic system where each element is directly or indirectly connected with the other. The meanings of specific cultural universals characterize national and ethnic features of a culture and its vision of space and time, good and evil, life and death, its attitude to nature, work, and personhood.

There are at least three interconnected functions that cultural universals perform with regard to human activity. First, they provide some selection of diverse and historically changeable social experience. This experience is evaluated and classified according to the meanings of cultural universals and is united in clusters. Due to such "categorical packing" this experience enters the process of translation and is transmitted from person to person, from one generation to another. Second, cultural universals act as the basic structure of human consciousness in each concrete historical period. Third, interrelations among cultural universals create the generalized picture of the human world,

120 *Global Studies Encyclopedic Dictionary*

which is usually called the "outlook" of the epoch. This picture, reflecting general concepts related to person and the world, introduces a scale of values accepted by this culture.

Stereotypes of group consciousness are specifically reflected in the consciousness of each individual. People always treat cultural universals in accordance with their life experience. As a result, the picture of the world that exists in their consciousness becomes very personal and serves as their individual outlook. There are several innovations specific for each outlook dominating this or that culture. Basic beliefs and concepts can be combined (frequently in a conflicting way) with very personal orientations and values, while the whole complex of individual beliefs can change during persons' lives.

The dialogue of cultures can play an important role in working out new orientations and strategies of growth. Alternatively, unification and destruction of cultural variety may result in degradation of cultures.

V. S. Stiopin

Cyberculture is the term denoting radical changes in mentality and sensory experience of the modern person, which have been caused by global expansion of new communication technologies at the end of the twentieth century. The term "cyberculture" first appeared on pages of various publications (from popular mass media to academic ones) in the early 1990s. A qualitative leap connected to transition from analog systems of information transfer to digital systems (the so called "Digital revolution") determined a transformation of a scale equal to the invention of the printing press by Guttenberg.

The role of mass communication in the process of formation and development of European civilization was first given scientifically significant theoretical grounds in the 1960s and 1970s, in works of the Canadian culturologist and philosopher Marshall MacLuhan (1911–1980). McLuhan's ideas had huge influence on public consciousness of the United States, preparing for a radical transformation of communicative media (and, probably, having partly initiated this transformation) connected to the invention of a personal computer in 1980s, and the Internet during the 1990s. While studying the phenomenon of electronic mass media, McLuhan foresaw many characteristic features of a global computer net. He generated the worldview from which "cyberculture" originated as the phenomenon of multidimensional reflection in modern consciousness of the role, place, prospects, and risks of the further development of the new technology of human dialogue.

The basic object of analysis, theme, and problem of cyberculture is a special dimension of existence, which is not physical but virtual, represents an arena for communication and interaction of hundreds of millions of people and has been named "cyberspace" (term given by writer William Gibson). The cyberspace is structured as a net, having features of nonlinearity, acentricity, self-referentiality, multi-agency, polycontextuality, synergism, and globality. Consequently, the process of studying cyberculture acquires an interdisciplinary character. Theoretical basics of poststructuralism, postmodernism, quantum mechanics, synergetics, and other philosophical and scientific trends of the twentieth and twenty-first century were used as initial material.

Capitalism – Cyberculture 121

The scientific stage of the development of cyberculture began in the mid-1990s, when the ubiquitous introduction of Internet services allowed millions of ordinary users, including those from universities and scientific institutions, to use global net resources. The education and existence of "virtual communities" and "online identities" occupied initial attention to this phenomenon. Basic studies in this sphere include works such as *The Virtual Community* (1993) by Howard Rheingold, giving the analysis of different sides of life of the online community, Whole Earth 'Lectronic Link (WELL), and *Life on the Screen* (1995) by Sherry Turkle, who carried out an ethnographic study of a group of communicative units of the global net; for example, Multi-User Domains (MUDs). Having received this initial impulse, the process of studying of radical transformation of the traditional image of a person and the world in cyberspace has developed widely and within few years. The range of themes includes issues of genders, racial and national identity in cyberspace, and even ideologically loaded types of discourse produced by cyberspace in addition to the design of the communication environment, forming the human dialogue within its frames. Among the variety of sociocultural and socioeconomic themes, the problem of "digital divide," or disparity in access to Internet resources, is of great concern, and which is not decreasing with the technology development, but, on the contrary, is continuing to worsen even in the United States. Under increasing net globalization this problem might only continue.

True bearers of cyberculture as a specific innovative attitude stand as its theorists, rather than the representatives of the "cyberpopculture" (or hackers, "geeks," "nerds," and "netizens"), who pose as cyberculture's only voice.

References: Rheingold, Howard. *The Virtual Community*. Reading, MA: Addison-Wesley, 1993. Turkle, Sherry. *Life on the Screen*. New York: Simon and Shuster, 1995. Whole Earth 'Lectronic Link. Accessed 13 October 2013. http://www.well.com/.

M. M. Kuznetsov

❧ D ☙

Debt Crisis refers to the rapid raising, since the second half of the 1970s, of the debt dependence of peripheral and semi-peripheral states of the world system on the capitalist center. Debt caused by structural disparity between developed and undeveloped countries of the world deepened the world split. Under conditions of international debt crisis, specific rules have been formed, which ensure the privileged position of creditors at the expense of debtors. Creditors have united in the London Club and the Paris Club. However, they could not prevent grouping of debtor countries into a club or a cartel to protect their interests. The International Monetary Fund (IMF), on basis of creditor-states support, established the international control over the banks of commerce and the international financial system. After the crisis in 1982, the Paris Club of creditor states and the London Club of creditor banks of commerce were engaged to negotiate as a united front with each debtor state to impose their own rules. The IMF also dictates its requirements to various states. The basic mission of the IMF is to achieve the collection of debts. The IMF "rewards" debt, depending on a nation's ability and readiness to demonstrate "progress" taking strict measures to restrict expenses and carrying out internal reforms.

Relations between creditors and debtors in the world system appear as the domination of the rich and the strong over the poor and the weak. Sometimes the rich render their assistance to the poor; sometimes they lose their money as a result of their debtor insolvency. But on the whole, they derive huge benefits from their position.

V. Prodanov

Demographic Policy is the purposeful practice of state services and other public institutions to control population and reproduction.

The Limits to Growth (Meadows, et al., 1972), a report to the Club of Rome, used computer modeling of exponential economic and population growth with finite resource supplies. The concept of zero population growth was discussed in that report. The goals and a system of measures of population control are determined by prevailing ideological concepts, by the peculiarities of existing social systems, by the governmental structure, by the level of economical development and resource potential, by the quality of life, and by cultural and religious standards and traditions.

Demographic policy should be based on analysis of population statistics, structure, and settling dynamics. A principal feature of demographic policy is its indirect influence on the demographic processes' dynamics, through demographic behavior and through decision-making concerning marriage, family, children birth, vocational choice, employment, and place of residence.

Demographic policy is implemented through a complex of various measures: (1) economic (such as paid leaves and various welfare payments for child birth, children's allowance according to their number, age and family type, loans and credits, taxes, and housing allotments); (2) administrative and legal (such as administrative deeds to regulate marriage, divorce, status of children in families, alimony, protection of maternity and childhood, abortions and contraception means' use, social security of disabled individuals,

Debt Crisis – Dynamic Srategy Theory

employment conditions and labor regiment for women that own children, and external and internal migration); (3) educational and propagandistic measures that seek to form public opinion, norms, and standards of a demographic behavior, fixing the demographic climate of a society.

Global threats from drastically increased population growth rates in the developing countries during the latter half of the twentieth century caused demographic policy to become a major priority for state interests in many countries of Asia, Africa, and Latin America, and an object for the efforts of international organizations such as the United Nations. Traditionally high birth rates in the Third World, almost unregulated at a family level along with decreasing death rates caused a phenomenon called the "population explosion." The idea of population growth limitation has been terms "neo-Malthusianism," after Robert Malthus, who first wrote how unchecked population growth is exponential (1798).

High population growth rates are obstacles to economic development. In dozens of developing countries, government programs supporting the distribution of family planning practices oriented toward the decrease of birth and population growth rates have been applied. The recommendations of World Population Conferences, especially those of the Programme on Population and Development, adopted at the World Population Conference in Cairo (1994), expressed basic guidelines regarding the implementation of state population policy. Recommendations were recited, and tasks and objectives were assigned. A special emphasis in the documents is laid on a sovereign right of each country to assign tasks in population reproduction and development and to concern means toward the tasks' achievement independently.

The global aspect of the demographical policy is relevant to the expressed suggestions on necessity of a world population policy that is to supplement national policies, to lay down general approaches and to adjust the tasks of their realization at a national level. An improvement in the standard of living inevitably leads to an increase in the expenditure of natural resources. This leads to the further population growth at a price of the degradation of conditions of life.

References: Malthus, R. An Essay on the Principle of Population. London: J. Johnson, 1798. Meadows, D. H., et al. *The Limits to Growth.* New York: Universe Books, 1972.

V. V. Yelizarov

Demographic Transition refers to the transition from high to low rates of birth and death as a country develops from a pre-industrial to an industrialized economic system. During development, all countries experience a rapid growth of population, followed by an even more rapid decrease in the relative growth rate. This demographic transition is a universal phenomenon, and countries of the developed world—Europe, North America, Russia, Japan—have already passed through it; their populations have now stabilized.

As birth rates decrease and life expectancy increases, the proportion of senior citizens in the population increases, which increases the burden on social and medical services. The profound change in the age profile, along with massive urbanization, leads to major changes in values, family life, and criteria for growth and success in society. Finally, the rate of change during

124   *Global Studies Encyclopedic Dictionary*

the transition is so rapid that neither the individual nor society has time to adapt to the new circumstances or to reach a new equilibrium. This is one of the reasons for much of the strain and strife of the modern world.

The developing world is now experiencing its most critical stage of rapid growth and even more rapid economic development. The main actors, the populations of India and China, are growing at a respective rate of 1.7 percent and 1.1 percent, but their economies are growing much faster—at 6–7 percent for India and 10 percent for China, although the economic crisis has led to a slowdown. These amazing statistics and their consequences should be assessed. At this stage in the demographic transition of Europe, between 150 and 200 years ago, large numbers moved from the rural areas to towns. Today, they are migrating to the great megalopolises of Asia and Latin America.

The transition is accompanied by an even faster economic development. During this period, large numbers of people move from rural areas to towns. Waves of immigration move people to less inhabited parts of the world. This pattern occurred during the nineteenth century in Europe, when North America and Australia were populated. In Russia, people emigrated to the New World or to Siberia, the frontier of development and growth.

As developing countries now experience this remarkable growth rate, its younger and active members can be either the instrument for economic development or a force for social instability. When the now developed countries passed through this stage, the rapid economic development of Europe culminated in the major instabilities that led to World War I.

The demographic indicators are a message for action. In the first place, appropriate demographic policies should be defined and pursued as part of national, social, and economic plans and development. At present, most governments do have such programs, but a detailed discussion of these issues is beyond the scope of this article. The cases of China, India, and Indonesia show the results of these programs, following somewhat different trends. The demographic situation in Asia is often explained as the result of the incomplete demographic transition.

A growing volume of evidence indicates that only by treating the human population as an evolving system can we obtain some insight into the reasons for the demographic transition. In considering the passage through and effects of the demographic transition in a broader historical perspective, we should emphasize that we are dealing with a global event never before experienced by humankind. Recent research has shown that, since the earliest human civilizations, the growth of the global population system has followed the same pattern. This is not linear or exponential growth, but it is a self-accelerated growth, described by the same law over an immense interval of time. This growth, when the rate is proportional to the square of the number of people, is nonlinear, and this rate can be related to the sum of all participating regions. Only if we refer to the whole world, to all people interacting in the demographic processes as members of the same species of the global demographic system, can we describe and assess growth throughout the whole world and over an immense time.

S. P. Kapitza

Depopulation (French *dépopulation*, "population decrease") is the systematic decrease in the number of inhabitants of a country or a region typically caused by imbalance between birth rates and death rates. Depopulation can occur, for example due to a long-term regime of a restricted reproduction at the same time as increased life expectancies, influence of migration patterns, or wars. Depopulation as a state of demographic development is characterized by net loss of the population seen when reproduction rates fail to match increases in life expectancy. Insignificant population growth can occur in different periods of depopulation process, while the natural decrease of population is often compensated by the migration growth. In the beginning of the twenty first century, depopulation can be observed in most of the industrialized countries of the world.

V. A. Iontsev

Determinism, Geographic: (Greek *geographia*, "description of the Earth"; and Latin *definire*, "to limit" "to determine") Geographic determinism is a world outlook and a scientific/philosophical school of thought whose proponents absolutize the influence of geographic conditions on human lives and actions, claiming that geographic environment is the determinative factor of social development. Under geographic environment is meant, as a rule, the part of terrestrial nature (the upper lithosphere, the atmosphere, the hydrosphere, flora and fauna) involved in human activities and being a necessary precondition for social existence and development. Opposed to geographic determinism, geographic indeterminism rejects causality as far as interactions between nature and society are concerned.

Although the French philosophers of the eighteenth century Anne-Robert-Jacques Turgot and Montesquieu are commonly thought of as the founders of geographic determinism, this concept dates back to antiquity. We can distinguish: real geographic determinism, developed by the Greek historians Herodotus and Strobon, aimed at revealing the dialectics of interrelations between nature and human development; physiological geographic determinism, founded by the Greek physician Hippocrates, that studies interconnections between human health and natural environment; and geopolitical geographic determinism, dating back to the Greek historian Thucydides, that studied the influence of nature on social development. Putting emphasis on climate variations among different regions, the Greeks considered Greece and the Mediterranean the friendliest environment for human life and activities.

In the eighteenth century, Montesquieu espoused a mechanistic geographic determinism that insisted on nearly total dependence of human activity on natural environment ([1748] 1989). He asserted that geographic and climatic environment directly causes human life, morals, laws, customs, and even political organization. At the same time, acknowledging that nature has created all people equal from their birth, Montesquieu stressed their differences from a geographic determinist viewpoint. He wrote that fruitless land makes people industrious, ascetic, hard working, fearless, and belligerent because they must get for themselves what their land is unable to provide; fertile land brings not only wealth but also femininity and an unwillingness to sacrifice human lives. He gives the same explanations when describing life of

126 *Global Studies Encyclopedic Dictionary*

insular nations. Accordingly, they are more prone to freedom than continental inhabitants because the small sizes characteristic of islands make oppression of one part of the population rather hard for the other part. Islands are separated from the big empires by a sea and they cannot support a tyranny; besides, a sea intercepts a conqueror. That is why, Montesquieu thought, insular nations are not in danger of being conquered and it is easier for them to preserve their laws.

A bit later, English economist Thomas Malthus put forward his prominent concept that there is some "natural law" regulating the level of population in accordance with the available food supply. According to this law, population growth always exceeds the growth of the means of subsistence because population increases in a geometrical ratio while the means of subsistence increase in an arithmetical ratio, which necessarily leads to "absolute overpopulation" that threatens many social disasters. Emphasizing strict determination of social development by the eternal laws of nature Malthus argued that the laws of nature are unchangeable from the beginning of the world.

German ethnographer and geographer Friedrich Ratzel (1897) was the foremost representative of geographic determinism at the end of the nineteenth century and the beginning of the twentieth century. Building upon ideas of Immanuel Kant, Alexander von Humboldt, Carl Ritter, and others, he was the first to link politics to geography in his attempts to explain the behavior of states by their geographic position. Arguing that human beings are a part of the whole world and that landscape and natural environment have a central impact on the character of nations living in various regions of the planet, Ratzel came to the conclusion that human beings must, like plants and animals, adapt to their natural environment. Moreover, even a state's wealth depends on to what extent it has adjusted itself to its natural environment, because a state, from Ratzel's point of view, is a biological organism acting according to the laws of biology and is rooted in the land like a tree. He concluded that expansion of living space (*Lebensraum*), or territorial expansion, is the most important means for a state to acquire power. Ratzel applied his ideas to Germany arguing that its central economic and political problems were caused by unjustly drawn and too narrow borders limiting the country's progressive development.

Territory, climate, resources, landscape, and the lay of the land, have an impact on social development because, while working and producing material goods, people use various natural materials, and the broader the sphere of their economic activity becomes, the more actively they use these materials. As a result, the sphere of direct interactions between humans and nature undergoes changes and expansion, meaning the role of the geographic environment in social life grows. While primitive people mostly used natural living sources (plants and animals) and made their labor tools from naturally available materials (stone and wood), at later stages of historical development the significance of mineral and energy resources grew and the geography of human activity expanded dramatically. Simultaneously, the complexity of labor was constantly growing due to geographic factors because human activity took place not only within friendly environments but also under severe, life-threatening natural conditions. Thus, moderate climate, fertile land, and optimal humidity provide an opportunity for bountiful harvest under relatively low labor costs, and natural resources' availability and their easy extraction make mining simpler and the production costs lower.

Until the twentieth century, human economic and social life depended on the geographic environment and the forces of nature less than when their economic and technical capabilities were higher. In the twentieth century, this trend changed and even reversed. That means now, after humankind has become a planetary phenomenon, its economic growth encounters natural limits of the geographic environment that is too limited in its size and resources to correspond to the growing human productive activity. It has been calculated that over the last three years the amount of raw materials consumed equaled the amount consumed during the whole of previous history. In the next decades, if the speed of economic growth remains the same, industrial production may increase by 200–300 percent, causing a demand for an additional huge amount of natural resources.

The image of Earth has been even more changed after the emergence of human-made environments, or so-called second nature. That means the presence of huge megalopolises and countless cities and towns covering nearly all the inhabitable territory of our planet's continents and islands. That also means a dense net of roads and railways, channels, pits, dumps, waste sites, and other objects made by humans that did not previously exist. Thus, nowadays, human beings play the major role in changing the geographic environment but, at the same time, after having encountered the above-mentioned natural constrains they have lost their relative independence from them. This does not just explain the long-lasting attempts to find theoretical foundations for the role of geographic environment in social development, but also makes many discussions concerning the process of globalization and the re-emergence of global problems especially significant.

References: Malthus, R. An Essay on the Principle of Population. London: J. Johnson, 1798. Ratzel, F. *Politische Geographie* (Political Geography). Munich and Leipzig: R. Oldenbourg, 1897. Montesquieu. *The Spirit of the Laws* (original French, *De l'esprit des loix*). Translated and edited by A. Cohler, M. Miller, and H. Samuel. New York: Cambridge University Press, (1748) 1989.

A. N. Chumakov

Development, Human: Human Development is a concept developed in the 1980s, within the framework of the "Program of Development" adopted by the United Nations, which aims at the integration of the solutions for economic problems related to the production and distribution of goods and services into the context of the formation and use of human abilities and skills, as the latter is regarded as the main goal of the social progress.

The principles underlying the concept of human development can be formulated as follows: (1) Human development aims at enlarging people's choices and building wealth. (2) There are no limits for human choices and their parameters can change as time passes; however, at any level the priorities are the same: living a long and healthy life, being educated, and access to the resources needed for a decent life. (3) Incomes are regarded as a means for extending options available for human beings, or a means allowing to make a free choice and increase the number of available options for tackling

128 *Global Studies Encyclopedic Dictionary*

particular tasks. (4) As incomes increase, the positive influence of this factor on human development is being diminished (according to the principle of diminishment of the maximal benefit).

In 1994, a new concept, "sustainable human development," was added, which signifies a development leading to economic growth and fair distribution of the results of such a growth, or, a development that leads to the restoration of the environment rather than to its deterioration, and which provides for enhancement rather than depletion of human abilities and capabilities. Such development gives priority to the human development of the poor and extension of their capabilities, and their participation in decision-making related to their life. Such kind of development also provides for the empowerment of women.

Since 1990, "Global Reports on Human Development" have been issued annually by the UN Development Program. Each report describes the current world situation and considers the behavior of social and economic parameters used for classification of countries into groups by the extent of human development and analyzes particular government policies and social and economic programs. "Cultural Liberty in Today's Diverse World" (2004) analyzes the problems of identity existing in various countries and societies, and various political approaches existing in multinational countries and societies.

I. A. Aleshkovsky

Development, Social: Social development refers to: (1) gradual ascension from undeveloped primordial forms to more advanced forms of social life accompanied by qualitative changes; (2) a complex process of interactive socio-economic changes; (3) a process of differentiation of social structures including full development and actualization of their potentialities. At any systemic level, development involves differentiation, the growing complexity of systemic organization, the increasing capability of adaptation to the environment, and the increasing independence from the environment. Development produces irreversible changes, evolving from a less to a more desirable state.

The UN considers development to be an integral process that includes economic, social, political, and cultural elements, aimed at the improvement of material well-being, including nutrition, habitation, education, public health protection, and culture. According to United Nations experts, development should not be reduced to purely economic factors; it should also include free choice of the development pattern, satisfaction of population needs, and solution to labor issues. Development is a feature of the system of international relations; it involves postcolonial societies into the optimum international division of labor, which should be the primary objective of the new international economic order. Development planning should be based on a realistic vision of national needs and on consumption models typical for the country in question.

Social development is not a movement from down up, increasing complexity and perfection, a transition from the simple to the complex, or, ascension. Complexity is not necessarily an attribute of development; simplicity is not its antipode. Systemic complexity is not always an indicator of development. Progressive development does not necessarily mean growing complexity of a system, since it may include dysfunction, non-efficiency, and decreasing reliability. Progress may mean organizational simplification presented as rationalization. The level of structural and functional integration, growing optimization and effectiveness, increasing autonomy of a system, and growing quality and reliability of operation systems, indicates social progress. Progress is a form of development, connected with increasing level of organization, the maintenance of evolutionary capacity and further development potential.

Two basic concepts of social development, linear (not repeating, unique, separate) and cyclical (naturally repeating, archetypal, inter-related with surroundings), have emerged. The notion of linear development goes back to the utopian socialists of the nineteenth century, who believed in constant improvement and perfection connecting development with growing material well-being, economic position improvement, and growth of per capita income. This idea spread all over the world. The notion of cyclical development existed for many centuries in Asian countries, and in recent years, this theory began to revive.

In linear thinking, exhaustion or death on any level is a failure. In cyclical thinking exhaustion and death are seen as natural, as letting go of old ideas and going within. Seen as a period of release and gestation, this "going within" is deeply integrated with the following inevitable "birth" phase. There is some synthesis of the notions of linear and cyclic development, assuming that while humanity succeeds in some spheres (such as material life, economics, knowledge, and technology).

M. Y. Pavlov

Dialogue is earnest conversation between two or more human beings for the sake of achieving understanding. Three twentieth-century accounts of dialogue, those of Paulo Friere, Albert Camus, and Paul Ricoeur, together elucidate the notion of dialogue. Friere presents dialogue as a sharing of ideas and views between individuals, cultures, and worldviews, wherein one begins with where the participants are, with no preconditions and no assumption of hierarchy. A goal of dialogue is to awaken and hear the voices of all, including the oppressed—to enlarge dialogue to encompass the totality of humanity.

Ricoeur's key contribution is his concept of "attestation." Each self in a dialogue "attests" to "well-considered convictions," convictions, which as starting points are not arbitrary, but are confident beliefs, but still held as "belief," and not as "the truth." Attestation is an ongoing felt commitment within fallibility, an urge to speak of where one is and what one has come to hold, from a history, a situation, and a set of circumstances, and, hence, open to error.

Attestation refers to an aspect of human evidence apart from the evidence of science and history, for attestation is evidence drawn from an earnest inspection of one's self; it is that which one claims upon self-examination. In attesting, one admits that the issue on which one is focusing is worth the struggle of thinking about it, even in the absence of the possibility of certainty. In attesting, one not only attests to some conclusions one has drawn, but attests at the same moment to some issues one has taken to be important enough to risk error and failure. Thus, for Ricoeur, attestation broadens into a stress on human action, effort, and commitment in the full knowledge of fallibility, with the error and finitude that is an ingredient in human action as a part of effort and commitment. Dialogue is a form of human action.

130 *Global Studies Encyclopedic Dictionary*

For Camus, it is not argument that is important, but human beings engaged in argument. The task is keeping this insight present and visible to human beings as they theorize. A sustained lucidity of "keeping voices alive" must be established, in the sense of staying aware that theories and idea systems are "sayings" put into practice, which come from human beings and affect human lives. Camus's first instinct is to see the human voice as the model for thought, rather than the text as the model for thought. There are voices within theories, in that humans are the authors of theories, *and* in the sense that human beings live out the consequences of theories.

For Camus, dialogue possesses no ease or guarantee of success. Dialogue is not a naive faith in words, but instead only the tempered faith that words are all we have. Our words are part of a weary and ongoing struggle, for to write and speak is akin to treating an illness or cultivating health in that each day the effort must be renewed. Finally, Camus's model for deliberation and action stresses finitude and limitation as elements to be kept in the forefront of decision-making and human awareness, never overcome.

The contributions of each of these thinkers to an understanding of dialogue are germane to how participants in global studies, especially ones from divergent traditions, can enhance their communication with one another. In this regard, such humane dialogue can improve international relations and promote positive peace.

References: Camus, A. *The Rebel.* Translated by A. Bower. New York: Vintage Books, 1956. Freire, P. *Pedagogy of the Oppressed.* Translated by M. B. Ramos. New York: Continuum, 2000. Ricoeur, P. *Oneself as Another.* Translated by K. Blamey. Chicago: University of Chicago Press, 1992.

D. L. Stegall

Dialogue among Civilizations and the Islamic Factor: Humanity is becoming increasingly more aware that contemporary global civilizations have to and may coexist, cooperate, and possibly complement each other. Based on that assumption and on the actual global situation related primarily to the contemporary level of technological progress, we can come to acknowledge the necessity of strengthening ties and relations among different cultures and, accordingly, among different civilizations.

The initiatives by Mohammed Khatami, Iranian scholar, Shia theologian, and Reformist politician, who served as the fifth President of Iran from 2 August 1997 to 3 August 2005, introduced the concept of dialogue among civilizations as the necessary and main factor in relationships in the global system. Civilization as a social system should not be understood as a mechanical aggregation of its components (countries, nations, or regions), but should be considered along with a new quality of the system that was acquired within the process as an "integral effect." For example, for Muslim civilization the Word of God was such an "integral factor."

The components of Islamic civilizations are the nations and countries that had striking ethnic, socioeconomic, and geographic differences and had been on different levels of cultural development before the Holy Koran was sent. The observance of the divine summons served as a powerful integration

Debt Crisis – Dynamic Srategy Theory 131

factor to allow formation of the Islamic civilization. At the same time, Muslim nations kept their cultural peculiarity and singularity, and every one of them feeds and enriches Islamic civilization in its own way, hereby generating many different interpretations and forms of expression. Thus, a valuable requirement of Islamic teaching is tolerance, which is prescribed for all members of the Islamic *umma* (community).

In Islam, this assumption originated from the teaching of the Holy Koran, where the representatives of all monotheistic religions (such as Muslims, Christians, Jews, and Zoroastrians) are promised all the blessings, while tolerance and the ability to listen to each other are prescribed. Learning is viewed as an important means in this process: it is not by chance that the first verse (*ayat*) sent to the Prophet of Islam starts with the word "read" (Surah 96, ayat 1). The Prophet of Islam emphasized the importance of science and instructed all his followers to be engaged in scientific research. Besides that, science is one of the most reliable means of intercivilizational relationships and dialogue. In terms of developing scientific and cultural values, Islam does not accept any boundaries.

Islamic scholars developed and introduced entire scientific courses new to the Middle Ages such as algebra (Mohammad bin Musa al-Khawarizmi), chemistry (Jabir ibn Hayyin, also known in Europe as "al-Geber"), and systematic medicine (Ibn Sina, known in the West as "Avicenna"). Abu-l-Qasim Mansur Firdausi, Umar Khayyam, Shams ad-Din Muhammad Hafez, and other representatives of Muslim literature had a great influence on the development of the European world of letters. The beginning of the European Renaissance was to a great extent due to the works by Eastern, first Muslim, scholars. That was the time of rapid and very fruitful exchange of scientific, ethical, and aesthetical values. The Renaissance was not a deviation from religious values, but, quite the contrary (Khatami, 2006). The mutual openness of the world and of the human being was the main idea of the Renaissance which itself is full of religious meaning since the Renaissance was aimed at strengthening, purifying, and expanding religion.

The globalization trend should be based on and should imply multipolarity of the modern world, taking into account the emergence of new morals, new collective justice based on international establishments and laws, enhanced sovereignty and democracy of all the nations, and their increased responsibility for international issues. This requires that all components of the international system (states and civil institutions) actively and equally participate both in development and decision making on the international level.

The events of 11 September 2001 cannot refute the doctrine of dialogue among civilizations, but they are a rather convincing argument for humankind's growing need for constructive and permanent dialogue of contemporary global civilizations to solve the accumulated problems. While condemning any acts of terrorism, prevention of such disasters requires greater involvement and deeper mutual understanding on the international scene of different religions, cultures, and civilizations of the East and the West.

132   *Global Studies Encyclopedic Dictionary*

References: Zarrinkub, A. *The Chronicle of Islam*. Teheran, 1996. Imam Ali, the Master of the Faihful. "The Green Memory. A Bunch of Flowers." In *Nakhj-ul-balaga*. In Russian. Moscow, 2001. Sanah, M. *Social and Political Problems of Iran's Relationships with the Countries of Central Asia. Political Scientific Analysis*. In Russian. Moscow, 2001. Khatami, M. Address given at Washington National Cathedral as part of the Interfaith Dialogue and the Role of Religion in Peace. 7 September 2006.

M. Sanah

Dialogue on Wealth and Welfare was the ninth report to the Club of Rome (Giarini, 1980). This report provides a deep analysis of the modern economic tendencies and pays specific attention to a mentality that has been destructive to the economic system. Making attempts to determine the ways of solving global problems at the present stage, the authors of this project turned to morality and consciousness issues encouraging a shift in "human characteristics" and the creation of "a new economic conception." They proposed revision of the concepts of "wealth" and "welfare," to meditate upon the problem of consumption and to construct a new economic conception oriented to the rational use of nature and environmental preservation. Among the urgent measures the report suggests limitation of consumption and of humanity's needs in order to use a minimum of energy. According to the recommendations, economic interests should correspond to ecological demands.

Reference: Giarini, O. *Dialogue on Wealth and Welfare*. New York: Pergamon, 1980.

A. N. Chumakov

Dignity: (Latin *dignitatem*, "worthiness") Etymological considerations alone do not define the nature of human dignity. If such a notion is to have an ethical employment, we must have a more precise definition.

The notion of dignity is one that recommends itself to ethical inquiry since connotes a kind of reserve, something that must be treated with appropriate discernment and respect. The dignity of the person is a kind of magnetic field that attracts some forms of treatment and repels others. However, the metaphor of the magnetic field is misleading in at least one important respect: it is mechanical and deterministic, whereas the notion of dignity connotes an ethical appeal to each and to all to accord appropriate respect, and so is the very opposite of the magnetic field that attracts or deflects by force. Human dignity is a fundamental moral principle; people deserve dignity merely because of their humanity. Emmanuel Kant speaks of this as an a priori, saying human beings "possesses a *dignity* (an absolute inner worth)" (1996, p. 558).

We could view the notion of human dignity as a kind of categorical imperative: act only so that your action always embodies treatment of human beings in a way that enhances sociability and facilitates freedom. Acts are morally evil when they compromise human freedom or inhibit sociability without warrant. Knowing how to judge acts in this way, being practiced, good, and efficient at such judgment, is what practical wisdom means.

Although much clarification still has to be done on the concept, we can suggest that human dignity is the best available formulation of the fundamental principle of action, coevally ethical and political, to which philosophy can contribute to steer globalization in the direction of greater humanity and solidarity. The philosophical contribution to this task of steering globalization is

Debt Crisis – Dynamic Srategy Theory 133

thus, precisely, the clarification of the notion of human dignity in a way, which makes this concept persuasive as an ultimate guide for action, planning, and judgment. To fail in this task seems to condemn the future of globalization to the victory of sectional interest over solidarity, of exploitation over fraternity, of inequality over equality, and of irrationality over the rational.

Reference: Kant, E. *The Metaphysics of Morals*. In *Practical Philosophy*. Translated by Mary J. Gregor, 353–604. New York: Cambridge University Press.

T. A. F. Kelly

Diplomacy: (French *diplomatie*, from Greek *diplōma*: *diplo*, "folded in two," and the suffix *-ma*, "an object") In Greece, *diplōma* was the practice of conducting negotiations between groups or states. Now, it usually referred to international relations conducted by diplomats with regard to issues of peace making, trade, war, economics, culture, environment, or human rights. Diplomats usually negotiate international treaties prior to endorsement by national politicians. Under conditions of peaceful international relations, diplomacy takes the leading role over other foreign policy means and methods.

In an informal or social sense, diplomacy refers to the use of tact to gain strategic advantage or to find mutually acceptable solutions to common problems in a non-confrontational manner.

Personal qualities and psychological characteristics conducive to diplomacy include mindfulness, tact, and adaptability, and communication skills. Communication in diplomacy is of such paramount importance that diplomacy is often defined in terms of communication as "a regulated process of communication" (Constantinou, 1996, quoted in Jönsson and Hall, 2002) or "the communication system of the international society" (James, 1980, quoted in ibid.).

Diplomacy as a social and historical phenomenon existed in the Ancient Orient, Greece, and Rome, where diplomacy represented a means by which to address issues of war and peace. The diplomacy of the ancient period was often spontaneous and chaotic, in that diplomatic envoys were only briefly deployed; once a diplomatic agreement was established, very little occurred to ensure that the agreement was upheld. The accomplishment of the mission was a public issue.

The period of feudalism can be characterized by a "diplomacy of force" or "diplomacy by pollaxe." The diplomacy of negotiation and compromise played a minor role, primarily in peace settlement after hostilities among states. The church and especially the papacy promoted diplomacy of this nature during medieval times.

Special attention has been paid to Renaissance diplomacy. During the thirteenth through fifteenth century, Italy is considered to have been "the cradle of permanent (or modern) diplomacy." Brisk commerce with many nation-states led to the necessity of the appointment of ambassadors (viscounts, Bauls, and consuls). Costly wars were common in this epoch, leading to the necessity of conflict resolution by means of negotiation. Precise information about economy and political situations as well supervision over concluded agreements became essential elements of diplomacy. Thus, in addition to maintaining military institutions, states also organized diplomatic services such as embassies, negotiations, treaties, and agreements. The first embassies in

134   *Global Studies Encyclopedic Dictionary*

foreign courtyards were established; Francesco Sfortza of Milan founded the first embassy in Florence in 1446. Diplomatic institutions were also implemented at that time along with diplomatic hierarchies, whose purpose was to loyally advocate the interests of a nation by sending its representative abroad.

In the sixteenth century, clerics continued to have an important role in diplomatic relations. Monarchies drew upon their connections to the Catholic hierarchy in forging international agreements. Thus, Latin came to be considered an international and diplomatic as the result of its use in drawing up treaties and agreements. There were cases when diplomatic correspondence was written in national languages, but Latin was the primary language of international relations. Many bishops, archbishops, and even cardinals performed diplomatic missions for kings and imperators, thus becoming the first professional international diplomats. By the mid-eighteenth century, the expression "diplomatic corps" had been introduced in Vienna, referring to all the ambassadors, envoys and other diplomatic personnel of foreign missions present at the seat of any given government. Another expression, "diplomatic service," also appears, meaning the branch the state that delivers personnel of the permanent missions of the state abroad.

The terms "old diplomacy" and "new diplomacy" have been in use since the end of World War II. The alliance of France, England, and Russia against Germany before 1914 is the old, while the system of international security that emerged with the League of Nations Covenant in 1919 is the new. That covenant was intended to regulate the relations of the then recognized states of the world. The new diplomacy was a break with bilateral alliances of the past to be replaced with a universal association of states pledged to compliance with a set of general principles embodied in international law, and the abandonment of power politics. Thus, the use of force to settle disagreements was to be replaced by diplomatic negotiations (Géraud, 1945). Woodrow Wilson was deeply committed to the new diplomacy, of which openness is a paramount feature. In his "Fourteen Points" address to a joint session of the US Congress on 8 January 1918, he called for "open covenants of peace, openly arrived at, after which there shall be no private international understandings of any kind but diplomacy shall proceed always frankly and in the public view.

"Open" diplomacy, or, "conference" diplomacy to use Dag Hammarskjold's term, refers to diplomatic negotiations that are carried out openly with transparency, and which are based on the open diplomacy, which grew out of Wilson's Fourteen Points. It succeeded the "French method" of "secret" diplomacy, which was conducted by professional diplomats, and which was originated by Richelieu and practiced for three centuries ("Open or Secret Diplomacy?," *The Milwaukee Journal*, 11 August 1955). In modern times, some critics speak of the "depreciation" of modern diplomacy, meaning that it does not pay proper attention to advantageous aspects of surreptitious diplomatic techniques.

"Bilateral" diplomacy refers to official relations between two states, while "multilateral" diplomacy is a set of official relations among more than states and their common activity as regulated by diplomatic methods and means, mainly within international conferences and organizations.

Debt Crisis – Dynamic Srategy Theory

Concerned with economic policy issues, the term "economic diplomacy" refers to the use of economic tools (e.g., trade agreements, import/export laws) to achieve state interests. Both states and non-state agents, such as NGOs, that are engaged in economic activities internationally are also players in economic diplomacy (Bayne and Woolcock, 2007). This term first appeared between 1950 and 1960 in Japan, when diplomatic means were employed to bolster the Japanese economy internationally. For economic diplomacy to work, the economic status of the country employing economic diplomacy is a decisive factor. Samuel P. Huntington considers it to be the most effective instrument of American foreign policy (1978).

"Trade" diplomacy refers to diplomatic efforts to support a nation's trade and financial sectors. The main goal of this work is to provide information regarding actual or possible exports and investments, contributing to trade relations and the national good through the promotion of domestic production in the world market.

Definitions of "public" or "popular" refer to communication with foreign populations to establish a dialogue designed to inform and influence. The appearance of new actors within international relations, such as nongovernmental organizations, international organizations, transnational corporations, mass media, and even some individuals has had much influence on the concerns and development of popular diplomacy.

Another aspect of public diplomacy is its role as an integral part of the United States' diplomatic policy regarding broad programs and projects of cultural exchange through the involvement of public figures (journalists, scientists, doctors, artists and other representatives) in efforts to accomplish foreign policy objectives. For instance, a special organ for the advancement of human rights, headed by the deputy Secretary of State has been created within the United States' State Department. This particular type of activity has been dubbed "cultural diplomacy."

The aim of cultural diplomacy is to promote the cultural, scientific, technical, social, and humanitarian achievements of the state abroad as part of a larger program of gaining allies and developing friendly relations with those countries. In some cases, special structures and organizations are employed to advance cultural diplomacy (e.g., the British Council or Peace Corps).

The widening of the diplomatic sphere has included many specialists other than professional diplomats. These specialists perform specific functions and their activity is often called "quasi-diplomacy." In international organizations the term "para-diplomacy" is used; a "para-diplomats" are persons who, in virtue of their competence, have been called upon to represent their government in international negotiations and in assemblies of international specialized agencies.

The modern practice of diplomacy should be based on understanding the problematic character and interdependence of world powers and should be aimed at international peace and cooperation, primarily by means of noncoercive dialogue.

References: Géraud, A. "Diplomacy, Old and New," *Foreign Affairs* (1 Jan. 1945). Accessed 12 November 2013. http://www.foreignaffairs.com/articles/70396/andregeraud-pertinax/diplomacy-old-and-new. Huntington, S. P. "Trade, Technology, and Leverage: Economic Diplomacy," *Foreign Policy* 32 (Autumn, 1978), pp. 63–106. Jönsson, C., and M. Hall. "Diplomatic Theory and Practice." Paper prepared for the 43rd Annual ISA Convention, New Orleans, LA, 23–27 March 2002. Accessed 11 November 2013. http://isanet.ccit.arizona.edu/noarchive/jonsson.htm. Woolcock, S., and Bayne, N., eds. *The New Economic Diplomacy*. Ashgate: Aldershot, 2007.

S. N. Pynzari

Diseases of Civilization (Lifestyle Diseases or Diseases of Longevity) are human diseases that appear to increase in frequency as countries become more industrialized and people live longer (e.g., Alzheimer's, stroke, depression, some forms of cancer, Type 2 diabetes). As such, these are seen as ill-fated forms of adaptation to unfavorable factors of the anthropogenically changed environment. Along with their undoubtable advantages and benefits, scientific and technological progress has also brought about damage and break-down of the gene pool of *Homo sapiens*, the consequences of which are unpredictable and possibly even catastrophic.

In response, worldwide efforts should be aimed at elimination of the most dangerous illnesses, defining the unfavorable consequences of the scientific and technological revolution, and resolving ecological crises. Ecological crises can cause changes that the threaten the natural foundation of human life, viz. the human gene pool.

Medicine has undoubtedly played a special role in shaping humankind's destiny. In many respects, medical achievements produced the considerable changes in demography. The diseases that used to cause mass epidemics (plague, smallpox) were conquered. Discovery of new therapeutic methods led to a substantial increase in the duration of human life. Treatments for ailments previously deemed incurable were found. Now however, even more virulent diseases supersede the conquered ones, some imitating other illnesses to deceive the immune system. A growing number of disease-causing organisms, also known as pathogens, are resistant to one or more antimicrobial drugs. A wide range of pathogens—including the bacteria that cause tuberculosis, viruses that causes influenza, parasites that cause malaria, and fungi that cause yeast infections—are becoming resistant to the antimicrobial agents used for treatment (Center for Disease Control and Prevention, 2013)

Leading among occupational diseases are ones caused by overstraining, including: (1) diseases of peripheral nerves and muscles, (2) of musculoskeletal system, (3) of the veins of the lower extremities, and (4) of the vocal apparatus.

Paradoxically, a number of drugs used to treat cancer have been shown to increase the occurrence of secondary cancers. In these instances, the benefits of exposure to the drugs for treatment or prevention of a specific disease have been determined by the Food and Drug Administration to outweigh the additional cancer risks associated with their use.

Before the 1920s, there were no fatalities among bronchial asthma patients. Wide use of adrenoceptor agonists and hormone inhalators for treating bronchial asthma sometimes results in fatalities. Medicine-induced etiology of some tumors (including cancer), leucoses, gastritises, stomach and duodenum ulcers, hepatites, bronchites, kidney problems, bronchial asthmas, and many other illnesses has now been established.

Modern civilization is characterized by a substantial growth in illnesses caused by human behavioral patterns and modern industrial technology. The Twelfth Report on Carcinogens, RoC, released by U.S. Department of Health and Human Services Secretary Kathleen Sebelius (10 June 2011), lists cancer-causing agents in two categories:"known to be human carcinogens" and "reasonably anticipated to be human carcinogens." The list includes a host of human behaviors and human-made products common to the modern industrial world: alcoholism; many chemicals including some dyes; smoking, smokeless tobacco, and environmental tobacco smoke; estrogens; gamma radiation; some glass wool fibers; Hepatitis B and C; lead; the artificial sweetener Saccharin; silica crystals; styrene (used to make Styrofoam); UV radiation; wood dust; and x-radiation (National Toxicology Program, US Department of Health and Human Services).

For many centuries, tuberculosis took innumerable lives. At the end of the twentieth century, after a long period of relative calm, tuberculosis was on the rise again; its acute progressive forms, sometimes fatal (called "galloping consumption" in the nineteenth century), have emerged. World Health Organization statistics indicate that TB is currently second only to HIV/AIDS as the greatest killer worldwide due to a single infectious agent. In 2012, 8.6 million people contracted TB, of which 1.3 million died.

Lev Zilber's virus-genetic theory (Shevlyaghin, 1979) is of merit, and, while initially rejected and subsequently modified, the dysontogenetic theory of Julius Konheim (defective anlage during embryogeny) turned out to be instructive since immune response to cancerous antigens is similar to that of an organism's response to an embryo (Columbus, 1994, p. 18). T. J. Svishcheva has put forward a theory of a trichomonad causitive agent (cancer of parasitic origin) (2006). None of these factors was present 200 years ago; all of them are a fruit of civilization.

One of the most common diseases in the developed world, especially in the United States, is obesity. As well, the food industry widely uses carcinogenic preservatives. The use of these, such as nitrites, is an issue of debate. Opponents believe that they should be banned, while the FDA takes the position that the substances are permitted as long as food manufacturers show that nitrosamines will not form in hazardous amounts in the product under the additive's intended conditions of use. The most recent National Health and Nutrition Examination Survey (1999–2004) shows that, while overall dental caries in teeth of children aged 2–11 years declined from the early 1970s until the mid 1990s (probably due to fluoride in drinking water), from the mid 1990s, this trend has reversed evinced by a small but significant increase in primary decay. As well, chemical treatment of fruit and vegetables can cause poisoning, which can be lethal.

This inventory by no means exhaust the list of pathologies typical of technogenic civilizations. There is a strong possibility that in the near future new, currently unknown illnesses will emerge. Health protection and the fight against the most dangerous diseases is one of the humankind's global objectives, being central to the preservation of life on Earth. Apart from a healthy

138   *Global Studies Encyclopedic Dictionary*

lifestyle, the condition of the environment, heredity, and quality of health protection, factors influencing the health of the population include the cultural and moral potential of society.

References: Shevlyaghin, V. "The Scientific Heritage of Professor Zilber and His Contribution to Virus-Genetic Theory of Malignant Tumours," *Neoplasma* 26:2 (1979) 113–123. Columbus, F. *Cancer Immunoembryotherapy*. Commack, NY: Nova, 1994. Svishcheva, T. *Prospective Diagnosis*. Dilya, 2006 (Russian). "Diseases/Pathogens Associated with Antimicrobial Resistance," *Center for Disease Control and Prevention*. (30 September 2013). Accessed 12 November 2013. http://www.cdc.gov/drugresistance/diseasesconnectedar.html.

N. A. Agadzhanyan, A. Y. Chizhov

Diseases, Global Threat of Infectious: The anthrax mailings after the 9/11 attacks, the SARS outbreak, and growing concern over avian flu have boosted public awareness of natural and human-made bio-security threats. According to the US government report *The Global Infectious Disease Threat and its Implications for the United States* (CIA, 2000), new and re-emerging infectious diseases pose increasing global health threats and will complicate United States and global security over the next twenty years. Since 1973, twenty existing diseases reemerged, spread geographically, or exhibited drug resistance. Over thirty new diseases have been identified. Advances in telecommunications and information technology have increased knowledge of, and possibly the availability of materials, about bio-weapons to potential adversaries. Globalization of trade and travel, environmental degradation, urbanization, population growth, sub-national conflict, poverty, development disparities, and microbial resistance to drugs have increased vulnerability to old diseases such as malaria and TB plus newer ones such as Ebola and AIDS/HIV. Information and telecommunications, and space technologies offer powerful tools to combat these threats. Surveillance and early warning are essential elements for preventing, containing, and responding to disease outbreaks. The World Health Organization (WHO) defines surveillance as "the ongoing systematic collection, collation, analysis, and interpretation of data; and the dissemination of information to those who need to know in order that action may be taken."

(1) Space: In 1994, NASA established the Center for Health Applications of Aerospace Related Technologies (CHAART). With the National Institutes of Health (NIH) and their National Institute of Allergy and Infectious Diseases, CHAART began investigating how disease studies could benefit from satellite imagery. Earlier, NASA researchers had worked on detection of environmental conditions related to habitats for mosquitoes that spread malaria and had used satellite imagery to identify areas at high risk for Lyme disease.

Other countries also recognize the potential for space-based surveillance. India began using its Indian Remote Sensing Satellite (IRS-1C and IRS-1D), plus commercial GIS (Geographic Information Systems) software, to battle malaria in the 1990s. Scientists prepare digitized thematic maps of variables like altitude, temperature, soil type, and rainfall. Since ideal ecosystems for particular malaria-spreading mosquito species are known, more accurate predictions of where species might be expected can be made and addressed.

Debt Crisis – Dynamic Srategy Theory 139

(2) Telecommunications and Information: Efficient communications are essential for effective disease surveillance. The Information Revolution has dramatically increased tools and capabilities (for example, Internet-based surveillance, telemedicine, and integrated surveillance) available for exchanging, comparing, and gathering information. E-mail, Internet publishing, fax, audio/video conferencing plus traditional telecommunications systems are all being employed for data transfer and communications.

Many national and global networks are being established. The United States Public Health Service operates the National Electronic Telecommunications System for Surveillance (NETSS) as part of its National Notifiable Disease Surveillance System (NNDSS). CDC is developing a National Electronic Disease Surveillance System (NEDSS) that will automatically capture real-time electronic data even though collected by various sources and methods. In 1998, PulseNet was established by the Food and Drug Administration (FDA), CDC, and United States Department of Agriculture (USDA). This national lab network performs DNA fingerprinting of bacteria. For example, molecular fingerprints of bacteria from contaminated foods/infected patients can be compared through a CDC database. If a pattern emerges suggesting an outbreak health, agencies receive computer-generated warnings. The United States Defense Department's Global Emerging Infections System (DOD-GEIS) monitors emerging diseases too.

The United Kingdom Public Health Laboratories Service and France's Pasteur Institute also maintain surveillance information networks. Electronic surveillance media is spreading to the developing world. Malaysia's Ministry of Health established a telemedicine applications project, proposed the development of a Health Management Information System, and plans to expand online computerization. India's Ministry of Information Technology is attempting to link hospitals through an intranet. Private and non-governmental actors also post disease information electronically. For instance, Health Canada maintains the Global Public Health Information Network (GPHIN) while Medecins sans Frontieres and the Red Cross disseminate information via the Internet.

WHO is planning to establish a global disease surveillance system linking existing systems in a "network of networks." Although a global network is a long way off, progress is being made in information exchange through expansion of networks, increasing network connectivity, and diffusion of communications equipment. For example, international networks have been established under the aegis of WHO such as Flunet, RABNET, and the Antimicrobial Resistance Information Bank. Others exist for yellow fever, HIV/AIDS, and sexually transmitted diseases. Networks addressing viral hemorrhagic fever outbreaks and African Trypanosomiasis are planned. WHO posts the Weekly Epidemiological Record and Disease Outbreak News on the Internet. ASEAN (Association of SE Asian Nations) is developing a web-based system to provide real-time regional infectious disease outbreak information and is proposing Tuberculosis control surveillance projects.

These developments create significant opportunities and challenges. Space, telecommunications, and information technology offer fast and efficient options for gathering, analyzing, and disseminating critical information that can save lives and contribute significantly to national and global security.

140 *Global Studies Encyclopedic Dictionary*

Considerable business opportunities and interests in health and surveillance technology exist in developed and developing countries.

Technical obstacles to network integration, data standardization and formatting, software compatibility, languages and search engines pose formidable challenges. Great gaps exist in technology and skills between developed and developing countries. The United States and Australia have over 40,000 Internet hosts per million inhabitants, Western Europe 20,000–40,000, but most of Africa and Asia: 1–100. Of fifty-two local Ugandan health facilities surveyed by the CDC only 27 percent had telephones and 14 percent radio call facilities. However, Africa and Asia are where conditions are ripest for the emergence of new diseases. The United States Government Accountability Office reports that "surveillance systems in all countries suffer from several common constraints" especially funding. Even in high-income countries, per capita spending health care amounts to only 3 percent of expenditure. Many investors perceive software and hardware development for computer surveillance systems as unprofitable because of its specificity and public sector orientation.

Information overload and quality control are persistent problems. The Internet allows quick access to hundreds of reports, but quality is not controllable. Premature or erroneous information can lead to inappropriate responses, panic, and expensive yet unwarranted countermeasures. The WHO has thus had to establish an outbreak verification system for identifying, extracting, and assessing outbreak reports from electronic media sources.

(3) Conclusion: The utilization of space and information technologies to fight infectious diseases and bio-terror is an area still in its incubation phase. A significant part challenge is that health care profession and space and information technology paradigms do not inherently overlap. It takes innovative, patient individuals to see the potential for cooperation and work toward implementation. International and interagency cooperation is essential. It also takes funding—that will mean that the burden will fall primarily on the developed countries. The potential for payback, especially considering the increased risks of infectious disease and bio-terror should provide clear impetus.

C. Jasparro, J. Johnson-Freese

Drug Addiction (Drug Dependence) is a group of diseases resulting from the non-medical use of drugs and psychotropic substances, characterized by mental and physical dependence on and tolerance to drugs. Drug addiction causes the mental, intellectual, physical, and moral degradation of personality. Drug addiction is the result of a complex of biological, social, and cultural determinants that cause pathological adaptation. The average life expectancy of an addict does not exceed thirty-five years. According to World Health Organization data, 95 percent of HIV-infected individuals are addicts. In addition, drug addiction directly and indirectly damages the economy of the majority of countries.

Presently, the term "drug addiction" is used in social and legal senses to define a negative mass social phenomenon of non-medical use of drugs and psychotropic substances by some groups of the population. Several experts offer the term "narcotism" to define drug abuse as a legal phenomenon. Drug addiction should be distinguished from addiction to toxic substances, or the

Debt Crisis – Dynamic Srategy Theory 141

abuse of substances that are not prohibited and are not recognized as narcotic ones; sometimes addiction to legal substances is considered a variety.

Narcology studies various forms of drug addiction, such as cannabis addiction, morphine addiction, cocaine addiction, amphetamine addiction, and hallucinogens addiction. Poly-addiction is defined as addiction to two or more narcotics simultaneously. Drug addiction is conditionally divided into three stages: initial-adaptation (change of the organism's reactivity, appearance of mental dependence), the middle stage (appearance of physical dependence in the form of abstinent syndrome), and severe (reduction of drug tolerance, lingering abstinent syndrome).

The *World Drugs Report* for 2012 shows that 230 million people worldwide (1 in 20) took illicit drugs within the last year. The report also says that problem drug users, mainly heroin—and cocaine-dependent—number about 27 million, roughly 0.6 percent of the world adult population (1 in 200). A majority of drug abuse is found in wealthy, developed countries (UN Office for Drugs and Crime, 2013).

B. F. Kalachev, K. V. Kharabet

Dynamic Strategy Theory was developed inductively by Graeme Donald Snooks (1996; 2003) to explain fluctuating fortunes of nature and society over the past 4,000 million years. The first to attempt this ambitious objective, it is unique for its endogenous and demand-side characteristics. It consists of four interrelated elements and one random force: (1) driving force of individual organisms to survive and prosper, or, "strategic desire," which provides the theory's self-starting and self-maintaining character; (2) "dynamic strategies," or genetic/technological change, family multiplication, commerce (symbiosis), and conquest employed by organisms to achieve their objectives; (3) "strategic struggle," through which organisms attempt to gain or retain control of the sources of prosperity; (4) constraining force of "strategic exhaustion" (not natural resource exhaustion), which leads to the stagnation and collapse of societies, species, and dynasties; (5) exogenous shocks, both physical and biological, that impact randomly and marginally on this exogenously-driven dynamic system. The Snooks-Panov vertical (viz., each biological/technological transformation—or paradigm shift—requires only one-third of the time taken by its predecessor) provides important evidence of the relevance of this theory in explaining and predicting the dynamics of life on Earth both before and since the emergence of humankind.

References: Snooks, G. D. *The Dynamic Society*. New York: Routledge,1996; ——. *The Collapse of Darwinism*. Lanham, MD: Lexington Books, 2003.

G. Snooks

०ॄ E ॄ०

East-West is the traditional Eurocentric (Western) us-them categorization of the world. The West refers to "our better, more civilized and more developed" culture or civilization, while the East denotes the "inferior world" of the others, of the rest, with whom we want nothing to do. This divide coincides with the ancient Greek and Roman perception of the "outer world" in the scope of which they pitted themselves against all other peoples, disparagingly labeled as "barbarians." In the late Middle Ages and early Modern times, this value-laden categorization was replicated in the image of "civilized Europe" as opposed to "barbarian Asia and Africa," which sometimes was clothed less judgmentally into the opposition between the Occident and the Orient. However, the category of "the Orient" reserved for the others with the tradition of literacy, excluded most of Africa.

Colonialism and the projection of European power worldwide seemingly confirmed the perception of European superiority. Interestingly, in the eighteenth and nineteenth century the concept of the West was extended to embrace the British colonies, consisting mostly of white settlers (Northern America, Australia, and New Zealand). This attribution displaced the former Spanish and Portuguese colonies in Latin America, which gained independence at the beginning of the nineteenth century. Their Creoles (mixed European-indigenous-Black) crossed the ideological threshold of "racial purity" so enviously guarded in the British colonies, where the perceived skin color determined one's social position. The top-notch rank was exclusively reserved for white Europeans. This situation also pushed the Russian Empire outside the boundaries of the West, although Russian imperialism drew heavily on the Western colonial model: Russia was never regarded as completely part of Europe (the Occident) or Asia (the Orient). Nineteenth century ideologues of racism perceived the Russians as an "impure mixture" of Slavs and Asiatic Mongols. Being a "pure" Slav did not help much either, because spurious etymological similarity of their self-ethnonym with the word "slave," Westerners believed Slavs to be descendants of slaves and, thus, inferior.

In terms of European geopolitics, by the eighteenth century, the North-South opposition was more pronounced; e.g., the war involving Muscovy (Russia), Denmark, Sweden, Poland-Lithuania, Brandenburg-Prussia and Saxony, among others, became known as the "Great Northern War" *not* "Eastern" or "East-Western War." The militarily and economically successful states of Scandinavia and Central Europe, along with the United Kingdom and France, belonged to the "superior North" pitted against the "inferior South" associated with Spain, Portugal, the southern section of the Apennine Peninsula and the Ottoman Empire. Russia only then entering the European politics, it was not included in this scheme, thus, still relegated to the non-European Orient.

The Russian Empire "moved" into Europe through the radical westward shift of its borders in the wake of the partition of Poland-Lithuania at the close of the eighteenth century. Between 1813 and 1815, Russian armies allied with Napoleon's Western European enemies pushed as far west as Paris

East-West – Evolutionism, Global 143

and defeated imperial France. After the Napoleonic Wars, the East-West axis replaced the North-South one in European geopolitics. Subsequently the category of "the East" coincided with that of the Orient associated with Russia and the Ottoman Empire and its successor states in the Balkans. The eastern half of Austria-Hungary was also subsumed in this category. By extension, after World War I, the nation-states that emerged in the area from the Balkans to the Baltic Sea, between Germany and Russia, were usually perceived as part of the East. Other categorizations cast them as a separate entity of Central Europe, or as a deficient part of (Western) Europe visible in the disparaging sobriquet of "New Europe" for this region. When Bolshevik Russia transformed itself into the Soviet Union the pejorative label of "the East" stuck to it, leaving Central Europe and the Balkans as an anomaly, or a buffer zone between the West and the East.

World War II, having degenerated into the Cold War, split the globe between the United States-led West and the Soviet-dominated East. By the 1950s, French statesmen and scholars disagreed with this bipolar vision of the world and proposed the triple world scheme, in which the West was the "First World," the Soviet bloc and other communist states in Europe and Asia—the "Second World," while the colonies and postcolonial states were lumped together into the "Third World." The logic of the Cold War being global, the Third World states had no choice but to align with the communist or capitalist bloc. The Nonaligned Movement emerged during the 1960s, and allowed a degree of reassertion for the separate existence of the Third World. However, the concepts of the "First and Second Worlds" never gained wide currency. In their stead the categories of "the West" and "the East" denoted the United States and Soviet enemy blocs, respectively. Hence, after 1945, the traditionally Eurocentric East-West divide was projected global-wide. In the last two decades of the Cold War the concept of the "Third World" waned too and became synonymous with the category of "developing or underdeveloped states." Staunch political alliance with the United States or rapidly increasing prosperity made Japan, South Korea, South Africa, Singapore, and Hong Kong into part of the West.

After the collapse of the Soviet bloc and following the end of the Cold War, new global categorization schemes were proposed. Initially, a tendency emerged that grouped the world's states into democratic and non-democratic ones. In 1993, American political scientist Samuel P Huntington proposed that the world consists of some ten distinctive civilizations classified as such on the basis of religion, ethnicity, history, geography, or convenience. The dynamics of the bipolar world still alive, he concluded that it is still "us"—the West against "them"—the Rest. Unfortunately, this diagnosis materialized at the beginning of the twenty-first century when the United States-led West seems to have found the replacement enemy for the defunct Soviet Union in the form of the amorphous (hence, even more unpredictable and dangerous) Islamic world.

When it comes to economic development since the end of the Cold War the rhetoric of globalization has ruled supreme. However, with few exceptions, this process seems to be benefiting the rich West and putting the poor states at an even greater disadvantage. Critics say that the term "globalization"

144 *Global Studies Encyclopedic Dictionary*

is a new euphemism for "(economic) colonization"; the dynamics between the erstwhile colonizers and the colonized replaced by a similar one between the globalizers and the globalized. This situation prompted the United Nations to introduce the developmental categories of the rich North and the poor South, which are *not* geographically determined, but correlated with the general level of prosperity in a state. On top of that the older opposition of the West versus the East still appears in Western intellectual discourse as a political shorthand currency for denoting us (rich North) and the rest (poor South).

T. Kamusella

East-West Problem resulted from the traditional historical, political, and cultural division of world civilization into two independent, though interacting, and worlds whose origins and particularities are determined by a different geographic position, a different history, and different cultural historical types (local civilizations).

The Western and the Eastern have always interacted closely and effectively, even though there were often wars between them. East and West have been related since ancient times. From time immemorial, the coming-to-be Europe (Western civilization) realized a deep interest in Egypt, Phoenicia, Carthage, and Persia and, later, toward India, China, and Japan. Europe received many cultural impulses for its development from the East. First, this refers to the Judeo-Christian doctrine. Alexander the Great's campaigns in Persia and India, the invasions of Huns, Mongols, and Tamerlane (Timur) into Europe, and the invasion of Arabs into Spain related East and West many times in the military, cultural, commercial, and biological (genetic) aspect. Relations between West and East include the Silk Road, spice trade, round-the-world marine travel, and many other issues. Arabs and Jews helped Europe keep and develop the Greek scientific heritage, which, together with the Christian ideology, gave Europe its spiritual form, promoted the Renaissance, the Enlightenment, and scientific, technical, and political progress (the formation and consolidation of democracy).

While the West was implementing its expansionist (and colonization) policy (initiated by the crusades), the East was inundated with new cultural and technical impulses, and with new political ideas. The economic and scientific technical effect of this influence by the end of the twentieth century proved to be so obvious that there is already no sense in mentioning the traditional backwardness of the East (the examples of modernization of Japan, China, South-West Asian countries, India, and several Islamic countries). Now we can only point to the cultural originality that still makes the East very different. Perhaps Niels Bohr was right when, in assessing the eternal historical and geographic correlation of West and East, he used his methodological subsidiarity principle: West and East neither abolish nor absorb each other; they only supplement each other. In the modern globalizing world, the East-West problem can also be viewed as Hermann Hesse did, namely, not as hostile and opposed positions, but as poles between which life swings.

A. V. Katsura

East-West as a Social Model: The Western world has reached its high level of well-being due to the perfection of social relations. A rational division of labor, human specialization in different spheres, and optimal variants of specialized activity contribute to the general progress. The Eastern world preferred the development of separate individuals and their moral perfection and, thus, could not find an optimal formula of social progress. Only in the West is the principle of cumulativeness used efficiently.

The East and the West can be, in some aspects, compared to art and science; the East is art and the West is science. The Eastern way of thinking is compared to art because in the East, like in art, the moral world of an individual is put in the forefront. The value of a human being lies in their moral perfection. The spiritual world of an individual determines their social characteristics. In the West differences among people do not hinder the cumulating of their activities; their unity and joint activities are basic conditions of preservation of the social structure.

The Western style plays the same music using different musical instruments—complementing each other, producing a symphonic effect. When playing identical instruments, people can only obtain synchronism. The East strives to uniformity, while the West turns to unity in diversity. In the first case the human will dominates; in the second case the social will dominates. Since human activity in the West is carried out within the frame of common social principles, it is possible to involve many people in the process of the realization of a single idea. Separate individuals become particles of society. In the East society is a totality of people, while in the West a person is a part of society. In the East person predominates; in the West society prevails.

In the age of globalization, mainly seen as Westernization, ethnic groups and minorities tend to merge into a single featureless structure. They lose their individuality. Their particularity, only recently viewed as their beauty, becomes a sign of backwardness. Whatever was original and unique is now depreciated because it does not correspond to Western patterns.

In the Western world, much talk about human rights and freedoms, democracy, and political pluralism now occurs; for several centuries individual happiness and well-being have been considered the major task of society and the state, and many actions have been undertaken in this direction. Eastern countries are criticized for violating human rights, limiting individual freedoms, and subordinating all people to one person (a shah or a monarch).

A specific feature of the West is that human self-realization takes place mainly not in the society, but within our own selves and within a family. People are not prepared to become a part of something bigger. Relationships with others take place at the level of personalities, as bilateral relations. These relations are not related to national and state interests. This type of relationship is mainly based on morality and traditions. Every person creates their own local social environment to enjoy freedoms.

However, in the West, rights and freedoms, possibilities for creative work, and free competition are guaranteed to all only by the society. To become free, one must first partly refuse freedom, find mutual understanding with others, accept joint activity, and become dependent on the state and the society. In exchange, the state and the society protect the remaining individual freedom, which has not fallen under legal limits. Thus, in the West, individual freedom is mediated by state and society. Human freedom and individuality

146   *Global Studies Encyclopedic Dictionary*

are realized through professional activity, creativity, economy, and science, but everyone follows single legal regulations. In the East, human freedom is often manifested through non-observance of legal regulations (civil disobedience).

S. Khalilov

Ecology, Deep: While shallow ecology refers to the fight against pollution and resource depletion, the central objective of which is the health and affluence of people in the developed countries, deep ecology (biospherical egalitarianism), refers to the rejection of the man-in-environment image in favor of a "relational, total field image" (Næss, 1975, p. 96). Through a process of enlightenment or "awakening," one recognizes one's ecological connectedness to the biosphere. In gaining this insight, the deep ecologist squarely repudiates the anthropocentric (human-centered) orientation of the Western (Occidental) tradition.

Norwegian Arne Næss is a philosopher and naturalist who coined the term "deep ecology" in "The Shallow and the Deep, Long-Range Ecology Movement" (1973). Næss criticizes Western civilization for arrogant human-centeredness and a related instrumentalization and subjugation of non-human nature by contrasting his new "deep" environmental ethic with "shallow" ("reform") environmentalism. Shallow environmentalism is simply an extension of the anthropocentric Western paradigm, because the reasons for preserving wilderness or biodiversity are inevitably couched in terms of human welfare. Shallow environmentalism falls short of valuing non-humans apart from their use-value. Deep ecology, in contrast, asserts that all organisms have intrinsic value. In this way, deep ecology is fundamentally non-anthropocentric.

Two interrelated underpinnings support deep ecology's non-anthropocentrism. The first principle, "biocentric equality," asserts that all biota have equal intrinsic value. The second, "expansionary holism," asserts that by a process of "self-realization" one comes to the understanding that the biosphere does not consist of metaphysically discrete individuals, but ontologically-interconnected individuals comprising one unbroken whole. Thus deep ecology is an egalitarian and holistic environmental philosophy.

Biospherical egalitarianism is the view that all biota have equal intrinsic value (or, to put it another way, it denies differential valuation among living things) (ibid.). In this sense, deep ecology is not merely non-anthropocentric, but anti-anthropocentric. In terms of moral considerability, human beings have absolutely no priority over non-human beings. Næss, Bill Devall, and George Sessions have all affirmed this way of thinking: "The equal right to live and blossom is an intuitively clear and obvious value axiom" (ibid., p. 96). "All organisms and entities in the ecosphere, as parts of the interrelated whole, are equal in intrinsic worth" (Devall and Sessions, 1985, p. 67). The target of biocentric equality is Western anthropocentrism. Deep ecologists contend that organisms have equal intrinsic value, with the implication that no form of life (even *Homo sapiens*) carries more weight in adjudicating conflicts of interests.

The second underpinning of deep ecology's non-anthropocentrism is expansionary holism. Some deep ecologists (notably, Devall, Sessions, and Warwick Fox) elaborate Aldo Leopold's holism by arguing for a breakdown of the ontological boundaries between self and other. This breakdown is achieved through self-realization: "It is the idea that we can make no firm ontological divide in reality between the human and the non-human realms . . . to the extent that we perceive boundaries, we fall short of deep ecological consciousness" (Fox, 1990). The ontological boundaries of the self are extended outward, including increasingly more of the lifeworld in it. Thus, this particular formulation of metaphysical holism can be correctly understood as "expansionary" holism: there is in reality only one Self, the lifeworld.

When ontological boundaries are overcome, one realizes nature's interests are one's own interests. Devall and Sessions believe, "if we harm the rest of Nature, then we are harming ourselves. There are no boundaries and everything is interrelated" (1985, p. 68). John Seed, an Australian environmental activist, nicely illustrates this attitude:

> I am protecting the rain forest" develops into "I am part of the rain forest protecting myself." I am that part of the rain forest recently emerged into thinking. . . . the change is a spiritual one, thinking like a mountain, sometimes referred to as "deep ecology. (1988)

Since the rain forest is part of him, he has the moral obligation to look after its welfare. The rain forest's well-being is indistinguishable from his well-being, so its needs become Seed's needs.

Environmental philosophers have criticized these two principles for being vague and even incompatible. David Rothenberg, however, has pointed out that a vague yet original idea like deep ecology may have more influence by stimulating new ways of thinking than a precisely delineated system. By this criterion, deep ecology has been enormously successful.

The contribution of deep ecology to environmental philosophy is the recognition that non-humans have intrinsic value. Deep ecologists are right to excoriate the Modern Western view of nature that views organisms as biomachines. From this standpoint, the only value fauna, flora, fungi, protista, prokaryotae, and inanimate matter have is instrumental value for humankind. Therefore, the more natural resources used by humans, the more value nature has. The importance of deep ecology is the rejection of this instrumental view of nature and the realization that non-humans have value above and beyond use-value for humans.

References: Næss, A. "The Shallow and the Deep, Long-Range Ecology Movement: A Summary," *Inquiry* 16 (1973), 95–100. Devall, B., and G. Sessions. *Deep Ecology*. Salt Lake City, UT: GM Smith, 1985. Fox, W. *Toward a Transpersonal Ecology*. Boston: Shambhala, 1990. Seed, J. "Beyond Anthropocentrism." In *Thinking Like a Mountain*. Philadelphia: New Society Publishers, 1988.

D. R. Keller

Ecology, Engineering: Engineering ecology is a section of industrial ecology connected to the development and practical application of technological and technical methods of regulation of the influence on the environment, including sewage and waste gases treatment, recycling, recuperation, and organized waste placement. Human survival requires the reasonable management of global technogenesis in rigidly regulated limits. A goal of engineering ecology is the research and implementation of reliable ways and means of maintenance–the necessary and sufficient conditions for this survival. The achieve-

148 *Global Studies Encyclopedic Dictionary*

ment of that goal is probable only from a position of a system approach, using integrated decision-making processes in all spheres of the production of goods and life functions.

Despite the obvious differentiation of nature protection functions involved in various branches of the national economy, heterogeneity of organizational-administrative structures, and difference in the specificity of labor processes and the conditions of their implementation, there is a need to develop a uniform scientific methodology that unites the following basic directions of engineering ecological maintenance:

(1) The analysis of reversible and irreversible processes of degradation in areas with complex engineering—geological, hydrological, geocryopedological, and other natural—climatic conditions;

(2) A quantitative estimation of general and local losses of the environment (correlation of these losses at regional, nation-wide, and planetary levels);

(3) Working out and classification of objective criteria of the balance condition of ecosystems, "person-natural object," "industrial object-environment," "person-machine-natural landscape";

(4) Regulation of ways and means of receiving of ecological information on concrete natural-technical geosystems;

(5) Working out local and regional ecological scales of various kinds of industrial influence;

(6) Creation of ecologically clean materials, manufactured products, machines, and technologies;

(7) Working out methods of engineering-ecological preventive maintenance, advance planning of nature protection actions, and the restoration and complex reconstruction of anthropogenic landscapes;

(8) The introduction of economic methods nature protection activity at all stages of the production of manufacturing of industrial or construction products;

The listed directions are complex in their character and based on the research strictly adequate to the specificity of a concrete ecosystem. It is necessary to speed up consolidation of all scientific, engineering-technical and industrial forces on a uniform methodological basis in the direction of nature protection and reduction of ecological risk to the Earth.

The main goal of engineering ecology is overcoming ecological antagonism in the system "person-nature."

Implementation of concepts of engineering ecology is the system of engineering-ecological maintenance of production (SEEMP). SEEMP is a complex of the interconnected cooperating elements (subsystems) functioning in optimum management modes. Management in the engineering-ecological sense is understood as a system of constant control and purposeful influence on conditions and factors effecting the ecological situation of the natural—technical geosystem with the purpose of the establishment, maintenance, and support of the necessary level of environmental safety during designing, production (including construction), and operation of artificial objects.

SEEMP functions in the development of the following subsystems: scientific-methodological maintenance (general principles of making decisions, normative regulations, and optimization of SEEMP management criteria); design maintenance (working out of calculated models and structures, and designing of ecologically clean objects); technological maintenance (ways and means of ecologically rational usage of constructive decision-technological processes, and regulating-technological schedule of ecological restoration of natural-technical geosystems); organizational-methodical maintenance (optimum organizational-methodical structures of production, principles of maintenance of ecological efficiency of production, and ecologically optimum forms of organization of labor processes); complex ecological control (ecological examination of scientific-methodical, design and organizational-technological decisions, industrial ecometry, and monitoring); informational maintenance (principle of accumulation, transfer, storage, and usage of ecological information, criteria of information quality and parameters of its productivity); quantitative estimation and forecasting (methodology of objective estimation of ecological situations in regional and planetary scale, multilevel identification, and engineering-technological aspects of limit forecasts); optimum management (substantiation of allowable limits of regulation of labor processes and management of natural-technical geosystems, social-methodological aspects of formation of ecological knowledge and culture of labor collectives, and general principles of ecologically optimum management).

Engineering ecology creates the necessary conditions for the mobilization of all actions for the protection of nature in the sphere of material-technical production and formulates technical decisions on the maintenance of true environmental safety on Earth.

I. I. Mazour

Ecology, Global: Global ecology describes the ecological condition of the whole planet. In a more precise sense, it has two meanings. First, it is a branch of biology dealing with the aggregate of living organisms of the whole planet in the context of interaction with their environment, that is, the discipline that deals with the entire biosphere as a whole from the point of view of its productive, adaptive, and recreational capabilities. This area of knowledge integrates all environmental research directions related to natural sciences, and projects them to the global level. Secondly, Global Ecology is the sphere of scientific knowledge that deals with the issues of the interaction of society with the environment. More precisely, it is a special branch of knowledge about the biosphere transformed by the human beings who are a part of it and about human beings as the main component of the biosphere.

As a Branch of Scientific Knowledge, global ecology is a sub-section of ecology, which focuses on anthropogenic influences on the biosphere as a whole and processes induced there by these influences; devising a forecast for the processes' after-effects and defining the direction of the activity necessary to prevent or weaken the negative after-effects of this kind. Global ecology to a wide extent includes knowledge about regulation mechanisms, peculiar to the biosphere, about interconnections between its inner processes, whether they have anthropogenic origin or natural ones, especially about the circulation of substance and energy in the biosphere and their change under the influence of different disturbances. The central problem of global ecology is the stability of life and the biosphere as a whole.

150 *Global Studies Encyclopedic Dictionary*

The key to global ecology is the idea of the biosphere as an integrated system. Vladimir I. Vernadsky (1926) and Alfred Lotka formed this view between 1920 and 1930 (we can consider Charles Darwin to be their predecessor, who was undoubtedly the first to realize the integrity of all living organisms or biota as an integral system). Vernadsky defined the biosphere concept itself and made a conclusion, that humans developed and extended the influence of their economy on nature to such a scope that the influence became comparable to geological forces. Lotka was occupied with studying the biosystems' participation in the circulation of substance and energy in the biosphere, including the quantitative level, using modeling tools (Begon, Harper, and Townsend, 1996).

In the middle of the twentieth century, the word environment, which had been roughly used in different terms and contexts long before, came into scientific use. Simultaneously, ecology, which had been a section of biology, started a rapid extension of its subject field, almost taking into its field of view the whole environment, its every level and every aspect of its consideration. Consequently, global ecology can be defined as a section of ecology, the subject of which is the global environment. The attempt to systematically synthesize knowledge, acquired by the global ecology by the end of the twentieth century, is accomplished in the theory of biotic regulation of the environment. In the center of attention of the global ecology stand global changes of the environment, causing the raising of the global ecological problems.

References: Vernadsky, V. I. "La multiplication des organismes et son rôle dans le mécanisme de la biosphère" (Multiplication of Organisms and Their Role in the Mecanism of the Biosphere), *le Bulletin de l'Académie des Sciences de l'Union des Républ. Soviét.* 1926. Accessed 12 November 2013. http://www.larecherchedubonheur. com/article-31047519.html. Begon, M., J. L. Harper, and C. R. Townsend. *Ecology.* 3rd ed. Cambridge, MA: Blackwell, 1996.

A. N. Chumakov, V. I. Danilov-Danilyan

Ecology, Industrial: Industrial ecology addresses the direct influence of economic activities on the environment; basic areas include: monitoring, regulation, control and management of the influence on the environment both at the level of a separate production site and at the territorial level.

I. I. Mazour

Ecology, Social: Social ecology is an interdisciplinary area of knowledge that studies the laws of optimization (compatibility) of society and nature. Social ecology is a further development of general ecology whose most fundamental theses about interaction of organisms and populations with their environment are mostly applicable to the interaction of society with nature.

Chicago school sociologists laid the foundation for social ecology in the 1920s. The term was coined in response to the need to consider the specifics of the impact of urbanized environment on human life and behavior. Social ecology has come to mean the study of the interaction between society and the global natural environment with all the variety of anthropogenic transformations it undergoes. It develops a theory of normalization of this interaction and of prevention of destruction of the biosphere in ways dangerous for society to the extent of losing the living qualities of the biosphere. One of the main purposes of social ecology is the development of maximum allowable nature management standards from the point of view of the biosphere, starting from emission of pollutants and produced heat and up to the quantity of resources withdrawn from nature.

People must observe the basic laws of conservation and support of life as strictly as any organism on the planet, even the smallest one. The fact that people interact with nature mainly with the help of mechanical devices does not give them the right to violate common laws of the conservation of life, and, furthermore, they make people even more dependable on the condition of the nature technically transformed by them. Until recently, the human transformation of nature from the point of view of the biosphere was rather casual, that is, it was not correlated with the condition of the biosphere and the logic of the processes of self-regulation intrinsic to it. Continuation of such ecologically careless practice is becoming dangerous for people themselves, because the scale of the aggregate of their activities has become quite comparable with the scale of biospherical processes, and in many aspects, has even significantly surpassed them. That is why the further practice of nature management should be correlated with the forecast of consequences of nature transformation caused by all kinds of human activity. As any type of activity involves interaction with natural objects, social ecology becomes a concrete aspect of several special areas of knowledge. Apart from the general theoretical topics that social ecology studies, it also has application areas: human ecology, which studies mainly medical aspects in relation to environment contamination with xenobiotics; and engineering ecology, which studies technical aspects of the treatment of the environment by people. Other types of social ecology that have emerged are urban, industrial, agricultural, and transport, legal, recreational, and chemical ecology, and geoecology, ecology of culture, problems of ecological education and attitude development.

Ecological knowledge will be developed in accordance with the variety of types of human activity. In the future, people with their mind, ideology, culture, and habits are expected to undergo changes so fundamental that they can result in the formation of a new kind of human being who, unlike the currently existing one belonging to the *Homo sapiens* species, can be classified as belonging to the *Homo ecologus* species. But, however much the social ecological science can develop, its central concept will always be the complex concept of the society-nature system, or socioecosystem, and the main purpose of social ecology will be research on laws of the self-conservation of the biosphere as they are specifically used by human beings whose activity should from now on support the biosphere.

E. V. Girusov

Ecological Audit is the business activity of ecological auditors or auditing organizations that perform independent professional analysis and assessment of economic and other activities affecting the environment and work out recommendations aimed at minimization of the negative influence on the environment and human health.

152   *Global Studies Encyclopedic Dictionary*

The need for the procedure of ecological audit becomes especially evident: (1) in the process of privatization of the state and municipal enterprises; (2) in the course of the activities of realtors, during the auditing and certification of real estate objects, technological processes, and industries over the entire "life cycle"; (3) when defining the responsibility for the environmental damage inflicted by industrial and consumption waste; (4) when developing the system of ecological insurance; (5) in the ecological expertise of investment projects and programs.

The main types of ecological audit include:

(1) Regulative audit that defines whether economic or other activity of the enterprise correlates with environmental legislation and ecological regulations;

(2) Management audit assessing the efficiency of the internal administrative system and corporate policy, and ecological risk related to the enterprise functioning;

(3) Supply audit that studies the existing practices of purchasing raw materials and looks for alternative production and equipment, which reduces the environmental impact and, at the same time, does not increase the production costs of goods and services and, in the ideal, even diminishes the cost price;

(4) Real estate (basic assets) audit aimed at estimating past economic damage caused by pollution and possible ecological consequences of critical wear of basic assets, identifying potential ecological responsibility and defining the real value of enterprises, land plots with industrial and residential constructions on them, and necessary volume of ecologically oriented investments;

(5) Waste-processing audit aimed at reducing the quantity and danger of industrial waste and its optimal utilizing, burying, and disposing;

(6) Energy-saving audit that monitors the energy consumption at an enterprise and reduces, as a result, environmental damage;

(7) Audit of business partners aimed at assessing whether their activity conforms to the ecological policy of the enterprise;

(8) Strategic audit aimed at working out a long-term business strategy taking into account the ecological characteristics of the enterprise;

(9) Ecological insurance audit preceding ecological insurance aimed at the protection of property interests of citizens, juridical persons, and the state that suffer as a result of environmental damage in case of technological disaster or fault, non-rational utilization of nature, or spontaneous natural force.

According to the requirements of the World Bank for Reconstruction and Development, the system of classification of investment projects by the nature of ecological assessment (audit) needed includes these categories: (1) A complete ecological audit is necessary. A project whose realization could inflict an irreversible environmental damage or create a precedent of environmental effect. (2) A partial ecological audit is needed in accordance with the specific ecological problems of the project. The project realization does not strongly affect the environment and the population and does not change the ecologically important regions of the ecosystem. (3) Ecological audit is not needed. The project realization affects the environment just minimally.

Recently, the importance of the nature-utilization audit has increased considerably in management systems of various enterprises and entire branches of economy. This type of audit involves business activity of auditors or auditing organizations that perform independent professional analysis and assessment of economic and other activity affecting natural resources and that work out recommendations aimed at minimization of the negative influence on these resources and the higher efficiency of their utilization. The main directions of nature-utilization audit are audit of mineral resources, water resources, land resources, forests, and production sharing.

V. N. Lopatin

Ecological Balance is a quasi-stationary, quasi-equilibrium ecological system condition, which involves a balance of natural and artificially modified environmental components and natural processes that contribute to the protracted existence of an ecosystem. Ecological balance also entails the dynamic exchange of energy, substance, and information, maintaining the integrity of an ecosystem. Component equilibrium, based on the balance of the components of an ecological ecosystem, is distinguished from territory equilibrium, which occurs in the juxtaposition of intensively (such as agrocoenoses and urban complexes) and extensively exploited areas (e.g., pastures, natural woods, forest reserves), providing stability in larger geographic areas. Ecological balance should also be distinguished from evolutionary balance, as the alteration of the latter results in an irreversible shift of the ecosystem and its attributes.

The concept of ecological balance is crucial to our understanding of the natural environment, since it is based on systematic scientific knowledge and conceptions of the function and properties of biogeocoenoses. The natural balance resulting from regular human impact tends to alternate, as displayed in singular or communal forms, explicit or implicit consequences, or considerable disproportion. One of the fundamental tasks in the investigation of the ecosystem involves the accurate forecasting of the possible imbalances in accordance with the virtual and planned parameters of such imbalances, thus bringing about shifts in equilibrium values

The following criteria are used to assess industrial technogeneous impact upon ecological balance:

(1) absolute environmental losses conveyed in the biogeocoenosis;

(2) ecosystem compensation capability reflecting maintainability for natural and compulsory regimes;

(3) the ecological imbalance threat designating the probability of irreversible loss and local ecological modification;

(4) the ecological loss concentration level describing the scale of the industrial techogeneous impact to the environment.

I. I. Mazour

Ecological Crisis occurs when the environment of a species or a population changes in a way that destabilizes its continued survival. It is a break of the balance between nature and a society or changes of concentration of biogenes (substances necessary for life) and of other substances in the environment.

Human economic activity changes the environment, but environmental changes affect humanity. Some ancient states of the Middle East ceased to exist because of the salinity and low fertility of the soil. When Athens diversified its economy because of low fertility, reducing the area of arable lands, the subsequent shortage of wheat required a policy of colonization for acquir-

154 *Global Studies Encyclopedic Dictionary*

ing bread supplies. These and other historical instances testify that, as a rule, chemical, biological, and physical degradation of soils leads to resource crises. We distinguish local, regional, and global ecological crises. Local ecological crises can be noted in almost every large city, and in small cities with one enterprise, around which a city was formed and which is usually a source of environmental disturbance. Regional ecological crises happen, for example, in regions of acid fall-outs because of the outburst of sulfur dioxide and nitric oxides, which mixes with droplets of water in atmosphere. Scandinavia and some other areas in Europe, and the northeast of the United States and the southeast of Canada are the largest regions that suffer from acid fall-outs. Lake fish die and forests are extinguished because of acid fall-outs. However, a crisis, as opposed to catastrophe, is a reversible state.

Only in the twentieth century did persistent global environmental changes surpass the natural level of fluctuations, causing global ecological crisis. That signified that human economic activities had gone beyond the carrying capacity of the global ecosystem (biosphere) and had disrupted the stability of the global environment. Overcoming the global ecological crisis is a priority problem on which the survival of humans as a species depends.

K. S. Losev

Ecological Expert: (Latin *expertus*, "experienced") Ecological experts forecast the ecological consequences of a project, with the intent to reduce its environmental impact. The main principles that should guide the practice are: obligation; presumption of potential ecological risk arising from any planned economic or other activity; open discussion of projects; independence; scientific basis and complex (multi-criteria) nature. Russian law regulates two types of ecological expertise (state and public):

(1) State Ecological Experts work under the auspices of authorized state structures aimed at evaluating ecological efficiency of projects and their compliance with existing ecological standards and regulations; ecological efficiency is defined by exposure, analysis, and comparison of all the real and rational alternatives including a refusal to act. This activity must be performed in accordance with the principles of obligatory realization, scientific foundation and legitimacy of conclusions, independence in organization and realization, openness, and citizen's participation.

(2) Public Ecological Experts evaluate and revise (by stating appropriate requests and limitations) projects related to the environment and natural resources. Their activity is performed in relation to civil initiatives (including initiatives of research centers) with participation of any person interested. The recommendations of civil ecological experts become legally obligatory only after the appropriate state institutions have approved them.

To perform expert ecological analysis, the following data is required: standard documents on preservation of the environment for working out ecological criteria; statistics on the reliability of objects similar to the ones under expertise; data on the condition of the environment: (a) field and laboratory research data, and (b) calculated data (methods of calculating the amount of injections of hazardous substances into the environment).

East-West – Evolutionism, Global

Considering the amount of initial data and the difficulty of evaluating the level of preparatory paper work (scientific and technological validity of decisions), it is necessary to create and use an ecological expert system specializing in different branches of the economy, which should automated for making expert decisions. Projects presented for ecological analysis should have a section on evaluation of environmental impact. The nature of the documents depends upon the type of the project and the possible consequences of its influence on the environment. For example, in documents concerning engineering and geological landscape, the following questions should be considered, which consider the preservation of nature, the rational management of resources, and ecological security: (1) geological, hydrogeological, and cryogeological data on the developing territory; field and laboratory research results; (2) initial data and requirements necessary for working out nature preserving activities; (3) data on natural flora, soil, landscape, subsoil waters, and the hydrographic web; (4) materials on the spread of plants and animals in the construction area—migration routes and spawning places; (5) quantitative evaluations of the natural influence on the engineering and geological characteristics of the territory, such as surface processes, cryogenic, erosion, and other processes; (6) characteristics of natural relief-forming processes that emerge or increase due to the construction; (7) levels of influence of various (regular or supernumerary) technological interventions (removing top-soil, filling roadbeds, and transportation passing) on geological and landscape characteristics and on animals and plants; (8) possibilities of natural and artificial restoration of damaged natural landscapes; (9) prognostication of changes of the natural environment after the interference (construction, exploitation, and industrial activity); (10) identification of dangerous regions and zones of extreme ecological vulnerability which require intense protection activities.

Expert ecological analysis of projects should be based on: (1) the priority of a society's right for friendly environment; (2) balance of the economic and ecological interests of a society; expedience of the project realization from a territorial, industrial, and ecological viewpoint; (3) ecological compatibility of the existing and planned projects with environmental protection requirements; (4) observance of projecting, building, and exploitation standards and environmental norms. The results of the analysis are documented in a written report that details the expert opinion. This should be complete and clear enough to exclude multiple interpretations of the materials studied; it should contain objective and valid conclusions about the completeness and legitimacy of a project; it should reflect the whole process of ecological expertise and the main questions studied at every stage. The structure of a decision is as follows: introduction, statement (description), and conclusion (evaluating and generalizing). Documentation is complete without the signature of the head of the ecological expert's institution.

I. I. Mazour

Ecological Information: (Latin *information-*, "to inform") is any information about the state of waters, air, soil, fauna, flora, and native zones and about activity influencing the natural environment, including conservation and administrative measures for the rational management of nature. The UN con-

156   *Global Studies Encyclopedic Dictionary*

vention about access to information and public participation in the decision-making process and access of the justice system to what concerns the environment (Orkhuss Convention) gives the following definition:

> Ecological information means any information, written, audiovisual, electronic or of any other material form, about: (1) the state of environmental elements, such as air and atmosphere, water, soil, ground, landscape and natural objects, biological diversity and its components, including genetically changed organisms, and interaction of these elements; (2) factors, such as substances, energy, noise, radiation, and activity or measures, including administrative measures, arrangements in the field of environment, politics, legislation, plans and programs, having influence or able to have influence on the environmental elements, covered in the "a" item, and analysis of costs and results and other economic analysis and assumptions, used during making decisions concerning the issues dealing with environment; (3) the state of health of people and the safety of people, living conditions of people, the state of cultural objects and buildings and constructions to the extent, that they are influenced or can be influenced by the state of the environmental elements or via these elements, factors, activity or measures, mentioned in (2).

The Orkhuss Convention came into force in November 2001, but neither Russia nor the United States has ratified the convention.

According to the Orkhuss Convention and the legislation of Russia, any physical or juridical person should be provided with free access to ecological information. Such access is an important guarantee, providing for valid decision-making concerning economic and other issues dealing with influence on the environment, and the basis of social and ecological public welfare.

Ecological informing or ecological propaganda means the spread of exhaustive authentic ecological information, ideas about the conservation of the environment and the rational use of natural resources with the purpose of forming public ecological awareness (raising the level of ecological culture). Taking into consideration that in the twenty-first century information becomes the determinative factor of human development and global activity, and the looming biosphere switches to the state of noosphere, the aggregate of the information structures, systems and processes in science and society is described by the infosphere concept.

The essential and rapidly developing component of the information supply in the environmental conservation field is the Internet. A stellar example of ecological information is the *Red Data Book of the Russian Federation* (RDBRF), a state document established for documenting rare and endangered species of animals, plants, and fungi, and some local subspecies that exist within the territory of the Russian Federation and its continental shelf and marine economic zone.

N. G. Rybalsky, V. V. Snakin

Ecological Law, International (IEL) (International Environmental Law): IEL is a branch of international law representing the aggregate of norms and principles of international law that regulate the activity of its subjects in the direction of preventing and removing damage to the environment from different sources and in the direction of the rational management of natural resources. The object of IEL concerns the conservation and rational management of the environment for the sake of the present and future human generations.

IEL has been recognized since the nineteenth century and has passed through several stages in its development: (1) 1839–1948, which is linked to the first attempts of "civilized" states to solve regional and local ecological problems; (2) 1948–1972, which is connected with the start of the UN activity; (3) 1972–the present, which marks the holding of global international conferences concerning these questions.

The sources of IEL are the norms of the international ecological agreements and international practice. It has not been codified. In the source system, the norms of regional international agreements are the dominant ones. The most important sources are such statements, as the Biological Diversity Convention (1992), Climate Fluctuation Frame Convention (1992), Ozone Layer Conservation Convention (1985), and Migrating Species of Wild Animals Conservation Convention (1970).

IEL development and functioning is built on specific fundamental regulations that are kind of juridical axioms in the relatively mobile substance of international law. It has two types of basic principles, fundamental principles of international law, and those specific to IEL.

The fundamental principles of international law include the principles stated in the UN Charter, UN Declaration of Principles (1970), the Helsinki Summit Final Page (1975), and the principles worked out by international legal practice. These are basically the fundamental principles of international law: sovereign equality, the non-usage of force and threats, the inviolability of state boundaries, territorial integrity of the states, peaceful settlement of arguments, non-interference into internal affairs, respect for human rights and fundamental freedoms, national self-determination, collaboration, and the conscientious executing of international legal liabilities.

Specific international ecological law principles are a developing category. These principles have not been fixed in any full-codified form yet; they are scattered over several international legal statements having a mandatory and recommendatory character. Such diversity causes some uncertainty in the positions of international lawyers concerning the question about the number of IEL principles.

Usually the following IEL principles are emphasized: (1) The environment is a common concern of humankind. (2) The environment beyond the frontiers is the common property of humankind. (3) Everyone has the freedom to conduct research and use the environment and its components. (4) We are responsible for the rational management of nature. (5) We hold an ideal of international legal collaboration in research and use of the environment; the interdependency of environmental conservation, peace, and security development, ensuring of human rights and fundamental freedoms. (6) A careful approach toward the environment is ideal. (7) Development is a right. (8) We should prevent harm and environmental pollution. (9) State liability is recog-

158 *Global Studies Encyclopedic Dictionary*

nized. (10) Immunity from the jurisdiction of international or foreign judicial organs is rejected.

International legal regulation of environmental conservation varies from component to component of the environment: conservation of waters, air, soils, forests, flora, fauna, and so on. Consequently, different international legal institutions within the IEL sphere exist: international legal air conservation, international legal animal conservation, and others.

P. A. Kalinichenko

Ecological Medicine is an area of medicine that deals with all aspects of the influence that the environment surrounding human beings has upon their health. Particular attention is paid to factors that directly result in deterioration of the environment. Ecological medicine became an independent branch in June 1986 (at a conference in Cleveland), and includes the following divisions: human biology, medicine (hygiene, toxicology, including genetic toxicology, and epidemiology), chemistry, physics, sociology, and technology of various types of production. Ecological medicine detects factors that influence human health (and studies the dose:effect ratio), studies cellular and molecular mechanisms of their action, develops theory, and implements practices of sanitary control over the purity of the environment. It controls the use of drugs, chemicals in industry and in domestic conditions, the use of construction materials, and pesticides. It studies the impact of poisonous industrial and agricultural waste, reasons and mechanisms of formation of congenital anatomic and physiological abnormalities, formation of cancer (ecological oncology), and immunological and pulmonary diseases. It develops methods of their treatment, controls the purity of the living environment (in centers of population, living quarters, and production facilities), and so on.

N. A. Agadzhanyan

Ecological Monitoring refers to the inspection of environment preservation; a system of measures established to prevent, reveal, and suppress violations of environmental regulations, to control the maintenance of the requirements designated in environmental preservation standards and normative documents by the subjects of economic and other activities. In accordance with the possible level on which they are conducted, we distinguish state, manufacturing, municipal, and public ecological monitoring.

The general purpose of ecological monitoring, or environmental inspection, is to ensure the maintenance of existing environmental and resource-saving regulations, requirements, and standards at every stage of manufacturing, constructing, or any other human activity connected with direct or oblique environment or its components' modification (including humanity). Ecological monitoring ought to be multilateral, meaning that it should take into consideration every human activity capable of influencing the environment.

The main ecological monitoring tasks embrace the construction of an information data base on the environmental condition and its modifications, collection of the required and sufficient information according to the criteria of completeness, accuracy and authenticity; revelation of harmful effects, or precautions for abnormal ecological damage.

The manufacturing inspection service officials may accomplish environmental monitoring in case it does not require special monitoring or analytical equipment. The responsibilities of such services usually include identification of the ecological modification attributes, efficient culprit exposure, and in some cases the authority to bring in the inspecting ecologists in order to implement instrumental survey, to assess the damage rate and to assign the appropriate sanctions.

The objects of ecological monitoring include harmful technogeneous impacts to the environment, on the one hand, and the environment itself prone or not prone to these impacts (background monitoring), on the other hand. The gained results are then compared to the standards.

The ecological monitoring parameters and indexes and the environment preservation standards are established by the state's system for environmental preservation; recently, more and more attention is also being paid to the regulations of the International Standardization Organization (ISO). Thus, a range of ecological monitoring regulations is determined by the state legislation.

The fundamental criteria of ecological monitoring are information precision and authenticity, which could adequately determine the efficacy of the solutions. Due to this circumstance, an actively developing applicable area of ecometry becomes highly valued.

One of the following solutions may be accepted. If the situation is assessed as landscape degradation or destruction, the reconstructing and stabilizing tasks are to be conducted while taking into account the predictable speed of harmful modifications; if the situation is steady and sufficient and the landscape ecological balance is preserved, there is no need for special operations.

The diversity and complexity of the ecological monitoring process and ecometry principles pose a many problems for specialists. The real occasions and circumstances accompanying the ecological inspection make it essentially inconsistent due to the ambiguity of the monitoring process and of the environment itself. The other reasons for disorder embrace the formality of regulations and numerical uncertainty, absence of a concept of environmental standardization of ecological balance and maintenance stipulations, the lack of inter-consistency in the regulation of information and monitoring processes, or the inadequacy of solutions accepted due to the qualitative and quantitative assessments of the current environmental state. These circumstances generate occasions when the same ecological situation, assessed by different monitoring services, obtains diametrically opposite evaluations on the subject of its usefulness or harmfulness for the further environment evolution.

The monitoring of technological items is continuous in the environmental monitoring services of the various ministries and departments. In spite of their methodological disconnection and lack of coordination, a considerable stock was assembled that contains exemplary facilities for specific and universal monitoring and analytical techniques and for analytical, controlling, assessing, and processing technologies. Thus, the question of technological support was turned to the point of the optimal monitoring facilities' complex selection from the list of those serially produced by various manufactures and departments on the standard base of ecological inspection.

160   *Global Studies Encyclopedic Dictionary*

The inspection results constitute the database of the environmental preservation service, where the information is collected, processed, preserved, and analyzed. The information database of the service forms, in its turn, the ground for environmental preservation management and resource-saving politics. The monitoring self-descriptiveness primarily depends on the technical equipment level; therefore, in order to establish the gadget stock it is necessary to keep in mind the complete monitoring standard complex which includes the parameters of volume, of periodicity, of the required accuracy and authenticity, and of plenitude.

The ecological manufacture monitoring service and the environmental preservation departments majoring in one or two ecological inspection trends have different goals. The ecological manufacture monitoring service has to have a complete set of the necessary technical facilities in order to monitor the basic environment pollution parameters.

I. I. Mazour

Ecological Optimism: (Latin *optimus*, "best") Ecological optimism is a part of the optimistic philosophical tradition. It accepts the concept of progress and recognizes progressive stages in evolution; it reflects a positive side of the complex and difficult relationship between man and nature, stating that a harmonious relationship with nature is possible, and it confirms that there always is a way out of any hopeless, desperate situation. It opposes the tradition of ecological pessimism prevailing in the environmental movement of the twentieth century. It is based on the long-established tradition of ecological wisdom and the ideas of Pierre Teilhard de Chardin and Vladimir Vernadsky regarding the transition of the biosphere into the noosphere. Ecological optimism fits the optimistic perspective in philosophy (Aristotle, Epicures, and Leibniz) and reflects a conventional positive attitude pre-existing in social consciousness in the form of myths, legends, fairy-tales, maxims, and proverbs where Good always overcomes the Evil. Through cultures and eras, ecological optimism manifested itself in different ways, growing from utopos (seeking a return to Eden) to topos (seeking an achievable state of harmony with nature). The notion of ecological optimism correlates with Native American beliefs, with Shamanism, Taoism and Buddhism. Recently it is associated, among others, with Gaia hypothesis of Lovelock and Margulis, with practical environmentalism, and with the work of natural scientists.

Ecological optimism embraces such concepts as progress and progressive evolution of the complex self-organizing systems, and the possibility of persistent human advancement. It incorporates intellectual and emotional components. Ecological optimism represents a positive side in the complexity of human/nature relationships, and it views environmental crises as valid transitional steps in human/nature co-evolution. Ecological optimism suggests that the harmonious coexistence of nature and humanity is possible as a result of joint effort, work, and intellect of awakened and responsible humankind. As a method of practical orientation and as an emotional tuning, ecological optimism is an important component of ecological culture. By anticipating a better condition, ecological optimism provides hope and forms a constructive ideal; it provokes, motivates, and initiates practical steps; it stimulates ecological perception and intensifies involvement; it re-programs the individual and collective mentality and frees creative energy for immediate actions toward unconventional solutions of environmental problems. Ecological optimism can be attained by intellectual training through cultivating an optimistic set of cognitive skills and re-programming long-standing mental habits.

Naïve ecological optimism is based on a lack of knowledge and blind faith; it reflects interrelations of the micro- and macro-cosmos sustained by traditions and taboos. Utopian ecological optimism is based on an unreal and mistaken prognosis and wrong interpretation of scientific facts. It denies the real danger of environmental crisis and redirects intellectual and practical resources from the reality of existing dangers to idealistic dreaming. Mature ecological optimism derives from scientific data, understands accurately the complexity of the ecological situation, and proposes appropriate solutions for environmental problems.

I. Y. Tuuli

Ecological Portrait of Human Beings (Ecoportrait): An ecoportrait is the aggregate of genetically conditioned qualities and inherited features characterizing a specific adaptation of individuals to a specific set of particular factors of their habitat. Characteristics of the ecological portrait are retained in migration to new regions, and determine the functional condition and the general efficiency of the organism in a new habitat. This factor must be taken into consideration in genetic and demographic analysis of migration and in development of a settling evolution concept. The final purpose of individual selection of people for work in various natural climatic zones, so long as the health of those people needs to be conserved, should be the determination of the ecoportrait of each individual and, on this basis, the detection of the region where the endogenous, internal environment of an organism will be in harmony with the exogenous, external ecological environment. Significant phenotypic flexibility related to some degree to the morphofunctional adaptation is most clearly seen in regions with extreme environmental conditions (such as Arctic, equatorial regions, and mountains). Biological adaptation of a human being is specific, because it involves, apart from the biological, also social functions and is performed in conditions of the growing importance of the social factor.

N. A. Agadzhanyan

Ecological Problems, Global: Global ecological problems are a complex of transformations of the planetary natural environment caused by rapid population growth, depletion of natural resources, and increasing pollution. The following factors are considered among the most important ecological problems:

Global climate changes: part of global (general) changes of the natural environment caused by changes in the balance of atmospheric heat and circulation of ocean water and fresh water. The greenhouse effect is believed to be very important owing to the registered relationship between carbon dioxide content and average planetary temperature that in recent years tends to grow. The chronology of paleoclimate shows substantial variability of climatic system independent of human influence with fluctuations having exceeded possible climate changes due to greenhouse effect of gases emitted to atmosphere

162 *Global Studies Encyclopedic Dictionary*

over the last two centuries on account of human activities. One of the aspects of global changes, the global warming, is the increase of average atmospheric temperature caused both by anthropogenic (greenhouse effect) and natural factors.

According to the data of the Hydrometeorological Center of the Russian Federation, in 1997, air temperature over the entire territory of the Northern hemisphere exceeded the average annual temperature by 1 °C. More significant temperature increases are registered in a near surface layer of the world's oceans. According to estimates of some scholars, the ocean level may rise over 4 m due to global warming, thermal expansion of waters, and partial deglaciation, and substantial parts of such cities as St. Petersburg, Amsterdam, Shanghai, and New Orleans may find themselves under water. Not only many island countries, but also the greater part of Bangladesh and Senegal will disappear. The warming of oceans, which is a potential threat to coral reefs, materially affects the environment of shelf areas.

The other aspect of the problem in question is aridization of land (Latin *aridus*, "dry"), or xerotization of soil (Greek *xeros*, "dry") occurring with the growth of average temperature, which is a complex and multifaceted set of processes leading to the decreasing of soil moistening and the ensuing reduction of the productivity of ecosystems. Aridization of land occurs due to both natural (cyclic changes of climate) and anthropogenic (depletion of subterranean waters, soil erosion, and dust storms) causes. One of the factors of aridization is the afore-mentioned increase of the average atmospheric temperature. The consequence of aridization is desertification and further aridization of the arid territories, which has become an object of special care of the UN Convention on Aridization.

Emission of greenhouse gases (carbon dioxide, water vapor, nitrogen oxide and sulfur, and chlorofluorocarbon or Freon) is presumably one of the main causes of global warming, since they arrest the heat radiation of the Earth creating the so-called greenhouse effect. A steady growth of greenhouse gases concentration in the atmosphere has been recorded in recent history. The contribution of different countries to emitting greenhouse gases is illustrated by data on CO_2 emission accounting for 80–90 percent of the total amount of greenhouse gases: United States, 22 percent; Russia, 11 percent; China, 10 percent; Germany, 4.8 percent; Japan, 4.4 percent; other countries, 48.8 percent. In 1990, the share of Russia in the global emission of gases was as follows: methane, 7.2 percent; nitrous oxide, 10 percent. This problem received special treatment in the Kyoto Protocol.

The use of renewable natural resources exceeds the ability of nature to recover. Such a disparity between rates of mining (production) of natural resources and the size of its reserves both in consequence of economic reasons (mining costs drawing near to the cost of the product obtained) and ecological considerations (disruption of productivity or over-production in excess of the rate of renewability of reserves) leads to depletion of water. A steady decrease of reserves and a deterioration of the quality of the surface and subterranean waters and exhaustion of soil; depletion of nutrients in the soil as a result of abuse or in the course of natural evolution leading to diminished fertility of soil.

Depletion of natural resources along with overall pollution of elements of the biosphere with substances alien to nature (xenobiotics), and concentra-

East-West – Evolutionism, Global 163

tion of ordinary substances over and in excess of maximum permissible concentration (MPC) leads to degradation of natural systems (top-soil, natural waters, landscapes). Degradation of soils, which is a stable deterioration of soil properties as environment for biota, and diminishing of fertility due to the impact of natural and anthropogenic factors can be divided into physical (deterioration of hydrophysical properties of soil, depletion of nutrient reserves, resalinization, realkalization, xenobiotic pollution) and biological (loss of species diversity, loss of the balance between various species of soil mesofauna and microorganisms, pollution of soil with pathogenic and alien microorganisms, and deterioration of sanitary and epidemiological parameters). Agricultural activities, overgrazing, and forest clearing can cause degradation of soil. It should be noted that undesirable acidification of soils results from the use of physiologically acidic mineral fertilizers and acid rains, containing considerable amounts of acidifying sulfur and nitric oxides, and eutrophication (blooming of water) on account of the excess supply of fertilizers (especially excess nitrogen and phosphorus) being dumped into water.

Loss of biodiversity includes the irreversible loss of many species (especially of large animals) as a consequence of excessive extermination (including hunting), and the destruction of natural ecotopes of animals and plants. An important means of registering and developing measures to protect rare and endangered species are international, national, and regional versons of the *Red Data Book of the Russian Federation.*

Clearing of forests, which are figuratively speaking the "lungs of the planet," has a tremendous negative impact on the planetary atmospheric balance. Having an important function in regulating climate and protecting soil and water, the forest cover of the Earth is one of the factors of biospheric stability needing conservation and renewal. Forests once used to occupy over half of the land territory of the Earth; at present they cover a little over a quarter of Earth's surface. According to some estimates, one hectare of forest absorbs around ten tons of carbon dioxide annually. The reason for forest clearing is relentless exploitation of timber and other resources, and diseases and subsequent death of forest vegetation because of environment pollution. Forest diseases, related to the development of big industry, have been registered in Europe since the middle of the twentieth century. Their main causes are acid precipitations (due to emission of sulfur and nitric oxide) and the impact of ozone. To prevent it, reduction of emissions of pollutants harmful for plants and soil is needed.

Depletion of the ozone layer has recently become a source of special concern for the international community. The ozone layer (ozonesphere), a stratospheric layer (at the altitude of ten to fifteen km) arrests the greater part of cosmic radiation (including ultraviolet radiation), which is fatal. The destruction of this layer observed recently (appearance of so-called ozone holes) can bring about significant changes in the biosphere and cause a higher level of ultra-violet radiation that is harmful for organisms. In the last ten years the average annual concentration of ozone in middle and high latitudes at the altitude of approximately 20 km has diminished by about 10 percent. The "ozone holes" are believed to be of anthropogenic origin due to the interaction of ozone with so-called ozone-destroying substances (e.g., Freon).

164 *Global Studies Encyclopedic Dictionary*

Overpopulation is the state of the ecosystem when the number of individuals of a species exceeds the capacity of the environment. It is often accompanied by intensive death of individuals (due to self-regulation of the population number) leading to relative stabilization of their number. Human population is regulated in a different way. Traditional methods like wars, ecological expansion, unequal wealth distribution, and hunger should give way to new humane methods.

V. V. Snakin

Ecometry is a complex engineering discipline of a physical and chemical nature embracing measurements for determining and evaluating the parameters characterizing the environment in all the diversity of its constituent elements and complexes from the viewpoint of their benefit or harmfulness for the functioning of biota. From the standpoint of environmental control, ecometry can be regarded as an information circuit of engineering ecology; as a branch of science and engineering, the measuring status, which determines the level of the objectivity of solutions formulated within the framework of the tasks of engineering ecology.

Ecometry is based on natural and artificial ways and means of obtaining, presenting, and using information on the natural environment. The issues of the optimization of measurement according to such parameters as precision, authenticity, and adequacy are in each specific case solved in the direct connection with a given ecological situation. For instance, when analyzing the qualitative and quantitative composition of atmospheric air, the required and adequate precision, authenticity and information capacity of a measurement cycle can be calculated based on the physical characteristics of the controlled object, like the size of particles polluting the air, with regard to climatic factors. All controlled parameters and characteristics will acquire clearly expressed quantitative criteria. The simplest case of ecological control from the viewpoint of the methodology of standardization is thus characterized. An alternative case is that of "landscape control" which becomes more and more widespread and involves a ground, airborne, or space photographic survey. Pictures of one and the same territory taken at different times are processed or deciphered, and the conclusions about the ecological vector of that territory are made upon the registered landscape changes. The changes thus registered are qualified as "beneficial" or "harmful" for a landscape.

Modern ecometry is based on the following logical and philosophical interpretations of opportunity: unconditional opportunity determined by univocal premises; accidental opportunity presupposing the symmetry of the initial premises; bilateral ("ciliary") opportunity or "opportunity of opportunity." The first interpretation of opportunity is implemented under classical (a priori) theory, based on the apparatus of mathematical analysis. Such a case is the most clear and simple, but it is extremely rare. The second interpretation is implemented under modern probability (relative frequency) theory and mathematical statistics. This method is used most often in the interpretation of ecological control data and prognostication, although statistical models seldom meet authenticity requirements.

The third interpretation has not been theoretically defined up to now yet; however, the bilateral opportunity corresponds most fully to the real situation with the environment, natural and technical systems, situations of ecological control, and the logic of their engineering and ecological study. The use of the "ciliary opportunity" concept for the purpose of long-term evaluation of ecological data array requires relevant fundamental studies and is another factor making ecometry an independent subject within the framework of science and industry. Ecometry can be regarded as a branch of science, the measurement status of which makes it responsible for the objectivity of engineering ecology solutions. Ecometry rests on natural and artificial ways and means of obtaining, presenting, and using information concerning elements of natural and technogenic geosystems.

I. I. Mazour

Economic Crisis, World: The world economic crisis is comprised of the disorganization and disintegration of economic ties between various countries and nations, accompanied by large-scale negative consequences, such as the stagnation and destruction of specific industrial spheres, the decline of business activity, the degradation of the most national economies, the growth of unemployment, political instability, and demoralization of the population. The industrial crisis of 1825 that took place in many European countries was a forerunner of such a crisis. However, the first true world economic crisis occurred in 1857. It overcame the economies of the most advanced countries of Europe and America that by that time had in many aspects become part of the emerging world economic system. Between 1873 and 1895, in Europe and overseas a second, more profound economic crisis occurred and bankrupted many enterprises in Germany, the United States, Great Britain and Austria.

The largest economic crisis in human history took place in 1920s and 1930s. The economic upsurge and financial movement in Europe and America that had occurred after the end of World War I were replaced in 1927, with a decline of manufacturing. In 1929, it ended with the Great Depression that developed in the US, which produced a world economic crisis. This crisis revealed those tight connections between national economies, having influenced nearly all capitalist countries and provoking an unprecedented level of unemployment, pauperization and, as a result, social unrest.

Upon the advent of modern globalization, conditions some segments of world economic system periodically find themselves in a situation of crisis. For example, the first world oil crisis that burst out in 1973 demonstrated the increasing dependence of the world economy on fuels and the growth of world oil prices engendered in most of developed countries of the world an abrupt industrial decline and an increase in unemployment and inflation. Instability of foreign currency markets, bankruptcy of the largest commercial banks, "dirty money laundering," and the criminalization of economic activities all imply real danger of large-scale financial shocks that make the coming of the new world economic crisis a very topical problem for the world community.

A. N. Chumakov

Economic Relations, International (IER): International Economic Relations (IER) is an economic discipline that explores forms, relations, and cor-

166   *Global Studies Encyclopedic Dictionary*

relations occurring along with the functioning of the world economy, concerning the economy as a whole and its separate links. These relations and interconnections, manifested in various forms as the conditions of international labor are more rapidly divided (especially increasing the dynamism of intensification and internationalization processes), lead to the acceleration of the world economy movement as a unified global system.

IER is based, first, on the methodological principles of fundamental political and economical statements (such as the base of economical theory), and, second, on the theoretical and methodological base of the discipline of "world economy." The concept of international economic relations is close to the concept of world economy, as it is based on the relations, connections, and correlations occurring in the functioning system in the world economy.

IER Key Problems: Conventionally, we can select the following key problems of IER: analysis of trade and economic flows, assistance and regulation in this area; assistance in determination and implementation of crediting world states and international organizations; elaboration, application, and regulation of customs rules as important components of the foreign economic complex of all the world's states; regulation of leading international monetary and financial, trade and other supranational institutions' and organizations' activities; working-out of the international economic (commercial) legal rules regulating IER, and of fundamental "standards," established and recommended by UN institutions (such as the UNCTAD, WTO, ILO, IMF, World Bank, and Basle Bank) within the sphere of UN international economic activity.

International economic relations is a system of economical, scientific and technological, industrial and administrative, commercial, monetary and financial, credit and currency relations and connections among various states' national economies; international collaboration based on the international division of labor; and international organizations' trade, business services, labor force and financial funds' regulation activities. Thus, examination of the two correlated trends, international economic collaboration and international economic regulation, is emphasized in the concept of international economic relations. As a research subject, IER appears to be very dynamic. It reveals how the connections and interconnections are arranged in the world economy. It analyzes instruments, establishments, and institutions forming the basis of these connections, and the methodology (diplomacy-negotiations, agreements, unilateral and multilateral activities of governmental and international organizations authorized by the International Law powers).

The Subject of International Economy: The study of international industry, as shaped by the modern multinational corporations and tending to slip out of the national economic sphere, is of a primary importance. This is the field of knowledge currently regarded as the "world economy," though its modern interpretation seems excessively broad (that is, the international "economy"). The international economy as a concept and entity reflects the aggregated totality of transnational economic agents' economies and thus forms just a part of the world economy. The dynamism of this particular part of the world economy and world finances-international economy assists globalization processes of the whole contemporary worldwide economy.

External Economic Links as a Subject of Research: This category is often used in an unwarrantedly broad manner in the scientific literature, as a synonym of "international economic relations," that tends to be substantially incorrect and leads to further methodological confusion. The character of the very concept of "external economic links" (or "external economic activity") is derived from the subject of these links: the state (or its official body; for example, the Ministry or Department), company (bank), or private person. This definition corresponds to another definition, which is the concept of "the foreign economic activity of the state," or the "external economic activity of an enterprise (company, organization)." The system of state external economic links includes various types of international collaboration of this state and of its managing organizations (or individuals) beyond its boundaries. Thus, it is a complex of relations and connections between a state and other states and international financial and economic organizations.

World Market Equilibrium Conditions: The main condition of world economic development is a growing world market. Hence, world market stability, steadiness, and growth, that is, the attainment and maintenance of world market equilibrium, are its strategic quality providing stable development of the modern world economy. The formula of such world market development is simple: adjustment of world demand and world output. At this point, the present formula does not differ from the national market equilibrium formula, but its implementation in the scale of the global world market constitutes a huge difficulty. Well known to the former socialist countries, when the conditions of world output lag behind world demand, a destructive deficit category (market demand exceeding the supply) appears on the world economic stage, the market equilibrium is disturbed even on specialized world markets or the surplus (such as oil and metals) shows up.

The aggregated world supply is the amount of goods that the producers of different countries are willing to offer on the consumer world market at the existing price level. It is provided for by the world countries' home production and by the import of mutual goods and services.

Thus, to guarantee world market equilibrium, the aggregated world demand must equal to the aggregated world supply, arranging a stable interdependency between these two gigantic volumes of goods and services. From the viewpoint of individual countries' interests and routines such equilibrium on the world market is considerably rare; the common rule is deviation from one side to another of the balance depending on the expected advantages for the major economic agents dominating the world's specialized markets.

Worldwide Regulation: There is no practical or theoretical system of regulating world market processes that considers the actual significance of the separate WTO, IMF, or IBRD measures and suggestions. Depending on the largest world states' internal and foreign policies or home situations, unilateral actions undermining the delicate market balance are frequently practiced. The world community (its largest and most influential partners) makes attempts to affect the world market by indirect means, at the world level (by establishing GATT-WTO trade rules), at regional levels and through integration institutions (EU, NAFTA, ASEAN, CIS and the others), and during the states' bilateral trade negotiations.

168   *Global Studies Encyclopedic Dictionary*

Removal of the artificial protective and customs barriers on the world market would, apparently, have led to the growth of the market's significance as a balanced and self-regulated system. But such an evolution is improbable insofar as the contemporary world market stays extremely politicized; powerful states striving to solve "their" problems exercise unrestrained expansion on the world market, thus stimulating and urging on national agents (multinational corporations and financial flows primarily), freezing the world market balance and increasing the disproportion of supply and demand volumes and of price policy.

World Economic Policy: The concept of "economic policy" is applicable to national states and authorities and to the states integrated into a union, for instance, EU states. At the less extent it concerns the leading world countries ("the Seven") joint policy, although annual congresses of their leaders negotiating solely world economic problems and their solutions have influenced international economic relations in specific periods.

Every time the ways of implementing these basic tasks of national economic policy are discussed in international forums, disagreements appear in the opinions of the various countries' leaders. Many disputants look for the problems' culprits outside their states and, what is especially worrying, correlate the successes or failures of their governments with international assistance or interference, without realizing that the finances for which they strive are not treasures found in ancient burials, but in the money of some countries' taxpayers, earned by their productive labor.

Being distracted by the complexity of national economic problems, the majority of the authorities "forget" about the accumulating global menaces: the danger of the possible collapse of the international and financial system, the irregularity of regional economic growth, the threat of dispersion of the various types of weapons among different regions and countries, the intensification of local military conflicts, famine and diseases and ecological catastrophes. In such a complicated international situation, the role of the United Nations—the only universal world organization—is declining in discussions of social and economic world development and in peacemaking missions.

R. I. Khasbulatov

Economic Security is a system of measures to protect national interests in the economic sphere from external and internal threats that is based on developing the national economy. At the national level, solving the basic tasks of economic security is possible due to conducting long-term socioeconomic policy that secures dynamic economic development, stability of financial systems, and social harmony. The high pace of economic growth, low inflation, reduction of the degree of social differentiation of the population, and gaps in socioeconomic development between various regions are the main criteria of the success of such a policy. Eventually, the economic security measures are directed toward protecting the constitutional order, sovereignty, and territorial integrity of a country.

Under globalization the watershed between external and internal threats blurs more and more. Leaving domestic socioeconomic systemic problems unsolved contributes to sensitivity to global threats. For example, global fi-

East-West – Evolutionism, Global 169

nancial crises under the circumstance of unbalanced state finances threaten to destabilize the macroeconomic situation within a country. Increasing competition between the main world centers of economic influence for access to strategic raw resources and for expanding sales markets is a feature of the modern world. In the beginning of the twenty-first century the global threat of international terrorism has become of particular importance. Thus, to strengthen economic security measures, it is necessary to develop international cooperation in the sphere of combating "legalization" (laundering) of illegally acquired money that can be used for financing terrorism.

S. V. Stepashin

Economy, World: The world economy refers to an integral world economic system. International economic relations, first just a natural exchange, and later which adopted the form of commodity trade, date to the most ancient times. Many millennia had passed before the separated and sporadic economic links among different tribes and later among States developed into a stable international division of labor. As a result, its participants obtained the possibility of concentrating their labor resources and capital on the production of those commodities and services that could be created, in the conditions of a particular country, with the maximum economic efficiency, others being obtained by means of international exchange. The national economies of the countries participating in this division of labor began to complement each other to some degree. Such interaction, mediated by money circulation and various credit and financial links, united step by step the national economies into a more or less integral economic system-the world economy.

Like any complex system, the world economy has new properties that do not pertain to the constituent national economies. These specific characteristics include: (1) the appearance of world markets for commodities and services that have a special significance for all countries (such as markets of oil, grain, and air traffic; (2) the formation of world prices on commodities and services that determine, to a great extent, the policies of national producers; (3) the development of the world market of credit capital, which determines not only its international price, but also the preferable directions of its flow; (4) the appearance of the resonance effect, i.e., transfer of economic growth or depression from one country to other ones closely connected with it.

The establishment of the world economy dates to the epoch of industrial revolution and developing capitalism, i.e., to the eighteenth-nineteenth century. Originally it comprised a relatively small number of countries involved in active trade and other interactions with Britain and other pioneers of industrialization, that is, Western and Eastern Europe, North America, and a small number in Asia, Latin America, and Africa. Most of the world community had not yet passed the stage of natural or semi-natural agrarian production and crafting, and hardly participated, in the above sense, in the international division of labor.

However, along with the development of capitalist commodity relations in such peripheral regions, on one hand, and with their involvement, by the industrial countries, into international trade and financial relations, on the other, the limits of the world economy gradually expanded. This process was

170 *Global Studies Encyclopedic Dictionary*

either voluntary or compulsory, in the form of colonization of agrarian countries by the industrial ones. Eventually, increasingly more countries joined the active international division of labor, and the world economy continued to grow until it comprised, around the middle of the twentieth century, the entire globe, thus reaching its spatial limits. In this sense, it had become global.

On the stage of globalization, the world economy acquires several qualitatively new peculiarities making it different from the preceding stages of the internationalization of economic life. They are caused by the extent of economic, technological, legal, and informational interdependence of the countries reaching the stage when three new phenomena become possible.

First, the world economy, being previously only a more or less integral economic system, becomes a united global economic organism cemented not just with the international division of labor, but with gigantic and even global production and marketing structures, a global financial system, and the planetary information network. The world economic space becomes a single field for the commercial games of big business, when the geography of location of productive forces and the branch structure of investments, production, and marketing are defined by the economic subjects taking into account the state of the global market, while the economic booms and recessions acquire global scale. By the end of the twentieth century, the globalization of the economy became an overwhelming self-developing process, a kind of chain reaction, which breaks all obstacles, constantly amplifying and accelerating.

Beginning from the middle of the twentieth century, such economic organisms began to form, on a small scale, in the most developed regions of the world. These were the highly integrated blocs of countries within the framework of customs or economic unions. The most developed is the European Union (EU), whose member countries carry out about two-thirds of their trade operations inside the region, have introduced, at the beginning of the twenty-first century, a single currency, invest the crucial part of foreign investment into the each other's economy, and intensively interact in the field of industrial and technological cooperation. In the economic, political, and legal aspect, they have turned, over the last half-century, into a rather integral economic organism that, for a long time, has been a single subject of international trade and political relations.

Other regional economic blocs, such as NAFTA, MERCOSUR, and ASEAN, are rather inferior to EU in the extent of a real integration, though they are moving in the same direction. As for the tens of other "free trade zones" and "customs unions" that try to follow the EU example, they just remain on paper. The practical experience accumulated over the last half century clearly demonstrates that turning national economies into a solid regional economic organism proves possible only at a very high level of industrialization of the member countries and an equally high development level of democracy and a legal state system. It is only then that growing national societies into a larger organism becomes not only objectively necessary, but also sufficiently safe for the interests of these societies in the economic, political, cultural, and other aspects.

This law of integration, approved at the regional level, is valid also on a global scale. The process of forming a united self-regulating economic organism, born in the highly developed core of the world community, is more and more expanding over the planet along with the technological, economic, social, and political maturing of the countries on the world's periphery. This process will take many decades, possibly extending beyond the limits of the twenty-first century.

New qualities of the world economy in the epoch of globalization are not only favorable, but also involve serious troubles for humanity. Economic globalization opens the widest prospects of truly global progress of industry, science, culture, and growing welfare of people all over the world. The entering of the world economy into the globalization stage in the last two to three decades was accompanied not only by a noticeable acceleration of growth of the world income as a whole, but also by a considerable shift to its more uniform distribution among the various groups of countries. In the new conditions, the less developed countries got the possibility of overcoming their technological, economic, social, and cultural backwardness much sooner than was possible before.

However, the globalization of the economy, as any boon, has its dark side. Thus, an accelerated economic growth is accompanied by the increasing consumption of fossilized hydrocarbon fuel and, correspondingly, by atmospheric pollution, which leads to dangerous climatic changes, hurricanes, floods, and other natural disasters. The globalization of financial markets created conditions for large-scale speculations capable of causing economic shocks on a regional or even global scale. Global information technologies facilitated organized crime, providing it with international range and new methods. The list of such dangerous consequences of economic globalization is not at all limited by these examples.

Thus, the main line of progress of the world economy in the twenty-first century will be inevitably accompanied by a sort of shadow-the parallel line of a permanent struggle against the negative consequences of globalization and of a constant search for such balance of positive and negative aspects of this process that would provide a stable positive result.

Y. V. Shishkov

Ecotoxicology is a sphere of ecology studying modern concepts of toxicity of elements and their compounds, specific biogeochemical features of behavior of toxicants in the environment, the mechanism of their diffusion and metabolism, and interrelations between the toxicity of elements. It prioritizes the localization of carcinogenic ions and a quantitative assessment of threshold effect of the toxicological influence affecting systems' "toxicant-environment" and the "toxicant-living organism."

The threshold concept assumes a high quality of environment and full safety for the person and any species if the environment contamination is below a specific level.

Objects of toxicological research are extremely various. These are waters, soils, pharmaceutical drugs, biological objects of animal origin, foodstuffs and drinks, pesticides, household chemical goods, vegetation, and waste products. Therefore the complex of applied tasks of ecotoxicology is very specific. The priorities list includes the following tasks:

172   *Global Studies Encyclopedic Dictionary*

(1) working out the modern methodology of ecotoxicological research, allowing it to carry out reliable assessment of the environmental quality in conditions of nature management and the complex influence of its basic components on living organisms;

(2) carrying out early diagnostics of changes in an organism, revealed before morphological, genetic, population and other changes;

(3) working out the applied bases of chemical-toxicological analysis of the priority contaminants including various methods for their detection, isolation, and quantitative definition in objects of the environment;

(4) creation of purposeful monitoring of toxicants, invoking those or other deflections in living organisms which facilitate finding a new approach to identification of the most active factor, as the specificity of the biochemical answer of an organism will facilitate tracking a route from consequence to cause, that is to find the corresponding toxic agent or a smaller group of agents distinguishing them from the general body of contaminants.

N. N. Roeva

Education is one of the major spheres of human activity. Developed countries invest 5–8 percent of their gross domestic product in education.

In the twentieth century, a serious attack on illiteracy was undertaken all over the world. The most important actions in this direction were carried out after the Second World War. Secondary education is now compulsory in every developed country, and higher education becomes a normal and frequent thing. Presently, over 20 percent of students in the developed countries continue their education in high schools. The literacy of the population has sharply increased. At the end of the twentieth century for the first time in the history of humankind a decrease began in the absolute number of illiterate people on Earth. At present, this number does not exceed 15 percent. Society today clearly understands that the sphere of education is the locus for the fundamentals of the development strategy of a country and of the whole world. Yet education in the modern world is considered to be in a crisis, which is evinced by its low quality; in the weak stability of the knowledge and skills of students; in the unsatisfactory state of students' morale and civic consciousness; in the decline of students' health while they are in school; and in the drug addiction and criminality widely spread among young people.

The great disparities in the educational levels in developed and developing countries are especially disturbing. Overall, developing countries allocate twenty-five times less to the education of one citizen than do developed countries. Very alarming is the "brain drain" which is common in almost all countries (but especially in developing countries) and which increases the differentiation among states in a larger scale.

It is possible to point out the following major directions of modern reforms in the sphere of education: fast development of systems of continuous education with wide use of all modern computer technologies first; democratization, fundamentalization, humanization, and humanitarianization of education; maintenance of high-level natural-science and mathematical and computer literacy of students; introduction of state standards for all educational levels; provision for an organic connection of the educational system to public institutions and mass-media, representing the major sources of informal education of citizens; internationalization of education. Education should become an integral part of life and an intrinsic need of every individual person.

V. I. Kuptsov

Educational Globalism is a complete modification of the order of knowledge distribution through educational institutions, which leads to the emergence of a society of learning. In the process of educational globalization the academic activity of every social and age group of the population becomes the basic instrument of social development. Educational globalization is stipulated mainly by the economic integration of various states, by the activities of multinational corporations, by the processes of the redistribution of financial flows and labor force migration. Educational globalization is directly connected with the increasing tension between the developed industrial countries and the Third World countries.

One line of educational globalism is the market orientation determined by investments of large corporations and international financial institutions into educational standardization and elaboration of teaching modules for employees meeting the requirements of technology. The other line of educational globalization stems from the program of general illiteracy eradication (in the beginning of the 1990s, there were more than 113 million children not attending schools, and almost 1 billion illiterate adults, mostly women). UNESCO regards gradual illiteracy eradication at the global level, including functional illiteracy, as the main humanitarian (humane) method of fighting chronic poverty and economic disparity and as an instrument of conflict prevention and HIV/AIDS eradication.

G. A. Klucharev

Education, Global Philosophy of: Against the historical background of the twentieth century we must critically analyze the educational philosophies and systems that define and control human cultures and civilizations.

Let us first look at the basic purpose of education. It should impart information; it should help the candidate to acquire necessary skills with which material production should increase qualitatively and quantitatively besides enhancing all creative potentials; and it should create definite attitudinal changes in the individual leading to better human relationships and collective achievements in all spheres of life.

(1) Experiential Education: In the emerging world context experience itself should be treated as a major educational component for the individual's total development and academic side of a person's performance and achievement. The experiential and technical aspects of a person's achievements must be given equal value and importance at all levels.

(2) Relationship Change: Another important change is bound to occur in the teacher-student relationship and traditional psychological hang-ups. A new sense of partnership and friendship should dominate the inter-personal and institutional relationships.

E. P. Menon

174 *Global Studies Encyclopedic Dictionary*

Education, Globalization of: Globalization of Education refers to a process of gradual adjustment of the educational system to the demands of the growing world market economy that transcends the borders of nation-states.

Globalization of education has contradictory consequences. The demand for education, especially higher education, increases. The number of students in the world has grown from thirteen million in 1960, to eighty-two million in 2000, and now surpasses 100 million. What is negative in mass higher education is the abrupt growth of paid education, the limited access to quality education for everyone, narrower specialization, and decreasing quality.

Globalization of education requires looking for the optimal balance between a strong state-oriented educational system and a strong system of high quality private education.

I. M. Iliinskii

Education, Philosophy of: The philosophy of education is the interdisciplinary study of the philosophic aspects of theoretical models and methods of research regarding processes of education and learning and training technologies. Philosophy of education also researches practical problems related to the challenges facing humankind.

Since ancient times issues of education and learning and training technologies were considered within the frameworks of specific philosophical traditions. In the second half of the twentieth century education obtained the status of a separate branch of knowledge, accepting Anglo-American traditional values in order to reveal the fundamental problems associated with education, seeking solutions for the refinement of academic programs and the training of teachers for colleges and universities.

Contemporary philosophy of education is characterized by a wide spectrum of approaches and conceptual schemes. Such directions in educational philosophy are informed by analytic, psychological, hermeneutic, axiological, existential, "culturologist," and postmodernist approaches. This diversity of perspectives facilitates the search for new ideas in the province of education, reviewing traditional views and opinions in response to the rapid transformation of the world.

In the 1990s, the philosophy of education gained more widespread recognition. The achievements generated by information and computer technologies, the challenges of globalization, and the dynamics of the transformations and contradictions associated with them lead to significant changes in the concept of education.

L. Apsite, Z. Rubene

Education in the Postmodern Age: In the framework of globalization, education gets new social functions. For instance, in the conditions of the global market, education becomes an economic force. The global world is characterized by the processes of its people and culture migration. Cultural templates and traditional values collapse; people's identity is lost; and a log jam grows between ethnic groups. Education as the means for social unity can become a force capable of overcoming these contradictions, of leading people to mutual understanding, and of smoothing cultural distinctions. The aging of the population demands an expansion of the framework of higher education and involvement of adults in acquisition of higher education.

I. Nekvapilova

Education, Transnational (Trans-Boundary Education): Transnational education refers to types of higher education programs, courses, or educational services, including distance education, where trainees do not reside in the country of eventual employment or where the educational institution awarding certificates, degrees, and qualifications are situated. Transnational educational programs can either belong to the educational system of a foreign country or be provided regardless of any national educational system. Because of internationalization, transnational education represents a transition to globalized educational markets.

Education crosses national boundaries because the higher education institutions in the most developed countries use different methods to promote their educational programs abroad both in the traditional form and through the full-scale usage of information technology and network engineering. The UNESCO European Center for Higher Education, the Council of Europe, and the European Commission are now working on the resolution of international legal problems involving transnational education development.

V. A. Galichin

E-Governance, the Case of India: As per the World Bank, "E-Government refers to the use by government agencies of information technologies (such as Wide Area Networks, the Internet, and mobile computing) that have the ability to transform relations with citizens, businesses, and other arms of government. These technologies can serve a variety of different ends: better delivery of government services to citizens, improved interactions with business and industry, citizen empowerment through access to information, or more efficient government management. The resulting benefits can be less corruption, increased transparency, greater convenience, revenue growth, or cost reductions."

Governments around the world try to achieve these goals. When governments tend to become smaller and closer to citizens, it is necessary to rely more heavily on the use of technology. In a survey on Government and the Internet, *The Economist* has pointed out that the same continuous availability and convenience, fast delivery, customer focus and personalization the norm in the public sector (as offered by the best web retailers and service providers) would not just make life easier, but would fundamentally change the way that people view government itself.

E-Governance Experiments by Local Governments in Kerala: A few local governments in India have taken small but important steps in this direction. Notably, the Five Village Panchayats in the Thiruvananthapuram district of Kerala, namely, Amboori, Vilavoorkal, Kattakada, Madavur, and Vellanad have been equipped with software for the implementation of a decentralized plan. Similarly, the Thiruvananthapuram Corporation has made efforts to introduce e-governance with an aim to decentralize administration and improve the services. The computerized system is expected to help in monitoring plan projects, administration of many welfare schemes, and management

176 *Global Studies Encyclopedic Dictionary*

of revenue, finance, and accounts systems among many other things. Packages for registration and issue of certificates have also been acquired.

The Ernakulam district panchayat has also taken up a pilot project in e-governance in three of the village panchayats in the district. It has been pointed out that in the existing system each government department collected data relevant to its functions. Studies show that in such a compartmentalized collection of data, integration of data collected by various departments became difficult. In the new system, the local governments would be custodians of a computer-aided data base server of all primary data relevant to any government office in the state. Thus, a database for any micro level planning activity would be available in the local government office itself. A scientific society called Electronic Industrialization Infrastructure Development (EIID) has been created by the Ernakulam district panchayat in this regard. Considering the problems faced by the researchers in the data systems used and in the generation, storing and collection of data at the local level in a relatively developed state like Karnataka, this effort is worth studying and replicating wherever possible.

Apart from the above, the efforts taken by Information Kerala Mission (IKM) of the Department of Information Technology in bringing e-governance to local governments are worth lauding. Sevana, a software application developed by IKM, was used on a pilot basis and it was found out that rampant corruption prevailed in welfare benefit disbursal and there were several instances of poor service delivery due to various reasons such as poor bookkeeping practices, duplicate registration by the beneficiaries, and incorrect calculation of pensions. Most of these things were eliminated after the introduction of Sevana. Sanchitha is another effort in e-governance in Local Governments, wherein an encyclopedic repository consisting of information on all legislation and executive orders and promulgations relevant to decentralization and daily operations of the Local Governments, has been prepared on a compact disk (CD). The IKM further aims at: (1) providing connectivity between the State Planning Board, District Planning Officer and 1214 Local Governments in the state of Kerala; (2) developing a mechanism to regularly monitor the plan targets achieved by the local governments; (3) automating various operations at the local government level such as accounting, finance, project management, public services, statutory functions, and general administration using appropriately developed information systems, among many other things. It is expected that there would be a quantum leap in accountability, transparency, and efficiency in public services due to these efforts.

E-Governance Experiments by Local Governments in Karnataka: Karnataka has been a pioneer in both local government reforms and information technology. In decentralization it has gone well ahead of many other states in India. Nevertheless, it has not lived up to its own image in using information technology in the local governments, despite serious efforts that have been taken to utilize it extensively at the state government level. The Karnataka e-governance policy, Mahithi policy, outlines various aspects of e-governance in this state. A passing reference should be made to the action to computerize municipal corporations. This computerization is aimed at simplification of the payment of property tax, the early issue of birth and death certificates, and

East-West – Evolutionism, Global 177

redressing of grievances. Unfortunately, efforts to highlight the achievements have not yet been initiated.

E-Governance Experiments by Local Governments in Other States: Okha Village Panchayat, Gujarat, has been claiming that it is the first Village Panchayat on the Internet from India. This is a comparatively better designed website because it provides more information for promoting Okha as an industrial and tourist destination. Details about its vision, investment opportunities, infrastructure facilities, administrative support, investment incentives to Okha by the central and state governments, and possible investment sectors conducive to Okha have been provided with an interest to bring new investments in this locality. Feedback mechanisms such as seeking visitor comments, among other things, have made it more of a two-way communication.

Gyandoot is another experiment that has received wide attention. This project was implemented in the Dhar district of Madhya Pradesh; it received the Stockholm Challenge IT Award 2000, described as "a unique government-to-citizen Intranet project . . . with many benefits to the region." Internet kiosks were established in the Village Panchayat buildings; the server hub was a remote access server in the District Panchayat. The kiosks offer various services, such as (1) agriculture produce auction centers' rates; (2) copies of land records; (3) registration of applications for income/caste/domicile certificates; (4) public grievance redress, (5) village auction site for land, agricultural machinery, equipment, and other durable commodities; (5) transparency in government by providing information on beneficiaries of developmental schemes. Questions about the viability of the kiosks and connectivity related problems need to be addressed to make this a model e-governance for the world.

Based on these cases, it can be concluded that efforts taken so far in India to adopting e-governance at the local level are worth appreciation. At the same time, intermittent efforts by a few local government institutions are not sufficient to change the face of the country that is facing many problems, such as poverty, inadequate governance, and lack of transparency in major functions of governments at all levels. Unless similar efforts are made all over the country with the laudable objectives of less corruption, increased transparency, greater convenience, revenue growth, or cost reduction leading to better public management, such isolated efforts will not yield much result. If concerted efforts are made to scale up faster, plug loopholes, and deliver value to citizens by adopting complete e-governance, the local governments in India will become a truly adoptable model for the rest of the developing world.

References: "Survey: Government and the Internet," *The Economist* (Special Section) 355:8176 (2000), 9–19. Indira, A. "Data Systems at Local Levels: Experience in an IT State," *Economic and Political Weekly* 37:13 (2002), 1188–1189. Malick, M. H., and A. V. K. Murthy. "The Challenges of E-Governance," *The Indian Journal of Public Administration* 47:2 (2001), 237–253. Raghunandan, T. R. "New Governance: Decentralization with Infotech," *Economic and Political Weekly* 37:12 (2002). The World Bank. "Definition of E-Government." http://go.worldbank.org.

V. Venkatakrishnan

Elite: (Latin *eligere*, "to choose"; old French *elire*, *elisre*, "pick out" "choose"; French, *élite*, "selection" "choice") The elite have come to be un-

178   *Global Studies Encyclopedic Dictionary*

derstood as the supreme stratum in a system of social stratification (irrespective of the bases of this stratification; in a political context there is a political elite). Since the seventeenth century, in English, the word "elite" has been used to define goods of first-rate quality. Since the eighteenth century, it referred to selected military units. The term has also been used in genetics and seed-growing for designation of the best species. Since the nineteenth century, the term has been applied to the supreme strata in the system of social hierarchy (synonym of aristocracy). Vilfredo Pareto applied the term in sociology.

Predecessors of modern theories of the elite were Plato, Machiavelli, and Nietzsche. Gaetano Mosca, Vilfredo Pareto, and Robert Michels formulated more comprehensive theories of the elite. An initial postulate of these theories was the affirmation that there are two necessary elements of social and political structure—a narrow supreme privileged layer with functions of social and political management and development of culture or creative functions and the other population executing noncreative, reproductive functions. For defining the elite Pareto suggested a statistical method. "In each area of human activity an individual is somehow given an index, which is an indicator of his capabilities. . . . A group where each of the individuals has received the best mark in his area, we shall name elite" (1935). Mosca considers the elite as a ruling or political class:

> In all societies, starting from those that are just approaching civilization and finishing with the most advanced and powerful, there are two classes of people-ruling and controlled. The first, always less many, is responsible for all political functions, monopolizes power and enjoys advantages which are given by this power (1939)

Many definitions of the elite, given by modern political scientists and sociologists (Giovani Sartori fairly notices the plurality of senses of the term), may be reduced to two main interpretations: value (moralizing) and functional. Supporters of the first explain the existence of elite by "superiority" (first intellectual or moral) of some people over the others. Supporters of the second approach explain the necessity of the elite by the exclusive importance of functions of management for the society that thusly determines an exclusive role of people executing these functions (successfully implemented by the qualified minority). Both approaches suffer essential defects. Suzanne Keller (1991) rightly notes that "the moralizing approach degrades to mysticism whereas functional turns out to be a tautology." Indeed, being asked the question of who has the power, an elitist of the functional orientation answers: the one who has the power, who heads the basic institutions of authority. But the true problem is in explaining why a specific elite group has seized imperious positions. Besides, both interpretations of the elite easily degenerate into an apology for those in power. Some elitists (José Ortega y Gasset, Theodor W. Adorno), in contrast to definition of the elite as a group in power (political elite generally is pseudo-elite or vulgar elite, subject to mass influences), think that elite is a value in itself regardless of the power it possesses. C, Wright Mills (1956) expressed an interesting point of view when, distinguishing the elite in power and the spiritual elite, searched for ways of establishing accountability of the first to the second.

There are narrow (elite is the supreme echelon of political power) and broad definitions of the political elite (the total hierarchy of managers making the important decisions for the country, a middle group of political-administrative elite making decisions that are important for some regions, spheres of social activity, and, finally, bureaucratic personnel). Suzanne Keller introduces the concept of "strategic elites." The political elite consists of the ruling elite, the opposition elite (if it is a "system" opposition, contending over power within the framework of the given political structure), and the counter-elite, which aims to change the whole political system. There are also economic, cultural, military, religious, and other kinds of elites.

In Russia during the Soviet times, all discussions of the elite were taboo. Official ideology affirmed that in the Soviet Union, no exploitation of one person by another existed; there were no exploiter classes. Hence, there was not and could not be any elite. Nonetheless, the Soviet authority included a supreme social stratum that carried out administrative functions and possessed institutional privileges, thereby having all the attributes of an elite.

References: Pareto, V. and A. Livingston. *The Mind and Society*. Translated by A. Bongiorno. New York: Harcourt Brace, 1935. Mosca, G. *The Ruling Class*. Translated by H. D. Kahn. New York: McGraw Hill, 1939. C. W. Mills. *The Power Elite*. New York: Oxford University Press, 1956. Dye, T. *Who's Running America?*. Upper Saddle River, NJ: Prentice Hall, 2002. Keller, S. *Beyond the Ruling Class*. New Brunswick, NJ: Transaction, 1991.

G. K. Ashin

Energy, The Countdown (1979) was the sixth report to the Club of Rome presented by French researcher, professor of National School of Arts and Crafts Thierry de Montbrial. The report showed the limitation of the power resources of the planet and called for active international cooperation in order to prevent possible energy crises.

Reference: De Montbrial, T. *Energy, the Countdown*. Elmsford, NY: Pergamon, 1979.

A. N. Chumakov

Energy Crisis means an imbalance of the energy economy, which is characterized by depressive processes affecting the generation, transformation, and utilization of energy. Such processes lead to the instability of the trends of development and negative consequences for the global and regional economy, the conservation of resources and the preservation the environment. The energy crisis has exacerbated international and interstate controversies and competition for fuel and energy resources.

I. I. Mazour

Energy Paradigm is a system of ideological and strategic imperatives based on an adequate analysis of general trends and tendencies in the development of the global economy (with regard to the efficiency of utilization of natural resources in the social or ecological context), and on scientific arguments for the adherence to the established principles of optimization and management of the use of energy resources at local, regional and global levels, taking into account the priorities of energy conservation, innovative technologies, and

180   *Global Studies Encyclopedic Dictionary*

renewable sources of energy. The energy paradigm determines the vector of the development of the global energy economy in the interests of the present and future generations.

I. I. Mazour

Environmental Changes, Global: Global environmental changes are observable changes in atmosphere, water, and soils and in the flora and fauna associated with human economic activity that have negative impact on the economy, human health, and the ecology. Exponential industrial and population growth has brought about changes in all natural environments.

The twentieth century proved to be a century of the most intensive growth, when many parameters surpassed every possible level achieved during the preceding period. The enormous changes that occurred on the Earth's surface surpassed the changes that took place during the last Ice Age, eighteen thousand years ago. Economic expansion resulted in the destruction of natural ecosystems on 63 percent of the Earth's surface and their replacement with anthropogenic systems. The number of forests halved. The concentration of gases in the atmosphere changed, including appearance of formerly absent ones, namely, chlorofluorocarbons (Freon).

Concentration of organics in fresh water grows on all continents, which entails accelerating the eutrophication of water. We can see physical, chemical, and biological soil degradation named by the Russian soil scientist, G. V. Dobrovolsky, a "quiet planetary crisis." Mass-scale extinction of species takes place. Species extinction is a natural part of evolutionary process, but this phenomenon is barely observable, as under natural conditions approximately one to ten species disappear annually. Currently, according to biologists, the rate of species extinction has accelerated up to 1,000 species a year. Thus, resources, formerly considered renewable, ceased to be such, because they are not renewed as before within the natural variation range.

K. S. Losev

Environmental Liability is a public law instrument, by means of which potential polluters are made liable for the costs of restoring the environmental damage they may cause.

"Civil liability" is covers *personal injury and damage to goods and property* (traditional damage). Civil liability is not, however, well suited for tackling environmental damage since such assets are not usually privately owned. Even if the damaged environment is privately owned, the so-called pure ecological damage is hardly covered by civil liability since this does not constitute a damage that personally affects the owner, but society as a whole. There is usually no duty on the person who gets the compensation to spend it to restore the environment.

Environmental liability does not cover "traditional damage," such as economic loss, personal injury and property damage. For example, in the case of an organic farmer whose crop cannot be sold as "biologically produced" or "organic" due to contamination (e.g., by genetically modified organisms), the damage caused to the farmer is a purely economic one (crop cannot be sold) not an environmental one: this is therefore a traditional damage which is to be dealt with in accordance with national law.

The notion of environmental liability has the following characteristics:

(1) It is based on the polluter pays principle, thus, polluters should bear the cost of remediation of the damage they cause to the environment, or of measures to prevent imminent threat of damage.

(2) Polluters would meet their liability by remediation of the damaged environment directly, or by taking measures to prevent imminent damage, or by reimbursing competent authorities that, in default, remediate the damage or take action to prevent damage.

(3) Competent authorities would be responsible for enforcing the regime in the public interest, including determining remediation standards, or taking action to remediate or prevent damage and recover the costs from the operator.

(4) Strict liability shall apply in respect of damage to land, water and biodiversity from activities regulated by specified legislation; fault-based liability would apply in respect of biodiversity damage from any other activity.

(5) Defenses would exist for damage caused by an act of armed conflict, natural phenomenon, or from compliance with a permit, and emissions which at the time they were authorized were not considered to be harmful according to the best available scientific and technical knowledge.

(6) When an operator is not liable, the state would have subsidiary responsibility for remediating that damage (under EU liability regime).

(7) Individuals and others who may be directly affected by actual or possible damage, and qualified entities (non-governmental organizations) may request action by a competent authority, and seek judicial review of the authority's action or inaction.

A "strict liability system" where there will be few defenses is rather the exception than the rule in liability laws. Legally, a liability system can vary according to the *defenses* available under the liability regimen. Strict liability is exceptional for activities that do not present some potential risk to humankind or the environment. The starting point has usually always been fault-based liability. Over time, liability has become strict in some areas because of the hazardous character of the activities or the potential consequences for human health (a classic example is pharmaceuticals). National laws differ with respect to which activities should be subject to strict liability.

"Liability for biodiversity damage" is something new in legal order. It is important to have a very precise and workable definition of biodiversity adapted to available legal systems. For example, the regime of environmental liability currently existing in European Union will be reviewed after five years and, if appropriate, the definition of biodiversity will be reviewed then.

The definition of "biological diversity" in Article 2 of the Convention on Biological Diversity cannot be considered at this stage as providing a suitable basis for the proposed regime. This includes the liability to be attached to genetically modified organisms). The Convention's definition goes beyond habitats and species and subsumes the idea of "variability." Using this definition, it could be argued that damage to biological diversity would encompass injury to "variability among living organisms." Such an approach raises the delicate question of how such damage would be quantified, and what would

182   *Global Studies Encyclopedic Dictionary*

be the threshold of damage entailing liability. Biodiversity should be defined by reference to areas of protection or conservation that have been designated in pursuance of national or sub-national legislation on nature conservation.

"Genetically Modified Organisms" would cause environmental damage when they cause damage to biodiversity, water or soil (in this latter case, the soil contamination should create a potential or actual serious harm to human health). The environmental regime established in European Union covers both contained use and release into the environment of GMOs. When the release of the GMO has been specifically authorized or when it was not possible to foresee the damaging effect of the GMO on the basis of the best science, there would be no strict liability. In case of negligence (for example, when the operator does not follow the instructions given by the manufacturer on how to use the GMO), the operator would still be liable.

Environmental liability implies *non-limited financial costs* liable parties can be required to pay for damage restoration. Limiting the amount of damages would reduce private parties' *incentives to take due care and prevent damage* since they would know that, whatever the consequences of their actions, their financial responsibility would not be greater than the limit.

Setting a high limit for environmental damage risks would lead many private parties to insure against the full limit, which is likely to be excessive in many cases. This would unnecessarily raise insurance costs for a typical liable party instead of lowering them. Moreover, setting a limit does not in itself facilitate the supply of insurance by limiting insurers' exposure to liability. Given that insurers are not potentially liable parties under the existing legal regimes, insurers' exposure to risk is limited by the amount of insurance coverage they provide. Irrespective of whether there is a financial limit to the liabilities of the insured or not, insurers generally provide coverage only for specified, contractual amounts.

In the EU in cases when damage is "orphaned," that is, the responsible polluter cannot be identified or has no money to pay for the damage, Member States are required to ensure that the contamination is cleaned up or the damage restored. To this effect, they may rely on funding mechanisms set up specifically for this purpose, for example, pollution fees collected from the activities most likely to cause the damage.

"Prevention of damage" plays an important role in the environmental liability scheme. Holding potential polluters liable for the costs of restoring environmental damage they cause creates incentives for avoiding damage. When 1€ spent on prevention is likely to avoid damage whose cleanup and restoration costs more than 1€, the parties responsible for the potential damage are encouraged to invest in prevention rather than shouldering restoration costs. Therefore, the environmental liability regime leads the economy toward socially efficient prevention levels environment-wise.

Environmental liability covers *all kinds of pollution except diffuse pollution*. Diffuse pollution is not covered since in this case, liability is not an efficient policy instrument. With diffuse pollution, it is either practically impossible or socially inefficient to make individual polluters pay for environmental damage they may cause. Imagine, for instance, using liability to make each car driver pay for the diffuse pollution generated by his or her driving, or holding one farmer joint and severally liable for all the nitrates pollution caused by farming in a river basin. Therefore, other legal instruments cover diffuse pollution.

D. Ratsiborinskaya

Environmental Psychology is a complex branch of knowledge about the psychological aspects of the interaction between human beings and their geographic, social, or cultural environment. The environment is a part of human activities; it is an important factor in the regulation of human behavior and social interaction. Environmental psychology unites psychology with ecology and studies many social and humanitarian problems concerning interactions between human beings and their environment. This branch of psychology is based on studies in psychological aspects of architecture, industry, and habitat. Environmental psychology has collected some important factual material, but so far it has no theoretical structure. This research field is very topical with regard to attempts to find effective ways to overcome the ecological crisis. Environmental psychology is focused on the following issues:

(1) the study of ecological consciousness through understanding human perception of the environment and its negative aspects;

(2) the study of the motivation for ecological behavior, when people destroy or maintain their environment;

(3) the analysis of the psychological consequences of ecological crises (mental deviations, crimes, demographic changes);

(4) the elaboration of psychological methods of propaganda in order to form ecologically adequate worldviews; ecological expertise in new scientific and technical projects.

O. Y. Baksansky

Environmental Safety and Security: Advantages of Pebble-Bed Nuclear Reactors: A growing consensus in governmental, scientific, and business communities has emerged that conventional forms of industrial and consumer power are approaching obsolescence, probably within two or three decades. The basic reasons are twofold: worldwide depletion of oil and natural gas fuels, and atmospheric contamination with global-warming waste gases produced by burning hydrocarbons. The inevitable conclusion is that non-global warming nuclear fission power facilities, currently meeting 16 percent of worldwide electricity demand, will have to be substantially expanded. However, environmentalists have raised objections to nuclear power on safety and security grounds citing the Three-Mile Island and Chernobyl accidents as cases in point.

An advanced nuclear reactor that redresses environmentalist arguments is the pebble-bed variety of high-temperature gas reactor (HTGR) that uses triple coated carbon-silicon carbide-carbon (TRISO) fuel. TRISO-fueled reactors are not only environmentally desirable and relatively secure, but they are also cost competitive and solve waste disposal and weapon proliferation concerns. Several pebble-bed nuclear power facilities are either operational or under construction.

Pebble-bed fuel consists of tennis-ball sized spheres, each loaded with about 15,000 TRISO-coated particles in a graphite matrix, every particle con-

184   *Global Studies Encyclopedic Dictionary*

taining a fuel kernel. Several hundred thousand TRISO spheres are in a typical reactor at any given time. The fuel kernels are uranium dioxide enriched to 8 percent. The spontaneous fission of the uranium atoms releases neutrons and heat energy, and it is this energy that is absorbed by the coolant. The fuel, which is fed continuously into the top of the reactor and removed at the bottom, is cooled by gaseous helium flowing through the pebble bed. The heated helium serves as working fluid for the turbo-generator, and is ultimately recycled back to the reactor in a Brayton Cycle.

Unlike the fuel loading of light water reactors, the TRISO design permits efficient utilization of uranium oxide. Typically, each fuel sphere is recycled more than ten times and is depleted to where very little fissionable material is left. Additionally, although they permit the transmission of heat, the silicon carbide/carbon layers that surround the fuel are relatively impermeable. Thus, the contamination of the plant is minimal with a favorable impact on decommissioning costs. Also, the thermodynamic design of a TRISO fuel load provides inherent safety; that is, shutdown in the absence of coolant can be achieved without reliance on human intervention. Additional advantages include: (1) simplified waste disposal; (2) non-incentive for nuclear weapon proliferation; (3) resistance to terrorist attack; (4) no excess heat disposal; (5) pollution-free electricity and heat; (6) pollution-free manufacture of hydrogen; (7) competitive cost.

TRISO-fueled nuclear plants have either been built or are planned in four countries, Japan, China, Russia, and South Africa. The only existing pebble-bed reactor is the Chinese HTR-10, a ten MW plant near Beijing that achieved criticality in December 2000. The South African pebble-bed is under construction. The Japanese reactor, the High-Temperature Test Reactor (HTTR), is a batch-fueled 30 MW facility that achieved criticality in November 1998. The Russian reactor is planned for licensing in 2010, and will have a 600 MW output.

HTGRs using TRISO fuel are especially suitable for hydrogen manufacture because they provide the high-temperature heat needed in methane reforming, currently the industry standard. Environmentally clean techniques for hydrogen manufacture should provide the basis for future generations of hydrogen-fueled cars.

M. B. Schaffer

Environmentalism, Western: Western environmentalism constitutes a critique of modernity. At the core of modernity are the ideas of a fundamental ontological divide between humans and non-human nature (metaphysical dualism), that non-human nature operating according to the deterministic laws of physics (mechanistic materialism), and that non-human nature has no value above and beyond use-value for humankind (anthropocentrism).

Modern philosophers, scientists, and theologians articulated a similar vision: the universe is a superlatively exquisite machine, the handiwork of God. Nature can be likened to an immense machine, created by God and designed to operate according to the laws of physics. All natural motions—including life—can be explained in terms of material and efficient causation without needing to refer to formal or final causes. Ultimate causes refer to God, and

God is outside nature. Thus, the undergirding of the modern view of nature is mechanistic materialism, which Francis Bacon, Galileo, William Harvey, Thomas Hobbes, René Descartes, Sir Thomas Newton, and others advocated. In the words of Johannes Kepler, "I am much occupied with the investigation of physical causes. My aim is to show that the celestial machine is to be likened not to a divine organism but to a clockwork" (letter, Kepler to Herwart, 10 February 1605). The modern view of nature conveniently bifurcates the subject matter of science and religion: the domain of science is the efficient causation of material substance; the domain of religion is the ultimate purpose of transcendent God. Science investigates the predictable and clockwork-like operations of nature, without having to infringe on theology's search for why there is any purposiveness in nature at all.

Environmental thinking, as it developed over the course of the twentieth century in the United States, is significantly different from non-American environmentalism, especially European environmentalism. Given the influence of Karl Marx (and by extension, the Frankfort School of Critical Theory), European environmentalism has been based in a social and political critique of class hierarchy inherent to capitalism, and the implications of these hierarchies for the human domination of non-human nature. American environmentalism, on the other hand, has been based in the ideals of the conservation movement and wilderness preservation. These distinctions are not absolute; fortunately, mainstream American environmentalism over the last decade of the twentieth century has widened its previously narrow focus on wilderness.

Finding inspiration in Marx, the fundamental premise of European environmentalism is that a necessary condition of solving environmental problems is transforming social structures. Both social and environmental problems result from unjust hierarchies of domination; attaining a more just ecological relationship with non-human nature entails securing just political structures.

American environmentalism has a totally different genealogy. With its emphasis on wilderness preservation and natural resource conservation, no Marxian critique of class hierarchy has been involved. Henry David Thoreau articulated a preservation ethic as early as 1851, when he wrote, "in Wildness is the preservation of the World" (2007, p. 25).

Aldo Leopold and Rachel Carson have largely set the tone of contemporary environmentalism. Both excoriated the idea that *Homo sapiens* is separate and superior, and painted a picture of the fundamental ontological interconnectedness of all things. Leopold, wrote, "a land ethic changes the role of Homo sapiens from conqueror of the land-community to plain member and citizen of it." (1949, pp. 219–220). In her seminal work, *The Sea around Us* (1961), Carson decried the arrogance of thinking that humans can manipulate and control nature. Before Leopold and Carson, the only critics of extreme anthropocentrism were a smattering of Romantics and other fringe characters.

At the end of the twentieth-century, critics of extreme humanism and nature-as-a-machine populate the ranks of the environmental movement. However, the degree of reaction against anthropocentrism within environmentalism varies drastically. In all of these environmentalists, we see a common emphasis on maintaining the integrity and stability of ecological systems, and none of the European emphasis on the need to substantially revamp existing social

186 *Global Studies Encyclopedic Dictionary*

and political structures. During the 1990s, however, American environmentalism widened its scope. Historian William Cronon notoriously claimed that wilderness is no more than a human social construction (1995). Others have criticized mainstream environmental groups as being primarily lobbies aimed at ensuring outdoor playgrounds for physically fit middleclass whites.

There has been a growing European-like awareness in the United States of the social and political elements of environmental problems. Interestingly, however, this growing awareness finds its inspiration not in a Marxian critique of the hierarchies of domination inherent to capitalism, but in the Civil Rights Movement. Whereas European environmentalists seek to mitigate ecological problems by tempering capitalist economics with socialist remedies, American environmentalists tend to iterate and reiterate each individual's right to a healthy habitat.

References: Leopold, A. "The Land Ethic." In *A Sand County Almanac*. New York: Oxford University Press, 1949. Caron, R. *The Sea around Us*. New York: Oxford University Press, 1961.—. *Silent Spring*. Boston: Houghton, Mifflin, 1962. Cronon, W. "The Trouble with the Wilderness." In *Uncommon Ground*. Edited by Cronon, pp. 65–90. New York: W.W. Norton, 1995. Thoreau, H. D. *Walking*. Rockville, MD: ARC Manor, 2007.

D. R. Keller

Epogenesis (Greek *epoche*, "hold-up," and *genesis*, "conceiving") refers to the formation and subsequent development of epochs and the transition from one epoch to another. As a result, the scale and territorial limits of the ongoing events expand. Within these limits, events preserve their wholeness, and reveal general trends and logical development. Human history can be divided into periods based on the scale of ongoing events. We can speak of four basic epochs in the formation of global connections (relations) accompanying historical development from the emergence of society to the foreseeable future:

(1) Epoch of fragmented events and local social connection that starts at the moment of the emergence of human beings five to three million years ago and lasts till the end of the Neolithic revolution, or the emergence and formation of the first states (circa the seventh to the third millennia BCE).

(2) Epoch of regional events and territorially limited international relations, from the end of the Neolithic revolution, when states had already been formed and stabilized (from the seventh to the third millennia BCE), to the beginning of the Great geographic discoveries, that is, to the Renaissance.

(3) The epoch of global events and of the universal economic and socio-political interdependence that lasted from the Great Geographic Discoveries to the middle of the twentieth century, when the world united with regard to geography, economy, politics, and ecology.

(4) The epoch of space explorations and space conflicts, which began in 1957, with a launch of the first Sputnik and which will continue in the foreseeable future. In this period, the world becomes informationally closed.

The transition from one epoch to the next was always accompanied by the expansion of the scale, of the territorial limits, within which events preserve their wholeness and reveal general trends and some logic of development.

A. N. Chumakov

Epometamorphosis (Epomorthosis): (Greek *epoche*, "hold-up," and *meta-morphosis*, "transformation") Epometamorphosis refers to an epochal transformation, a transition to the next stage of development of a geo-bio-socio-system accompanied by fundamental changes in this system in terms of its form, content, essence and manifestation are concerned. This notion provides a better understanding of macro-historical and macro-social transformations influencing the interchange of historical epochs, that is, the changes that cause the transition from one epoch of global connections to another.

Globalization involves the structural transformation of an entire society as a part of planetary biological and geographic conditions. At this moment, the fundamentals of social life and the perspectives of individual people within these structures change radically. As a result, the whole system of relations between human beings, the biosphere and the geographic environment around them is changing.

Fundamental changes (transformations such as metamorphism) caused by globalization take place at the level of social consciousness and worldview. The most important stages of the development of human worldviews or, to be more precise, the basic historical forms of such development are: (1) the mythological and religious (chronologically appearing approximately at the same time), (2) the philosophical, (3) the scientific, and (4) the global. Given these divisions, one may speak about five turning points in human history, each of which may be regarded as an "epometamorphosis."

The first epometamorphosis was connected with the emergence of *Homo sapiens* and with the formation of religious and mythological outlook. This period transpired from forty to sixty millennia ago to the beginning of the first millennium BCE. The Neolithic revolution that took place between seven and ten millennia ago marked the culmination of the first epometamorphosis. Because of this epometamorphosis, human beings finally distinguished themselves from the animal world and the creation of material and spiritual culture (in the form of musical instruments, dances, songs, pictograms, and verbal communication). Hence, began the epoch of fragmented events and local social connections. This epometamorphosis also corresponded with the beginning of history.

The second epometamorphosis was characterized by the formation and the emergence of philosophy as an historical form of world outlook in the middle of the first millennium BCE. Karl Jaspers referred to this era as "axial time." The transformations that took place during this time were essential characteristics of the epoch of regional events and territorially limited international relations. The term "culture," which describes the results of human activity against the background of "wild" nature, and the first symptoms of globalization emerges during this period.

The third epometamorphosis was directly linked to the partition of science as a form of social consciousness separate from philosophy and to the beginning of scientific and technological progress. These transformations characterize the epoch of global events and the emergence of universal economic and sociopolitical interdependence. At that time the term "civilization" emerges, expanding linguistic capabilities for describing complex social realities in terms of their structure, organizational forms, and scientific and technological

188 *Global Studies Encyclopedic Dictionary*

achievements. The third epometamorphosis is marked by the beginning of actual globalization and with the fundamental stage of its development.

The fourth epometamorphosis is presently taking place and is linked to the formation of global consciousness that dates back to the nineteenth century, but which has only become visible since the second half of the twentieth century, when the era of space exploration began. It is connected with multi-aspect globalization.

After the fifth (hypothetical) epometamorphosis begins, or even in a more distant prospect, a radical change leading to the understanding of human essence should happen. Most likely, the term "humanization" will become the main category of social and individual consciousness.

A. N. Chumakov

Equality: As understood in global studies, equality refers to the state of being equal with regard to status, rights, and opportunities. A central topic in contemporary political discussions, most contemporary and political theories hold that human beings are essentially equal—of equal worth—and that this ideal should be reflected in the economic, social, and political structure of society. Ronald Dworkin has emphasized equality as the basis of all political discussion.

Egalitarian theories maintain that justice consists in treating people equally. The identification of justice with equality has a long tradition in political thought since the time of Aristotle (*Politics*). People are different in physical and mental ability and have different interests and aims. It is generally acknowledged that some inequalities are morally neutral. Differences in ethnic origin, interests, abilities, intelligence, and conceptions of the good are generally thought to be not morally repugnant, but they are clearly hindrances in achieving equality as the social and political goal.

A wide-ranging variety of conflicting conceptions exist regarding equality. The theories of equality differ about what precisely is to be the object of equality: welfare, preference satisfaction, primary goods, economic resources, social status, political power, capacity of personal fulfillment, opportunity for welfare, and opportunity for scarce resources and social positions. While egalitarians uphold equality as the "sovereign virtue," in Robert Dworkin's phrase, non-egalitarians dispute the significance of equality as an ethical ideal or a political goal. Some egalitarians maintain that equality involves equal consideration of interests. But this does not specify how to rank competing interests or whether the scope of interests should include animals and plants. It is also disputed whether interest satisfaction is an appropriate good to be distributed. Regarding the thesis that there is always a presumption in favor of treating people equally unless some relevant difference has been found, it has been objected that this view of equality is empty unless the determining criteria of relevant differences can be specified. Aristotle's conception of justice as equality under law is open to the similar objection: what the laws are needs to be specified.

Primary goods, resources, economic benefits, wealth, power, prestige, class, welfare, satisfaction of desire, satisfaction of interest, need, and opportunity are generally considered to be relevant in determining equitable distribution. Resource egalitarians maintain that all are entitled to equally valuable shares of resources (wealth and other primary goods). Welfare egalitarians argue that people should receive equal welfare--interpreted in terms of preference satisfaction. Those who uphold equality of welfare as the goal claim that justice calls for arranging social institutions to ensure that every person would find welfare equally. The common core of humanity (common needs and sympathy) is also thought to provide a justification for equal treatment unless there are reasons for treating them differently.

Governments by their very nature necessitate hierarchies of power. That those who govern are subject to the same laws does not mitigate the essential inequality, since the laws are made by their authority and they are in a privileged position to interpret the laws. Marxist and socialist egalitarians argue that franchise is insufficient for political equality. Wealth, education, and leisure determine who can participate effectively in the political process. Marx and Engels have argued that so long as the state exists, political equality is impossible: the rulers always have more power than the individual citizens.

Two lines of criticisms against equality exist. The first upholds equality as the ultimate social goal and advocates unequal treatment as a means to this end. It is claimed that when some are worse off than others, treating people equally will only lead to further inequalities. Existing inequalities need to be remedied by affirmative action policies and preferential treatment of the underprivileged to provide opportunities historically denied to the minorities.

The second line of criticism rejects equality as a social or a political goal. Hume argued that people are not equal and that only the most tyrannical forms of governments can ensure equality. The criticism by Friedrich August Hayek and Robert Nozik of promotion of equality by governments on the grounds that it interferes with individual liberty is not original, since David Hume anticipated it. Hayek argues that people are not equal and, given equality before the law and freedom to develop their talents, some people will do better than others. Inequalities are unavoidable: even given equal resources, differences will arise. He maintains that inequalities in wealth, including inheritance, cannot be altered by government policies without seriously jeopardizing individual freedom.

Equality as a social goal seems to be ambiguous. When is it just to treat people unequally? Merit, desert, ability, or capacity to do greater good for the greater numbers is sometimes regarded as a strength that justifies a larger share of limited goods and resources. Deficiency, inability, or propensity to do harm to a larger group would require more of different kinds of goods and resources to be incorporated into the communal life and to minimize harm.

If equality is seen as having an intrinsic value, it is a moral value to be pursued for its own sake and not because of beneficial consequences or to avoid harmful effects of inequality. In response to the criticism that egalitarian views fail to account for different interests and claims, we invoke Kantian and utilitarian bases of justice. In Immanuel Kant's imperative presumes equality of all human beings in human worth and dignity as rational moral beings. In the utilitarian goal of general well-being, there is also an assumption of equality of all in contrast to the promotion of individual interests as constituting the goal of justice. How do we arrive at the criteria for distinguishing the areas of equality that are in the jurisdiction of government? They

190 *Global Studies Encyclopedic Dictionary*

are to be worked out, not within the limitations of specific existing forms of governments at present, ranging from theocratic states and constitutional monarchies to democratic states and market driven economies, but with a conception of justice as equality before law, equal treatment of all, and not favoring those privileged in natural endowments, social standing, or wealth.

References: Hayek, F. A. *Constitution of Liberty.* Chicago, IL: University of Chicago Press, 1960. Dworkin, R. *Taking Rights Seriously.* Cambridge, MA: Harvard University Press, 1977.—. *Sovereign Virtue.* Cambridge, MA: Harvard University Press, 2002.

H. Höchsmann

Equality of Gender (Sexual Equality): In conditions of globalization gender equality, meaning equal rights for men and women, is one of the main goals of social policy of democratic countries and the world community of the twenty-first century. This principle is affirmed in the UN Declaration of Human Rights (1948). At the sixth International Conference of the United Nations on improvement of female status (Peking, 1995), two documents were adopted—the Peking Declaration and the Peking Platform of actions. Those documents confirm the orientation of the world community to values of gender equality. The Declaration puts new guiding lines for old and new democracies, stating that "equal rights, opportunities and access to resources, equal distribution of family duties between men and women and harmonious partnership between them have the key value for their well-being and well-being of their families, and for strengthening of democracy" (article 15).

The European Convention for the Protection of Human Rights and Basic Freedoms (1950) and the European Social Charter (1989) are, in many respects, based on the ideas and principles of the UN Universal Declaration on Human Rights and other documents accepted by the United Nations. Both have played a significant role in unification of the legislation on human rights in this region, also from the point of view of equality of women and men.

The White Book of Social Policy, which frames the European Union actions on conservation and development of the European social model in situation of globalization of trade and production, also includes a special section, "Equality of Opportunities for Men and Women."

Starting from 1986, regular European conferences of ministers on issues of equality between women and men are held (Strasbourg, 1986; Vienna, 1989; Rome, 1993; Istanbul, 1997). At the Istanbul meeting on equality issues (1997) ministers adopted a politically important document-Declaration on Equality between Women and Men as the Main Criterion of Democracy. In 1995, Strasbourg held the European conference "Equality and Democracy: Utopia or Possible Achievement?" where issues of gender equality were discussed. At the second summit (Strasbourg, 1997), members of the European Council stressed "the importance of more balanced representation of men and women in all social spheres, including political life" and called to "advance and achievement of actual equality of opportunities of women and men

In 1995, under the initiative of the European Council, a group of experts was formed, whose main priority was a integrated approach to the problem of gender equality (gender mainstreaming). For the first time, such approach appeared in various international documents after the 3rd International Con-

East-West – Evolutionism, Global 191

ference of the United Nations on Problems of Women (Nairobi, 1985). "Gender mainstreaming" implies the necessity of incorporation of women and men equality criteria into the general system of social organization called "mainstream." The integrated approach to the problem of gender equality is perceived as regular integration of people's priorities and needs into social policy programs to achieve equality between women and men. Development of special social policy measures for achievement of this equality, accompanied with estimation of their effect on women and men, with subsequent monitoring and evaluation is already at the planning stage.

The integrated approach to a problem of gender equality also consists in organization (reorganization), improvement, perfection, and evaluation of decision-making processes made by persons involved in implementation of this policy with the purpose of integration of equality issues into all spheres and all levels. It is important to note that the integrated approach assumes wider and more detailed definition of equality estimating variety of existing differences. An introduction of this approach is connected to a new understanding of equality from the point of view of interrelation of concepts of equality and difference. Equality today means an equal position, independence, responsibility, and universal participation of women and men in all spheres of public and private life. According to experts of gender mainstreaming, the issue is not in confirming that this difference exists, but in preventing of the situation when difference leads to discrimination. The concept of equality of the new approach includes "rights of women and men to difference and diversification." Existing difference between women and men should be taken into account, and the purpose should be not the establishment and preservation of hierarchy, but overcoming of its negative consequences.

O. A. Voronina

Esperanto: Created by Ludwig M. Zamenhof in the late nineteenth century, Esperanto is the only international language communicatively realized and permanently functioning within a significant international community.

Zamenhof suggested that a nationally neutral and easy-to-learn language of international communication (ideally, second after everyone's native tongue) was not only the most reasonable solution for the problem of interethnic communication both between states and within multiethnic states, but would also promote mutual understanding among peoples and preserve peace. Practice has demonstrated that only an a posteriori autonomous, planned, international language can be communicatively realized. Due to Esperanto's linguistic structure, task-oriented cultural development, and ideological organization of the community speaking it, it has reached the level of an elaborate language and has the potential to be used globally.

During the first years after being invented, Esperanto spread very rapidly. The first Esperanto clubs, books, and periodicals existed already in the 1890s. In 1905, the First World Congress was held in France. In the first half of the twentieth century Esperanto and its ideology attracted great attention and support from outstanding intellectuals, many of whom were learning this language (Leo Tolstoy, Maxim Gorky, R. Rollan, Richard Bach, and Russian poet Mikhail Isakovskii who wrote poetry in Esperanto).

192 *Global Studies Encyclopedic Dictionary*

In 1921, the League of Nations discussed the issue of making Esperanto its official language, but this project was rejected by the French delegation. Since then, such a project has never been officially discussed and the Esperanto movement developed exclusively on a non-governmental level (although, for instance, I. Bros Tito in Yugoslavia and Todor Zhivkov in Bulgaria knew Esperanto and supported the movement).

In the 1920s, the movement of Esperantist workers grew, especially in France, the Soviet Union, and Germany. However, from 1933, in Germany, and from 1937–1938, in the Soviet Union, the Esperanto movement was, in fact, forbidden. Many Esperantists were executed or imprisoned on the charge of espionage, "international Trotskyist conspiracy." In Germany, the movement was recreated after the World War II, while in the Soviet Union only after 1956. Esperantists and their organizations exist in most countries, but the Esperanto movement is especially well developed in Bulgaria, Brazil, Hungary, Vietnam, Germany, Iran, China, South Korea, Lithuania, the Netherlands, Poland, Serbia, Slovenia, France, Croatia, Sweden, and Japan.

After the World War II, however, interest began to wither away. In the 1990s, Esperanto gained some popularity on the Internet; it is the second language (after English) used during international e-conferences. Currently, estimates range from 100,000 to 2 million active or fluent users worldwide, with about 1,000 native speakers who learned it with their native languages. With a notable presence in approximately 100 countries, usage is highest in Europe, East Asia, and South America.

The Universal Esperanto Association (Universala Esperanto Asocio; UEA) to make practical language use easier for the world Esperanto community. During the First World War, the UEA helped to connect prisoners of war with their relatives in 300 thousand cases using Esperantist channels. As a nongovernmental organization it has consultative status at the United Nations and UNESCO. In 1954 and 1985, UNESCO adopted special resolutions in support of Esperanto. The Universal Non-National Association (Sennacieca Asocio Tutmonda is a nonpartisan workers' Esperanto organization. There are also many youth, academic, professional, and cultural international organizations of Esperantists holding annual congresses, festivals, meetings, and seminars. Other local organizations have also been established in various countries.

N. L. Gudskov

Ethics, Global: Global ethics is a branch of philosophy that studies the ways in which different thought systems advocate ethical action and organize their moral universes in relation to metaphysical and other avenues of exploration. It also studies ways in which intellectual and moral pressure can bear on solving pressing global problems affecting the contemporary world as a whole.

T. C. Daffern

Ethnic Cleansing: The process or policy of eliminating unwanted ethnic or religious groups by deportation, forcible displacement, mass murder, or by threats of such acts, with the intent of creating a territory inhabited by people of a homogeneous or pure ethnicity, religion, culture, and history. The term entered English usage in 1992, in the context of the post-Yugoslav wars, es-

East-West – Evolutionism, Global

pecially in Bosnia. The term had emerged earlier in the Serbo-Croatian press reports on the goals of this warfare. Warring parties sought to "cleanse" areas that were perceived to belong to ethnic nation-states from populations perceived as not belonging to the ethnic nation who claimed such domains. The word "ethnic" made it from obscure scholarly works into mainstream usage during the 1940s and 1950s. Earlier, the noun "cleansing" or adjective "cleansed" or "clean" were employed to denote what we now recognize as an act of ethnic cleansing. The earliest such usages of these terms of which the author is award occurred in the name of the Polish organization entitled The Organization for Cleansing Poznań from Jews and Germans, established in 1918, in the city of Poznań. This organization postulated the removal of the targeted populaces from the city and its vicinity. Similarly, in the Third Reich the areas where Jews had already been exterminated were dubbed as "*Judenrein*," or cleansed of Jews.

Ethnic cleansings have taken place since times immemorial. Examples include: (1) the removal of the Jews from Palestine to Babylon in the sixth century BCE; (2) from the fourth to the tenth century CE the Germanic, Slavic, Turkic and Finno-Ugric ethnic groups that moved between Central Asia, Western Europe and northern Africa continually displaced one another; (3) in the eighth century, the Byzantine army resettled the members of the Paulician religious sect from Armenia to the Balkans; (4) intermittent expulsions of Jews from Western and Central Europe commenced in the twelfth century; (5) between 1609–1614, the Moriscos (Muslims converted to Catholicism) were expelled from Spain; (6) following the Thirty Years War, Protestants were often removed from Catholic realms and Catholics from areas controlled by Protestant rulers; (7) in the mid-eighteenth century, French-speaking Acadians were removed from the British colony of Canada; (8) between the 1820s and 1880s, indigenous Americans were forcefully relocated from their lands to reservations in the West; (9) in the mid-nineteenth century, British settlers exterminated all the Aborigens living in Tasmania; (10) in 1907, the German army almost exterminated all the Herero ethnic group in South West Africa.

Although all these instances involve ethnically or religiously differentiated groups, their removal or extermination was either the result of imperial policy (empires, by definition, are multi-ethnic), occurred in areas not controlled by any state, or stemmed from the quest for religious homogeneity (religions tend to disregard ethnic boundaries). Hence, it is incorrect to subsume these instances under the rubric of ethnic cleansing, which entails the existence of or endeavor for the creation of a homogenous ethnic nation-state. Only such a state needs to be 'cleansed' of all those persons and groups who do not conform or decline to assimilate with a given ethnicity that constitutes the basis for the commonality of an ethnic nation living in this state. Due to the logic of ethnic nationalism the ethnic nation, which strives for founding its own ethnic nation-state, or to whom such a state already belongs, becomes involved in ethnic cleansing so as to 'purify' its state from "foreign" (that is, ethnically incongruous) elements.

Ethnicity is an arbitrary and ascriptive label, which can be adopted by an individual who identifies with a group, or forced on the individual by a group. The group self-defined as a nation, and recognized as such by other nations,

194 *Global Studies Encyclopedic Dictionary*

defines which pre-selected elements of broadly defined cultural reality constitute the ethnicity of the thus formed nation. Obviously, the elites are instrumental in defining this ethnic core of the ethnic nation. Eventually, these elites decide who is to be included in the pale of the ethnic nation. The individual is the passive object of this process. In the Third Reich Germans of Jewish ancestry whose grandparents had converted to Protestantism or Catholicism and spoke no other language but German, were denied commonality with the German nation and exterminated together with other European Jews in the Holocaust. The 1923, Lausanne Treaty that provided for exchange of Greeks from Turkey for Turks from Greece oftentimes forced Greek-speaking Muslims from Greece to Turkey and Turkish-speaking Orthodox Christians from Turkey to Greece.

Italy in 1860, and Germany in 1971, were established as ethnic nation-states in stark opposition to the model of the civic nation-state widespread in Western Europe and the Americas. In the civic nation-state, citizenship equals or replaces ethnicity as the basis for one's commonality with the civic nation. (But even in ideologically civic nation-states ethnicity has played a role too. For example, until the 1960s, Afro-Americans and Native Americans were excluded from the United States' civic nation. In France, during the nineteenth and twentieth century, languages other than French and non-standard dialects were suppressed and liquidated, while their speakers were treated as second-class citizens.) In the last three decades of the nineteenth century most of the ethnic states that started emerging at the beginning of this century in the Balkans gained or confirmed their independence. Another wave of ethnic nation-states arrived in Central Europe and the Middle East after World War I. These states disregarded any existing state borders in pursuit of national territories that would "truly" reflect the settlement extent of their ethnically defined nations. On the other hand, they wished to remove from their territories persons labeled as belonging to other nations. Inevitably, ethnonationally induced and legitimized conflicts flared up among the neighboring ethnic nation-states. Invariably, instances of ethnic cleansing followed.

The twentieth century was the age of ethnic cleansing. In the Boer War (1899–1902) the British troops displaced and kept in concentration camps one-third of the Afrikaners. In the course of the Balkan Wars (1912–1913) mainly Muslims/Turks were expelled from this area but the Balkan nation-states involved in mutual ethnic cleansing too. In 1915, the Ottoman Empire exterminated Anatolian Armenians and expelled most of them to Mesopotamia. These examples still belonging more to the pre-national era, the first clear-cut instance of ethnic cleansing occurred after 1918, when Germans, Germanophone Jews, and other German-speakers chose to flee or leave the new Central European nation-states of Poland, Czechoslovakia or Hungary.

The administrative division of the Soviet Union was organized on the ethnonational principle, which spread the model of ethnic nation-state throughout Eurasia. During World War II Berlin and Moscow perpetrated many acts of ethnic cleansing including, respectively, the Holocaust, forced resettlement of entire "enemy nations" (for instance, of Chechens and Crimean Tatars) or parts of nations (for example, Poles or Estonians). Between 1945 and 1950, ethnic cleansings coded as "population transfers" involved over 30 million people in Central Europe alone. The forced resettlement of over 10 million people in the wake of splitting British India into India and Pakistan (1948) was of similar magnitude. Further ethnic cleansings occurred invariably when new ethnic nation-states emerged in Eurasia, recently, after the breakups of the Soviet Union and Yugoslavia.

Curiously, the usually civic nation-states of Africa have not been immune to the phenomenon of ethnic cleansing as demonstrated by the genocide of the Tutsis in Rwanda (1994). This has to do with the general failure of these multiethnic states in the credible replacement of ethnicity with citizenship and civic values as the basis for nation- and nation-state-building.

Methods employed to carry out an ethnic cleansing can be ranked according to the degree of their lethality: genocide, unilateral expulsion, multilateral and negotiated population transfer or exchange under pressure, forced emigration, and forced assimilation. Significantly, depending on circumstances these instruments of ethnic cleansing may be applied against a section of or the entire ethnic group or nation.

T. Kamusella

Ethnic Group: During the 1960s and 1950s, "ethnic group" replaced the term "tribe," which was tainted with the association of primitivism and split human societies into two worlds, namely, that of the "modern West" and of the "backward Rest," to be studied by sociologists and anthropologists, respectively. The former was composed of "societies" while the latter of "tribes." "Ethnic group" bridged this conceptual gap allowing for comparisons of "Western societies" and "tribes" as entities of the same rank and without any a priori preconception that one of them is somehow "better" than the other.

Taken as the defining quality of the ethnic group, ethnicity is said to be comprised of a variety of elements that include: a collective name (ethnonym), a common myth of descent, a shared history, a distinctive shared culture, an association with specific territory, a sense of solidarity, legitimizing symbolisms (mythomoteur), and the group's capacity for biological (endogamic) self-reproduction. This static approach presents the ethnic group as a clear-cut entity that does not change or interact with other ethnic groups.

From the vantage of analyzing social and inter-group relations as dynamic processes, the ethnic group constitutes itself only in relation to other ethnic groups. In 1969, the Norwegian scholar Frederik Barth introduced the concept of "ethnic boundary," which exists "in the head" of a member of an ethnic group who identifies himself or herself as a member of his or her ethnic group in contrast to other groups, through remembering and expressing the elements that constitute this group's specific ethnicity. This explains the durability of ethnic groups even in view of the fact that individuals tend to leave one ethnic group for another. Individuals cross ethnic borders by exchanging one ethnicity for the ethnicity of a chosen, new group.

Currently, in social sciences, the concept "ethnic group" possesses two basic meanings. The broader one treats all the human groups founded on a specific ethnicity as ethnic groups. The other contrasts the ethnic group vis-à-vis the nation. In the latter view, the difference between nation and ethnic group is constituted by several factors: (1) The nation has its nationalism that

196 *Global Studies Encyclopedic Dictionary*

forms the basis of the nation's political program in search of national autonomy or separate national statehood. The ethnic group does not have such a program. (2) Nationalism allows the nation to formulate its interests at the level of the whole nation. A common ethnicity gives the ethnic group a sociocultural coherence but does not restrain various subgroups from expressing their own interests sometimes in conflict with the interests of other subgroups and even of the whole ethnic group. (3) In the case of the nation, the ethnic group's ethnic boundary is reinforced with the political boundary delimited by the nation's nationalism, and, eventually, by the geographic border of the nation-state itself. (4) Nationalism and the institutions of the nation-state determine the continuation of national ethnicity. The reproduction of the ethnic group's ethnicity depends on the extant "traditional" and non-institutional structures. (5) The nation's ethnicity can be defined by ethnic elements and citizenship to a varying degree. By definition, the ethnic group has no recourse to defining its ethnicity via citizenship.

Ethnic minorities, this is ethnic groups and stateless nations, fare better in the civic nation-states that adopted the policy of multiculturalism (Canada, the United States, Australia, South Africa, Spain, the United Kingdom, and Russia). In this model the ethnic or ethno-national preoccupations and character of various native and immigrant groups are defined as the civic nation-state's "cultural richness." Simultaneously, they are subsumed in the commonality of citizenship. Despite their various ethnicities, citizenship determines the nationality of all citizens vis-à-vis the nation-state. On the basis of various ethnicities some concessions are granted in the public or administrative sphere, but, generally, ethnicity is considered a private matter that should not intrude on the mainstream politics of the civic nation-state. Obviously, this cannot be so in ethnic nation-states, where an ethnicity of a group of citizens that varies from the national standard poses an ideological danger to the state and its ethnically defined nation. Allowing unrestrained expression of their ethnic difference, minorities would undermine the legitimizing force of the ethnic nation-state's nationalism and its nation's dominant position.

The broader meaning of the concept "ethnic group" can become a potent category of comparative analysis in the future development of social sciences. Leaving aside the politically discriminatory distinction between the nation and the ethnic group, it may allow for fruitful comparison of variegated human groups steeped in ethnicity, be they non-national ethnic groups, nations in their ethnic or civic guise, stateless nations and ethnic and national minorities. Globalization necessitates this objective and non-judgmental approach to the problematic of ethnic groups. The Global State System, perhaps, will not accommodate more than 200 states; so, not more than the corresponding number of nations can count on securing and maintaining their own nation-states. But going only by the guideline of already classified ethnolects (languages that constitute or are part of an ethnic group's ethnicity), 4–7,000 extant ethnic groups can be identified. With the increasing ideologization of ethnicity at the turn of the twentieth and twenty-first century, facilitated by the growth of global transportation, communication, and media interconnectedness, many of these will turn their ethnicity into an instrument of political struggle, especial-

East-West – Evolutionism, Global 197

ly when the interests and economic sustenance of the ethnic group's members are endangered by the host nation-state or by international companies.

T. Kamusella

Ethos, Global: (Greek, *ethnos*, "people" "nation" "class" "caste" "tribe" "a number of people accustomed to live together") Ethnos denotes people of the same race or nationality who share a distinctive culture. "Global ethnos is a concept designating one of the probable tendencies of the development of public morals and manners in the post-industrial epoch; i.e., is expected that, as time goes by, humankind will ultimately work out a single ethos, the way it has been done by separate peoples, confessions, estates, professions, etc.

In the modern world, cosmopolitanism vies with regionalism and even calls for isolationism. To answer questions about the possibility of the global ethos, and about spiritual and moral processes in light of globalization, it is essential to distinguish the two levels of the human identity: social (group) and spiritual. Cultural diversity and pluralism oppose cultural standardization. Social and group particularism and spiritual-and-moral universalism obviously conflict and need to be accounted for in any concept of global ethnos. Therefore, a unified ethic and cultural scheme of activity, being transformed into a global ethos is a long and complicated historical process.

A. A. Guseynov

Euroislam is a conventional term signifying, in a general sense, the integration of Islam with the ideas and practice of democracy and liberalism. In the narrow sense it means the integration of Islam into European culture and politics. Some contemporary Muslim reformers, for instance Tariq Ramadan, understand Euroislam (*al-islam al-urubi*) as Shariat adapted to the conditions of the modern Europe. Some others, for instance Rafael Khakimov, consider Euroislam as an addition or an alternative to Eurasian ideas. The term "Euroislam" appeared in the 1990s, and was first used to designate "secular Islam," that is, Islam quite consistent with secular organization of western states. With time, Euroislam come to be understood not as the Islam of Europe, but as a process of inscribing Islam into social order and state structure of European countries. Such an interpretation of Euroislam finds many adherents, since the problem of the Islamic integration into European communities becomes increasingly real in view of the increase of the number of Muslims in the European countries and owing to the need for constructive interconfessional dialogue to reduce the tension between the East and the West.

In its many codes, Shariat collides with democratic norms of western community. Therefore Euroislam demands the removal of the most odious norms of Shariat, such as beating unfaithful wives with stones, and cutting off thieves' hands. More complicated conceptual problems also arise. For instance, to what extent may Islam exist within by secular culture and a secular state while continuing to be a religion of strict monotheism? European Muslims would not accept loss of the religious content of Islam. Besides, there is no unified Islam in Europe: most Muslims in Germany are Turks, in France, natives of Magrib, in Holland, natives of Surinam and Indonesia, in Great Britain, and natives of India and Pakistan. They all have their own traditions of practicing Islam and their own specific manner of entering the "European house."

198   *Global Studies Encyclopedic Dictionary*

Globalization leads to the integration of various communities, in particular, to the integration of Muslims into European communities. Leaders of European states further the integration and Muslim communities in Europe as they make reciprocal steps.

A. G. Kosichenko

Europe: Etymology of "Europe" remains obscure, but common wisdom holds that it comes either from the Phoenician or Sumerian word for "sunset," "evening," or "twilight" (which, coincidentally, are also the meaning of the term "Occident," or "West") The current usage stems from the Greek *Europa*.

The Romans took over the Greeks' mental categorization of the known world and its inhabitants. The most significant division was that between the Romans (lumped together with the Greeks) and the barbarians. Geographic entities remained ideologically neutral as the Roman Empire extended in Europe, Asia, and Africa. The renewed ideologization of geography came with the rise of Christianity on the ruins of the Roman Empire.

In Antiquity, Europe was perceived as separated from Africa by the Mediterranean Sea and from Asia by the River Don, the Sea of Azov, the Black Sea, and the Bosporus Strait. Some perceived the region between the Black and Caspian Seas as the European-Asian boundary. The border between Asia and Africa was set on either the Nile River or the Red Sea. The first extant map of Europe, dating to the early twelfth century, schematically depicts the continent with the lands of the Slavs and the Magyars in the east. The successor states of Charlemagne's Roman-Frankish Empire dominate the map's center. Soon Europe was identified with Western Christendom that paid allegiance to the Pope at Rome, wrote in Latin, and used the Latin alphabet for noting vernaculars. The idea of the *Respublica Christiana* (Christian Commonwealth) was born.

Due to the reforms of Peter the Great, Muscovy was transformed into the Russian Empire that dynamically expanded eastward and southward. This empire could not be disregarded in European politics any more. The pro-European aspirations at premium in St Petersburg, in 1730, Philip Johan von Strahlenberg, a Swedish officer in the Russian service, proposed that the Europe's eastern boundary should be moved from the Don to the Ural Mountains and the Ural River. In a century, this would become the accepted view. In the nineteenth century, with Russia's expansion to the Pacific and Northern America, it became obvious that Europe is not a continent, but one of Eurasia's significant peninsulas or subcontinents on par with India or Indochina. The perceived geographic uniqueness of Europe and the idea of "civilizational supremacy" of Europe (construed as the West) continued unabated. This posed Russia with the question whether it were part of Europe.

In the nineteenth century, Western Europe accepted Russia as a power, but the popular image portrayed it as part of "wild and autocratic Asia." Famously, in the first half of the nineteenth century, the Austrian Chancellor Clemens Metternich opined that the East (as a synonym for Asia) extends eastward of Vienna. This introduced into the popular thinking the division between the West (Europe) and the East (Asia), construing the line extending from Danzig (Gdańsk) via Vienna to Trieste. After 1945, this mental divide was reinforced by the descent of the Iron Curtain that ran along the western reaches of the Soviet bloc.

As of the beginning of the 1950s, the name "Europe" was used elliptically for denoting the six members of the European Communities: Belgium, France, Italy, Luxembourg, the Netherlands, and West Germany. This narrowest of usages excluded the rest of Western European states. The situation changed in the years 1981–1995, when the European Union (EU) extended to coincide with entire Western Europe (significantly excluding Iceland, Norway, and Switzerland) and Greece. Now, in the popular political and journalistic usage, Europe is synonymous with the EU.

It also seems that the larger and growing EU of the future may not be as homogenous as one may hope. Since the 2004, expansion the old divide between the West and the former Soviet bloc has been replicated in the opposition of Old versus New Europe. In the context of globalization, the EU may not be able to face increasing economic and political competition, which could push it into a closer alliance with the United States and Canada. Would then such a pan-Western or Euro-Atlantic alliance remain "European" or would it be transformed into an altogether different entity?

Western critics of extending EU membership to post communist, post-Soviet, and post-Yugoslav states utilize the concept of a "return to Europe" also, arguing that decades (or longer) may transpire before these states will approach "Western (Western European) standards" in terms of the economic and democratic progress which would make them eligible for accession into the EU. In these critics' eyes, any "return to Europe" of states separated from Western Europe for forty years by the Iron Curtain is next to impossible. Some Central European intellectuals and politicians criticize this negative stance, arguing that their states and societies do not need to "return to Europe" because they have *always been* in Europe.

T. Kamusella

European Constitution: The first European Constitution was adopted at the EU summit, 18 June 2004, the culmination of a complicated and intensive two and a half-year "constitutional marathon" consisting of two consecutive stages: (1) the calling and holding of the European Convention (February 2002 through July 2003). Valeri Jiskar d'Esten, the former President of France, acted as the Convent Chairman, who the media called "a father of European constitution." (2) The project of the constitution, prepared by the Convent, was presented to Intergovernmental Conference of EU member states for discussion and revision (October 2003 through June 2004), the final point of which were the summit talks on 17–18 June 2004.

The full title of the document, "Treaty Establishing a Constitution for Europe," was intended to emphasize the content, form, and field of this source of law and the ambiguous legal nature of the European Union as a formation containing both the elements of an international (intergovernmental) organization and the elements of a federal state:

The treaty form of the European Constitution leaves to the member states "the final word" in its text approval and in making amendments to it. The treaty form of the Constitution is not a precedent: the Constitution of the

200 *Global Studies Encyclopedic Dictionary*

United States, the oldest one in the world, was nothing more than a treaty, judging from the order of its adoption (it could come into force only in those states that agreed to ratify it). The experience of the Soviet Union should also be remembered. It was established in 1922, by Treaty of Alliance (the treaty became the main part of the first Constitution of the Soviet Union of 1924) and was dismissed in 1991, by treaty order.

Components of a state system in the organization of the European Union are noticeable in the use of the word "constitution," which, in fact, forms the content of the document ("Treaty establishing a Constitution for Europe"). This very term is generally used in the text, starting from the first article: "Reflecting the will of the citizens and States of Europe to build a common future, this Constitution establishes the European Union . . . ."

The most enigmatic element in the title of the new Constitution is its last part—"for Europe." It is unprecedented too: In 1949, when the text of the future constitution of Germany was being developed in its western lands, the document was entitled "Basic Law for Federal Republic of Germany" having in view the prospective formation of the united German state.

Presumably, in the case of the Constitution "for Europe" it was meant to "invite" other countries of the continent to join the European Union and, accordingly, to subject themselves to the force of its "basic law." Because of its objectives and prospective nature, the new European constitution is destined geographically not only to the Union within its present limits. It is as if the constitution anticipates its further expansion throughout Europe.

Another cause of the above-mentioned formula is the fact that during the draft constitution development the European Convent discussed the opportunity of renaming the European Union as "United Europe." However, it was decided in the end to stand by the old (accustomed and familiar) name of the organization, its constitution still bears traces of these initiatives and retains them for future generations.

What changes introduce the new Constitution for Europe into the political and legal "architecture" of the European Union? How will the "new" Union differ from the "old" one, and who shall assign it?

First, the structure of the organization will be much simpler. The "new" European Union will have a single legal personality, including an international one. The present dichotomy between the "European Union" and the "European Community" will disappear.

The structure, competency and functioning of the new Union will be controlled by the common Constitution instead of the three present constituent acts and many other sources of "original right."

For the first time the Constitution carries through the "distribution of powers" of the European Union, on basis of the experience of federal states.

Innovation of the Constitution is the acknowledged right of the member states to leave the Union voluntarily. This is the first time that this right has been secured in the Constitution.

A system of institutions will be essentially reconstructed in the "new" Union to ensure better efficiency and democracy.

Similar to the basic laws of most states, the new Constitution for Europe determines the official symbolism of the European Union. Along with the present flag (a circle of twelve gold stars on a blue background), anthem (a fragment of the "Ode to Joy" by L. v. Beethoven) and Day of Europe (9 May), the Union will adopt a motto ("United in diversity"). The monetary unit of the Union, the Euro, will also be recognized as its symbol.

The last item in the list of the Constitution attainments, but not the least one by significance, is the final settlement of the problem of human rights protection at the level of the Union. For that purpose, the Constitution incorporates in Part II the full text of the Charter of Fundamental Rights of the European Union developed in 2000, by a special convention chaired by R. Herzog (a former President of FRG). The Constitution also solves the problem regarding the Union's membership as an individual party within the European Convention for the Protection of Human Rights and Fundamental Freedoms. (In this case acts of supranational institutions of the Union may be appealed at the European Court for Human Rights in Strasburg.) It demands the Union to join the Convention without reserve. This, however, will become possible only after necessary amendments are made in the text of the treaty.

By volume, the Constitution for Europe may be compared with the most extensive and detailed constitution act—the 1950 Constitution of India—and even surpasses it. Taken together with protocols and declarations, the final text of the Treaty Establishing a Constitution for Europe makes about 1000 pages.

Approval of the Constitution of the European Union reveals the beginning of a fundamentally new tendency of regional constitutional right formation in the greater part of the European continent.

S. Y. Kashkin, A. O. Chetverikov

European Union (EU): The EU is an economic and political union of states located primarily in Europe. The EU operates through a system of supranational independent institutions and intergovernmental negotiated decisions by the member states. The organization's principal aim is "creating an ever closer union among the peoples of Europe" (Article 1 of the Treaty on European Union).

Any European state that respects the democratic principles of social structure, "the principles of liberty, democracy, respect for human rights and fundamental freedoms, and the rule of law" may apply to become a member of the Union (Articles 6 and 49 of the Treaty on European Union).

To enter the European Union, a country should also have a sufficiently high level of economic development comparable with the average indices of the EU and should carry out the legal reforms to bring in advance its home legislation into conformity with legal norms of the European Union.

The European Union was formed gradually along with the intensification of integration processes between its member states and peoples. The step-like character of forming the European Union also shows itself in the modern structure of this organization.

Thus, from the viewpoint of its internal design, the European Union is an organization with a complex structure consisting of three components ("pillars"): European communities, CFSP, and PJC. The functioning of each "pillar" is ruled by separate legal sources of the European Union, which are different in nature and content.

202 *Global Studies Encyclopedic Dictionary*

As concerns the character of its competence, the European Union is a supra-state (supranational) organization of political power, in favor of which the member states have voluntarily limited their sovereignty. The current European Union is a state-like formation, which is gradually evolving toward a full-fledged federation. At the same time, the European Union is still keeping some common features of an international (intergovernmental) organization and a confederation of states.

S. Y. Kashkin

European Union Charter on Fundamental Rights is the fundamental act in the area of humanitarian rights, adopted by the European Union (EU) on 7 December 2000; it is a kind of European Union "Bill of Rights." The Charter sums up the entire previous development of the ideas of human rights in constitutional and international law, and not only in Europe, but all over the world. The contents and spirit of this important document were influenced by the European Council and by the law regulating human rights that was formed on its basis. The EU Charter has incorporated the basic achievements of the humanitarian law already developed by humankind and at the same time has approached these achievements creatively and in its own way.

The Charter is the first document to defend a person not only against the state and its bodies, but also against supra-state organizations and their bodies. The Western doctrine of human rights, traditionally divides rights and freedoms into major ones, which include civil and political rights, and "minor" ones, which include the group of socioeconomic rights. By contrast, the Charter as a whole considers the legal status of a person and a citizen of the European Union in unity and equality of rights and freedoms, thus demonstrating the principle of unity and nondiscrimination of rights and freedoms in the "European variant."

The Charter brings all personal, political, and socioeconomic rights together and recognizes them as basic ones. The main idea of the Charter is the declaration of the principle that the European Union "places a person at the center of its activity by means of the introduction of citizenship in the Union and creation of a space of freedom, safety, and justice." The Charter fixes the rights and freedoms of a person with the help of seven basic principles—values constituting a uniform complex: the legal status of a person, the principle of respect for human dignity, the principle of the maintenance of the rights and freedoms of the person and the citizen, the principle of equality, the principle of solidarity, the principle of democracy, and the principle of a lawful state. It is especially emphasized that these principles are based on the spiritual, moral, and historical heritage of the people of Europe.

One of the main features of the Charter is a new comprehension of the principle of nonviolence, shown in the aspiration to evolutionary creation without destruction, groundless denying, and violence.

The Charter has also reflected the tendency of the increasing interconnection between humanitarian regulations of constitutional law and international law. This demonstrates the universal tendency of the unification of humanitarian law and its consistent advancement to the leading positions. Together with the planned expansion of the EU, further development will be given to the tendency of territorial expansion of "the European humanitarian law," and to the expansion of its influence as an example or as "the world humanitarian standard."

The Charter firmly fixes an extended character of securing the rights and freedoms of a person. First it demands unconditional observance of the rights and freedoms that are already fixed by current legislation. Besides, whenever possible it provides for the constant addition of rights arising as a result of social progress and scientific and technological development; at the same time, it aspires to fix not the minimum level as it is usually accepted in international legal practice, but the highest achievable maximum level of their number and implementation. At the same time the Charter works as an instrument of acceleration in the creation of civil societies within countries that are members of the EU and the establishment of a civil society within whole European integration group. The democratic globalism of the future can be constructed only on the basis of humanistic regionalism. In this regard states should inevitably transfer some of their powers to a supranational civil society in the course of its development. Supranational civil societies will expand their regional borders in a planetary direction.

The Charter of the European Union on fundamental rights has strengthened and expanded the opportunities of judicial defense of the rights and freedoms of citizens of the EU and others who stay within the territory of the EU.

The Charter evinces the aspiration to provide internal stability and security with the help of the instruments of the rights and freedoms created mostly as a result of the activity of the European Council, to reach a consensus within the framework of the integration group, reducing class, national, and other confrontations to a minimum, whenever possible avoiding rough supranational or state compulsion, peaceful resolution of conflicts arising out of the interests of citizens, balancing interests of individuals, places, regions, and countries within the framework of the Union, providing thus the support to the weak and prospects to the strong, finding effective stimulus and instruments for progress and ways to allow the majority of the population of the European Union to use its fruits.

The character of the Charter allows us to see an essentially new tendency of development of the humanitarian law in the twenty-first century, searching for legal means of restricting the negative consequences of integration, internationalization, and globalization. The Charter undertakes an attempt to find a new, more harmonious combination of rights and freedoms with duties and civil responsibilities to "human society and future generations."

S. Y. Kashkin

European Union Law is a body of treaties and legislation, such as Regulations and Directives, which have direct or indirect effect on the laws of European Union member states. The three sources of European Union law are primary law, the Treaties establishing the European Union; secondary law, regulations and directives based on the Treaties, which are established by the European Parliament and the Council of the European Union; and supplementary law. The functional principles of the European Union law include supremacy of the EC law—priority of the norms of the EC law over the norms of the na-

204 *Global Studies Encyclopedic Dictionary*

tional legislation of member states—and direct action—direct application of European Union law on the territories of member states; its norms are implanted into national legal systems without any transformation. These principles have been developed by the court practice in interpreting the constituent documents of the organization. General principles of the European Union law include protection of rights and personal freedoms, legal certainty, proportionality, non-discrimination, subsidiarity, and several procedural principles.

When defining the sources of the secondary law, the continental and Anglo-Saxon legal systems sometimes collide over the issue of the recognition of jurisdictional acts as sources; international law has influence.

The first category of secondary law acts is statutory acts, including court rules, directions, frame decisions, general decisions of the ECSC, and recommendations of the ECSC. The second category is individual acts, including other decisions. The third category is recommendatory acts, including other recommendations and conclusions. The next category of secondary law acts is acts on coordination of the Common Foreign and Security Policy (CFSP), and on cooperation of the police institutions and judicial bodies in the sphere of criminal law. This category of acts includes principles and general guidelines, general position, joint actions, and general strategy. A separate category of acts is made out of jurisdiction acts—decisions of the Court. Secondary law sources include sui generis acts–the "informal" legal acts adopted by bodies of the Union (usually decisions or resolutions of a concrete body) that have not been stipulated by constituent treaties. The last category of secondary law sources, defined as international acts, includes: decisions and acts of representatives of member states; conventions between member states made on the basis of constituent treaties; international treaties of the European Union.

The specifics of the European Union also predetermine structural features of European Union law. European Union law also has regulations on human rights and freedoms, norms adopted according to the CFSP, the European community law.

Currently, the European Union law is characterized by codification and enforcement. The Laaken Declaration of 2001, adopted at the summit of heads of the states/governments within the framework of the European Council, emphasizes the necessity of reform of sources of the original and secondary law of the European Union, simplification of legal forms, and establishment of the full-value Constitution of the European Union on the basis of constituent treaties of the European Union and the Charter of the European Union on the fundamental laws of 2000.

S. Y. Kashkin, P. A. Kalinichenko

Evolutionism, Global: Global evolutionism is a research trend, representatives of which aim at developing of new possibilities for the analysis of the strategic dynamics of life and activities of humankind with the help of the systematic development of the image of global evolution. It is an integral evolutionary process, in the course of which there is a succession of creation and development of objects in the inorganic world, then objects in the organic world, and then in the world of social systems. This trend began in the 1980s, when there was a rise in interest in global problems and interdisciplinary research. At that time, it was realized that the vagueness and complexity of global problems could be lessened by studying them in a broader context, which is naturally set by the image of global evolution. This trend was named with a special term because evolutionism approached the qualitatively new step and form of its development. It is no longer philosophical evolutionism, but a more detailed and systematic one based on continuous consideration of the synthetic theory of evolution, modern achievements in studies of social evolution, social and biological ecology, and interdisciplinary approaches and concepts. Another reason for self-determination of the new trend was that the traditional scientific structure could not incorporate the newly formed trans-disciplinary theory of evolution.

The basis for this developing trend is the hope of discovering common evolutionary rules, which can be used for studies and updates of social dynamics. The possibility of this is supported by cybernetics, systemology, rhythmology, synergetics, and other universal, trans-disciplinary research that shows the high degree of similarity between the spheres of life of different origin. This can also be confirmed by the fact that development of non-equilibrium thermodynamics permitted removal of a barrier between former evolutionary antipodes-animate nature and abiocoen—and inclusion of them in the uniform conceptual frameworks successfully set by synergetics. The worsening of the ecological situation in the last decades is a reminder of the fact that a person and society cannot be separated from nature, and consequently an approach is adequate only when it includes a person in fundamental structures and processes of the Universe, that is, they should initially be considered together.

An important precondition of taking evolutionary views to a new level in the middle of the twentieth century was the intensive development of various forms of complex activity that demanded close and interested interaction of researchers of different specializations, thus promoting removal of professional barriers and the transmission of ideas and images from one scientific sphere into another. Thus, there was a gradually grown tendency to carry common evolutionary views up to essentially more developed levels on the basis of wider usage of modern conceptual means and approaches. So far global evolutionism has been at the developing stage; therefore, research in this field may have different terminological description ("Universal evolutionism," "Big history," and "Study of the self-organizing Universe"). Self-determination of the global evolutionism puts forward two important and generally valid groups of questions. The first group has a gnoseological character and deals with the clarification of ways of construction and the organic inclusion of this new trend into the structure of science. It is more and more obvious that the traditional dichotomy between "particular scientific knowledge" and "philosophy" does not express the realities of research practice anymore and demands development and additions aiming for more precise identification of modern sciences of trans-disciplinary cycles.

The second group of questions, ontological, is connected to the necessity of correct description of many objects of different natures that science knows, with the help of a uniform model of scaled evolutionary process. It is no longer enough to be guided only by a traditional philosophical model of one-directional and linear evolution. There is the aim to learn to feature the

206 Global Studies Encyclopedic Dictionary

unity of system organization of structural levels of life in order to make external analogies visible between, for example, an organism and a society, and to study in its entirety the assumed fundamental homogeneity of space within the evolutionary process. The new evolutionary paradigm is developed within the framework of two major gradually approaching points of view. One of them, synergetic, relies upon solid models of self-organization, developed first by abiocoen sciences, and extrapolates them into other spheres of scientific knowledge. The second approach, evolutionary in essence, assumes the system of biological concepts as a basis, taking into account the high level of their development and the natural proximity of biology, on the one hand, to social knowledge, and, on the other hand, to knowledge of abiocoen that facilitates translation of ideas.

The prospects for the development of the Russian version of global evolutionism are connected to continuous concrete definition of some of the general ideas defining its specificity: (1) The historical existence of the Universe is not considered any more as a one-directional movement toward complication or "thermal death." It is more important to perceive interrelations of these two major types of changes, forming together the process of "cosmogenesis." Global evolutionism is focused on the theoretical reproduction of mainly "historical" constituents of cosmogenesis that are connected to the creation of the increasing number of different objects, with the originating of more and more complex forms. (2) The initial objects of the analysis are "evolutions"-complex systems with evolutionary self-determination; e.g., the noosphere and the biosphere.) Evolutionary changes are clearly revealed in their dynamics and that permits consideration of global evolution first as a continuous creation of more and more new evolutions. The unity of the structure and dynamics of objects of this class still requires more study. (3) The historical development of evolutions is considered as a co-evolutionary process that occurs within more scaled systems and is largely predetermined by limitations and peculiar features of this medium. Lastly, while studying global evolution, special attention is drawn to its stages that can supersede one another in a spasmodic and very ambiguous way.

A. A. Krushanov

⊰ F ⊱

***Factor Four*: *Halving Resource Use*:** (1997) is the report to the Club of Rome prepared by Ernst U. Weizsäcker, Amory B. Lovins, and L. Hunter Lovins, who all prominent experts in the field of the preservation of the environment. The report substantiated the concept of "resource productivity," understood as an opportunity to live twice as well while spending half as much. According to the report's authors, since the time of the industrial revolution, progress has meant an increase in labor productivity. But the issue is an increase in the efficiency of resources. This goal can be achieved by using electric power, fuel, materials, fertile grounds, and water more effectively, frequently without additional expenses and even with benefit. The report contains a lot of concrete examples, graphs, tables, and figures showing that the majority of technical solutions to many of our problems already exist and should only be applied. The authors also used experience and a great number of actual materials to show the way of making "the revolution of efficiency" a really profitable business.

Reference:Weizsäcker, E. U., A. B. Lovins, and L. H. Lovins. *Factor Four*. London: Earthscan, 1997.

A. N. Chumakov

Financial Market, Global: The global financial market is a system of global financial flows, mediated mostly by stock exchanges of industrially advanced countries, where purchase and sale deals with securities (mostly shares of leading transnational corporations and companies) are held. The forming of the world financial market is the main manifestation of globalization at the turn of the twentieth and twenty-first century. The global financial market per se is a system of nerve knots and endings of the modern world economy that has some independence concerning flows of commodity and services, circulating in the system of world economic ties. The analogy to the nervous system, holographically presented at the global level, is not a simple metaphor; it is quite a precise description of structural-functional features of the global financial market that has central and peripheral subsystems.

The central system of the global financial market is represented by the stock exchanges of the leading industrially developed countries functioning on the basis of hard currencies, and peripherally by security markets of countries with non-convertible currencies. The largest stock exchanges of the world are: the New York exchange (founded in 1792), located on Wall Street—a symbol of the global financial market, the Tokyo exchange, the London exchange (founded in 1773), the Frankfurt exchange, the Amsterdam exchange (the oldest exchange in the world, founded in 1602), the Zurich exchange, the Hong Kong exchange, and the exchanges of some other countries.

The basis of the global financial market is the construction of financial pyramids, which defines sustainability limits of this market within world economy. Stable functioning of pyramidal structures presupposes successive mobilization of financial resources on the global scale in order to construct a limited number of financial pyramids, mainly in the United States and Western Europe (on the basis of the world currencies—the American dollar and

208   *Global Studies Encyclopedic Dictionary*

the euro). The periphery of the global financial market therefore is characterized by immanently inherent stresses (instability and unsteadiness), resulting in recurrent defaults and devaluations (depreciation) of national (non-convertible) currencies.

Globalization promoted the forming of the global financial market. Starting from the end of the World War II and up to the beginning of the 1970s, world capital markets almost did not function, because to a large extent national governments regulated and even hampered international financial flows. Liberalization of external economic regimes at the end of the 1970s and the beginning of the 1980s, politically stipulated by the coming into power of conservatives and right-wing conservatives led by Margaret Thatcher in Great Britain, Ronald Reagan in the United States, and Herbert Kohl in Germany, has predetermined the subsequent enormous growth of global flows of financial resources. As calculated by the International Monetary Fund (IMF), between 1970 and 2000, the volume of international flows of financial resources in relation to GDP of industrially developed countries has grown ten times, and in relation to GDP of developing countries it has grown five times. Between 1990 and 2000, the gross volume of financial flows, circulating within the limits of twenty-one industrially developed countries of the world, has grown fourfold, from $1.5 up to almost $6.5 billion USD.

The net-effect of global circulation (mobility) of global financial resources is the construction of the next levels of financial pyramids, among which the financial pyramid of Wall Street is the most noticeable. Thus, in 2000, net-inflow of financial resources into the global economy has amounted to 424 billion dollars, from which the share of the United States was $407 billion (96 percent), the share of other industrially developed countries was $8 billion (2 percent), and the share of "developing markets," that is, all developing countries, republics of the former Soviet Union, countries of the Eastern Europe, and Israel, Singapore, Taiwan, and South Korea, whose total population is about five billion people, was $9 billion (also 2 percent).

Globalization of the world financial markets also predetermines the forming of a specific rhythm of their development in the form of paraboloidal dynamics of indexes of market financial activity, when every twenty years there is a steady increase in the activity of stock exchanges, which is followed by either a large market collapse or gradual "cooling" of exchange activity, and this period as well lasts for about twenty years. After the Second World War stock markets, especially those of the United States, developed intensively during the period from the middle of the 1940s, up to the beginning of the 1960, then they were in a state of "hibernation" up to the beginning of the 1980s. After that, they developed quite rapidly for twenty years until 2000, and then entered the state of continuous sluggish recession of activity of the financial markets. In particular, from 1980 until 2000, the mid-annual Dow Jones index of industrial activity has grown twelve times—from 891.5 points up to 10,735 points, the high technology corporations and companies (blue chips) index, first calculated in 1971, when its initial level was taken as 100 points, has grown 22.5 times (from 168.6 points to 3,788 points).

The process of globalization of the financial markets creates a potential for the "collapse" of global financial systems, similar to the Wall Street financial collapse of 1929. Many analysts and leading economists believe that the situation in the global financial market is reminiscent of the situation at the end of the 1920s, which also was a direct consequence of "miniglobalization" of the world financial system and the sharp strengthening of the U.S. dollar in global economics after the First World War, the result of which was the well-known "decline of Europe" and the collapse of the Europocentrism model of world economics and civilization.

The global financial market executes the major function for the establishment of the "market price" of world civilizations, which, however, does not always coincide with their actual "cost" and the significance for current and future planetary development.

V. S. Vasiliev

The First Global Revolution (1991) was a Club of Rome report prepared by its Alexander King and Bertram Schneider. The decision to prepare it was approved in 1989, at the Hanover Club of Rome session, where it was noted that the situation that had arisen in the world required reinterpretation, owing to which the new Club tasks were to be defined.

The report noted that since the first report to the Club of Rome was published in 1972, the roots of world problems had stayed the same while, the range of questions and problems demanding special attention had changed. Among the variety of different problems they selected that required urgent solution: military economic conversion and a switch to a peaceful course; reorganization of power consumption in order to prevent global warming; the impact of processes of world development and overcoming the backwardness of developing nations. The report emphasizes the role of personality in history, ethical issues, and social management problems. The authors critically analyzed various economic systems, mentioning the poor efficacy of the existing state and international management structures, and presenting the valuable role of informal organizations and public associations. When discussing the merits and demerits of a market economy and the related democratic administration forms, they agreed with Winston Churchill who, in their opinion, was absolutely right in stating, "Democracy is the worst among the systems, except for the rest." Thus, the choice is between the bad and the worst, and being aware of that the world community has also to realize that it will be able to survive only if it unites in the face of common troubles.

Reference: King, A., and Schneider, B. *The First Global Revolution.* New York: Pantheon Books, 1991.

A. N. Chumakov

Formation Theory was developed by Karl Marx, who presented world historical and social dynamics as a logical change of stages or socioeconomic formations based on modes of production. Marx distinguishes four (or in some works five) types of these formations in world history: primitive communal, feudal (Asian), capitalist, and communist. The structural kernel of formation is displayed as dialectical interaction of productive forces and production relations that stipulates the political superstructure (the type of state)

210 *Global Studies Encyclopedic Dictionary*

and the forms of social consciousness (ideology, culture, or religion) in the formation.

At a designated stage of historical development, unavoidable aggravation of productive forces and production relations occurs because the former naturally develop faster. The productive forces, having gone far beyond the production relations in their development, generate a conflict. This conflict can be settled by means of a revolutionary outburst or by a leap, after which a new proportion of the productive forces and of production relations arises, building up a new political superstructure and new forms of social consciousness.

The method of the articulation of world history into the stages, although based on a different principle and causing different corollaries, was already used before Marx, for example, in the systems of G. W. F. Hegel and Charles Fourier. But Marx was interested in two formations more than in the whole of world history: the capitalist one contemporary to Marx (he devoted the entire cycle of wide and fundamental economic investigations to its analysis) and the communist one, which he thought would inevitably emerge. The most important aspect of formation theory is Marx's interpretation of the transitional process from capitalist to communist formation, based on two main theses:

(1) The transition from capitalism to communism is a worldwide process or at least a process that should involve all the civilized and developed countries. That is why the communist revolution is necessarily a world revolution. Certainly, such a process cannot be galloping and, moreover, it cannot be completed by one act, but it consists of a series of communist revolutions that would be performed primarily in the countries like France, Germany, and England and then expanded to the periphery of capitalist formation.

(2) Communist society naturally rises up from the capitalist one, and acts as anti-capitalism and post capitalism. The world capitalist market is surmounted by the communist world economy, thus eliminating the competence of world economic relations, producing a homogeneous system, and striving for the confluence of national economic structures. During this period, according to Marx, the worldwide character of the transitional processes from capitalism to communism correlates to the line of social, economic and political reformations and cannot make sense without them. The model of social dynamics that substantiates formation theory constitutes a synthesis of the linear-stage progress (here we obviously notice the influence of the eighteenth century Enlightenment thought on Marx). This ascension of humanity from one formation to the other is understood as a manifestation of the social life spheres' increasing progress and the social systems' cyclical circulation (here Marx makes use of the ideas of ancient philosophers, Hegel, and Vico). In this ascension each formation runs its natural historical way of rise, maturing, and fall. During the transition from one formation to the other the similar repetitive situations occur: the conflict of productive forces and production relations, ruined then by revolution, or by the switch of the leading class and of the system of property relations.

Y. N. Moshchelkov

Freedom: (Old English, *freodom*, "freedom," "state of free will," "charter," "emancipation," "deliverance") Freedom denotes the power or right to act, speak, or think as one wants without hindrance or restraint. In social terms, the problem of freedom may be interpreted from the standpoint of the following alternative approaches, the individualistic and the socio-collectivistic (communitarian) models. The first model is based on the position of isolated individuals who possess natural rights and who are oriented toward wholly rational behavior. The logical implication of this position is the assumption that a social contract should be concluded between individuals that would limit everyone's unrestrained urges. The second position states that there are no such abstract isolated individuals in nature, and that freedom cannot be individual at all, since it originates from social institutions, practice, and social traditions.

V. V. Lyakh

Frolov, Ivan Timofeevich (1929–1999) was a Soviet and Russian philosopher, political and social activist, and one of the founders of global studies in the Soviet Union. Under his leadership in the Scientific Council on Philosophical and Social Problems of Science and Technology at the Presidium of the Soviet Union Academy of Sciences (1980), the first scholarly conferences on global problems were planned and held. After a series of his works on global studies had been published in the late 1980s, his philosophical definition of global problems became widely accepted by the Soviet academic community. He was also one of the first scholars to basic characteristics of global problems and formulated criteria that could discriminate and classify them.

A. N. Chumakov

ℂ G ℬ

Gandhism refers to the teachings and social practices associated with Mohandas Karamchand (aka Mahatma, or "a great soul") (1869–1948), preeminent leader of Indian nationalism in the then British-ruled India.

After studying law in London, Gandhi went to South Africa in 1893, where he spent twenty years working against discriminatory practices against Indians. There, he organized Satyagraha ("insistence on truth")—a campaign of peaceful civil disobedience to oppose the crudest manifestations of racism, which his compatriots faced in British Pretoria (1908–1914). Satyagraha attracted the attention of the international community. Among those who supported Gandhi were Leo Tolstoy, George Bernard Shaw, Albert Einstein, Bertrand Russell, and Romain Rolland. After his return to India in 1915, Gandhi again employed nonviolent civil disobedience. In 1920, at an extraordinary session of the Indian National Congress, Satyagraha was acknowledged as a national method of struggle against colonialism.

Gandhi considered nonviolence (*ahimsa*) as "one's highest religious duty." He consistently followed the principle of *ahimsa* all his life by conducting experiments with different forms of nonviolence on both individual and social levels. Gandhi based on ahimsa his visions for the future independent India and for world civilization as such. Hinduism (the Bhagavad Gita, in particular) and Jainism (primarily *ahimsa*) were influential in shaping Gandhi's world outlook. He also repeatedly acknowledged the influence of three Western writers, Henry David Thoreau, Leo Tolstoy, and John Ruskin.

Being a reformer, Gandhi reinterpreted several fundamental notions of the Indian spiritual tradition: he revised his previous formula "God is Truth" into "Truth is God"; interpreted *moksha* ("liberation from the chain of rebirths") as *swaraj*, self-rule, liberation from the dictates of government, a nonviolent political structure; called upon people to undergo *tapas* ("penance") of a new type-experiments with collective forms of asceticism at special farm-settlements; turned *ashrama* (a center for meditation and ascetic practices) into a kind of laboratory for social experiments to actualize collectively the principle of nonviolence.

Gandhi saw the national liberation of India and the radical reorganization of Indian society on nonviolent principles as the first step to achievement of a worldwide nonviolent civilization. He considered Western civilization to be an incarnation of "evil." He hated machine production since it was destroying cottage industries and the traditional Indian way of life, which was organically interconnected with nature.

Gandhi viewed nonviolent civilization as based on the principles of conscious and voluntary self-restraint. In the economics, it was to rely on the villages and cottage industries, on decentralization of production basically through cooperatives, on removal of exploitation and private property through trusteeship. He regarded the ideal political structure as a confederation of free and voluntary interacting villages governed through panchayats— five men and women, elected by the villages. He supported a concept of "direct democracy" based on labor status as the only electoral qualification, apart from age; opposed caste or religious discrimination; strongly voiced against imperialism and military aggression of any kind.

In his outer appearance and in his use of terms, concepts, and symbols of Hinduism, Gandhi constantly demonstrated an inseparable link with Indian culture. At the same time, he never tired of insisting on the unity of humankind and its ultimate aim, that is, a quest for Truth-God. Though some elements of Gandhism are utopian and even conservative, it had a great positive impact on world public consciousness, in particular, on the struggle against racism and militarism.

M. Stepanyants

Gender Oppression: Global (Mis)treatment of Women: Women in many places around the world experience various forms of oppression that have become issues of concern worldwide. These include sex trafficking, female genital mutilation, domestic violence, women being separated from their children while serving prison sentences, females being denied education and vocational training, forced and child marriage, sexual molestation and incest, and inequitable treatment in the job market.

Gender can be defined as the sociocultural interpretation of the significance of sex. Whereas "sex" designates our biological (determinist) conception of self, "gender" points to the socially-defined understanding of what it means to be a woman or a man. The sex-gender distinction implies that, unlike sex, gender is contingent upon social norms. Two competing ideological frameworks can be noted. The homogeneous model stresses women's sameness, whereas the heterogeneous model focuses on women's differences. The heterogeneous model, situated in the multicultural and global feminist agenda, is more promising because it provides feminist theorists with an honest portrayal of what it means for women to suffer from gender oppression.

Proponents of the homogeneous model insist that feminists must focus on women's sameness, because this theoretical approach provides a greater sense of political solidarity—a vital component for reframing legislation. However, critics question whether feminists need a unified and static definition of gender or "woman." On their view, the homogeneous model can universalize the perspectives of the dominant group and, as a consequence, veil the concerns, grievances, and specific needs of others who are also marginalized in society.

Global feminism focuses on the implication of dividing the modern world into first and third world countries, namely, the oppressive results of colonialism, imperialism, and big world government. Fostering a transnational feminist practice, global feminists ask that feminism broaden its scope of feminist thought to include the needs and concerns of women from all countries and not just first world countries. However, critics of the heterogeneous model warn that spotlighting women's differences can work against feminist/political solidarity and hamper social, political, and moral progress. On their view, pointing out differences risks one group or characteristic to claim superiority to another group or characteristic.

For a truly inclusive theory, a feminist agenda cannot claim that all women are alike even if they live in the same country, state, or neighborhood.

214 *Global Studies Encyclopedic Dictionary*

That is why the heterogeneous model, couched in the tenets of multicultural and global feminism, may provide feminists with a more critical understanding of gender oppression and how to go about ending that injustice. Proponents of the multicultural/global camp are correct to recognize that political marginalization is contextual in essence, that is, a woman who thinks of herself as oppressed or wrongly discriminated against in her society may discover that upon traveling to another country she nevertheless has more power than many other women who do not share her privilege of class and nationality.

We must be vigilant about overstating the differences among women. All over the world, women are more likely than men to be sexually exploited, suffer from discriminatory employment policies and practices, and endure unjust laws and practices concerning marriages, reproductive rights/ideologies, and child custody. We cannot let color, race, class, ethnicity or sexual orientation separate women from their underlying kindred spirits and goals

N. M. Williams

Geneva Conventions are the multilateral agreements in the field of the legislation of war aimed at protecting war victims. The Red Cross first put humane rules for the prosecution of war crimes forward in 1863. The, the first Geneva Convention, for the Amelioration of the Condition of the Wounded in Armies in the Field was adopted in 1864. It was aimed at improvement of the fate of soldiers, wounded or fallen ill, during military operations. This first effort provided only for: (1) the immunity from capture and destruction of all establishments for the treatment of wounded and sick soldiers; (2) the impartial reception and treatment of all combatants; (3) the protection of civilians providing aid to the wounded; and (4) the recognition of the Red Cross symbol as a means of identifying persons and equipment covered by the agreement. Despite its basic mandates, it was successful in effecting significant and rapid reforms. Due to ambiguities in the terms and the rapidly changing nature of warfare, subsequent conventions were established.

The second Geneva Convention (1906) had to do with the behavior of armed forces both at battlefield and at sea. The third Geneva Convention (1929) settled that conditions of the agreements concerning not only citizens of the countries that ratified the Convention, but all the people, irrespective of their citizenship (at that, not only military personnel, but also civilians). The fourth Geneva Convention (1949) set strict standards on protection of civilians and military personnel during war in places of military operations and in occupied territories (it also demanded to prohibit the war crimes). In 1977, two additional protocols to the 1949 Convention were adopted, which concern protection of the victims of the international armed conflicts and the victims of the internal conflicts and in 2005, a third protocol was adopted, which concerns a new international symbol that can be worn by medical or religious personnel in addition to the traditional Red Cross and Red Crescent. The four 1949 Conventions have been ratified by 195 states, including all UN member states as well as the Holy See and Cook Islands. As of June 2013, Protocol I had been ratified by 173 states, with the United States, Israel, Iran, Pakistan, India, and Turkey being notable exceptions. As of April 2013, the Protocol had been ratified by 167 countries, with the United States, Turkey, Israel, Iran, Pakistan, and Iraq being notable exceptions.

V. N. Kuznetsov

Genocide: International jurist Raphael Lemkin coined the term "genocide" to designate one of the greatest crimes against humanity, namely the Holocaust perpetrated by Nazi Germany. The UN Convention on the Prevention and Punishment of Genocide was adopted 9 December 1948 (General Assembly Resolution 260, entering into force 12 January 1951) adopted this term in identifying a crime punishable by international law whether in times of war or in peace (Article 1). The Holocaust represented what many intuitively found to be most heinous about the crime designated as genocide. This was the utter and irreversible attempt to annihilate a people, including the destruction with them of their culture and any genetic links with the future.

Most commentators insist on the necessity of intention as evidence of purposive, premeditated massacres organized or coordinated by government or military functionaries. Without stringent conditions for intentionality, say analysts, it would not be possible to distinguish genocide specifically from other forms of state sponsored mass murder, such as the use of death squads in El Salvador or the disappearances under the junta of the colonels in Argentina.

An alternative argument is that conclusive evidence of premeditated and planned state policy is too stringent a standard, especially since governments can lie about their intentions and obstruct efforts to uncover them. For this reason, it is sufficient to be able to impute or infer intent from consequences. Genocide occurs when the foreseeable and cumulative results of a course of action are the extermination of an out-group, and a state either produces this outcome or acquiesces by consistently refusing or failing to protect victims.

Concerns about the restrictiveness of the definition have led some commentators to move further away from traditional notions of intent to emphasize systematic patterns in which political, economic, and social changes, together with the negligence of a governing regime, collude in the massacre of a group, although the destruction of the group was not explicitly intended. For instance, in his analysis of the fate of the Australian Aborigines, Tony Barta argues that genocide can result from "relations of destruction" inherent within a political and social system (Wallimann and Dobkowski, 1987).

As with genocide more generally, questions have arisen about the relevance of acts of omission leading to mass death in contrast to intentional and active killing. It can be argued that when knowledge of the outcome is known, and there is both the capacity to avert the catastrophe and an intention not to do so, then the responsible regime may be guilty of autogenocide. Thus, one might argue that in presiding over the decimation by starvation of the people of North Korea, Kim Jong Il is guilty of autogenocide, as was the Chinese government of the 1960s, in deciding to permit tens of millions to perish from famine. These cases contrast clearly with tragedies in which the ravages of famine or disease decimate a population though the governing regime seeks to end the tragedy.

Disagreements over the definition of genocide often reflect underlying concerns about the best way the international community can respond to gen-

216 *Global Studies Encyclopedic Dictionary*

ocidal states or developmental tendencies that have genocidal effects. The Genocide Convention stated in Article 1 that genocide is "a crime under international law" and in 1951, the International Court of Justice reaffirmed its status as international law. The Convention has been ratified by such a large majority of states that it is now considered a rule of customary international law, binding on all states (whether a signatory or not) and requiring them to prosecute acts of genocide (Robertson, 1999).

Much effort has gone into explaining how human beings are capable of the atrocities of genocide. The rapidity with which genocidal hatreds can explode into orgies of killing, as in Rwanda in 1994, suggests to some theorists that genocide may be linked to innately murderous or aggressive drives that occasionally break through the veneer of civilization. Also cited as supporting these views are the facts that genocide was neither a modern nor especially European phenomenon.

Sociobiological explanations, along with some psychoanalytical theories, are deterministic in purporting to establish direct causal links between aggression that is genetically or psychologically based and genocide. However, since presumably all human beings possess potentially the same murderous drives, such explanations do not account for the critical differential factors leading some to participate in genocide while others do not. Moreover, plausibly testable hypotheses generated by determinist theories have failed to be supported by available evidence. The notion that the Holocaust had been perpetrated by psychotic killers or fanatics who were abnormal in their cruelty or aggression was refuted by psychological studies of Nazis at the Nuremberg war crime trials (Borofsky and Brand, 1980; Dicks, 1972) and by philosopher Hannah Arendt's study of Eichmann, resulting in her theory of the "banality of evil" (1963).

It was the original hope of the United Nations' delegates in 1948 (reaffirmed at subsequent conventions) that universal agreement on the protection of human rights would prevent future genocides. This has not yet occurred, but more effective protection of human rights, a better understanding of "early warning" symptoms, together with decisive intervention and swift and sure justice for criminals, provide the best prospects for the future.

References: Arendt, H. *Eichmann in Jerusalem*. New York: Viking, 1963. Dicks, H. V. *Licensed Mass Murder*. New York: Basic Books, 1972. Borofsky, G. L., and D. J. Brand. "Personality Organization and Psychological Functioning of the Nuremberg War Criminals." In *Survivors, Victims, and Perpetrators*. Edited by J. E. Dimsdale. Washington, DC: Hemisphere, 1980. Walliman, Isidor, and Michael N. Dobkowski, eds. *Genocide and the Modern Age*. New York: Greenwood Press, 1987. Staub, E. *The Roots of Evil*. New York: Cambridge University Press, 1989. Robertson, G. *Crimes against Humanity*. London: Allen Lane, 1999.

R. P. Churchill

Geocivilization is a part of the globe inhabited by a community of people with distinct history, ethnic origins, tradition, way of life, rituals, mentality, and values. Among modern geocivilizations we can list: Western, Islamic, Orthodox, Confucian-Buddhist, Hindu, Latin American, Russian, Japanese, and, perhaps, African (Samuel Huntington). Culture is the backbone a geocivilization. Initially, when geocivilizations sprang up, religion was the determining factor within culture itself, playing the role of a "cocoon," as Arthur J. Toynbee put it. Ethnic identity can also be a backbone feature. For Russian civilization that is a unique community of people belonging to different ethnic groups and different religious faiths (Orthodox, Catholic, Protestant, Muslim, Buddhist and Jewish), the determining feature is its history of co-habitation. This implies the use of a common language, norms of behavior, and value system. The backbone feature of the Russian geocivilization is historical destiny.

Geocivilizations may contain sub-civilizations. For instance, within the Western (otherwise called West-Christian, West-European, Catholic-Protestant) civilization, can be distinguished European, North American, and Australian sub-civilizations. Within the Islamic civilization, we can distinguish a Sunni, Shi'a, African Islamic, and, probably Turkish, Indonesian, and Malaysian sub-civilizations. Tibeto-Buddhist and Japanese sub-civilizations within the Confucian-Buddhist civilization.

Russia is rightly called an independent Russian civilization. Within the overall Russian culture can be found elements of different sub-cultures that have become closely interwoven. A unity was formed containing the Christian Orthodox nucleus and Islamic, Buddhist, and other threads in the original fabric of the Orthodox culture. Elements of different cultures co-existed peacefully for quite a while. Only recently have contradictions arisen in the intercultural interaction that it is extremely important to consolidate, while being conscious of cultural variations and the integrity of the Russian culture. The existing diversity should contribute to richness of the common culture, not to hostility and conflicts.

Geocivilizations, especially neighboring ones, are in close interaction. They can cooperate, when the achievements of one geocivilization become the property of other civilizations. At the same time, every local civilization seeks to exert pressure on neighboring civilizations, especially if it is strong enough. Expansion is a natural demand of a civilization (Toynbee), including territorial expansion, attracting a labor force, and assimilating neighboring peoples. Geocivilizations undertake expansion in order to control the material and labor resources of other societies and to disseminate their values. The main forms of expansion are external pressure (political, economic, cultural, demographic, military, and informational) and military attacks. The history of civilizations is an endless chain of wars waged for global domination.

S. G. Kiselyov

Geoeconomics is a branch of the social sciences that emerged in the 1990s, at the interface between economics and political science. Geoeconomics combines studies in economic history, economic geography, world economics, political science, conflict studies, and theory of managerial systems. The field of geoeconomic study includes:

(1) Geographic issues related to the organic links between economy and space, to the impact of climate and landscape on economic activities.

218 *Global Studies Encyclopedic Dictionary*

(2) "Power and its instruments;" the transformation of power from military-political to an economic one generating special type of conflicts: global geoeconomic collisions.

(3) Politics and strategy of enhancing competitive power of states within the framework of economic globalization (the world economy's unification and its absorption by the "universal market").

(4) Spatial localization (geographic and transgeographic) of different types of economic activity within the new global reality; new typology of the international division of labor.

(5) Politics and economics merged in the field of international relations; formation of strategic interrelationships and foundations for global governance.

A continuity of approaches and instruments of geopolitics is traceable in the genesis of present-day geoeconomics that emerged at the end of the twentieth century. Edward Luttwak (1990), compared geopolitics with its emphasis on the use of military force to achieve foreign policy goals to geoeconomics as a policy oriented toward winning economic competition.

In Italy, General Carlo Jean developed the concept of geoeconomics. According to him, "geoeconomics is based not only on the logic, but on the syntax of geopolitics and geostrategy, and in a wider sense on the whole praxis of conflict situations" (1991).

In Russia, scholars and governmental institutions assimilated the geoeconomic methodology and approach right after the disintegration of the Soviet Union. The Russian concept of geoeconomics interpreted the issue in a way different from the mainstream ideas of geopolitics and conflict studies. Geoeconomics is understood as a spatial localization of types of economic activities in a global context and a new type of world division of labor, and the merging of international politics and economy in the sphere of international relations (Alexander I. Neklessa).

Presently, the following two research models predominate in Russian geoeconomics: (1) A hexagonal structure of the global geoeconomic realm (Neklessa), and (2) An understanding of geoeconomics as a policy and strategy of enhancing the competitiveness of states in the new global context (Ernest G. Kochetov).

The hexagonal model is based on the phenomenon of merging politics with economics and reflects not so much the hierarchy of states, as "geoeconomic integrities"—economic activities (financial and legal technologies, high technologies, industry, and raw materials production), typical for a specific group of states or geographic region. The matrix of geoeconomic order, according to this approach, is formed by four geographically localized spaces, including: (1) The North-Atlantic West involved in the production of high technology commodities and services, (2) the raw materials-producing South, localized in the area of the "Indian Ocean Arch," (3) the least geoeconomically defined "Inland Ocean" of Northern Eurasia, and (4) two geoeconomic spaces without fixed geographic localization: (a) the transnational quasi-North originated in the North Atlantic region and related to the financial and legal regulation of the global economy, and (b) the Deep South archipelago—the "world underground" with a destructive economy based on plundering the resources of civilization.

Currently, the geoeconomic realm is dominated by the New North, leaning upon a strategic consensus between the North-Atlantic and transnational blocks with respect to ways and means of redistribution of the world revenue and collection of global "quasi-rent." Geoeconomic instruments for achieving these goals are global financial and legal technologies, such as world reserve currency, global debt, programs of structural adaptation and financial stabilization, the "Washington consensus," the emerging system of national and regional risk management, and the prospects of a global emission-taxation system.

Reference: Luttwak, E. "From Geopolitics to Geoeconomics." (1990). In *The Geopolitics Reader*. Edited by S. Dalby, P. Routledge, G. Ó Tuathail, pp. 125–130. New York: Routledge, 2003.

A. I. Ageev, A. I. Neklessa, V. I. Yurtaev

Geoethics (Greek *ge*, "Earth" and *ethike*, "morality") Geoethics is a set of moral standards to be used when exploring the geosphere, which emerged as an independent field of scientific research in 1992, when, in the Czech town Pribram, representatives from five countries participated in the symposium "The Mining Problem in Science and Technology." Czech scientist Vaclav Němec coined the term "geoethics" is considered the founder of geoethics.

The scientific community has accepted geoethics, born at the interface of ethics and geology, because any correct use of natural resources is inconceivable without a proper ethical attitude to it and to the whole geosphere. Geoethics, according to Němec, should be developed with regard to all specificities of Earth sciences and to the social responsibility of their representatives.

The danger of exhausting Earth's resources (and especially mineral reserves) is among the most important global problems. From this perspective, geoethics is of substantial importance. For instance, when evaluating mineral reserves in individual deposits geologists demonstrate for how many years the explored reserves will be available considering the given scale of exploitation. Here, an ethical (geoethical) problem arises: How much mineral raw materials can be exploited in order to preserve the resources for future generations? When limiting the solution of this problem to economic and ecological approaches, the freedom of choice for future generations becomes restricted. Geoethics may serve as the means of influencing the consciousness of people promoting the creation of resource-saving technologies; sustainable development and reasonable use of resources; the search for alternative (renewable) sources of raw materials and energy; preventing ecological crises. Geoethics is based on education, cultivation, and culture that lead to the emergence of harmonically developed personalities, able to find responsible solutions and to make responsible decisions based on the priority of universal spiritual and moral values.

G. S. Senatskaya, V. Němec

Geology (Greek *geo*, "earth" and *logis*, "study") Geology is the study of the components, structure, and developmental history of the Earth's crust and the deeper layers of the Earth.

A qualitative leap forward in the transformation of geology into a complex of sciences is linked to the introduction of physical, chemical, and mathematical methods of investigation at the end of the nineteenth and the begin-

220 *Global Studies Encyclopedic Dictionary*

ning of the twentieth century. According to the contemporary classification of the forms of the movement of matter (physical, chemical, biological, and social) the geological form is located as an offshoot proceeding from the chemical and biological forms of the movement of matter.

In terms of structure, modern geology includes stratigraphy, tectonics, geodynamics, regional geology, sea geology, mineralogy, petrography, lithology, natural resources, and several cross-disciplinary branches: geophysics, geochemistry, and geoecology.

These sciences being independent, the components of geology have their own objects and methods of inquiry, and, at the same time, contribute to the accomplishment of the theoretical and practical tasks of geology. Geophysics, for example, is a complex of physical sciences dealing with the physical properties of the entire Earth and with the physical processes in its solid (lithosphere), liquid (hydrosphere), and gaseous (atmosphere) covers. In compliance with the three constantly interacting covers, geophysics has the following subdivisions: earth physics, hydrophysics, and atmosphere physics. In its turn, earth physics embraces the following basic disciplines: gravimetry, seismology, terrestrial magnetism study (magnetometry), electrometry, radiometry; each of the indicated disciplines has essential practical significance being a theoretical ground for the corresponding geophysical methods of mineral exploration (seismo-, gravi-, electro-, magneto-exploring).

I. T. Gavrilov

Geopolitics: (from Greek *geo*, "earth" and *politika*, "affairs of state") Geopolitics refers to: (1) the interdisciplinary discipline synthesizing elements of geography, history, political science, sociology, demography, ethnology, and economics, and having the purpose to study the historical-geographic dynamics of states; or (2) actual political, economic, and military activity of states, directed at influence expansion, the original power "hydraulics" between countries, regions, and continents.

The primary circle of geopolitical ideas developed on the basis of works by Friedrich Ratzel (anthropogeography, geographic fatalism), Jean Brunhes, Paul Vidal de la Blanche (geographic possibilism), and André Siegfried (ecological and electoral geography). Johan Rudolf Kjellén coined the term "geopolitics" in the beginning of the twentieth century and it was soon picked up by the English scholars Halford Mackinder and German Karl Haushofer; their works are the recognized classics of this discipline. In the United States, Alfred Thayer Mahan and Nicholas J. Spykman developed geopolitical ideas. Geopolitical ideas were also studied by several twentieth century Russian representatives of Eurasian orientation (Nicolai Trubetskoj, Alexander P. Savitsky, Nikolay Alekseev, and Lev N. Gumilev).

The rise and rapid development of geopolitics occurred in the epoch when the power and international influence of states obviously depended on the geographic location of a country, its size, population, and supplies of natural resources. The founders of this new trend saw the territory of a country, the extension of its borders, and their natural or official movement as the key and significant political resource for ensuring the development and success of the state. Ratzel, Earnst Haeckel, and Nikolai Danilevsky all conceived of cultures and states as a living organisms: they are born, grow, and die. Ratzel (1897) defined land and soil as basic elements of state activity; land and soil act as a constant reality, a source of people's constant interest. Since the state develops not only in time, but also in space, geographic expansion becomes a natural process at a particular stage of its growth. He listed seven laws of expansion (1896): territory growth in accordance with culture growth, absorption of state and national units of smaller significance, understanding of borders as an important peripheral organ of the state, and tendency to occupy the most important and richest geographic regions. Influenced by Ratzel, Kjellen furthered the concept of geopolitics, defining it as a science about the state as the geographic organism embodied in space (1917), attempting to justify the natural tendency of Germany, a middle-continental European nation, "juvenile" and developing, to political and military expansion. Mackinder, introduced the concepts of geographic axis of history, the heart of the land (heartland), and the global island. Later, Haushofer introduced the concepts of moveable borders and life space.

References: Ratzel, F. *Politische Geographie* (Political Geography). Munich: Oldenbourg, 1897.—. "Die Gesetze des raumlichen Wachstums der Staaten" (The Laws of Spatial Growth of States). 1896. Kjellen, R. *Der Staat als Lebensform* (The State as a Form of Life). Leipzig, S. Hirzel, 1917.

A. V. Katsura

Geosystem, Natural and Technical (NTG): This term refers to a manifold of the earth's crust elements and anthropogenous components (constructions, transport systems, and recultivated sites) correlating and interrelating with each other and constituting a unity and integrity. The concept is close to the concepts of anthropogenous landscape, cultural landscape, technogenous landscape, and agrolandscape.

One of the basic distinctive NTG features is social and economic ground. Each natural and technical geosystem is established in order to satisfy specific needs of modern society, such as individuals' place of residence, irrigation of fields, transportation of fuel-energy resources, feedstock recycling, and the manufacturing of industrial and food products.

Thus, a natural and technical geosystem is an assembly of natural and artificial objects constituted in the process of construction and exploitation of engineering buildings, complexes, and technical facilities that interact with natural objects (geological solids, soil, vegetation, relief, water springs and the atmosphere, fauna, and the communities). Therefore, NTG is the kind of formation that indispensably appears in the region experiencing any economic activity and replaces the local natural geosystems.

Contemporary NTG development tendencies evince the disproportion between engineering design-theoretical grounds and the experimental bases concerning the factor of technogenous impact on the environment. The models' inadequacy to the real ecological situation of the industrially developing territories often leads to irreplaceable loss of natural landscapes.

I. I. Mazour

Global Modeling is a field of scientific research that studies various scenarios of world development by constructing models using a complex of mathe-

222 *Global Studies Encyclopedic Dictionary*

matical, socio-economic, and other methods based on information technologies applied to global studies, aimed at solving global problems. Over twenty global models have been made for solving different variants of problems. Notable is *Mankind at the Turning Point*, prepared by Eduard Pestel and Mihajlo Mesarovic, which describes computer modeling of the development of humanity with regional specifics taken into account. The report warned about potential large-scale financial costs and human causalities unless immediate actions are taken and the existing world problems solved.

Global modeling examines multifaceted activities of human beings who exert a deliberate and active influence on the functioning of the entire system to achieve multiple and often contradictory goals. Therefore, the issues connected with global systemic management and mathematically based studies of socioeconomic systems are of special significance. These issues predetermine to a large extent the choice of modeling methods, the methods of formal description of individual blocks, and the area of applicability of a model and potential results.

Upon formalization of socioeconomic and ecological processes, one often applies empirical correlations obtained based on the observation data of physical statistics without adequate analysis setting the limits of their use. For instance, the key problem of describing an economic mechanism-determination of the relationship between a system's economic benefits and available resources-is normally solved by setting the so-called production function.

In describing a process of production, one should keep in mind that the process is manageable, that is, occurs upon designated control actions. These control actions, and objectives of control systems are different for socialist and capitalist economies. Besides this, control actions change with time. The actual experience in determining parameter values of production functions shows that, firstly, these values often differ even for countries at the same level of economic development; second, they differ at different stages of development of one and the same country. Thus, there is no certainty that economic development of the future world would be described by a production function close to the present-day production function of the United States (Meadows's hypothesis), or that production functions of regions in the future world would remain the same as now (Mesarovic-Pestel's hypothesis).

Therefore, for the correct consideration of the process of economic development, it is necessary to have a differently formalized description of the production process with obvious isolation and description of control actions, the objectives of the system, and the level of scientific and technological progress.

In a wider theoretical context, the issues of formalizing socioeconomic and ecological processes are closely linked to the issue of the adequacy of the models. When we cannot rely on "precise" laws of natural science alone, one needs a system of hypotheses often using this line of reasoning: the model is adequate, as, firstly, the system of hypotheses applied at the moment looks plausible, and, secondly, by quantitative and qualitative characteristics at a known time segment the model and the real process are similar.

Research methods based on global models can be categorized into one of two groups: imitative and optimization models. (1) Imitative models are described by closed equations sets, that is, all functional relations, parameters, and exogenous values (including control actions) are set prior to the functioning of the model. The study of a system based on an imitative model consists in identifying how various selected propositions about functional relations and numerical parameter values and control actions will affect the behavior of the system. (2) The equation set of an optimization model is not a closed one, as a part of exogenous variables (controlling actions) has not been set. The study of an object based on an optimization model will consist in finding values of these variables that ensure the achievement of a pre-set specific goal, optimizing a pre-set function.

Globalization and the rise of antiglobalist movements set new tasks for global modeling. At the same time if the first stage of global modeling revealed global problems and set these before politicians, currently politicians and life itself are setting specific tasks before global modeling.

V. A. Gelovani, V. B. Britkov

Global Studies is an interdisciplinary field of scientific investigations aimed at revealing the nature, tendencies and reasons for globalization processes, the global problems it generates, and the search for ways of furthering positive and overcoming negative consequences of these processes for human beings and the biosphere. In a wider sense, the term is used to signify a body of scientific, philosophical, culturological, and applied investigations on various aspects of globalization and global problems, including the results of such investigations, and practical activities toward their implementation in economic, social, and political areas both at the level of individual states and internationally. Global studies is a result of integration processes characteristic of modern science. It is an area of investigation wherein various scientific disciplines and philosophy closely interact with each other (each from its own standpoint and method); analyze various aspects of globalization, offering solutions to global problems seen both separately and as an integral system.

In the United Kingdom, the phrase "global studies" first became popular during the 1960s and 1970s (e.g., Derek Heater, Dave Hicks), along with the related phrase "global education." The idea grew that the world system had to be studied interdependently: that the educators of different countries could no longer review their own intellectual world in isolation, but that the different subjects of the curriculum in schools and universities should form a new inter-disciplinary field.

In the United States, the phrase "world order studies" was more commonly used, associated with the work of Saul Mendlovitz and Richard Falk and the New York-based World Order Models Project, which began in the late 1960s. This was an attempt by American liberal intellectuals to consider all aspects of global problematics as an interrelated whole, including environmental problems, peace and conflict resolution, policy, population and resources, ideology, and social problems.

Never before has our planet been so overburdened, and human beings have come into collision with the results of their labor that made them critically dependent on scientific and technological achievements and unprotected against the power human beings themselves have created. In the wake of the unprecedented pollution of the environment there appeared alarming tendencies of population growth, arms races, and depletion of natural resources, which pose serious threats to social progress and even life on Earth. An active development of global studies was notably affected by the imbalance between society and nature that by that time had reached the maximum permissible proportions; furthermore, fragmentation and disunity in the face of global problems had become so evident not only to specialists, but to public consciousness as well. Consequently, the concepts of "ecology," "ecological crisis," "global problems of modernity," "globalization," and "anti-globalism," became widely used in scientific parlance and fairly quickly were added to the vocabulary of almost all languages in the world. They have become part of ordinary consciousness, political vocabulary, and attributes of the modern worldview. The number of publications, scientific conferences, and discussions on these topics has been growing annually all around the world and the results of such activities generate an ever-increasing scientific and public response.

Thanks to global studies, in recent years ideas have expanded significantly, in scope and depth, on the tendencies in the development of the global economy as an integral system and on the ensuing global problems. The nature and genesis of global problems were revealed; the profound relationship was exposed not only between natural and social processes, but also between the contradictions thereof and their dependence on social, economic, political, ideological, and scientific-technological circumstances. The most important achievement of global studies was the creation of a language for interdisciplinary communication acceptable for different sciences, and the development and upgrading of fundamental key concepts and categories, such as "globalization," "global problem," "ecological crisis," "ecologization of production," "demographic explosion," "global dependency," "world community," "new thinking," and "new humanism." Consequently, world outlook underwent significant changes; people gained a much greater understanding of the dependence of human beings on nature, the environment of the earth and space, and the resulting relations and balance of forces in the world. It becomes obvious that the interdependence of all aspects of social life is steadily growing, that different states increasingly interact with each other, and that while states were asserting their own national interests and sovereignty they were generating radically new contradictions in international relations.

From the standpoint of modern global studies, current global problems are not the result of someone's miscalculation, someone's fatal flaw or deliberately chosen strategy of social and economic development. They are neither whims of history nor the result of natural anomaly. Global changes and the ensuing problems common to humankind have been the consequence of centuries-old quantitative and qualitative transformations in both social development and the "society-nature" system. The root causes go back to the history of the formation of modern civilization, giving rise to an all-embracing crisis of industrial society, of technocratically oriented culture in general. This crisis embraced the whole set of interpersonal relations and relations with society and nature and infringed on vital interests of the entire world community. Such a development resulted first in a degradation of the human environment, and quite soon it revealed a tendency toward a degradation of human beings themselves, as their behavior, ideas, and thinking were found to be unable to adequately adapt to the changes that were occurring around them with an ever-increasing speed. The rapid development of social and economic processes were attributed to human beings themselves, and the purposeful transforming activities in which they engaged were enhanced many times by ever new achievements in science and technology.

Understanding global tendencies and the radical solution of the ensuing problems requires not only theoretical research, but also successful practical actions. Thus, global studies is objectively performing an integrating role, helping many scholars take a fresh view of the modern world and realize its participation in the common destiny of humankind. The results of the World Congresses of Philosophy (Brighton, 1988; Moscow, 1993; Boston, 1998), and especially the latest Eleventh World Congress of Philosophy (Istanbul, 2003) under the main theme of "Philosophy Facing World Problems," provide further evidence. These results indicate that in the modern world scholars of various orientations are increasingly interested in globalization processes, are concerned about problems common to humanity, and, in their professional activities, search for practical solutions. Global tendencies and problems do not leave humankind options other than striving for unity, while preserving the originality of cultures, age-old traditions, and the peculiarities of individual nations and peoples who are overcoming fragmentation and discord. Only in the light of the knowledge that is being worked out and formulated in global studies can we achieve an adequate understanding of the processes and events transpiring in the modern world.

I. I. Mazour, A. N. Chumakov

Global Studies in Philosophy seeks to solve, from the point of view of philosophy, the worldview and the methodological, culturological, and other aspects of globalization processes and the consequences they generate. The existence of this area is a result of the fact that modern science cannot do without specific worldviews and principles that reflect universal values when it comes to solving difficult complex problems. Philosophy, forming this outlook, influences the process of making economic, political, and other decisions. Without such a wide view of the research object, spreading beyond concrete disciplines and reflecting every modern achievement in other areas of knowledge, neither fundamental discoveries, nor the development of science are possible. For example, the wide view of the world (compared to the view that dominated for a long time in classical physics) once allowed Albert Einstein to develop the general theory of relativity, which included classical (Newtonian) physics. Any scientific discipline may find itself in a situation similar to Newtonian physics if it tries to solve any problem of a global character from its own position. Modern global problems form a very complex system dealing with people, society, and nature in their many interrelations, and consequently frameworks of concrete sciences are too narrow to see such objects of research as a whole, as a uniform system, in the context of modern global tendencies and the contradictions generated by them. Philosophy contains potential opportunities for development of planetary consciousness, humanization of international relations, and solutions for worldviews and theoretic-cognitive and methodological problems in the field of global studies.

226   *Global Studies Encyclopedic Dictionary*

Within the framework of global studies in philosophy several basic problems are being solved:

(1) Forming the outlook, a specific view of the world and a person's place in it, global studies in philosophy set corresponding estimation tasks, which in many respects determine the direction of human activity; thus their worldview and estimating functions are implemented.

(2) The methodological function of philosophy and generalizing theories that it generates, turn out to be extremely necessary to modern science as they promote integration of scientific knowledge.

(3) Philosophy helps to explain social phenomena and processes in their historical context; it formulates the most general laws of the development of society and nature and consequently in the course of studying of global processes it aims at understanding them as a natural phenomenon integrally connected to social progress. The phenomenon of globalization and its consequences are thus considered not as an accident or demonstration of blind fate dooming humankind to destruction in advance, but as a result of an objective process of the conflicting development of the history of humankind.

(4) From the point of view of philosophy it is possible to see the general tendency and dynamics of the development of world processes, and the correlation and interaction of the problems generated by them.

(5) Philosophy also carries out a culturological function as it enables us to develop a culture of theoretical thinking. Another aspect of this function is that studying the history of philosophy of various nations allows us to get acquainted with their customs, traditions, and culture, and none of the problems that this or that nation faces can be solved without this knowledge.

(6) The result of the whole vision of the natural-historical process and a complex approach to its interpretation is the opportunity for a more precise orientation in promptly increasing the flow of scientific information on global problems.

(7) Philosophy deals with issues of human life, death, and immortality, and that becomes of special value and urgency when confronting the threats posed by global problems.

Finally, the important methodological function of philosophy is the development of such categories as global studies, global problems, nature, society, civilization, social progress, scientific and technical revolution, globalization, and globalism, which are directly connected to the actual modern problems of humankind and are very important for comprehending the objective tendencies of the world development.

A. N. Chumakov

Global World Outlook refers to perspectives regarding the social subject's place in the world as the spiritual epicenter of life in society over the course of history, and who forms a unity in relation to other social subjects representing different regional cultures.

Historically, the first worldviews to emerge were mythological, giving birth to philosophy and providing genesis for science and monotheism. The place of the human being in the world, then, has been dominated by a one-sided perception: anthropomorphic, in the mythological worldview; conscientious, in the philosophical one; theoretical, in the scientific worldview and irrational, based on revelation and faith in a single God, in the religious one. Perspectives that are more global in scope have not yet been widely adopted, though their recognition is vital for the modern human being.

These global perspectives include the speculative approach that was dominant up to the industrial revolution of the eighteenth and nineteenth century and the activity-related outlook that has been developing since the industrial revolution. The industrial revolution marks a significant reorientation of global perspective and a simultaneous qualitative leap in the entire system of productive forces and relations of production, that is, in the mode of production as the material foundation of the life of the society. The consequences of the industrial revolution are diverse: the mastering artificial sources of energy, transition from small-scale manufacturing to production driven by large machinery, industrialization, urbanization and the appearance of "white collar" workers. However, the most significant result of the industrial revolution has been the growing power of human activity, to the point that humans now possess power on par with that of the forces of nature. The consequences of a nuclear explosion (Hiroshima and Nagasaki) are not to be compared with consequences of natural cataclysms. The place and role of the human being in the Universe is changing dramatically with a corresponding reconstruction of the foundation of philosophical and scientific knowledge; from the mid-nineteenth century onward modern philosophy and science have supplanted the entire previous tradition of philosophy and science.

The principal difference between the activity-related perspective and the speculative involves a more acute concern for the consequences of human behavior. Having become a significant technological force, humans are now forced to consciously assume complete responsibility for the results of their own activity. A carrier of the speculative perspective may be quite an active person; Don Quixote, for instance, was an energetic hidalgo ready to fight windmills for the sake of some higher cause, however, he did not worry much about the results of his actions. Before the industrial revolution the impact of human activity upon nature and the surrounding world in general was similar to that of a mosquito bite; therefore classical science may allow some inaccuracy and ignore the resulting impact. The mechanistic picture of the world takes no account of the subject of cognition while in the modern relativistic picture of the world the subject is the benchmark of cognitive and transforming activity.

Traditional philosophy from antiquity to classical German thought does not reflect purposefully upon the responsibility of a social subject for the results of their activity. Modern philosophy (philosophical anthropology, personalism, neo-Freudianism, existentialism, and logical positivism) has had to give priority to the problem of personal responsibility in the sphere of politics, science, art and religious conscience, as these concerns have a significant bearing on the world's destiny. However, modern philosophy is rooted in classical philosophy and these global perspectives, like all worldviews, originate in antiquity. The encyclopedic intellect of antiquity, such as that of Aristotle, was the first to classify virtues by identifying dianoethical and ethical virtues. Dianoethical virtues help us achieve the golden mean between two extremities. These virtues are a function of the speculative rational part of the

228 *Global Studies Encyclopedic Dictionary*

soul. Ethical virtues are virtues of character produced by the active rational part of the soul. For example, courage is a dianoethical virtue actualized as the golden mean between cowardice and reckless bravery; as an ethical virtue it is the ability to behave courageously. Thus, Aristotle singled out two components in the rational part of the soul identified by Plato: speculative and active, each having its own consistent logic. The active component relies on the speculative one; otherwise the subject will immediately perish from cowardice or reckless bravery.

Further epochs deepened these ideas in their own ways: the middle ages mastered the phenomenon of self-conscience, the Renaissance, the phenomenon of personality and individuality, the New Age, the scientific mind. As a bridge between the previous traditional and modern philosophy, classical German thought studied activity, though still in the former speculative way. Thinkers expressed the transition to the activity-related worldview of the nineteenth century in a variety of forms. In the 1840s, in his "Theses on Feuerbach" Karl Marx writes, "philosophers just *explained* the world in various ways but the idea was *to change* it." Two decades later, independently of Marx, Russian organicist Nikolay N. Strakhov in his work "World as a Whole" stressed that "activity is a more difficult notion than being" and revealed the inevitability of a transition to a more complex level of world cognition: from the descriptive study of being to the examination of its dynamics, that is the activity of all world forces. The theoretical-methodological fundamentals of the activity-related type of the world outlook are developed most completely and constructively by Russian organicism and cosmism.

The theoretical apparatus of the speculative perspective is comprised of such major categories as being, spirit, nature, substance, mechanism, objective and subjective reality, dynamic laws, and simple mechanical systems. The activity-related type of the world outlook shifts the emphasis to categories of activity, organism, organization, subjective and objective factors, statistical laws, complex dynamic systems, freedom, and responsibility. The development of one's activity-related worldview entails learning to organize one's activity so as to ensure optimal correlation of objectives, means and results. This is important for the human being in all spheres of his life: social, interpersonal, professional.

Today *the spontaneous variant* of transition from the speculative to the activity-related world view dominates; the social subject "plays" with his technological, political-economic and spiritual-intellectual force, which results in negative consequences: ecological problems, social crises, postmodernism and the mechanization of the human being.

The conscious aspect of an activity-related perspective is manifested to the extent that people are aware of their responsibility they reflect upon the unpredictability of their own activity and are eager to prevent good intentions from having quite the opposite results.

Reference: Marx, K. "Theses on Feuerbach." In *Selected Writings*. Edited by David McClellan. New York: Oxford University Press, 2000.

O. D. Masloboeva

Globalization is a process of development of a united interrelated world where protectionist barriers that restrain nations' communications and protect them from chaotic external influences do not separate them from each other in matters of mutual benefit. Various nations entered the new open world of globalization starting from different stages of preparation, with a variety of economic, military, and information potentials.

Many researchers comment about the vagueness of the term "globalization," but no longer can disregard it when discussing modernity and the future of human civilization. Globalization as general concept contains many specific meanings and connotations, and it is necessary to figure out its central features. It is important to avoid descriptiveness and inconsistency.

Interrelations of the developed and the less developed nations in the emerging global world are fraught with new shocks and collisions. It is not by chance that the most developed and powerful countries that perceive the weakening of former sovereignties as a new opportunity for their economic, geopolitical, and sociocultural expansion became the most consistent adopters of the global world propounding the idea of a united open society without barriers and boundaries. These countries are inclined toward Social Darwinism in their interpretation of the global world in terms of new natural selection destined to expand the boundaries of habitation and opportunities for the more adoptive ones at the expense of the less adoptive.

The less developed and secure countries demonstrate in response a suspicious attitude toward globalization and the liberal vision of the global open society, using all kinds of defensive and protective mechanisms. This provides evidence of the asymmetry of globalization. Reacting to the Western concept of the global "electronic village" being a united information space that responds to events in the remotest corners of the world "in a neighborly way," Third-World countries raise the problem of "informational imperialism," meaning unequal exchange of information between the North and the South, developed and the developing countries.

Today, not only are informational relations asymmetric. In the modern global world there appeared new financial and economic, political, and military technologies capable of undermining national independence when the foundations of human existence, daily well-being and the security of people are concerned. A sort of de-embodiment of many conditions of human existence occurred under the impact of the capability of remote control across state borders. The manipulation of currency exchange rates and short-term speculative capital, the expropriation of national savings, and the depreciation of the labor of millions are among the most striking examples of these new technologies in application.

Political pressure on the behavior of local political and intellectual elite, ethnic groups, and decision-makers is less studied. Nevertheless, these spheres could not avoid the impact of globalization that needs to be analyzed not only as a spontaneous increase of global interdependence, but also as a technology of global asymmetric interactions which structures the world in a new manner. In the field of information the world is often divided into donor cultures and recipient cultures.

230 *Global Studies Encyclopedic Dictionary*

Along with global information space, allowing users to affect human consciousness over state borders, there emerge other global spaces opening up opportunities for similar actions with respect to the material factors of human existence. This means that we are witnessing a process of the emergence of a global authority, differing from the traditional forms of government due to radically new technologies of distant control and the latent forms of its existence.

Humankind has yet to realize the significance of this new phenomenon and its long-term consequences. Ever since the American and French Revolutions of the eighteenth century nations have been solving two problems:

(1) Gaining national sovereignty and independence-freedom from external oppression.

(2) Establishing democratic control over its own rulers-their subordination to the will of voters and constitutional legal norms.

Admittedly, many realities of the modern global world call into question these achievements of the epoch of democratic modernity. Today neither guarantees from unsanctioned external action, nor is democratic control over the forces that execute such actions secured. We are living at the late hour of history when humanity is facing a dilemma: it will either open the door into a qualitatively different future, or there will be no future at all.

In conclusion, for at least three reasons demonstrate the process of globalization is preparing the transition of humanity into an essentially different quality, when continuation of the present by extrapolation of the existing parameters and tendencies is no longer possible. The first is connected with ecological "limits to growth." It calls for a change in the very paradigm of development of the modern technological civilization and its relationship with nature. The second is related to the no less dangerous tendencies of moral degradation that are manifested not only by catastrophic deterioration of mass morality, but also by significant deterioration of decisions made by the modern political, economic, and administrative elites. There arises a need to change the sociocultural paradigm shaping the moral and behavioral code of modern humanity. The third is connected with the increasing social polarization between the adapted (successful) and non-adapted (not successful) parts of humankind. Recently, it seemed that global modernization influenced all humankind, adjusting the less developed strata, countries, and regions to a single standard that embodied the long-cherished historical goal of humanity.

Today, we are facing the threat of losing the common human perspective, the split of the human race into the adapted cultural race ("the golden billion") and the non-adapted, to which the majority of the planetary population belongs. This split of the world is already rapidly destroying the mechanism of our planetary civilization, replacing solidarity and trust with ruthless social Darwinism, the war of all against all, and ubiquitous suspicion.

Humankind cannot survive on such a basis for a long time. The very paradigms of relations between the West and the East, the North and the South, the Sea and the Continent, the "growth poles," and the deprived periphery must be changed. The issue of a qualitatively different future is not another utopia, but a vital necessity, since apparently we are not able to stay forever in our present, even if someone is completely satisfied with it.

Here arises the issue of methodology related to finding ways to reach this qualitatively different future. Who discovers this future and by what means and procedures?

Modern secularized social thought has taught us that the most advanced countries had opened new horizons for the less developed ones. However, today, when we hear the world vanguard's statements about securing the future for themselves, which the less adapted and the less worthy ones will hardly have access to, we have an obvious breakdown of the modernization mechanism based on the messianic activities of some vanguard groups. The happy minority obviously wants to privatize the future instead of making it a common property.

Under such conditions, prognostication needs to reconsider the paradigm of modernity. The world elite feels comfortable in the present and welcomes "the end of history" preserving the status quo forever. A search for a qualitatively different future is a task for the "stepchildren of progress," who failed to accommodate themselves to the present. A special type of consciousness that is characterized by both social pessimism (concerning the present) and historical optimism will bring forth the required historical transformation.

A. S. Panarin

Globalization from Below: *Globalisering underifrån* ("Globalization from Below") is a Swedish organization that consists of radical groups and individuals, mostly on the extra parliamentarian left. It is a part of the worldwide network Peoples Global Action. The organization is not anti-globalistic, but, in contrast to "globalization from above," they are in favor of a worldwide globalization that includes freedom and democracy for all countries.

Therefore they are against the European Union (EU), partly for the following reasons: (1) They regard EU as a parody on true democracy. (2) They consider EU as a capitalistic project that has contributed to a growing gap between the rich and the poor, both inside the union and on a global level. (3) They are of the opinion that EU's foreign policy is becoming more and more militarized.

Globalization from below wants an alternative globalization and rejects capitalism, imperialism and feudalism. Their philosophy is based on decentralization and autonomy. The means they use are, among others, civil disobedience and direct action.

L. Knutsson

Globalization, Fundamental: Fundamental globalization is the emergence of structures, connections, and relations of planetary scope. As a result the world in all its aspects finally becomes a single whole, that is, a single political and economic system. Humankind becomes global in the full sense, capable of interacting with bio- and geo-sphere as a "geological force." The first signs of fundamental globalization became noticeable in the second half of the nineteenth century. In the middle of the twentieth century, it grew into multi-aspect globalization. Fundamental globalization's distinguishing characteristics are manifested in the fact that the world that has become geographically closed, thus becoming a whole in the economic and political sense. It

232 *Global Studies Encyclopedic Dictionary*

became global economically after the emergence of the world capital flows and the formation of the multinational corporations.

A. N. Chumakov

Globalization, Hyperglobalist and Transformational: Based on the degree of revolutionary extremism in globalist reshaping of the world, two approaches can be distinguished: hyperglobalist and transformational, opposition to which—hyperglobalism and transformativism—is lodged by a group of specialists skeptical about globalization.

Robert Keohane and Joseph Nye (hyperglobalists) argued (1977) that simple interdependence grew into complex interdependence, connecting economic and political interests so tightly that a conflict between the largest states was really out of the question. Ken'ichi Omae (1990) held that the significance of people, companies, and markets increases while prerogatives of a state weaken; in the new era of globalization all peoples and all basic processes turn out to be subdued to the global market space. This is a new stage of human history where traditional nation-states lose their natural foundations and become inappropriate business partners. Globalization is understood as the source of future prosperity, pacification, rules shared by everyone; as the means for survival, for increasing living standards, social stability, political influence; as the reason why conquering the neighboring states becomes no longer necessary. This wave will follow the rounds of world trade negotiations; it will lead to working out a new position on making trade barriers, quotas, tariffs, and subsidies for your own industry.

Theodore Friedman champions hyperglobalization; he addresses market capitalism and liberal democracy, allowing capital to move at lightning speed to the countries where the political situation is stable, the economy is effective and profits are the most promising. The supporters of globalization think it the only way to close the gap between the rich (Western) part of the world and its poor part.

Proponents of the transformational approach, headed by James Rosenau and Anthony Giddens, think that globalization in its present form is historically unprecedented and consider irrelevant the comparison with the pre-World War I period. Globalization requires from states and societies adaptation to a more interdependent but at the same time highly unstable world, characterized by swift social and political changes constituting the essence of modern societies and world order. Globalization is a mighty force transforming the world and responsible for the large-scale revolutionary changes in societies and economies, transformations of the forms of government, and the entire world order. It destroys differences between the domestic and foreign, internal and external. Rosenau stresses the emergence of a new political, economic, and social space within a traditional society to which states must adjust themselves on the macro level while local communities must do so on the micro level.

At the same time, transformists (unlike hyperglobalists) try not to specify the direction of this process, the very essence of which is unpredictable changes and whose main characteristic is the emergence of new contradictions. They see globalization as a long-term process full of controversies, subject to many situational changes and do not pretend to know the future of

Gandhism – Gumilev, Lev Nikolaevich 233

the world development, considering as idle talk any prediction of the future world parameters, any clear definition of the demands of the world market, or any complete description of the emerging world civilization.

Transformists demonstrate caution and "scientific modesty" and circumspection because they do not want to draw clear pictures of the kaleidoscopically changing world. They do not forecast the emergence of a single world community, not to mention any single world state. For them, globalization is associated with the formation of new world stratification when some countries become an integral part of the "typhoon epicenter," or the world development center, while the others are hopelessly marginalized. But even given this clear gap between two kinds of countries, there will be no distinction between the "first" and the "third" world; the distinction will be more complex. In fact, all the three worlds will exist in nearly each big city as "three circumferences": the rich, those loyal to the existing order, and outcasts.

Globalization ideologists are far from insisting that the process of globalization is complete, but they assert that this process cannot be stopped.

References: Keohane, R., and J. Nye. *Power and Interdependence*. Boston: Little Brown, 1977. Omae, K. *The Borderless World*. New York: HarperBusiness, 1990.

A. I. Utkin

Globalization, Limits of: Globalization faces many obstacles resulting from inertia of the past and from intrinsic contradictions within globalization itself. Militarist and neoliberal forms and methods of globalization are strongly opposed. Among the emerging restraints are political, economic, technological and cultural ones. In some areas, these (and other) restraints substantially hamper globalization, and sometimes even turn it back. It is due to their effect that globalization increasingly takes the form of regionalization.

Current globalization processes are generated mostly by economic considerations such as the ongoing social division of labor and progress of science and technology under the dominance of free market economies. At the same time, secondary (derivative) manifestations of globalization are gaining strength, while becoming more independent and self-contained: these are not only international political integration processes, international non-governmental associations and NGOs, various forms of international cultural exchange and communication and internet and individual communicational capabilities, but also protest (anti-globalist) movements, organized crime and international terrorism. All these phenomena are ambiguous and generate new conflicts and controversies, which impose limitations on prospects of further globalization, while constantly transforming its nature.

Political interests of national economies (protectionism) and technical and economical constraints limit the economic foundations of globalization. Under conditions of industrial development, interrelated economic units were located close to each other (city, economic clusters, national economic complexes), which brought substantial profits due to cutting infrastructure (both material and social one) and other costs. Modern means of communication provide some relief of this territorial imperative, yet they do not overcome it altogether. Moreover, in post-industrial society, with its emphasis on creative (intellectual) factors, a new imperative is emerging related to formation of local clus-

234   *Global Studies Encyclopedic Dictionary*

ters: direct communication between intellectual personalities is quite essential for development of creative ideas (computer-based communication is much less efficient here). As a result, we now see the advent of techno-police, financial centers, centers of business services and other kinds of clusters. New centers of power emerge, with global markets becoming oligopolistic instead of free.

Political imperatives of globalization are mostly determined by economic considerations: striving for strengthening positions of one's "own" national economy and "own" entrepreneurs in the world, and pressure from stronger powers and transnational corporations. It underlies active participation of countries in processes of forming the legal basis of global markets and investment flows. The inherent logic of political life in most of the countries is hostile, rather than favorable to globalization. Politicians generally see refusal to protect national interests as a political suicide. Globalization is fraught with downsizing of bureaucracy threatening quite a few state officials with unemployment (hence, their open or veiled resistance). In a broader aspect, globalization threatens with job losses for many groups of people living in developed nations (especially, in view of competition from cheaper labor force of developing countries); therefore, it is likely to cause large-scale protest movements in the capitalist world.

Influence produced by globalization upon various cultural aspects of social life is just as controversial. On the one hand, globalization expands significantly the area of cultural communication, exchange of concepts and intellectual achievements, gives an unparalleled impetus to new creative initiatives, development of science and technology. On the other hand, globalization destroys cultural foundations of existing communities, their historical heritage and singularity. Culture is an immune system of society, the basis of its successful opposition to destructive external forces. At the same time, an essential means of social consolidation is often created through the so-called enemy image, a concept penetrating through many aspects of spiritual and cultural life (religion, ethnos, national ideology). The ideology of globalization has so far been unable to offer a concept that would consolidate modern humans while being adequate in its effects.

V. M. Kollontay

Globalization, Historical Stages of: Globalization is an amalgamation of national economies into a united world system based on rapid capital movement, new informational openness of the world, technological revolution, and adherence of the developed industrialized countries to liberalization of the movement of goods and capital, communicational integration, planetary scientific revolution, international social movements, new means of transportation, telecommunication technologies, and internationalized education.

Gradual integration of countries and continents has occurred throughout human history. However, only twice did the rapidity of such integration become revolutionarily:

(1) The First Stage of Globalization: Between the nineteenth and the twentieth century, the world entered the phase of active integration based on the global spread of trade and investment due to the emergence of steamboats, telephones, and factory lines.

(2) The Modern Stage of Globalization: By the beginning of the twenty-first century, agreement on information technologies had been worked out, countless important agreements on telecommunications and financial services had been completed, and meaningful agreements, such as China's admission to the World Trade Organization, had been arranged.

While the first stage of globalization leaned upon the global British Empire, now the process of globalization is influenced primarily by the United States. To some extent, it has circumscribed its incomparable might and virtual hegemony for the sake of the process of opening the world economy and building multilateral institutions, participating actively in multilateral rounds of trade negotiations, opening its own markets for import and taking steps to bring trade liberalism into being.

According to Thomas Friedman, globalization has replaced the Cold War system. Not every country can consider itself a part of this system but virtually all states (and industrial companies) are under pressure to adjust to the challenge of globalization. However, political and economic choice of the majority of governments is restricted because there is only one superpower in the world and because capitalism is predominant worldwide.

The previous system of international division of labor based on the relationship between "the developed industrial world core," the semi-periphery of the industrializing economies, and the periphery of the underdeveloped countries is evolving toward building a single global economy dominated by "the global triad": North America, the European Union, and East/West Asia. Here the central productive forces of the world and the "megamarkets" of the global economy, where the central role is played by the globalized multinational corporations, are located.

Reference: Angel, N. *The Great Illusion*. London: William Heinemann, 1909.

A. I. Utkin

Globalization, Local Characteristics of (The Example of Georgia): A characteristic feature of globalization is movement toward an open society and reliance on the most advanced technologies and intellect-intensive labor. This speaks in favor of not being isolated from the worldwide process of globalization. At the same time, globalization does not mean the same for the developed countries of the West and for the underdeveloped South. Peoples of the developing countries and countries in transition often have to choose between national independence and international isolation. One of the current examples is Georgia; it does not give up hope of being integrated into the process of globalization. Important prerequisites are available in that respect. Georgia contributed much to globalization, especially in the Christian world, possesses human resources with high intellectual potential in science and the humanities, and occupies an important geopolitical niche. It is one of the central sections of the short-cut route connecting two continents and has significant natural resources that make it less dependent on the external world for its supply of energy or strategic raw materials. Given these resources, Georgia is able to overcome the difficulties that arose due to the breakdown of the Soviet economy and to domestic problems.

236 *Global Studies Encyclopedic Dictionary*

Other nations of the South Caucasus (Azerbaijan and Armenia) do not differ much from Georgia with its problems and potential, as these nations have lived in the neighborhood for centuries. They are facing the need to settle their territorial disputes, restore territorial integrity, and overcome separatism. The Caucasus has always been the apple of discord for the great powers and is no less explosive than the Balkans. For many centuries, Georgia has been located at the crossroads of civilizations. Its geopolitical situation against the background of the permanent military-political and economic confrontation of the great powers has largely determined the historical destinies of her people.

The beginnings of the ancient Georgian states of Diaohi and Colchis date back to the thirteenth and fourteenth century BCE One finds many references in this regard in Assyrian and later in Urartu sources. The legend about the Argonauts, one of the finest works of ancient Greek epic literature, contains interesting data on the country of Colchis and the magical "Golden Fleece." Ever since that time, the Georgian tribes have had to struggle continuously for survival and defend their independence and culture against the background of the sharp confrontation of the hostile great powers. In different moments of history, Georgia found itself in the spheres of influence of Persia, the Hellenic world, Rome, Byzantium, the Ottoman Empire, Russia, and the countries of Western Europe. From the era of ancient colonization up to the tenth century CE, the West-Georgian state of Colchis was closely integrated into the Western world (Greece, Rome, and Byzantium). The East-Georgian state—the Cartali (Iberian) kingdom—cooperated with the eastern Hellenic countries, later with Rome, and then with Sasanidian Iran.

The great powers always had their eyes set on Georgia, as the Great Silk Road crossed its territory on the way from the Mediterranean through the Black Sea and Caspian Sea on to India and China. The adoption of Christianity as the state religion in the early fourth century had conclusively determined the Georgian political and cultural orientation toward the West (namely, Byzantium). Since that time Georgia regarded itself as a part of Europe. Between 1089 and 1125, Georgia became a powerful state that assumed the role of protector of Christianity. During that time, Georgia claimed to establish control over the Black Sea and the southern section of the Great Silk Road. Georgian culture absorbed the ideas of humanism (neo-Platonism) reflected in literature, theology, philosophy, historical and legal thinking, architecture, and art.

Following the Mongol domination in the thirteenth and fourteenth century and the invasion of Tamerlan at the turn of fourteenth and fifteenth century, the flourishing of Georgia was interrupted for a long period. The seizure of Constantinople by the Turks in 1453, meant for Georgia the loss of the only way to the culturally advanced countries of the Mediterranean area. The Ottoman state, powerful and aggressive, replaced Byzantium. Having found itself in dangerous proximity to Iran and the Ottoman Empire, Georgia did its best to break away from that hostile encirclement and appealed to European states for help. However, the latter were guided by their own interests and were not ready to confront Iran and the Ottomans. Then, the Georgian political leaders turned to Russia. The interests of the growing Russian state coincided with those of Georgia; Georgia hoped, with the help of Russia, to liberate itself from the Iranian-Ottoman yoke, while Russia counted on Georgia, its co-religionist ally, in promoting its designs in the East. These mutual aspirations resulted in the Georgian Treaty of 1783, but soon thereafter the interests of Georgia and Russia drifted apart. In 1801, Russia abolished the kingdom of Eastern Georgia; later, Western Georgia shared the same fate. Only in 1918 did Georgia regain state sovereignty to lose it again in 1921, and join the Soviet Union in 1922. The disintegration of the Soviet Union brought national independence to Georgia, but the disturbed territorial integrity and shattered economy made the country's way into the world extremely complicated. Financial and moral supports and Georgia's participation in implementing projects of world significance provide a chance for Georgia to restore in the twenty-first century the role it used to play in the Caucasus.

V. A. Kvaratskhelia

Globalization and Localization: Given the perpetuation and even exacerbation of historically established inequitable trends of development that seem to characterize globalization in its current form, under a wide-ranging diversity of local conditions throughout the global "South," some scholars say that globalization is not a natural, evolutionary, or inevitable phenomenon. It is the imposition of Western culture on all others. Similarly, some writers characterize globalization as the third phase of colonization, the second phase being neo-colonialism. On this view, Western countries are employing globalization to extend and strengthen the fundamentally exploitative relations established between colonial powers and the colonized over the past 400 years. Industrialized countries are essentially entrenching a global capitalist system and consumer culture by establishing a global market controlled by the most dominant interests within the ruling elites of these "Developed" nations, especially the interests of the largest transnational corporations (TNCs).

In spite of critical attitudes toward globalization, there seems to be a growing consensus that globalization, in one form or another, is here to stay. Paradoxically, it is the inevitable and global reach of globalization that has unleashed a new interest in local cultures, economies, environmental standards, and unique ways of being that must either effectively manage the forces of globalization or be swept away into the dustbins of history.

As some writers have noted, although conventional approaches perceive a local culture as the opposite of the global, the concept of local culture is a relational concept. Thus, there is a way in which we become more consciously aware of the local through the very process of globalization. Accordingly, there is no inherent contradiction between localizing and globalizing tendencies. Many scholars have acknowledged that local happenings are shaped by events occurring many miles away and vice versa. Globalization and localization imply each other. Local identities cannot stand alone. Localization does not mean that local people produce everything locally; they do benefit from global structures.

Globalization brings into focus how as the local responds to global intrusions we might find new ways of understanding globalization. Local communities have a lot to contribute to global society. Many living indigenous traditions in the global "South," and even among Indigenous

238 *Global Studies Encyclopedic Dictionary*

American communities of the United States, have developed diverse traditions that are environmentally friendly and sustainable. In various regions of the world, we find surviving and functioning systems of natural law ethics, relational views of the self and world, and visions of the common good that are expressed through various cultural concepts. It would seem that many indigenous people and their sages throughout the world have known for centuries and millennia, from a diversity of locally articulated perspectives, what some of the West's most brilliant scholars have only recently discovered: what the Western world tends to dichotomize as the natural environment and the life of human beings, or more starkly, "man" versus "nature," are, in reality, inextricably part of *one* interdependent and intricate web of being. From this perspective, globalization is merely the Western world's belated recognition of this fact as the consequences of human behavior on the whole of our being become ever increasingly apparent to the five limited sensibilities on which the Western mind is so utterly dependent. The danger, from this perspective, is that with this new insight many powerful forces of the "developed" world now seek to control our being rather than find a place within. Such attempts can only end in catastrophe.

Local people in various parts of the world are responding to global changes. I share the view of many scholars that local groups, far from being passive receivers of transnational conditions, actively shape the processes of constructing identities, social relations, and economic practice. However, active responses to globalization are threatened by a strong tendency for powerful organizations and nations to think they, in effect, know all there is to know about literally everything. Localization, as an active response to the more hegemonic and destructive aspects of globalization is absolutely crucial. However, the struggle to localize knowledge is a difficult political problem because the most powerful forces of globalization are primarily interested in exploitation and profits. The fact is that passive recipients are, in the short term, much more profitable than active participants. Given the severe disparities of power there is an urgent need to encourage, enhance, and empower active localized responses capable of warding off the negative trends of globalization based in the exploitation and commodification of our being.

Such concerns have led to a call for the "democratization" of knowledge production and application in which claims and counter-claims from diversity and inclusively representative range of perspectives can confront each other in a critical manner. The localization of knowledge through the democratization of knowledge has been recognized as one way of avoiding the danger of making disastrous mistakes by applying practices that may make perfectly good sense in one social or environmental setting, but are totally inappropriate in other local settings.

Globalization is indeed a multi-dimensional phenomenon that affects every aspect of our lives, and, it is here to stay—at least for the foreseeable future. It]brings us closer together,]strengthening our consciousness of the interdependent nature of life on Earth, but at the same time, it pits us against each other in competitive confrontation based in historically entrenched asymmetrical relations of social, cultural, economic, and political power. As

Gandhism – Gumilev, Lev Nikolaevich   239

such, it is required of us to ask ourselves the question "how might we so construct our lives together such that all life flourishes?"

Many people from various perspectives or based in particular interests argue for almost an infinite range of strategies to our shared globalize human condition. Some of these strategies are mutually reinforcing, some are compatible, while others are mutually exclusive and contradictory. At the very least, wisdom would seem to dictate that strategies which are totally foreign to local socio-political, cultural and historical realities should not be imposed by those who have gained a monopoly on coercive economic, political, and military power over the course of the last 500 years of human development and history. We must analyze the moral consequences of the economic and political systems that we institute. Local people should empower themselves through networks of globalize localities and solidarity, adopting and adapting principles suited to local needs, and work together to democratize and humanize the global structures of what we call "globalization."

W. Kelbessa

Globalization at Macro-, Meso, and Micro- Levels: At present, globalization takes place at the macro-, meso-, and micro-levels, of which the most socially important is the meso-level. The meso-sphere is not simply a new face of globalization; it is a basis that links human and social dimensions. At the meso-level, the new world of globalization meets the world of people and the world of life itself.

The concept of globalization is not reduced to only the economic aspect; it presupposes a specific relationship between economic and political aspects since both of them constitute a united and universal system. At the same time, their mutual penetration creates a paradoxical effect: it is impossible to describe integrity from the economic point of view irrespective of the political aspect, and vice versa. Most likely, the current relationship between the political and economic aspects of globalization creates some third quality, existing parallel to the economic and political description and as a specific example of globalization. From the political point of view globalization includes problems pertaining to the theory of democracy; in the economic aspect globalization is a theory of market.

E. Kiss

Globalization, Multiaspect: A process of total involvement of all spheres of material and spiritual life by the nations of the world in the formation of a single system unbreakably tied to and interacting with the bio-geosphere. Multiaspect globalization started in 1970s, and continues, characterized by the fact that most of the population understands the existence of global threats and globalization processes, and the formation of the global community and value orientations, culture and life styles associated with it. This period also involves the intense development of the informational and technological revolution and the emergence of global consciousness as one more form of social consciousness on par with myth, religion, philosophy, science, and ecology.

Multiaspect globalization has radically transformed global market, which has gradually become stronger and more significant than national economies and revealing specific qualities and characteristics. Of them the

240 *Global Studies Encyclopedic Dictionary*

most important are: (1) the emergence of a resonance effect, when economic upsurges or recessions are transmitted from one country to other countries and regions tightly connected with it, (2) the formation of world markets for those goods and services of special significance for functioning of all, without exception, countries of the world (such as the oil and gas market, air-passenger operations market, grain market, and the informational and tourist services market), (3) the formation of world prices for the above-mentioned goods and services, which often guide policies of their national manufacturers, and (4) development of world loan capital market where not only the international price of capital but also the directions of its flow are defined.

Along with internationalization of the economy and the unifying role of money, multiaspect globalization is characterized by and responsible for the formation of the mass society and mass culture corresponding to it. In the context of globalization mass culture is a primary image of a transnational, world and universal culture and is sometimes understood as a culture opposed to ethnic values and ethnic and cultural uniqueness. This is only partly true. Globalization necessarily leads to the universalization of all spheres of social life, including spiritual values and worldviews. But it does not rule out the preservation of the tradition, originality and cultural diversity of individual countries and peoples, which, though not without conflicts and contradictions, must be transformed and changed under the pressure of external conditions and under the influence of the objective factors of world development.

Apart from serious structural changes in the sphere of culture, international relations and international law, multiaspect globalization has engendered strong demand for radical changes in morality, behavioral norms, and value orientations and patterns.

A. N. Chumakov

Globalization and the New World: Although globalization has a deep connection with economics and communications, it arises from a cognitive and philosophical foundation. On the basis of a new philosophy and way of thinking a new world has appeared with basic features that are contrary to human interests. Its more obvious cognitive characters are the following:

(1) a skeptical picture derived from a naturalistic metaphysics and science against knowledge affirmed in earlier metaphysical and scientific worldviews;

(2) subjectivism opposes objectivism in the earlier worldview;

(3) analytical thinking against speculative claims of earlier worldview;

(4) practical thinking against theoretical knowledge of earlier worldview.

On the one hand, globalization decreases some of our suffering. On the other hand, it increases our suffering. The ideal of living in equality around the world can be found in the utopias of great philosophers and even in major religions, although it seems that this idea is not compatible with our needs for pluralism and tolerance. Namely, the same world facilitates the processes of acquaintance, friendship, and finding our similarity, but it does not support individual evolution, personal genius, and the appearance of self-realization (self-identity); so, very real mental, cultural, and political problems follow

S. H. Hussaini (Akhlaq)

Globalization, Social Effects of: Social effects of globalization are generally characterized by the intensification of communication processes and the redistribution of survival resources both in world political-economical systems and in the stratification of social space. Positive social effects of globalization consist in the expansion of the field of freedom and relevance, first professional relevance, and the integration of diverse world culture treasures. The contraction of chronotopic distances (or the "compression" of space and time) enables an intensifying interchange of experience, significant information, and scientific achievements. Thanks to globalization states get an opportunity to create supernational structures, promoting dynamical solutions of various urgent problems while taking into account interests of both developed and developing countries; these could be global problems, problems of labor distribution in the world, the economic labor market, social protection (rendering by rich states assistance to poorer ones), the struggle against world terrorism and global criminal networks, the prevention of the consequences of large-scale natural disasters, and the search for ways to solve long-term and deep social conflicts.

The main negative effect of globalization is the exacerbation of contradictions on all the levels of social development. On the megalevel, exacerbation of social contradictions in modern world is effected by the preservation of relative backwardness of the majority of countries, or those beyond the "golden billion" zone, thus widening the gap in wealth and welfare between people living in developed and developing countries. This entails the formation of an international system employing "global resources for private egoistic interests of the minority" (Alexander Panarin). On the macrolevel, the contradiction between the small, wealthy minority and the majority of the planet's inhabitants involves the ability to move (physically or virtually) around the world. Zygmunt Baumann states that today "access to global mobility" is becoming the factor of the public stratification. The modern "mobile elite" lives in time, while paupers live in the local space; the field of freedom for the elite is expanding, while that for non-elite is contracting. Moreover, on the macrolevel globalization of crime takes place, especially in the area of drug industry, and female "slave trade."

On the mesolevel the phenomenon of "new poverty" (Baumann) is being formed, in which, in contrast with the old one, "inherited" poverty appears as the result of the propagation of new industrial technologies and the actual liquidation of unsettled territories.

The main danger of globalization is the creation of preconditions for the appearance of a world totalitarian power, which, in the final analysis, will nullify the possibility of the embodiment of such human values as freedom, privacy, and human dignity.

V. V. Nikitina

Globalization as a Stage in the Transformation of the European Value System: Up to the middle of the nineteenth century, European civilization was based on Christian values with God being the measure of all things. By the end of the nineteenth century the traditional system of norms and rules had started to crumble. It was accompanied by the search for new models and

242   *Global Studies Encyclopedic Dictionary*

identities, but none of the proposed ideas was comparable with the original Christian orientation with respect to reliability and stability. Social orders replaced each other: Christian values were replaced by Fascism (more precisely, totalitarianism), then by the ideas of freedom and democracy.

I. B. Chubais

Globalization and Universal Values: The value-related side of globalization is the intensive exchange of material and cultural values created within local communities of people, and, at the same time, the detection and generation of the universal core of the world ethical heritage. This brings in the issue of universal values, which are important for all humankind.

During the twentieth century, many values of the previous centuries were destroyed. However, the tragic experience of the twentieth century also revealed the unity of humankind, having demonstrated that the value split threatens the existence of humankind.

Integration of national cultures into a world culture implies some "common denominator," universal values. Cultivation of universal values in each national group leads to narrowing the split. It is analogous to people who are not influenced by national biases.

Growth of integration processes in the modern world at the regional and planetary levels is stipulated by the vital necessity of solving problems of ecology and peaceful coexistence. Thus, the principle of the priority of universal values is not just a good intention or fine words, but an axiological imperative based on the presumption of universal values existing in human culture; this is an imperative that can ensure the continued existence of humankind.

L. N. Stolovich

Glocalization (portmanteau of globalization and localization) is a process of interlacing global trends in social development and local particularities in the cultural development of different peoples. The term is a combination of the words globalization and localization. In this context, globalization is viewed as something huge, external, imminent, and, in the final analysis, suppressing everything local, which, in its turn, is viewed discretely as something expressed in small specific manifestations on the local every-day level. From this point of view, local and global are not mutually exclusive. Moreover, the local is considered an aspect of the global. In addition, according to supporters of this approach, globalization means retraction, confrontation of local cultures that requires a new definition in this clash of localities. They also believe that cultural globalization makes identification of a national state with a national-state society null and void, generating and knocking together transcultural forms of communication and life, notions of responsibility and ethnicity, of the way individual groups and persons view themselves and others.

A. N. Chumakov

Goals for Mankind (1977) was the fifth report to the Club of Rome authored by a group of scientists headed by the well-known American philosopher Ervin Laszlo. The report formulated the concept of global solidarity, which, the authors opined, can be achieved by "revolution of global solidarity," connected with reorganization of consciousness and formation of a new "global

Gandhism – Gumilev, Lev Nikolaevich   243

ethos." The basis for such a revolution was four global goals of world development: global safety based on termination of the arms race and refusal of violence; solution of the food supplies problem on a global scale; global control over the usage of power and sources of raw materials; global development focused on qualitative growth. It set a paradigm for the subsequent research of the Club of Rome, which started to pay more attention to concrete global problems, rather than to the situation as a whole.

Reference: Laszlo, E. *Goals for Mankind*. New York: Dutton, 1977.

A. N. Chumakov

Golden Billion: (*золотой миллиард*, tr. *zolotoy milliard*) Golden Billion is a term in the Russian speaking world that designates the most wealthy people living predominantly in the most developed countries and having all that is needed for a secure and comfortable life.

The present stage of scientific and technological revolution that has grown into informational and technological revolution contributes, on the one hand, to the unity of humankind and, on the other hand, to its differentiation. For example, while the level of social, political and ecological intensity in the world is growing, the gap between rich and poor people, countries and regions widens; capital and the newest technologies are accumulated in the most developed countries, entailing the transfer of the intellectual potential and highly qualified specialists from the poor states to the rich ones. Informational and technological revolution translates into a more profitable and privileged position for the most advanced countries. For example, at the beginning of the twenty-first century the overwhelming majority (88 percent) of the 300 million Internet users live in "golden billion" countries.

The income gap between people in the rich and the poor countries is especially striking. For example, while in 2000, one sixth of the planetary population (mostly inhabitants of North America, Europe and Japan) were receiving nearly 80 percent of world income (about \$70/day for an average person), 57 percent of world population from the poorest countries had only 6 percent of world income (meaning less than \$2/day for person). At the same time, 1.2 billion people having less than \$1/day were even in a worse situation.

Under the conditions of globalization the world as a whole becomes more and more structured, first, in the field of communications and world trade. This, in its turn, increases the specialization of specific countries and regions within the world division of labor. For example, the percentage of the population of advanced countries working in the sphere of service, science, education, art, light processing and the electronics industry, that is, the ecologically secure industries, is much higher than in the least developed or even developing countries. Considering more severe environmental laws adopted by the most of developed Western countries, most environmentally dangerous industries, such as chemical, metallurgic, mining enterprises, are being moved to the less developed countries. In a very short time such industries destroy traditional ways of life and create serious disproportions in nature/society relations. The informational and technological revolution has also fundamentally changed the directions of world financial flows: now capital mostly goes to the developed countries, because high technologies require educated, high-

244 *Global Studies Encyclopedic Dictionary*

ly skilled specialists, high industrial culture and developed infrastructure. Golden billion countries also become the main consumers of knowledge-oriented, highly technological production leaving outsiders no chance to shorten the gap and to reach the level of sustainable development. This is a serious contribution into social and political instability in the modern world.

A. N. Chumakov

Gorbachev Foundation (The International Public Foundation for Socio-economic and Political Research) is an international, nongovernmental, noncommercial organization. The Gorbachev Foundation (GF) was founded in December 1991, by Mikhail S. Gorbachev, former President of the Soviet Union, Nobel Peace Prize winner, President of the International Green Cross, and leader of Social-Democratic Party of Russia. Gorbachev is currently the president of the Foundation.

According to the GF Charter, its primary goals are:

(1) Participation in social, economic and political research.

(2) Strengthening international cooperation and exchange in social, economic, and political areas.

(3) Revealing of and assistance to the development of the most important directions in social, economic, and political research and practice.

(4) Assistance to faculty and research staff development.

(5) Support and encouragement of the most important research work of individual scholars and students, research and pedagogical groups, and popularizes of humanitarian and social sciences.

(6) Organization of international exhibitions and teleconference bridges discussing the problems of modern civilization and the entering of humankind into the third millennium.

(7) Charitable activities.

In the age of globalization, the most important GF activity is development of the principles of a new, more fair and more perfect world order. The motto of the Foundation is "Toward a New Civilization." GF cooperates with the largest universities in various countries of the world. The basic directions of the Foundation's research work are as follows: global problems, economic development and social problems in the world, in Russia, and in CIS countries; problems of cultural and spiritual life; the process of European integration; international security and disarmament; history of perestroika in the Soviet Union. Both organizations and individual citizens of Russia and other states take part in the Foundation's activity. The GF representation offices exist in several large cities of the Russian Federation. The Gorbachev Foundation can also be found in the United States, Italy, Netherlands, and Germany.

GF devotes special attention to charity programs. In this area children's health services are a top priority for GF; in particular, the first Russian Center of bone marrow transplantation for children suffering from leukemia has been founded with GF financial support. Now GF is running two long-term projects: "The struggle against tuberculosis in Russia" and "Quality management of medical aid in the country."

Y. P. Zavarzina

Government, World: Federal world government refers to a vision of a system of world law in which the nations of the world would limit their sovereignty to create a legal authority over all persons on earth, with enforceable legislative, judicial, and executive powers. Although it is possible to trace the roots of the idea of world government back to ancient Stoic notions of a universal *logos*, Christian ideas of a natural moral law, or popular ancient notions that the ruler was a living embodiment of the sovereignty of the people, the concept of a democratic earth federation under universal enforceable laws has become prominent only since the early twentieth century. In particular, this idea, and the worldwide movement behind it, is conceptually a product of eighteenth century democratic theory and twentieth century awareness of global issues such as overpopulation, militarism and weapons of mass destruction, the threat of global environmental destruction, and ever-increasing global poverty and misery on the planet.

Several books advocating world government appeared after World War I, greatly increasing during the 1940s and early 1950s. Soon after the development of nuclear weapons and the chaos of World War II, many books and movements advocating world federalism appeared. Many prominent leaders (such as Albert Einstein) and organizations began working for federal world government. One fundamental issue that has divided world government advocates is whether the United Nations should be reformed from within to serve as the basis for future world government or whether the UN Charter must be replaced with a well written constitution for the planet.

Several substantial world government organizations persisted throughout the Cold War period and we can identify signs at the dawn of the twenty-first century of a revival of interest in world political unity. The largest and most important of these organizations are the World Constitution and Parliament Association, the World Federalist Movement, the World Mundialist Movement, World Union, and the World Citizens Movement. In addition, there have been over 150 proposed constitutions written for world government. Most significant among these is the *Constitution for the Federation of Earth*, a document that was written and refined through four constituent assemblies of world citizens between 1968 and 1991, and has been translated into twenty-two languages.

There have been several general philosophical arguments for world political unity, such as those based on the progress of civilization, the oneness of humankind, or emergent evolutionary processes. However, the most common and more specific set of arguments for world government focus on global issues. Global issues are understood by many world federalists to be those many issues that are planetary in scope and beyond the ability of individual nation-states to address.

Issues most commonly cited are the following: the need to create world peace through ending war and militarism; the need to save and preserve the global environment; the need to protect the human rights and freedom of all people; the need to create democracy for all people on earth; the need to eliminate poverty by regulating global trade for the benefit of all peoples and creating an economic order of planetary prosperity; the need to control international crime that prospers under the anarchic nation-state system; the need to

246   *Global Studies Encyclopedic Dictionary*

create a sense of world community to diminish ethnic, cultural, racial, and religious rivalries; the need to find a solution for the population explosion; and the need for truly planetary planning and vision for the good of all persons on the planet and future generations.

Several thinkers in the Western tradition have characterized the relation between autonomous sovereign nation-states that characterizes the modern world system as a state of war. Thomas Hobbes first pointed this out in *Leviathan* (1651), followed by John Locke in his *Second Treatise on Government* (1690) and Immanuel Kant in *Perpetual Peace* (1795). If "sovereignty" means absolute autonomy over internal affairs and absolute independence in external affairs, then this system of more than 190 territorial units (in today's terms) all pursuing their economic, political and military self-interest is necessarily a system of perpetual war or threat of war, a thesis apparently corroborated by the history of the modern era.

Proponents of world government argue that the only way to secure peace is *create a peace system* (in contrast to the present war system), the key to which is to limit the sovereignty of nations to substantial control over only their internal affairs and place an authority over all of them with the mandate to prevent war and create the conditions for permanent peace. Effective world government with laws enforceable over all individuals, including leaders of nations, is the only practical means of creating a peace system, promoting demilitarization, and ending the chaotic system of sovereign nations that necessarily depends on militarism and war-making capacity.

A similar argument applies to all of the remaining global issues itemized above. The system of sovereign nation-states intrinsically requires each nation to promote its own perceived interests above all the others and the common good of the planet. Therefore, the possible imminent collapse of our planetary environment from global warming, ozone depletion, scarcity of resources, and proliferation of toxic wastes can only be prevented by a world government that limits the sovereignty of nation-states and is mandated to protect the global environment for the welfare of all.

During the twentieth century, we have experienced a growing awareness of issues and problems truly global in scope, such as those mentioned above. Some proponents of world government argue that the historical-causal roots of these problems lie in the dominant institutions and attitudes of modernity that first developed in the Renaissance during the fifteenth, sixteenth, and seventeenth century. The paradigms on which our institutions and attitudes are based are not global or planetary in scope, but are fragmented—premised as they are on a worldview centuries old and extremely naive by contemporary standards. Solving global problems requires a conceptual shift to a planetary paradigm and institutions premised on the realities of our global situation.

Many argue that a paradigm shift has been effected by twentieth century science, a shift pointing toward a transformation of all these fragmented modern assumptions. All features of the universe are now seen as interrelated and interdependent, none comprehensible in themselves apart from their relations to the environing universe. A concurrent revolution has taken place in the biological sciences that now emphasize the interrelatedness of all living things, of the processes of evolution, and of organisms with their environment. Similar

Gandhism – Gumilev, Lev Nikolaevich 247

transformations have taken place in the social sciences suggesting a holism in which individuals cannot be understood apart from their social environment.

The required paradigm shift entails that the system of more than 190 territorially autonomous sovereign nation-states be replaced with democratic world government premised on the interrelatedness of all peoples, cultures, and persons on Earth. In this argument, the concept of democratic world government is a *logical consequence* of the new scientific paradigm developed by the most advanced minds of the twentieth century premised on the holistic relation of unity-in-diversity found within all the processes of nature.

G. T. Martin

Great Eight (Great-8, G-8): The G-8 is a forum for eight of the world's largest national economies as nominal GDP with higher Human Development Index; not included are India at ninth, Brazil at seventh, and China at second. Due to their financial and economic potential, these nations play a significant part in solving global problems and produce an enormous effect upon all humankind. Various assessments attribute up to three quarters of the world production and trade to these countries; collectively, the G-8 nations comprise over 50 percent of the 2012 global nominal GDP and 40.9 percent of global GDP (PPP). Beginning in 1975, the leaders of Great Britain, Germany, Italy, the United States, France and Japan (joined by Canada in 1976) have been gathering for their annual meetings and coordinate their positions and actions in the economic and political fields. The EU is represented within the G-8, but cannot host meetings. The meetings are of an informal and unofficial nature; no documents are signed in the form of contracts or agreements obligatory for all the participants; only general guidelines are created, which each of the member-states is free to follow in its own way. Russia has been invited for these meetings since 1992, and achieved full membership in 2002. Russia hosted the G-8 summit of 2006, at which its full membership was confirmed

V. N. Kuznetsov

Green Revolution: a process of introducing new high-yielding varieties of grain crops, irrigation, chemicalization and mechanization of agriculture in many countries with a view to increase food resources; "explosion-like growth" of agricultural production during the 1960s and 1970s, in several developing countries (such as India) due to new agrarian methods (effective cultures, and modern technology).

V. M. Smolkin

Green Taxes (Ecological Taxes; Ecotax): are taxes collected to support ecologically sustainable activities via economic incentives, calculated on the basis of environment impact. In the broad sense, green taxes are ecological fees and pollution charges, which constitute one of the basic state economical tools for environmental protection.

Green taxes were first applied in practice in the countries of the northern Europe. There was an ecological reform carried out in other European countries and in Japan in the middle of the 1990s. Along with pollution quotas,

248 *Global Studies Encyclopedic Dictionary*

green taxes represent an economic mechanism to implement the regulations of Kyoto Protocol to the UN Framework Convention on Climate Change.

P. A. Kalinichenko

Greenhouse Effect is a warming of the lower regions of the atmosphere, caused by the absorption of long-wave (infrared) radiation by water steam, carbon dioxide, methane, nitrogen oxide, ozone, and chlorofluorocarbons (Freon), which are contained in the atmosphere and absorb the Earth's thermal radiation. These gases are called "the greenhouse gases." Without these the temperature on Earth's surface would be about -18 °C. The greenhouse gases raise it by thirty-three degrees to 15 °C. The absolute value of the greenhouse effect is 160 W/ml, out of which 100W are provided by water steam and 50W by the carbon dioxide. Currently, due to human economic activity, the concentration of the greenhouse gases, primarily of the carbon dioxide, is increasing due to fossil fuel combustion.

This trend is believed by many researchers to entail the growth of the average global temperature on the Earth's surface. For instance, according to the observations of meteorological stations, the average global temperature from the middle of the nineteenth century to the end of the twentieth century has risen by 0.6 °C. There is other evidence of global warming; for example, general waning of most mountain glaciers, reduction of the ice-covered area of the Arctic Ocean, and retreat and reduction of permafrost. However, it is still impossible to single out the anthropogenic component, as there are no methods for determining it. The rise of temperature is most likely fully or partly caused by natural climatic fluctuations. There are other possible anthropogenic mechanisms of warming; for example, it can be connected with reflectance changes due to desertification and destruction of ecosystems on huge land areas. A more serious consequence of environmental changes under the influence of economic activity is not so much global warming as the violation of climatic stability leading to the growth in the frequency and spread of various abnormalities, including catastrophic ones. In the problem of the emission of greenhouse gases the emphasis is primarily put on industrial carbon emission, accounting for carbon emission through land tenure being insufficient, although the latter by it mass varies from one half to more than the total value of industrial emission.

According to the Kyoto Protocol (1997), the system of industrial emission abatement is being elaborated by many developed countries, which are the main sources of the greenhouse gases. This protocol stipulates a 5 percent cut in emissions by 2008–2012, while in 2000, the emission was about 1.6 billion tons a year. The magnitude of this cut is less than the calculation error of the greenhouse gases full emission. From the point of view of the theory of biotic regulation, the maintenance and development of natural ecosystems and the recovering of natural woods facilitates stabilization of the concentration of greenhouse gases, because natural ecosystems serve as accumulators and consumers of these gases.

K. S. Losev

Growth Limits: The Earth being spatially finite with limited natural resources necessarily entails limits to population growth and growth of industrial and agricultural production. Questions about growth limits are not new. The most outstanding examples include French Enlightenment philosophers Montesquieu, who argued that money turnover, income growth, and number of people must be balanced, Marquis de Condorcet, who asked whether a retreat would be provoked when humankind faced the limit caused when the growing number of people exceeded the quantity of means for their maintenance, and the English economist Thomas Malthus, who theorized about the negative consequences of population growth in *An Essay on the Principle of Population* (1798).

Jay W. Forrester discussed growth limits in *World Dynamics* (1971), and the concept became popular within a framework of computer models of global development during the 1970s. Scholarly debate around the issue began after the publication of *Beyond the Limits to Growth* (Pestel, 1987), which concluded that a continuation of the existing trends of population and industrial/agricultural production growth would mean that by the beginning of the twenty-first century the levels of natural resources and environmental pollution would reach a critical point fraught with the prospect for global catastrophe. According to the authors of the report, its prevention was only possible if the adequate measures were taken to reach "global balance."

Mathematical calculations of the limits to growth have debunked the widespread technocratic myth about unlimited economic growth as a means to resolve problems; these have directed international attention to crises resulting from polices that waste natural resources. At the extreme, the concept of growth limits has given rise to the extreme concepts of "zero growth" and the demands to stop all industrial activity. More moderately, it addresses issues related to the meaning of natural (extrinsic) and sociocultural (intrinsic) growth limits, and the issue of development objectives with regard to increasing the quality of life.

Speculations about growth limits have given rise to rethinking earlier concepts of social progress. Supporters of the growth idea, which was popular in the 1960s, were forced to defend their positions. Non-conforming theorists reconsidered their outlook, abandoned the concept of civilizational "growth stages," and turned their attention to the issue of ecology. New concepts began to emerge in political economy mediating between an orientation toward unlimited growth and a trend to zero it. It became evident that the real issue is not the choice between growth and antigrowth but defining the characteristics and quality of growth (e.g., Pestel, 1987).

Some pinned their hopes on space explorations, arguing that they can engender a higher type of human civilization free from environmental pollution and threats of war. However, while space exploration was developing, it turned out that near-earth space pollution had grown and "Star Wars" plans had been nurtured that, in their practical realization, would threaten human existence even more than conventional armaments.

Others counted on the new technologies providing minimal environmental pollution and saving natural resources. Nevertheless, this solution with all its accomplishments has led to even greater waste of energy potential. For example, due to the emergence of new car models made of completely new materials, air pollution and gasoline consumption decreased. This meant a relative saving of energy, but, at the same time, increased the waste of energy

250   *Global Studies Encyclopedic Dictionary*

related not only to the production of but also the providing of service for the increasing number of cars, roads renovation, and fast track construction.

Still others, including some specialists in cybernetics, argued that the processes of information and communication make unlimited growth possible because telecommunications development and information consumption have no limits. Nevertheless, as has been demonstrated by the informatization of societies and the worldwide spread of the Internet, time and people's wish to assimilate and absorb the annually growing flow of information set their own limits on informational consumption and utilization. In an informational society, conventional environmental pollution related to industrial waste has been replaced with informational pollution (e.g., computer viruses; disinformation), which are no less destructive.

The concept of intrinsic limits set not by nature but by humanity raises questions not about how is production, but what is produced and how resources are distributed. Growth for the sake of the growth contradicts the interests of humanity. The emphasis must not be on quantity but on quality. Many scholars have concluded that human civilization is now beyond growth limits. Hence, decisive measures must be taken to stabilize the global system before global catastrophe occurs.

References: Malthus, T. *An Essay on the Principle of Population*. London: Printed for J. Johnson, 1798. Forrester, J. *World Dynamics*. Cambridge, MA: Wright-Allen Press, 1971. Pestel, E.. *Oltre i limiti dello sviluppo* (Beyond the Limits to Growth). Torino: ISEDI, 1987.

V. M. Leibin

Gumilev, Lev Nikolaevich (1912–1992)–Russian historian, geographer, and ethnologist, Gumilev developed an original theory of ethnos (ethnic group). He coined the term, "passionarity," which means the non-resistible inner drive to conduct extremely fierce activity, the purpose of which is changing the ethnic or natural environment of the person who possesses this drive. He distinguished the following phases of ethnogenesis and ethnic evolution based on the presence of passionarity:

(1) Incubation phase. Some people possessing passionarity emerge as a result of "a passionarity impulse," thus giving rise to a new ethnic system.

(2) Development phase. Passionarity steadily increases; this phase is characterized by a wide expansion of the new ethnos and population explosion. Nearby ethnic groups perceive this phase as a period of emergence of an active ethnos asserting uncommon ideals and fighting for its territory, often at the expense of neighbors.

(3) The phase where passionarity reaches its extremes. Sacrificial people possessing passionarity dominate the ethnic system, but now (unlike the previous phase) they aspire rather to the utmost self-assertion than to asserting their ethnic group.

(4) Break-down phase is a period of drastic decrease of passionarity accompanied by ethnic splits and severe conflicts within this ethnic system. The number of people having not enough passionarity ("philistines," "bourgeoisie") grows. The ethnic system declines and merges with other ethnic groups.

(5) Inertia phase is a period of increase and then slow decline of the level of passionarity. Strengthening of state power and social institutes, intensive accumulation of material and cultural values, and active transformation of landscapes take place.

(6) The phase when passionarity falls down to zero. As a result, the social structure falls apart: corruption becomes a law of life, crime rates increase, the army becomes non-effective, and power is taken over by cynic adventurers who play on the emotions of the crowd. The population decreases; this process is somewhat hampered by the inflow of the alien ethnic groups, who often begin to play a leading role in public life.

(7) Regeneration phase is a possible reconstruction of the ethnic system thanks to remaining passionarity.

(8) Memorial phase is the final phase of the process: the ethnic system completely loses passionarity but preserves cultural traditions of the past. The memory of the heroic deeds of the predecessors lives in the form of folk tales and legends.

A. G. Ganzha

⊰ H ⊱

Hegemonism: (*haegemonia*, "domination" "primacy") Hegemonism is an ideology and politics aiming at establishing domination by one state, group of states, political system, or religion over the whole world or specific regions. The ideology of hegemonism represents a system of beliefs and concepts that uses a policy of world domination to justify coercive methods of achieving expansionist goals. The politics of hegemonism are based on force and imply military and other interventions into the domestic affairs of states and peoples. Throughout human history, each civilization or world power demonstrated hegemonic trends. The cause of hegemonism is a state's desire to enlarge its political power and influence at the expense of other members of the international community. Using threats or force, building coalitions, and establishing zones of influence, states aspire to gain control over the behavior of the other elements of the international system and to create such an international environment that would allow them to realize their interests and goals with an optimum effect. This system's organizational principles and the forms in which its interests are realized reflect, as a rule, the power and influence of the systemic elements over one another. Control over the international system is based on the allocation of power and resources among its elements. The structure of power distribution defines who, in fact, governs this international system and with whose interests this system is in congruence. Within an international system it is the dominating states that organize and control the interactions among its elements.

The history of international systems knows three types of control: (1) imperial, when one powerful state controls the other weaker states; (2) bipolar structure, when two powerful states control and regulate interactions within and among their spheres of influence; and (3) balance of power, when three or more states control each other's actions with the help of diplomatic maneuvering, re-shaping of unions, and open conflicts. The dominating states organize and maintain a net of political, economic, and other relations within the system and, especially, in their own spheres of influence. They work out and enforce basic rules and norms guiding their own behavior and the behavior of the weaker states.

The "right to hegemony" of a great power can be obtained in three ways: (1) It can win the last hegemonic war or demonstrate the apparent ability to impose its will on the other states. (2) A state's domination might be—often is—accepted because it provides the dependent states with benefits such as an advantageous economic environment or international security. (3) A dominating state's position may be backed by ideological, religious, or other values shared by several states.

The interests of separate elements of the international system may confront one another. Imbalance and a change of dominating powers lead to a crisis of the given international system. Thought peaceful resolution of the crisis is possible, resolution often occurs through a hegemonic war that defines which state or group of states will dominate the international system. The peace that follows a hegemonic war is characterized by the re-shaping of

Hegemonism – Hypothesis 253

political, territorial, and other foundations of the system. This cycle of changes ends when a hegemonic war and succeeding peace create a new status quo and balance reflecting a new allocation of power within the system. In periods of stability, the international system and the world order are in a state of homeostasis, or dynamic equilibrium, though at the level of interstate relations continuous tactical changes occur. The international community today condemns all forms of hegemonism. A resolution condemning hegemonism in international relations was adopted by the United Nations in 1979.

S. V. Moshkin

History, Global: Global history is a trend in contemporary historical science based on the ideas of the holistic nature and interconnectedness of world history, of the role that global factors such as climate changes play in social development, and the interactions of various civilizations. Sometimes global history is considered as a historical dimension of global studies. Its subject matter may be described as the processes of the emergence and development of transnational spheres of human living that have their own logic and dynamics, as compared to the processes typical in the development of separated tribes, nations, and states. It is especially focused on issues of interactions between trends toward unity and polyethnicity in history, and on the role of ecological, cultural, and political crises in both the evolution of separated communities and the international community as a whole. There are three basic conceptual frameworks within global history based on the synthesis of historical, economic, geographic and other approaches but different in their methodology and founding principles. A theory of contemporary (universal) evolutionism and self-organization tightly connected with anthropological, ecological, and synergetic ideas is the first of them. A modern world-system theory interacting with the civilizational approach to the study of history is the second conceptual framework. The third conceptual framework is represented by historical geopolitics: the history of the world powers as cross-points (centers) of the global system of political relations. Global History provides an opportunity to overcome the Eurocentric and West-centered position in the study of history that is of primary importance for analyzing modern globalization processes.

Global history waves are pulses of development and advancement of global processes in the sphere of economics, politics, and culture that relate to the global changes in human and social life, which emerge because historical processes of global development (including modern globalization processes) are evidently non-linear. Their nature depends on various natural resources, socio-psychological, and other factors. Analysis undertaken within the framework of global history demonstrates that, generally speaking, global economic, social, and political development has both progressive and wave-like (oscillatory) components. Waves and oscillations of global socio-economic and political development are of differing durations and amplitudes. Global waves of differentiation in general coincide with periods of the emergence and initial development of a new mode of production (for example, ancient, feudal and capitalist) and new economic, social, political, and cultural institutions related to it. Global waves of integration correspond to

254 *Global Studies Encyclopedic Dictionary*

periods of their further development, advancement, and expiration. As a result, a global wave of differentiation and a global wave of integration make a "life cycle" of the mode of production in question and its system of social institutions. Thus, global history waves allow us to structure world history and to divide it into various periods and turning points. At the same time, there are shorter global history waves: in the industrial period a wave of global integration initiated by the industrial revolution (1846–1914) was replaced by a wave of "disintegration" (1914–1946) and then by a new, more powerful wave of global integration (since 1946).

V. I. Pantin

Homeostasis: (Latinized from Greek *homoios*, "similar," and *stasis*, "standing still") Homeostasis is a state of dynamic equilibrium of a system, which is preserved due to its resistance to the external and internal factors that upset the equilibrium. The concept of homeostasis originated in physiology to explain the stability of the internal environment of an organism and its principal physiological functions that are achieved through the mechanisms of self-regulation. Soon the principle of homeostasis was extended to psychology (Jean Piaget), and, since the 1950s, to cybernetics (Norbert Wiener) and other sciences. Since then, the concept of homeostasis has been applied on a large scale not only to biological and psychological systems at various levels, but also to technological, economic, social, political, ecological, and other systems to characterize the mechanisms of their management and control on the basis of negative feedback between the system and the external environment.

The homeostatic interaction of an open system with the surrounding world gives rise to two types of adaptability. On the one hand, there is a system's adaptability to the external environment by means of internal modifications, and, on the other hand, there is active influence of a system upon its environment and its "adaptation" to the system's "needs" by extraction and assimilation of the necessary resources. Such adaptability grounds the mechanism of self-regulation and self-organization of a system, which, in its turn, predetermines such functional properties of a complex system as integrity, latency, and purposefulness. When studying complex systems, one uses, apart from homeostasis, the concept of heterostasis that reflects a hierarchy of homeostatic systems, the division of parameters into more or less important ones, and the concept of homeoresis, which is a set of internal organisms that provide alteration of some parameters of the system in the process of its evolution. In social and political science, the principle of homeostasis is used mainly for analyzing the functioning and sociodynamics of the social and political system in general and its subsystems.

Y. V. Irkhin

Human Development Index (HDI): The HDI is a composite development index developed by the UN Development Program (UNDP). It is a quantitative index used for a quantitative representation of each of the key aspects of human development: longevity, education, and material welfare (standard of living). These fundamental aspects correspond to the three key-issues: to live a long life, acquire knowledge, have an access to resources sufficient for a decent standard of living. The HDI is regularly updated and comparable for almost any country of the world. This includes life expectancy at birth for both sexes; adult literacy, and primary, secondary, and higher education coverage; and GDP per capita adjusted according to purchasing power parity (material welfare). At that, the index of material welfare determines availability of resources for human development, while indices of education and longevity reflect the very purposes of human development.

One of the major advantages of HDI is that the methods and indicators of human development proposed by UNDP enable comparative study of development progress both on the national and regional level. They are based upon universal methods of national and regional socio-economic and demographic development programs.

I. A. Aleshkovsky

Human Development Reports (HDR) are annual reports, published under the aegis of UN Development Program (UNDP) since 1990. The founder of the HDR, Mahbub ul Haq, considered its primary objective to be the presentation to the world public of a wider vision of the people's living conditions and ways of their improvement, which would differ from a more narrow idea of economic growth in terms of GDP alone. On his initiative, the aggregative FTSE All-World Developed Index was worked out as an alternative method of measuring living standards and quality of life, taking into consideration not only per capita GDP, but also expected duration of life and education level. Later, on the basis of this technique, the Human Poverty Index (HPI) was developed in two variants: for developing countries (HPI-1) and for industrialized countries (HPI-2) and Gender Influenced Development Index (GIDI) and others. These indices were based on the compilation of detailed statistical information from more than 170 countries and became the acknowledged principle of place-to-place comparison of human development. They also enable monitoring the dynamic of social development on the national, regional or global scale. Between 1993 and 2003, fourteen reports were published, each of them laying the emphasis on a particular aspect, taking into consideration the influence of different factors on the human development, such as poverty, consumption, high technology, globalization, security condition, people's rights, and democracy. Regional and national reports on human development published under the aegis of UNDP are also of a significant value.

A. B. Veber

Human Experience, Objective and Subjective Factors in: Categories involving the dilemma of the modern social subject, who, as a significant technological force, must consider the boundaries of freedom to avoid self-destruction. The subjective factor consists in the free and purposeful activity of a social subject, which integrates the theoretical and practical dimensions of societal development. The objective factor consists in the laws of nature and society and the concrete historical conditions of the life of a social subject independent of will or conscience, which determine the direction and conditions of one's life.

"The subjective and the objective" and "subjective and objective factors" are not identical, as each categorization reflects a specific global perspective in transition from a speculative type of outlook to a more activity-

256 *Global Studies Encyclopedic Dictionary*

related one. The categories of *"the subjective factor"* and *"the objective factor"* are more concrete and in this sense, richer than the category of "the objective" and "the subjective." Objective (from the Latin *objectivus*) is: (1) existing outside and independent of consciousness, inherent in the object itself or corresponding to it, and (2) corresponding to reality, impartial, unprejudiced. Subjective (from the Latin *subjectum*, meaning a subject or self-aware human being) is: (1) inherent only in a person, personal, and (2) one-sided, deprived of objectiveness, partial, prejudiced. The category of "the factor" (from the Latin *factor*, meaning acting, making) has two major meanings: (1) a driving force or reason for some process, and (2) the general conditions under which a process takes place.

The category of "the subjective factor" expresses the moment when ideas motivate actions, becoming the driving force of social processes. The category of "the objective factor" denotes the moment when a person's activity influences the conditions and circumstances that set the limits for the accomplishment of a goal. The subjective factor is the activity of a social subject. This activity may be spontaneous or deliberate, determinate or incidental, but in any case it must be at least fundamentally free. The subjective factor is operative only when people have freedom of choice. Freedom acts as a link between the subjective and objective factors as the subject chooses among the available conditions and circumstances available.

Despite their singularity of various historical conditions, the subjective factor is nevertheless subject to regularities. First, objective conditions and relations determine the subjective factor. Second, the subjective factor involves a relative independence; under the same objective conditions a particular subject's actions will differ from another's. Thirdly, the role of the subjective factor is a phenomenon in continual evolution.

Most directions of modern philosophy, such as existentialism, logical positivism, pragmatism and personalism, are characterized by a subjective-idealistic orientation and absolutize the inner freedom of the social subject, thus expressing the growing role of the subjective factor in conceptual formations. Postmodernism goes beyond absolutizing inner freedom, lapsing into indeterminism or voluntarism, in the language of social cognition.

In my opinion, the dialectic of the subjective and objective factors has been most fruitfully examined by Russian organicism and cosmism, especially in its concept of the God-humanity and noosphere: the goal of humanities' self-revival has been objectively pre-assigned but the accomplishment thereof depends on humanity itself.

O. D. Masloboeva

Human Potential refers to aptitude or competency to engage in different kinds of activity. It can refer to the aptitude of individuals, generally, by a group of people distinguished for some reason (social, cultural, professional, demographical, ethnic, territorial), by a country or an aggregate of countries, or, most generally, to humanity as a whole. The ability is not only possessed, but also realized. Human potential is determined by the characteristics of the individual or group in interaction with their surrounding conditions.

Hegemonism – Hypothesis   257

The concept of human potential is closely related to earlier terms, such as "human resource" and "human capital," which were in usage several decades ago. Those terms allowed us to see a human being not as an element of a production, social, technical, commercial, or other systems, but as a creature that cannot be implemented in a system and reduced to functionality. At first, additional human features were studied to make systems more efficient and controllable. Eventually, it became clear that the systems should be reconstructed: it was inevitable to consider the non-linear elements, bifurcation points, and so on. However, the concepts were limited to studying the relationship between a human being and a system, to viewing human beings as a resource.

The term "human potential" is meant to capture the notion that human beings are considered simultaneously as special resources, in that they are able to act independently from a system, and as persons who demand and consume natural and social resources. The concept reflects the immanent value of human existence, its openness at any point of human life.

Scientific research and practical activity in the field of human potential mean to solve different problems of its conservation, protection, development, and realization with special attention to risk factors under different social and natural conditions.

B. G. Yudin

Human Rights: While the concept of "rights" has been common parlance for several centuries, the notion of "human rights," rights that belong to people qua humans and because they are human, is more recent. A watershed event in the establishment and recognition of the significance of human rights was the drafting of the Universal Declaration of Human Rights, passed by the United Nations on 10 December 1948. Since then, many other human rights documents have been drafted and the protection of human rights has been a basic component of international political relations.

The concept of rights, including that of human rights, is related to other concepts connected to the regulation of behavior, concepts such as justice and law. Richard Tuck (1979) traces the concept of "rights" back to the medieval and earlier Roman notion of *ius*, from which we also derive the concept of "just" and "justice." *Ius* was held in contradistinction to *dominium*, the notion of "dominion" and ultimately of "property." Early on, then, there was the distinction between power, or ability, and justice, or what is right. Nonetheless, the notion of rights as pertaining to all moral agents and being held in rem, that is, against all others, does not figure in the moral or political works of the classical or medieval theorists, especially not in the context of what constitutes the good life or good society. There were the notions of justice and of natural law that were basic to understanding the place and role of moral agents in their social contexts. Even early documents such as the *Magna Carta* (1215) detailed policies and regulations that are seen today as precursors to both legal and human rights.

This transition from early conceptions of dominion, sovereignty, and natural law to natural rights and international law came in large part from the efforts of Hugo Grotius (1583–1645), John Selden (1584–1654), and Thomas Hobbes (1588–1679), but even more significantly from the writings of John

258 *Global Studies Encyclopedic Dictionary*

Locke (1632–1704), Baron de Montesquieu (1689–1755), and Thomas Jefferson (1743–1826). By the end of the eighteenth century, two crucial political documents, the French Declaration of the Rights of Man and the American Declaration of Independence, along with the Constitution of the United States, incorporated the concept of natural rights into the framework of governmental systems. With nineteenth-century criticisms of the Enlightenment's faith in natural reason, the concepts of natural law and natural rights that in large part rested on that faith evolved into the twentieth century notion of human rights, with a concern for the well-being of persons who suffered from political and economic oppressions.

Though the language of human rights did not become ubiquitous until the middle of the twentieth century, following the widespread recognition of the Holocaust, there were several earlier transnational efforts to implement actions and policies that focused on the rights of human beings, such as international efforts in the 1800s to abolish slavery and the slave trade, and efforts to regulate the treatment of civilians, wounded soldiers, and prisoners of war in the early 1900s. The International Court of Justice (the World Court), was established in 1946. Unquestionably, however, the Universal Declaration of Human Rights, drafted under the leadership of Eleanor Roosevelt, altered global consciousness on the issue of human rights as nothing before it did. It was truly international in scope, with significant contributions from Alexander Bogomolov and Alexei Pavlov from the Soviet Union, Peng Chun Chang of China, John Humphrey of Canada, Charles Malik of Lebanon, Hernan Santa Cruz of Chile, and many others. It was followed over the next several decades by many other international agreements: the UN Convention on the Prevention and Punishment of the Crime of Genocide (1951), the UN International Covenant on Civil and Political Rights (1966), the UN International Covenant on Economic, Social and Cultural Rights (1966), the Helsinki Agreement (1975), the Convention on the Elimination of All Forms of Discrimination against Women (1979), the UN Declaration on the Right to Development (1986), and a host of others.

Besides increased attention to human rights legislation from international bodies like the UN, agencies and organizations have emerged to protect and enhance human rights. These include Amnesty International, Human Rights Watch, the Economic and Social Council, the International League for Human Rights, the International Helsinki Federation for Human Rights, the Red Cross, and Red Crescent, and many others, both governmental and nongovernmental. Still, despite this attention, persistent violations persist, perpetrated by governments against their own citizens, by states against other states, by nongovernmental organizations against citizens of nearly all countries, across cultural, racial, religious, and political lines. The sustained implementation, protection, and enhancement of human rights has focused along the lines of national, international, and nongovernmental foci, and there appears to be an obvious need for a twenty-first century approach to the perennial issue of regulating the behavior of agents and enunciating

Hegemonism – Hypothesis 259

the identity of agents in the context of seeing them flourish in an ever-expanding global interconnectedness.

Reference: Tuck, Richard. *Natural Right Theories.* New York: Cambridge University Press, 1979.

D. Boersema

Human Rights from the Philosophical Point of View: The most significant achievement of humanity in the twentieth century is its bringing to the fore the idea of human rights. This occurred especially after the World War II, with the establishment of the UN. But toward the end of the century, human rights became a fashionable concept and an increasingly elastic term. Thereupon, we have observed an increase in the number of assumed human rights and an ever-increasing conceptual confusion in international human rights instruments.

Thus, despite everyone now speaking of human rights, flagrant human rights violations go on in many countries of the world; laws are enforced without taking into consideration the foreseeable consequences they bear for human rights; states violate the human rights not only of their citizens, but also those of the citizens of other countries.

One of the main reasons of this state of affairs, at least conceptually speaking, is our lack of sufficient knowledge of what human rights are. If people really knew what human rights are, most of them would be surely ready to do anything they can for their protection.

This is why before we start dealing with any theoretical or practical issue related to human rights, we need an epistemically justifiable answer to the question, "What are human rights?" This question is crucial, because its answer also constitutes the criterion used, or to be used, to distinguish a human right from other rights.

Norms of moral evaluation, or general propositions on what is good or bad, are expected to determine our evaluations, that is, to be used in order to determine the value of an evaluated object—whether an action, person, or situation; while norms of moral conduct or "ought-should" propositions are expected to determine our actions in life. Norms of evaluation are supposed to lead those who use them as criteria to the knowledge of the value of given actions of one (which is not the case); while norms of behavior—moral rules or principles—are expected to determine actions carried out in given situations.

The "ought" or "should" of these latter norms of conduct or behavior is deduced from premises of different epistemic specificities and by different ways of reasoning.

For my purpose, I shall confine myself here to pointing at only two kinds of such norms: those deduced in different given historical conditions from experience by induction and those deduced by the comparison of different given (human or historical) conditions in the light of the knowledge of the specificities of the human being, or of what is called human dignity. The first kind of norms may be justified statistically, the second by a kind of reasoning similar to *reductio ad absurdum.*

At the origin of the first kind of norms (behavior) are given natural-social conditions and the conceptions of different cultures concerning the human being. They are norms of behavior relative to existing conditions, possessing a

260 *Global Studies Encyclopedic Dictionary*

practical function, in view of establishing or safeguarding any order in those existing conditions at the moment they are deduced. So long as the conditions in which they are deduced prevail, if deduced with sagacity, they are functional. But when these conditions change, they lose their function and meaning, that is, the "ought" or "must" which they express lose their ground.

Norms of this kind, to which many proverbs or products of practical wisdom also belong, are deduced by evaluating the effects this or that way of behavior has had, that is, the benefit or harm they mostly (or often) caused to those who happened to behave in this or that manner. Norms of this kind tell us, in fact, this much: when someone behaves in this or that way, the probability of safeguarding his or her benefit or interests, and sometimes those of the others, increases. When, in a given case, one is unable to make a right evaluation, but possesses the will to protect what is considered to be in such a case his or her benefit or interest, or that of the group he or she belongs-it is more probable to protect it, if he behaves in the way the relevant norm demands. As to the other kinds of norms (or "ought" propositions) originating in the knowledge of the value of the human being, they are deduced from this knowledge, directly or indirectly, in the face of human or historical conditions doing harm to this value (for instance, "thou shall not kill," "no racial discrimination shall be made").

Rights, in general, are "rights of someone." The Platonic understanding of right is something due to somebody by somebody else. Human rights express the treatment that is due to each human being by other human beings.

Human rights, as they are worded in the international instruments, intend to express demands concerning what a human being and every human being should not suffer, that is, what direct and indirect treatment should or should not undergo, so that a human individual may actualize, as much as each one can, the human potentialities, ethical ones included, which constitute what we call human dignity.

Still, too often, individuals do not treat other individuals in the way that human rights demand. Then, who will do that?

If we look at the international human rights instruments, or, at the addressees in these instruments (with the exception of the Universal Declaration of Human Rights, whose addressees are "every individual and every organ of society") the answer appears to be clear: the instance that will safeguard-directly or indirectly: the treatment of individuals in the way that human rights demand is the State in each state.

Such legislation, implementation, and education depend first on clear knowledge about each human right, since they presuppose that those responsible for the implementation and education of human rights have sufficient philosophical knowledge of the conditions that human rights demand and have become able to put in connection this knowledge with the cases they will face.

I. Kuçuradi

Human Security entered the vocabulary of contemporary international relations with the publication of the *Human Development Report* by the UN Development Programme in 1993. When introduced, this idea of making "security" pertain to human beings rather than to states and nations was both inno-

Hegemonism – Hypothesis

261

vative and controversial. Human Security again became the main focus of the UNDP's annual report in 1994, wherein the authors urged "it is time to make a transition from the narrow concept of national security to the all-encompassing concept of human security."

Human Security, according to its proponents, is both a vision and an agenda, a projected picture of a finer world and a goal to be pursued via co-operation among states and peoples. According to the UN Development Pro-gramme, those who endorse and promote the idea of human security "stress the security of people, not only of nations." In the international environment of the twenty-first century, they say, there ought to be less "stress on national security" and "much greater stress on people's security," less emphasis on "security through armaments" and more on "security through human devel-opment," and less concern about "territorial security" and more about "food, employment and environmental security." Whereas national security has to do with protecting states from hostile neighbors, human security may have as much to do with protecting people from their own oppressive governments.

Advocates of human security assign its promotion and realization to the United Nations and associated international organizations. These are the set-tings within which the international community can choose to act collectively, and when the international community does act collectively it should act in the interest of human security. The United Nations has long embraced the idea of "collective security." So, then, why not extend this to the notion of "collective human security," which would call for international collective action in response to threats to social, economic, political, and cultural well-being, and to military threats? Everything that the United Nations does, advo-cates of human security say, should be related to improving human welfare in all of its aspects or dimensions.

Programmatically speaking, having the promotion of human security as a central objective of policy and action integrates the main areas of United Nations activity by relating peacekeeping directly to development and both of these directly to the promotion of human rights. Most often, peacekeeping, development, and promoting human rights are looked upon as separate kinds of undertakings. They are all assigned to the United Nations, but they are bureaucratically compartmentalized within the organization, because different agencies deal with peacekeeping, development, and human rights. Until re-cently, moreover, these activities have also been conceptually compartmental-ized. That is, keeping peace in troubled areas, encouraging development, and protecting human rights were essentially thought of as different kinds of mis-sions. But the idea of promoting human security combines the different as-pects of the United Nations assignment: promoting human security means keeping the peace; it means furthering economic and social development; it means protecting human rights. Human security gives the United Nations a unified mission directed toward improving the lives of human beings.

Much more than simply a change in international rhetoric, the new em-phasis on human security is prompting changes in policy and behavior at the United Nations. As noted, it has become the main focus of the work of the United Nations Development Programme. It has, in addition, become the principal justification for "humanitarian intervention" on the part of the international community. Promoting human security has extended the agenda of the Security Council, which has assigned to itself the responsibility of acting in response to threats to human security. Such reasoning, for example, underpinned United Nations missions like those undertaken in Kosovo, East Timor and Sierra Leone.

But what is perhaps most significant in the notion of human security is that it focuses international organizational attention directly on individuals and their circumstances. This implies that an appropriate role for international organizations, and the prescribed role for the United Nations, should henceforth be to relate immediately to people and to serve them directly in such ways as to enhance the quality of their lives.

Yet, herein also lies the core of today's opposition to human security in theory and practice: promoting human security via international organization collides with preserving and protecting national sovereignty. Most states remain highly sensitive about their sovereignty and the idea of humanitarian intervention by the international community smacks of United Nations supranationality, which has not been especially well received among member-governments of the United Nations. Particular objections have been raised by small states and by non-Western ones that perceive international intervention as usually coming from large states and Western ones. Those who take a skeptical view of this new concept of human security wonder from whence comes the international community's authority to intervene into national territory and national affairs, either for purposes of promoting human security or for any purposes whatever.

The international politics surrounding "human security" are intense because the stakes are perceived to be very high, as they surely would be where the question is one of constraining the authority of the state and elevating that of the international community and the United Nations as its agent. Therefore, within the United Nations today, and in contemporary international relations more generally, there are staunch opponents and enthusiastic proponents of using human security to justify collective international action. There are likewise as many critics as there are champions of humanitarian intervention, and as many interests and coalitions that would relegate human security to history's dustbin as there are forces that welcome it as the wave of the global future. The balance of these forces and the outcomes of interactions among them will decide the future of "human security" at the United Nations.

Many students and practitioners of world affairs view human security as a good idea. But the political contexts of the early twenty-first century suggest that human security may be an idea before its time, an idea perhaps that the world is not yet ready to embrace. A better world, that is, a humanly more secure world, will never be constructed until it is first imagined. The international civil servants at the United Nations Development Programme deserve great credit for compelling the governments of the world to at least imagine that the human condition can be improved through international cooperation.

D. J. Puchala

Humanist Manifestos: The Humanist Manifestos are declarations of humanist principles drafted by members of the international humanist community to the world community.

The first stated the necessity of reorienting essential values on the basis of common sense, principles of democracy, and a socially focused economic policy. The second advocated human rights and democratic values, and condemned totalitarianism, racism, and religious and class antagonisms. A distinctive feature of the third is its optimism and the belief that humankind is able to cope successfully with the unprecedented challenges it faces at the beginning of the new century.

The humanistic outlook is based on: respect for the dignity of all members of the world community; individual responsibility not only for traditional communities, but also for the destiny of humankind. The ethics of planetary humanism demands universal application of the principle to act in the way that will decrease the sum of human sufferings and increase the sum of happiness.

Humanists believe that the most urgent task of the twenty-first century is the creation of global institutions able to solve global problems. Unfortunately, there is a wide gap between the capabilities of international institutions like the United Nations or the World Health Organization and the needs of the world community. Turning the assembly of sovereign states into an assembly of sovereign peoples should increase the efficiency of the United Nations. Separate states will have to delegate part of their national sovereignty to a system of transnational governance. This would require creating: (1) A world parliament, elected by the people of the Earth, which will represent the interests of people, not governments. (2) An effective security system capable of eliminating international conflicts. (3) An effective world court and international legal system powerful enough to implement its resolutions. (4) A planetary agency of environmental control. (5) International taxation to help developing nations satisfy those social needs not satisfied through market mechanisms. (6) Global institutions regulating activities of corporations. (7) free exchange of ideas, respect for different opinions, and the right to disagree.

V. L. Ginzburg, V. A. Kuvakin

Humankind as a Family: The idea of the humankind as a family was first put forward by the prominent Indian thinker Vasudev Kutumbakam and became widely popular in India as an alternative to Americanization, Europeanization, and globalization. Its essence is struggle against cultural expansion and a transition to a dialogue among civilizations. Western civilization, burdened with its own contradictions, is not capable of coping with the complex problems of our time. The drive toward forceful expansion of material and cultural hegemony resulted in alienation among nations and the widening of the breach among civilizations. The era of territorial appropriation and colonialism, with its exploitation of material and human resources and alteration of traditional way of life in colonies, lasted until World War II. Post-war times saw further aggravation of inequality and instability in sociopolitical and technological-economical spheres throughout the world. The conflict stemming from the struggle for politicoeconomic advantages, which always are an instrument for restructuring cultures and a weapon against unsubmissive states,

264 *Global Studies Encyclopedic Dictionary*

is intensifying. Clashes in the cultural sphere are becoming more pronounced in the international arena. Formation of a single religious center and a single power and culture is a major reason for terrorist acts. Attempts to root out the differences in the biological, cultural, religious, and economic spheres are always accompanied by economic, psychic, or social coercion and can throw a civilization back to the Stone Age.

A new dialogue is called for that substantiates the necessity of the preservation of cultural and civilizational diversity, which is the basis of the stability of social institutions. Indian leader Mahatma Gandhi noted that if the world takes on the motto "an eye for an eye," it would become blind. In this blindness new civilizational structures must support harmonization of moral values. In the absence of such harmony violence and terrorism stand in the way of building a desirable future.

Greediness, the fight for power, craving for appropriation, and the violation of moral and ethical order demonstrated by nations contradict civilizational values. Wars between civilizations are unfolding. Attempts to define a civilization as Western, Hindu, Slav, or Latin American testify to the intention to express the deepest human essence of the ancient cultures in market terms. Human rights provisions, financial and trade instruments, and patent systems, used for the purpose of propagating consumerism, clash with deep-rooted limitations intrinsic to the majority of Eastern cultures. Propagation of cultural superiority, which is dictated by market needs and the interests of a specific single power, is channeled at so-called weak civilizations. In the situation of a monopolar world, civilization is becoming a field for market activity. Consumerism has had a reducing effect on culture and has weakened the influence of civilization. Human relationships are coming down to material relations, depriving people of the chance for self-actualization and ascending development. Market opposes culture, victimizing it. We have to search for new constructive and creative principles, to open our mind to new levels of reality. Human collective consciousness lies in its inherited experience, manifesting itself, among other things, in myths, legends and dreams of a human being.

Many European thinkers (e.g., Edmund Husserl and later, Pierre Teilhard de Chardin) proclaimed that Europeanization of humankind was inevitable. According to some retrograde concepts, the globalized world has to be Europeanized or Americanized. Some insist that the East must be Europeanized, because Europeans create history. But these claims to a special European mission and to universality of European values were undermined after World War II and the subsequent collapse of the colonial system.

Attempts are made to cultivate modern Western culture and its way of life in African, Asian, and Latin American villages. All of this undermines local traditions and cultural norms. The world, ruled by science and technology, is a world of blind (bare) rationality, of its supremacy over myths, legends and other spiritual and cosmic ties. A dialogue among cultures and religions has to be free from banalization of the sacred. The language of science, metaphysics, and historical references cannot provide the basis for such a dialogue, due to their inherent limitations. Such a dialogue has to employ the traditions of all kinds of thinking. We must make a reality of Vasudev Kutumbakam's dreams about the world as a family.

J. C. Kapur

Humus (Latin *humus*, "soil," "land") is a combination of all organic compounds, contained in soil, but not contained in living species that have an anatomic structure. Individual (including specific) organic compounds compose humus, as do products of their interaction, and organic units in the form of mineral formations. In the composition of humus, there are specific gumine substances (characteristic products of soil formation-synthesized in soils), non-specific organic compounds (synthesized in living species, enter the soil with remainders of floral and faunal origin), and intermediate products of decay and humus formation.

Difference in amounts of humus in different types of soils depends upon the geographic conditions (starting at .1–10 percent, sometimes more). The amount of humus is defined in tons per hectare in the upper layer of the soil (0–20 cm.). The reserve is considered to be large if the quantity of humus exceeds 200 tons per hectare. Now it is usual to assume that during the twentieth century arable lands lost about 20–30 percent of humus reserves.

T. V. Prokofieva

Hunger is the sensation of desiring food. When used by politicians, relief workers, or social scientists, the term "hunger" refers to people suffering from a chronic food insufficiency. There can be absolute hunger—total or nearly total absence of food (famine)—or relative hunger—constant or long-lasting lack of necessary nutrients in human food (undernourishment, malnutrition). Hunger as a social phenomenon can exist latently even when food is consumed every day. Large-scale famines (absolute hunger) are rare events occurring, as a rule, as a result of some emergency situation (war, drought, harvest failure). Systematic malnutrition (relative hunger) remains a phenomenon of global scope affecting large groups of people and resulting in weakening their health and decreasing their life expectancy. For example, in India the last large-scale famine took place in 1943, in Bengal (three million victims); meanwhile, a relatively large proportion of characterizes this country undernourished in the population (23 percent in 1997–1999).

Hunger as a Global Phenomenon: According to UN Food and Agriculture Association (FAO) estimates, about 826 million people (792 million in the developing countries, 37 million in developed countries) suffer from chronic malnutrition. Thirty million people die from hunger each year. The highest percentage of undernourished can be found in the eighteen African countries (including Somalia where about 75 percent of the population is undernourished), Afghanistan, Bangladesh, Haiti, North Korea, and Mongolia.

Hunger and malnutrition have devastating impact on all aspects of human social and political life. Protein-, mineral-, and vitamin deficiency-caused diseases (edema, rickets, beriberi, pellagra, scurvy) and general immunity decreases cause life expectancy reduction. However, the most damag-

266   *Global Studies Encyclopedic Dictionary*

ing consequence of hunger is personal degradation, making entire nations unable to implement socioeconomic and political development. The combination of hunger-provoked apathy and nervous breakdowns becomes a cause of political instability and the impossibility of democratic governance.

The Causes of Hunger and the Methods of Hunger Eradication: The need for eradication of hunger is reflected in the "world food security" concept. Approximately up to 1980s, a neo-Malthusian idea prevailed worldwide that an insufficient quantity of food was the main cause of hunger. Initially "food security" meant availability of a sufficient amount of food (in grain equivalent) at the world market. Increasing food production in the developing countries was considered the main hunger eradication strategy. After the "green revolution" that implied increasing land productivity through cultivating new food crops and introducing new agricultural techniques, many developing countries have reached impressive results. In India wheat production has grown from twelve million tons in 1966, to forty-seven million tons in 1986; in Nigeria and Ghana the number of undernourished due to cultivating new varieties of cassava has dropped from 40 to 10 percent of the population.

Nevertheless, the Green Revolution also had some negative effects: overdependence on fertilizers, melioration, and other ecologically harmful techniques; the use of genetically transformed crops; promoting one-crop agriculture. Besides, the growing quantity of food did not automatically bring hunger eradication because many people and whole states had no financial capacities to purchase food. Sometimes devastating famines coincided with years of peak food production (Bangladesh famine of 1974). Even in case of food scarcity some social groups suffer from hunger and some not (for example, the 1973 famine in Ethiopia affected only the province of Wollo, and the most interesting is that food prices in that province were not higher than in the rest of the country).

Currently, there is enough food on our planet to provide each of its inhabitants with at least 2700 kcal per day. In the developed countries, there is often food overproduction, but under-consumption may be prevail among the poor who have inadequate money to purchase food. Accordingly, it is of special importance to fight the social causes of hunger, that is, deficiencies in the food distribution system. Nowadays, food security is understood as the right of all people to access to food sufficient to support their physical, mental, emotional, and spiritual health. The right to food is recognized as one of the fundamental human rights; but a minimum level of income is its necessary precondition. Thus, the problem of hunger is closely related to the problem of eradication of poverty and underdevelopment worldwide.

Modern hunger eradication strategies are based on the ideas of the 1998 Nobel Prize winning economist Amartya Sen and consist of providing each human being with sufficient livelihood to prevent large-scale famines and eradicate malnutrition. Organization of public works allowing the majority of the population to have decent wages is much more effective for combating hunger than free food distribution because, on the one hand, it prevents mass starvation and, on the other hand, it strengthens human dignity and self-esteem. The other strategy successfully used in the developing world is providing the poor with mini-loans (about fifty dollars) to let them start their own small enterprises and get away from poverty. Direct food distribution is only used in a state of emergency, when a sudden famine outbreak must be suppressed.

Achieving political stability and promoting democracy plays an important role in combating hunger and malnutrition. The spread of hunger in the Sub-Saharan African states is directly linked to the endless civil wars undermining the economy and especially agriculture. The absence of democratic freedoms (such as freedom of press) allows neither to locate and prevent a famine outbreak in time nor to organize a society-wide campaign against it.

The world struggle against hunger is conducted by intergovernmental organizations (first, the FAO and the World Bank), separate states and non-governmental organizations (NGOs). In 1996, the World Food Summit adopted the Roman Declaration on World Food Security and the Action Plan with a goal of halving the number of undernourished in developing countries by 2015. However, according to FAO estimates, by 2015, the number of undernourished in the developing countries can decline to 580 million at the most and by 2030 to 400 million people. The intergovernmental organizations' anti-hunger activity is hampered by their dependence on national governments that often lack the political will needed to overcome hunger. In many countries suffering from hunger there are no national governments or they are not operating, which is why international or local NGOs often must fight against hunger.

Current common wisdom is that to overcome hunger, the issue should be understood not as a separate issue of food scarcity or abundance, but as a part of a broader global problem of poverty and underdevelopment eradication, of increasing the level of education and political stabilization.

A. V. Mitrofanova

Hydrosphere, Global: (Greek *hydro*, "water" and *sphaira*, "sphere") The global hydrosphere is a continuous envelope of the Earth, comprising all water in the solid, gaseous, and chemically or biologically bound state. The unity of the global hydrosphere is provided not only by its continuity, but also by the permanent water exchange between all its parts and by constant transformations from one state to another.

The water envelope of the Earth was formed in the course of the global geophysical processes resulting also in the formation of three other envelopes coupled with the hydrosphere: the mantle, the lithosphere, and the atmosphere.

The water of the hydrosphere affected and is still affecting greatly all the processes going on both on the Earth's surface and in its bowels, up to the most drastic geological transformations. There is actually no substance on Earth that does not contain water.

The global hydrosphere played in the past and is playing now a basic role in the geological history of the Earth, in forming the physical and chemical environment, climate and weather, and in the beginning and development of life on Earth.

The water of the hydrosphere affects rocks both mechanically and chemically. Freezing and expanding in the cracks, water acts as a destructive agent. Rivers develop broad valleys, carrying the fragmental material to the lower region. Depositing at the sea or river bottom, the solid material forms the sedimentary and detrital rocks. Rivers also carry great quantities of natu-

268 *Global Studies Encyclopedic Dictionary*

ral material in the dissolved state. Various salts precipitating from the water of the hydrosphere form rocks and minerals of chemical origin (such as gypsum and dolomite). Some aqueous organisms are able to absorb substances (calcium carbonate, silica) from ambient water and deposit them in their bodies. Accumulating at the bottom of water reservoirs, their skeletons (shells) form the thick layers of limestone and various siliceous rocks. The greater part of sedimentary rocks and such valuable minerals as oil, coal, bauxites, and manganese and iron sedimentary ores formed this way in the past geological epochs as a result of the influence of the hydrosphere and the processes inside it.

It was in the hydrosphere that the life on Earth first appeared. The evolution of living organisms proceeded in the marine environment all over the Precambrian, and it was only in early Paleozoic that the dry land began to be populated by various organisms.

The waters of the hydrosphere are comprised of complex chemical solutions of various substances. They are quite varied, differing not only by the chemical elements and their total concentration, but also by the quantitative relation between the components and the form of compounds. Water contains gases, mainly in the form of molecules and partially hydrated compounds, and salts, mainly in the ionic form. As far as high concentrations are concerned, it can contain molecular complexes and organic substances, both in the molecular and high-molecular form, and in the colloidal state.

The chemical composition of the hydrosphere determines the various processes going on in the aqueous environment.

The waters of the hydrosphere play a quite important role in human life. They are used in the water supply, navigation, fishing, and hydroenergetics. Mineral lakes are utilized to produce valuable chemical raw materials. Stocks of underground water are used for irrigation and water supply. The water from many mineral springs has a healing power.

The global hydrosphere actually runs through all the geospheres of our planet. All the Earth's crust down to its lowest boundary 35–45 km deep contains underground water. Moreover, even at the depth of 100 km, especially in the regions of deep tectonic breaks, traces of underground water can be found. At the great depths, the water from the mantle is supposed to come to the asthenosphere with the volatile hydrides of alkaline metals and fusible silicates of alkaline salts, which then dehydrogenate into the water vapor and drain through the breaks.

The upper boundary of the hydrosphere actually coincides with that of the atmosphere. Though the general body of water vapor is concentrated in the troposphere, there is a permanent mass exchange through the tropopause. In spite of the small quantity of water vapor in the stratosphere, condensation is still possible, resulting in the appearance of nacreous clouds. Above 80 km, the molecules of water are supposed to be completely dissociated. At the heights of more than 300 km, the exchange with cosmic space takes place.

Though the Earth's hydrosphere is an integral aqueous envelope, it can be subdivided into several parts, three of them being the basic ones. The first part comprises the Earth atmosphere from the Earth's surface to the height of 300 km. It consists mainly of vapor, along with the drops of liquid water and ice crystals, and, at the great heights, of separate molecules and hydrogen atoms. This part of the hydrosphere contains the greatest volume of water. The second part of the hydrosphere, and the second in its volume, comprises the surface waters of the Earth. They are distributed from the maximum depths of the ocean (11 km) up to the maximum heights of the mountainous snows (9 km). This water is mainly in the liquid state along with the solid and biologically bound one (oceans and seas, rivers and lakes, reservoirs, and glaciers, flora and fauna). The third part of the hydrosphere is the largest in its volume. It comprises the underground waters, which can be in a vaporous, liquid, solid, and chemically bound state. This is the soil moisture, gravitational waters of the upper layers of the Earth crust, the deep pressure waters, underground waters, water in the bound state in various rocks and sediments, water forming the part of minerals, and juvenile waters.

R. K. Klige

Hypothesis (Greek *hipothesis*, "basis" "suggestion") is an epistemological method consisting of setting up credible presumptions concerning the presence of laws that would explain the causes of phenomena and facts earlier unknown to science. A need for a scientific hypothesis emerges in a problem situation when new facts cannot be explained with the help of the old theoretical knowledge. There are several criteria of a hypothesis's veracity: the principle of verifiability, prognostic abilities, generality, and simplicity. A hypothesis should be verifiable, that is, the results it predicts must be empirically testable; in case they are not, the hypothesis is not scientifically valid.

A hypothesis should be free of formal logical contradictions and be internally harmonious. It should explain the maximal number of facts and conclusions drawn from it. A hypothesis's prognostic ability manifests itself in its ability to predict events not yet found empirically.

The simplicity criterion means that a hypothesis should explain a maximum of phenomena based on a minimum of presumptions.

In global studies, hypothesis mostly takes a form of prognostication; building ideal models of possible Earth and space development; or extrapolation of some hypothetically possible events that took place over the Earth's history (for example, the discovery of craters in various spots of the planet, that appeared perhaps as a result of meteorites, became a ground for a suggestion that there were many ecological catastrophes that radically changed the planet's surface and its biota).

N. F. Bouchilo

03 I 80

Imperatives to the Cooperation of North and South (1981) (*L'imperatif de cooperation nord-sud*) was the eleventh report to the Club of Rome, made by the director of the Department of Economic Forecasting of the Ministry of Finance of France Joan Saint-Geours, who continued the research of the basic issue of the eighth report, *Tiers-monde, trois quart du monde* (The Third World: Three Quarters of the World) and focused attention on the analysis of a problem of the interrelations between the North and the South, overcoming of backwardness of social and economic development and the gap between developed and developing countries.

Reference: Saint-Geours, J. *L'imperatif de cooperation nord-sud* (Imperatives to the Cooperation of North and South). Paris: Dunond, 1981.

A. N. Chumakov

Industrial Revolution: The Industrial Revolution was a radical change in the economic sphere that occurred during the nineteenth and twentieth century, when machines and mechanisms replaced manual labor. The most important invention of the industrial revolution was that of the steam engine based on the inventions of Thomas Savery, who patented the first crude steam engine (1698), based on Denis Papin's Digester or pressure cooker of 1679; Thomas Newcomen, who invented an atmospheric steam engine (1712), an improvement over Thomas Slavery's previous design; and James Watt, who patented an improvement of Newcomen's engine (1769). As the result of this invention, fundamentally new opportunities were opened for material production. The machine became the basis for technological process, while human beings became its appendix, a link or an element in a technological chain. Key events that influenced the globalization processes occurred in the nineteenth century, when the basic types of transportation and communication had been invented.

The second half of the nineteenth century was marked by the emergence of many engineering innovations, inventions, and discoveries that radically changed economic life, labor and living conditions. In 1860, industrial oil operations began, opening the "oil era" in human history. Oil has become the most important raw material, impacting the industrial development of many key nations and of the world as a whole while determining world historical development in the twentieth century.

A. N. Chumakov

Inequality, Digital: Digital inequality is inequality in the sphere of access to modern informational technologies. Digital inequality places those having no access to the world of modern communications at a disadvantage, depriving millions of education, medical assistance, and informational services. Causes of digital inequality include poverty, age, education, gender, territory (related to low mobility), and culture.

E. V. Mazour

Inequality, Global: Global inequality consists of inequalities among individuals within nations and between nations within the world system. Measurement of inequalities between social systems is made with the Gini Coefficient, while that of international inequality employs the Absolute Gap of Gross Domestic Product (GDP). Previous calculation of inequalities has focused only on the average incomes (GDP per capita) between countries with inattentiveness to the inequality of distribution within each country. At present, the evidence is unclear as to whether the two aspects of global inequality move in the same direction, such that when the gap between nations widens the gap between individuals also widens, and conversely, when it is lessened the other follows suit. There seems to be separate mechanisms underlying these two dimensions of global inequality, so that their interrelationship may not be about dependence, synchronicity or direct correlation.

Global inequality involves imbalance, disparity, and mockery of what is supposed to be an egalitarian global order. It raises the issue of global justice because it imposes a monopoly of benefits and burdens. It is about the irreconcilable imparity of human misery, suffering, impoverishment and deprivation experienced by many and amassing of wealth, power and income by the rich few. Hence, the oft-cited adage: the rich get richer and the poor bear more children. This brings to light the unnecessary chasm between North-South, the tri-partite global stratification of the world system into core, semi periphery, and periphery, or most simply, the divide between the rich and poor countries.

Multifarious factors explain such global inequality: transnational economic integration and national disintegration as a result of globalization, open capital markets, trade and investment liberalization, deregulation, overpopulation, colonialism and its attendant legacy of inequalities, neocolonialism and its concomitant exploitation, government decisions, a shift to new technological breakthroughs, lack of equitable distribution, lack of education and natural disasters. The unwanted effects of malnutrition, hunger, famine, income gaps, poverty, AIDS, illiteracy can be solved by education, democracy, Green Revolution, government action, gender sensitivity, technological transfers, more development assistance, just free trade and transnational economic policies, debt relief or reduction, and a deep sense of shared humanity.

A. S. Layug

Information: (Latin *informationem* "outline," "concept," "idea") In its most technical sense, information is a set of symbols that can be interpreted as a message. It commonly refers to data transferred from person to person via communication: oral, written, by technical means, or conventional signs. Since the mid-twentieth century, the avalanche increase of volume of information in social life, science, technology, and industry that was called "the information explosion," and the increase of requirements for its reliability, especially in the military sphere, gave a push to the development of various ways of measuring information and the parameters describing its quality; e.g., redundancy, reliability, value, and utility. Processes under study included ways of information exchange not only between people, but between living organisms, and technical devices (first between and within self-steered dynamic systems).

Information is especially important in the analysis of processes happening in modern society. Formation of a post-industrial society significantly

272 *Global Studies Encyclopedic Dictionary*

transformed the production, transformation, storage, and transfer of information. Especially revolutionary has been the development of communication via computer networks. Informatization overcame the limits of industry and became the integral attribute of various sides of modern human life: scientific inquiry, aesthetics, ethics, religion, and daily life. Manuel Catells has called the modern stage of social development "the information age."

V. M. Adrov

Information Field of the Universe: Modern science faces the problem of building a single theory of the development of the Universe that incorporates all major forces, including the social one. Thus scientific knowledge requires a new approach. The information field of the Universe (IFU) is a universal relation of reality, namely, the correlation of the entire foundation of the Universe with the results of its own evolution. The IFU provides a genetic link between the genus of thinking individuals and the foundation of the All. In conformity with the laws of the IFU the aggregate elements—material objects and forms of their interaction—turn into a relatively closed world. On this basis, the IFU should be considered as a self-developing integrity generating a system of concrete forms of movement, the supreme form being the social one.

The IFU is an entity whose self-development is ensured by the means generated by itself. The human being as a result of evolution of the Universe is the subject of its information field and a means of its reproduction. The human being thinks and acts using the information-energy possibilities of this single field of the Universe, human evolution being determined by its laws. The IFU is a necessary condition of development of thinking individuals; outside it they are just objects (things) on a level with a variety of other things while together with it they are elements forming cosmological integrity. Moreover, the IFU is a prerequisite for the existence of thinking individuals and is reproduced as a result of their own activity. As the human being is involved in the IFU through the process of formation, existing as thinking individuals means becoming an element of this field. The IFU is reproduced in the process of activity of the human being that makes his essence objective in the form of the necessary elements. Thus, the activity of thinking individuals acquires a cosmic significance.

Generating an information field in the Universe being observed they create a body common for all of them and different from the multitude of their discrete organisms, that is, the IFU as an endlessly reproducing process of interaction is seen as their universal continual body. This solves the contradiction of the human being—the subject of the IFU with his organism that is finite in space and time. Therefore, the objective of social development must be both the process of reproduction of the natural environment of existence of the human being and expanded reproduction of universal possibilities of the IFU. There will logically appear a type of society whose self-development will be based on the very process of IFU reproduction. Such a civilization can be logically called an information-cosmic one. By expressing the properties and laws of the IFU in the theoretical system the humanity will understand the sense of its own existence and prospects of its evolution, creating new means to solve practical problems of the modern time.

I. A. Safronov

Information Revolution: The newest telecommunications and information technologies have created supranational bridges, across which information easily overcomes physical barriers and state borders creating global cyberspace. Modern systems of telecommunications, that allow us to get necessary information from any distance in real time and make decisions faster, essentially facilitate the internationalization of economic processes. Development of the Internet and mobile systems of distant communications considerably deepens the international division of labor and intensifies international trade, allowing manufacturers of goods and services to study foreign markets of production faster and more intensely and to optimize their geographic structure, and allowing buyers to choose necessary goods and services with the help of electronic commerce not only in their own country, but abroad as well.

New information technologies allow investors to estimate more quickly the investment climate in any part of the global economic space, and in concrete projects. It essentially increases capital export, promoting the development of international industrial, trade, and financial ties both at the level of enterprises, banks, trade or insurance companies, and national economies as a whole. Since the mid-1980s, direct foreign investments have literally poured into the world arena, quickly surpassing investments in the national economy. This was the same period when the world computer net began to emerge. The share of foreign direct investments in the world gross domestic capital investments increased from 2 percent in 1982, to 16 percent in 1999, more than eight times. This means not only additional financing of the economies of countries (capital importers), but also the inflow of new technologies, advanced management, and marketing. This has become a powerful additional impulse accelerating the technical and economic development of countries-capital importers—and the world economy as a whole.

Modern information technologies contribute considerably to the further development of international industrial cooperation, that is, to the creation of technologically and economically complete industrial chains, the separate parts of which are located in different countries, but which function according to a uniform plan and in a uniform rhythm, similarly to the shops of one factory. Streams of details, units, and components move along these chains following a strict schedule and providing continuity of the whole technological process resulting in this or that end product.

In conditions of the new informational revolution, the global economic space is quickly covered with the dense net of direct and systematically organized cross-border industrial ties. This modifies the character of international economic relations. Spontaneous market relations become more closely bound with strict regulations and rigid technological discipline within the framework of cooperative alliances. National macroeconomic complexes become more deeply involved into the system of direct industrial ties, supported by domestic and foreign microeconomic actors. Not only do directly cooperating enterprises

274 *Global Studies Encyclopedic Dictionary*

become components of large international industrial systems, but also whole countries become more and more pulled together by stable and long-term ties.

The modern system of telecommunications will also speed up the transformation of world financial centers: London, New York, Frankfurt, Tokyo, and others, into a complete global market of financial resources functioning almost around the clock. Any fluctuations in any one such center become almost instantly known to the others and influence the whole global financial system. However, this new situation is fraught with considerable threats both for national economies and for the world economy as a whole. The international capital market that used to be an auxiliary mechanism called to facilitate commercial transactions and investments in the industrial sector of economy, turned by the end of the twentieth century into a self-sufficing global system independent of this sector and functioning according to its own rules.

In conditions of world informational integration, globalization of local processes occur, such as getting a higher education while living far from the best universities and educational centers. Developing countries get access to knowledge and an opportunity to raise the quality of their human capital. At the same time, the interactive exchange between national cultures is accelerated and extended. The world community becomes more and more integrated not only at the material and technical levels, but also at the spiritual level.

Y. V. Shishkov

Information Rights and Human Rights: The information age revolution has had significant globalizing impacts on space (the shrinking of physical distance), time (time compression), and the (declining) role of the state, politics, global economics, culture, and identity. The impact of the information age on global communication and community varies.

If we are to achieve a needed balance between individual rights and societal interests, then informational rights will require far greater specification, even as communal values of trust, social solidarity, and mutuality are increasingly vital to the promotion and protection of human dignity. For human rights are moral and legal, and it is their morality that underlies and justifies their legality. Human rights have as their objects the necessary goods of individuals to which they are entitled and which require duties on the part of others. Community protects and fulfills rights, especially for individuals who do not effectively have them, on the basis of mutuality and the social solidarity of individuals who recognize and fulfill common needs. Human rights require community for their implementation, while community requires human rights as the basis of its morally justified economic, political, and social operations and enactments. However imperfect and inevitably time bound ethical traditions may be, a proper understanding of the correlative nature of human rights and responsibilities is vital for ethical reflection on the information age (Walters).

Technology cannot provide an ethical "operating system" for the deeper problems that plague communicative reason among individuals. Nor does technology provide a response to the problem of balancing individual and communal goods, private and public interests. At the limits of technical design and moral norms, the challenges of the information age open on to the

Imperatives to Cooperation of North & South – *Issyk-Kul Forum* 275

need for a wider rationality or reasonableness and profound questions about the possibilities of human trust, mutuality, and social solidarity.

References: Walters, Gregory J. *Human Rights in an Information Age*. Toronto: University of Toronto Press, 2001.

G. J. Walters

Informatization is a socioeconomic and scientific-technological process of the creation of the optimum conditions for the satisfaction of informational needs and the implementation of the rights of citizens, government institutions, local authorities, and NGOs that require information resources.

The basic goals of informatization are: provision of conditions for development and protection of all forms of ownership of information resources; formation and protection of state information resources; creation and development of federal and regional information systems and networks; provision for their compatibility and interaction in the uniform information space of the Russian Federation; creation of conditions for a qualitative and effective information supply for citizens, government institutions, local authorities, organizations, and public associations on the basis of state information resources; maintenance of national informational security and provision for the implementation of the rights of citizens and organizations in conditions of informatization; assistance in the formation of the market of information resources, services, information systems, technologies, and the means of their maintenance; formation and implementation of a uniform scientific-technical and industrial policy in the sphere of information in view of the modern global level of the development of information technologies; support of projects and programs of informatization; creation and development of a system of attraction for investments and a mechanism for stimulation of development and implementation of informatization projects; development of legislation in the sphere of information processes, informatization, and information protection.

V. K. Loskutov

Integrity of Human Existence is one of the most important concepts, acquiring a special significance in the globalizing world. The importance of human existence is impossible to grasp by studying data about humans exclusively from the outside.

Language is the starting point for understanding human existence. The normal process of speaking includes three elements: (1) I, the speaker; (2) communication; and (3) the recipient of the message.

A concept of "possibility" constitutes the main axis on which modern philosophy revolves. However, it is necessary to admonish that: (1) Human life cannot be reduced to pure possibility or projection because projections are formal and a human is always greater than them (this is especially important for globalist-type concepts). (2) Possibility presumes fortuity, though it is not defined by it. (3) Every choice is made with consideration to normative limitations that are directly built into the essential order and, as a result, into the law of eternity, reflected in the moral law.

276 *Global Studies Encyclopedic Dictionary*

Human beings long for the integrity of existence and want to be protected from ontological loneliness. But "being in the world" more commonly takes place as an invisible union with loneliness, not wholeness.

We always hope. The desired can come true. For a desire to come true, hope includes an element of pleasure. But this pleasure is always mixed with anxiety, because what is most enticing is always beyond reach and volatile. However, this hope comprises confident waiting, which relies upon Someone. We do not trust objects, but we do trust people. It is impossible to build a life on the foundation of hopelessness. Only hope—adventurous hope—permeates through time and institutes life. Sadness that expresses everything that threatens us can never disappear for good, but we always try to head toward hope. Like courage, hope has to be of specific proportions. An excess of hope leads to complacency; an insufficiency of hope leads to despair. While complacency is an unnatural, premature integrity, despair is premature apprehension of a failure, of a verdict. Life per se always corresponds to life as hope. Complacency and despair freeze the current of existence on the way toward the ocean, making it impossible or turning it into an air-castle.

Without love, life is not worth living. Love imparts clear awareness to a person's destiny. Love leads misery to integrity. The yearning to give one's self, to expand and to delight in this expansion characterizes love. Only someone who is able to overflow can put himself or herself together. Love sheds light on the hidden best qualities of the loved one and arranges the integrity of values in a hierarchical unity. 276

We need to make our life complete, using our own calling for it. Calling is a means to approach integrity, a set of values directing our life. A human life has the structure of a Calling: a call, a response, and a mission. Calling cannot be inherited or contrived; it can only be discovered. Until we get this integral image, we are in debt to ourselves.

A. Basave

Intellectual Revolution, Global: The global intellectual revolution refers to the intellectualization and transformation of human thinking into collective professional activity and the formation of organizations, services, and networks which jointly process information. From the late twentieth to early twenty-first century, the global intellectual revolution has resulted from developments in science and technology, which have completely transformed the socio-political and socio-cultural appearance of the humankind. The phenomenon of global intellectual revolution is generated by the development of several historical processes. First, it was caused by the transition of people into the sphere of intellectual activity, which was related to the creation of powerful personal hardware and software and global networks of intellectual production, distribution and appropriation of knowledge and information, formation and "multiplication" of the number of generally accessible databases among all the field of human activity (that is, the "associated memory" and "collective intellect" of humankind). Second, a new information society is being formed, where the main and deciding component is personnel, which has been trained in software and is armed with most sophisticated information processing equipment. Third, giant "machines" aimed at collective processing of information are being constructed and employed. Fourth, the development of "mobile" intellectual property and the process of economy globalization demonstrate the powerful and controversial manifestation of the growing global intellectual transformation. The sphere of intellectual activity (info-sphere) is becoming the main source of wealth, overriding the "real econo-my" (technosphere).

Y. A. Vasilchuk

Interdisciplinarity, Types of: There is a rather encompassing trend toward cross-disciplinary systems in a progressively interlaced world. This development involves complex systemic trends, which are having progressively more comprehensive impact that manipulate and reshape, if not revolutionize, our environment and the social world. This is a sociotechnological world, viz. largely human-constructed technological systems and, thus, to a considerable and increasing degree, it is an "artificial" world.

Systems methods and methodologies prevail. This trend is to be found in all science-induced technological developments and in administrations. Besides systems theory in the narrow sense, these methods are characterized by operations technologies controlled by (methodical or even methodological) processes controlling and systems engineering, and by operations research. Methodological assessment as articulated in philosophy of science is needed.

In general, information, abstraction, formalization, and concentration on the articulated operational procedures are essential. It is by the way of computerization and information processing, and by using formal and functional operations technologies (for example, flow charts, and network approaches), that the formal essentials of increasingly comprehensive processes and organizations, and the interrelations of different fields and subfields, are integrated. Information technologies lead the way.

Characteristic of comprehensive systems engineering or systems technology are different technological developments, including economic and industrial changes that lead to system(at)ic interaction and generally to a kind of systems acceleration across different fields. (This is a trend which had been predicted by Friedrich von Gottl-Ottlilienfeld already in 1923, when he described mutually interactive spill-over effects, ramifications across traditional realms and a sort of what we nowadays would call positive feed-back processes.) All these ongoing processes necessarily require a far-reaching, if not encompassing interdisciplinary interaction and stimulation (interstimulation). Indeed, interdisciplinarity led by spillovers from science to science and, from there, to technological development and innovation implemented in society at large is prominent nowadays. Systems analyses and systems technologies require interdisciplinary approaches in practice. The pertinent challenges within this "world of systems," including technosystems, social systems, and ecosystems, require a thorough methodological study for the respective approaches and types of interdisciplinarity in research and development.

Short of providing such a methodological analysis here, it may suffice just to mention that one has to elaborate criteria for the methodological distinction of disciplines according to the objects and areas and scopes of the

278 *Global Studies Encyclopedic Dictionary*

research, development, and prospective implementation. Relevant (arsenals of) methods have to be articulated. Knowledge interests and the relationship between theory and practice have to be studied. Methodologically important is the difference between theories and their systematic and historical connections and contexts, substantivity versus operationality of theories (substantive versus operative or procedural theories). One has to specify from a philosophy of science perspective the extant patterns of explanation and systematization (descriptive versus explanatory, historical versus systematical) and questions of cognition and normativity (descriptive versus normative approaches and practical combinations).

These perspectives lead to different types of bi- and multilateral interdisciplinary relationships among the respective disciplines, as in Geomatics engineering (GIS), for example. Stages of the more or less strong formal and law-based interpenetration or merely aggregative coordination are reflected in the following types of interdisciplinarity: (1) interdisciplinary cooperation in well-defined projects; e.g., GIS in cartography or geology. (2) bidisciplinary or interdisciplinary research areas; e.g., satellite geodesy. (3) multidisciplinary aggregate fields of research; e.g., environmental research. (4) genuine interdisciplinary areas; e.g., physical chemistry or biochemistry. (5) multidisciplinary areas resulting from or relying on multidisciplinary theoretical integration. (6) abstract interdisciplinary theories; e.g., general systems theory. (7) mathematical theories of abstract and complex dynamical systems; e.g., deterministic probabilistic chaos theory. (8) supradisciplinary, abstract, structure-analytic, and operational disciplines; e.g., operations research. (9) methodological supradisciplines; e.g., philosophy of science and science of science. (10) philosophical and methodological epistemology as a metadisciplinary approach; e.g., methodological schema interpretationism.

At first, we saw just the simple cooperation of different experts for or within a developmental program, as land use, for instance, in coastal zone management planning where experts from different fields like geography, cartography, hydrography, geodesy, biology, ecology, limnology, or oceanography, and engineering in dike building and landscape planning, have to cooperate. Second, an interdisciplinary or bidisciplinary realm of research like satellite geodesy might evolve or, third, even a multidisciplinary aggregative research area as, such as environmental research (systematic ecology). The fourth level of cooperative integration would amount to a real interdisciplinary area (like molecular biology or population genetics) or, fifth, a multidisciplinary area in the more specific sense (multidisciplinary theoretical integration), for instance, the integration of natural and social science approaches in systems engineering of eco-techno-sociosystems. The sixth through eighth levels are formal theories of an abstract mathematical brand being used as instrumental vehicles of modeling real or constructed systems. Furthermore, the meta-theoretical levels nine and ten are addressed on a higher stage of methodological or epistemological (meta) analyses (for example, philosophical, social, and methodological assessments of the afore-mentioned approaches as in social impact analyses of geosystems engineering).

H. Lenk

Interfaith Studies is an important branch of global studies focusing on the interfaith movement, which analyzes the history of the interfaith movement since at least 1893 (the date of the first Parliament of World Religions), and the philosophical assumptions behind the major international interfaith organizations, including the World Congress of Faiths, the World Parliament of Religions, the World Conference on Religion and Peace, and various interfaith peace initiatives organized in the former Soviet Union and China. Additionally, this study involves a critical discussion and analysis of forces and movements opposed to the idea of interfaith harmony and coexistence, such as cults and religious bodies which advocate their own paths to spirituality as the only one possible for humanity, while decrying other such paths as heretical, deviant, and damnable. To some extent, "anti-interfaith" bodies have served as vehicles of terrorist violence, both in previous ages and in the contemporary world, advocated and justified by individual terrorist leaders on pseudo-theological grounds. Global Studies analyzes these aspects of interfaith opposition, as well as the extent to which opposition bodies have been supported and aided by various intelligence agencies worldwide for their own geo-strategic ends and the lack of sustainable moral controls on such agencies. Finally, Global Studies is concerned with the extent of positive communication and interaction among humanist and positivist philosophical traditions, interfaith communities, and cultures. Further attention is directed toward instances of such conversations being impeded or blocked and the reasons for those instances (such as unconscious, geo-political, or economic factors).

T. C. Daffern

International Federation of Philosophical Societies (Fédération Internationale des Sociétés de Philosophie, FISP) FISP is a global, nongovernmental organization founded in 1948. Its principal objectives are: to contribute directly to the development of professional relations among philosophers of all countries; to foster contacts among institutions, societies, and periodical publications dedicated to philosophy; to collect documentation useful for the development of philosophical studies; to sponsor the World Congress of Philosophy every five years (since 1900); to promote philosophical education; to make publications of global interest, and to contribute to the impact of philosophical knowledge on global problems. FISP also sponsors major international conferences and other philosophical meetings.

FISP publishes a biannual newsletter containing information on its activities and the activities of its members. FISP archives, containing documentation on the past activities of the Federation, are kept at the Institute of Philosophy, University of Düsseldorf (Germany). FISP carries out its activities through its General Assembly, Agenda Commission and Bureau. The General Assembly is composed of delegates from all societies-full members of FISP each having one delegate. The General Assembly is convened every five years during the World Congresses of Philosophy. The Agenda Commission is an administrative body, consisting, apart from the President and Ex-President, of 31 members elected by the General Assembly for ten years. The Bureau consists of the President and the Ex-President, three Vice-Presidents, the Secretary General, and the Treasurer. At present the Federation has four standing consulta-

280   *Global Studies Encyclopedic Dictionary*

tive committees: Committee on Philosophical Encounters and International Cooperation, Committee on Finance, Committee on General Policy, and Committee on Intercultural Research in Philosophy. In addition, there is a Committee for the World Decade of the Education of Human Rights, Committee on the Teaching of Philosophy and Committee on Bioethics. FISP is open for any kind of proposals concerning cooperation both from individual philosophers and philosophical institutions. See http://www.fisp.org.

I. Kuçuradi

International Framework for Collaboration in Legal Matters refers to a legal framework for the coordination of the activities of countries in law enforcement and the protection of human rights, civil liberties, property, public order, public security, the environment, and other human values based on international law and norms and national laws. As globalization progresses, international efforts to combat international terrorism and crime becomes more and more important. Over 500 million crimes are annually committed worldwide, a significant number of which stem from globalization. Often, dangerous crimes are committed on the territories of two or more countries. Common forms of collaboration in addressing international crime include aid in handling criminal, civil, and family cases; international treaties and agreements for combating crime, transnational crimes, and international terrorism; compliance with and enforcement of decisions of foreign law enforcement bodies in respect to criminal and civil cases; regulations on criminal, legal, and human rights matters in respect to public order; exchange of information that is of mutual interest for law enforcement bodies; exchange of experience in law enforcement; aid and assistance in training lawyers; technical aid; material support and consultations; and joint scientific research and development. Special international organizations have been established to combat crime. Among these organizations Interpol and Europol play an especially important role.

G. V. Dashkov

International Green Cross (IGC): The idea of the International Green Cross (IGC) was put forward by Mikhail Gorbachev at the Global forum of spiritual and parliamentary leaders on problems of the survival of humankind (January 1990, Moscow). In 1992, at the World Ecological Forum in Rio de Janeiro, heads of 158 states and parliamentary and spiritual leaders unanimously supported the creation of this global ecological organization, and it was established on 18 April 1993, when its first General Assembly was held in Kyoto, Japan. In his speech there, Gorbachev stated that the IGC aim—solving transnational ecological problems—should become a point of concern for all countries and achieved with the help of the powerful national organizations of the IGC. The Green Cross motto—compromise instead of confrontation—corresponds to the principles of civil society where ecological problems are solved from the position of being a good partner and a good neighbor. The major programs of universal importance were determined at the first IGC General Assembly: "Liquidation of the Consequences of the Cold War," "Ecological Education," "The Earth Charter," and "Pure Water for the Planet."

Imperatives to Cooperation of North & South – *Issyk-Kul Forum* 281

Twenty-six national organizations, together with their regional representations, comprise ICG. It is able to solve complex problems that involve political contradictions and difficult ecological situations. National organizations are an integral part of the IGC; they support IGC policy and its projects and programs in their countries; their practical activities combine the solution of ecological problems on their own territories with solution of the problems set by the IGC.

S. I. Baranovsky

International Law and Ethics Conference Series (ILECS), now in its sixth year, was envisaged as an opportunity for leading American and Western European scholars (initially philosophers, and now these are increasingly multidisciplinary gatherings) to meet with their Eastern European counterparts and discuss issues of global importance. The projects have been envisaged as pairs of annual conferences held in Belgrade in June with follow-ups in October in the United States. Many organizations have come together during last four years to make this possible: Serbian Philosophical Society, University of California at Santa Barbara Global Studies, The Center for Philosophical Education at Santa Barbara City College, University of San Francisco Philosophy Department and Law School, Portland State University Conflict Resolution, the Open Society Foundation, USAID, British Council, Goethe Institute, and Machette Foundation. It is safe to say, after five years, that the two-part model proved extremely successful.

Recent pairs of conferences include the following: (1) "War Crimes: Moral and Legal Issues" (Belgrade University, 21 June 1997; University of California at Santa Barbara, 14–16 November); (2) "War, Collective Responsibility, and Inter-Ethnic Reconciliation" (Belgrade University, 26–28 June 1998; Santa Barbara City College, 9–11 October); (3) "Secession, Transitional Justice, and Reconciliation" (Belgrade conference was cancelled, Follow-up took place at the University of San Francisco, 29–31October 1999); (4) "The Ethics of Humanitarian Intervention: Ways of Internationalizing of Internal Conflicts" (Belgrade University, 23–25 June 2000; Portland State University, 13–15 October); and (5) "Institutionalization of Human Rights and Globalization" (Belgrade University, 15–17 June 2001; Portland State University, 2–3 November).

The intended product of each year's meetings is a conference volume. For example, Aleksandar Jokic is the editor of *War Crimes and Collective Wrongdoing* (2001), *The Ethics of Humanitarian Interventions* (2002), and a special issue of *Peace Review* with fourteen articles on "Secession, Transitional Justice, and Reconciliation." Our collaborative work has resulted in publications such as "The Ethics of Economic Sanctions: The Case of Yugoslavia." The material from each year's conference was also published in Serbian, first in the journal *Theoria*, and then in the Institute's Yearbook. ILECS' 2002 theme was "Political and Religious Toleration in the Age of Globalization."

References: Jokic, A., and J. Babic. "The Ethics of Economic Sanctions," *Fletcher Forum of World Affairs* 24:1 (2000); reprinted in *Guild Practitioner* 57:1 (2000) and *Peace Review* 12:1 (2000).

J. Rajkovich

282 *Global Studies Encyclopedic Dictionary*

International Law, International Legal System, and Global Studies: An area of vital importance for global studies, international law, international legal systems, and global studies concerns a variety of overlapping fields of research. Internationally, global studies considers international legal mechanisms such as the International Declaration of Human Rights and its adopted covenants; the law of war and peace and the Geneva Protocols governing the use of armed force to settle international disputes; the legal protocols of the International Court of Justice in The Hague; the work, history, background, future, and politics of the new International Criminal Court established by the Rome Convention; and the work of international United Nations agencies and their legal fields of operation, such as environmental and human rights law and cultural issues. Comparatively, Global Studies analyzes national legal systems on a global basis, comparing and contrasting the ways different legal systems train their officers and judges and operate different rules of evidence, trials, and sentencing procedures.

This study further involves analysis of attitudes toward crime in different cultures, the work of legal professions and their adjunct agencies (police, intelligence, politicians) in pursing domestic and international criminals, issues of political corruption, and the extent to which different legal systems pursue or fail to pursue those who are suspected or known to be active in corrupt criminal practices.

Philosophy of law is comparatively studied on a global basis, considering which cultures and nations advocate or base their legal systems on which philosophical or spiritual principles and traditions with which results in practice. Additionally, this study compares the legal bases of political constitutions, the legality of different political parties in different legal systems, the laws governing various aspects (like financing) of political parties, and whether such mechanisms lead to corruption or transparency. The status of women in different legal systems, the rules governing women's rights, and the roles of women in the legal profession are compared as well.

Globally and comparatively, Global Studies considers questions of the (il)legality of nuclear weapons and weapons of mass destruction generally and the prospects for genuine multilateral and universal nuclear disarmament long-term. Another topic of analysis is academic work to reconcile legal systems derived from religious law and secular or philosophical traditions, including Chinese, Japanese, Christian, Jewish, Islamic, Hindu, Buddhist, Roman, Russian, Marxist, Napoleonic, Common law, and other legal traditions.

T. C. Daffern

International Organizations are unions of states (or their agencies); of NGOs, ethnic communities, and private individuals from various states. These unions aim at achieving common objectives in various spheres (political, economic, social, cultural, and scientific) and represent the primary form of international cooperation.

International organizations emerged in the second half of the nineteenth century, when economic and sociopolitical relations crossed the borders of nations giving rise to an objective demand for cooperation and coordination of interstate efforts to solve new transnational tasks. The first mass international nongovernmental organizations were the Red Cross (1863), founded by Swiss Henry Dunant, and the First International (1864), an international fellowship of workers founded in London by Karl Marx and Friedrich Engels. The first international intergovernmental organization was the Universal Postal Union, founded in 1874, to provide organization and functioning of the international postal service (since 1878, the World Postal Union).

World wars, especially World War II, gave a new impulse to the building of international organizations to prevent new wars and to create an effective system of international security. Thus, in 1919, the League of Nations was founded (officially disbanded in 1946); it was an international organization that proclaimed development of cooperation among nations and the promotion of peace and security as its central goals. In 1945, the United Nations Charter was adopted. The United Nations Organization was created to strengthen security and peace and to develop cooperation among nations. In the second half of twentieth century globalization and the growing interdependence of nations have led to the emergence of an increasing number of international organizations, extension of their functions and agenda and their growing influence and visibility in the international arena.

Presently, there are tens of thousands of international organizations in the world diverse in their activities, organizational forms, goals and objectives, number of participants, agenda and functions. Depending on organizational form, we can distinguish two types of international organizations: intergovernmental and nongovernmental.

Intergovernmental international organizations are founded and act in accordance with treaties and agreements among states at the level of governments or governmental agencies. They have adequate authorities and administrative structures and are considered subjects of international law, that is, they possess international rights and duties, sign international treaties with states and among themselves. Such organizations include the United Nations Organization (the United Nations, 1945), the International Monetary Fund (the IMF, 1945), the World Trade Organization (WTO, 1995), the North Atlantic Treaty Organization (NATO, 1949), the Oil Producing and Exporting Countries (OPEC, 1960), and others.

Nongovernmental international organizations unite public unions and scientific societies, private companies and enterprises, and private individuals from different countries. Unlike governmental organizations, the nongovernmental ones are not subjects of international law. Among the largest and the most powerful of them are the Inter-Parliamentary Union (1889), the International Olympic Committee (1950), the World Federation of Trade Unions (1945), the World Peace Council (1950), the International Union of Students (1946), the Club of Rome (1968), Greenpeace (1971), and others.

Depending on their participants and the scale of their activity, international organizations can be universal and regional.

Universal organizations conduct their activity on all territories, states, and continents and embrace the world community as a whole. As a rule, they are open to all states, and to organizations or private individuals regardless of the socioeconomic and political systems of these countries.

284   *Global Studies Encyclopedic Dictionary*

Regional international organizations unite states (or intrastate agencies), communities, or private individuals of a specific geographic region forming a territory-based structure.

The largest universal organizations, taking an appreciable and essential part in the process of globalization are:

(1) the United Nations Organization (UN);

(2) the UN specialized agencies: World Meteorological Organization (WMO), World Health Organization (WHO), World Intellectual Property Organization, World Postal Union, United Nations International Children's Emergency Fund (UNICEF), International Development Association (IDA), International Maritime Organization (IMO), International Civil Aviation Organization (ICAO), International Labor Organization (ILO), International Finance Corporation, International Bank for Reconstruction and Development (IBRD), International Monetary Fund (IMF), International Telecommunications Union (ITU), International Court of Justice (ICJ), International Fund for Agricultural Development (IFAD), United Nations Industrial Development Organization (UNIDO), Food and Agricultural Organization of the United Nations (FAO), and United Nations Development Program;

(3) UN Educational, Scientific, and Cultural Organization (UNESCO);

(4) autonomous organizations under the aegis of the UN: World Trade Organization (WTO), United Nations Conference on Environment and Development (UNCED), International Atomic Energy Agency (IAEA).

Other key international organizations include: the World Confederation of Labor (WCL), the World Federation of Scientific Workers (WFSW), the World Federation of Trade Unions (WFTU), the World Bank, the World Council of Churches (WCC), the World Wildlife Fund (WWF), Interpol, The London Club, the International Association of Political Science, the International Organization of Journalists (IOJ), the International Federation of Philosophical Societies (IFPS), the International Military Tribunal, The International Green Cross, the International Olympic Committee (IOC), the International Council of Scientific Unions (ICSU), the International Union for the Conservation of Nature and Natural Resources, the Organization for Economic Cooperation and Development (OECD), the Pugwash Movement, the Paris Club, and The Great Seven (since 1997, the Great Eight).

The most well-known international regional organizations are: the Asian Development Bank (ADB), the Association of South-East Asia Nations (ASEAN), the South Asian Association for Regional Cooperation (SAARC), the African Development Bank (AfDB), the Baltic Assembly, The GUUAM (Georgia, Uzbekistan, Ukraine, Azerbaijan, Moldova), The European Union (EU), the European Investment Bank (EIB), the Eurasian Economic Community, the Latin American Integration Association (LAIA), the Latin American Economic System (LAES), the League of Arab States, the International Association of Philosophy Professors (AIPhP), the MERCOSUR (the Southern Common Market), the Organization of American States (OAS), the Organization for African Unity (OAU), the Organization of the Islamic Conference (OIC), the Organization for Security and Cooperation in Europe (OSCE), the North Atlantic Treaty Organization (NATO), the Oil Producing and Exporting Countries (OPEC), the Organization of Central-American States, the Parliamentary Assembly of Council of Europe, the Colombo Plan, the Nordic Council, the Council of Europe (COE), the Commonwealth of Independent States (CIS), the Council of Persian Gulf Arab States Cooperation, the Council for Mutual Economic Aid (CMEA), the North American Free Trade Agreement (NAFTA), the Asian-Pacific Economic Cooperation Forum, and the Shanghai Cooperation Organization (SCO).

A. N. Chumakov

International Philosophers for the Prevention of Nuclear Omnicide (IPPNO): IPPNO is a worldwide organization of philosophers dedicated to the elimination of nuclear weapons, took shape in a series of discussions among participants in the Colloquium on World Peace held at the seventeenth World Congress of Philosophy in Montreal, Canada, August 1983. John Somerville, the driving force in the group, had proposed earlier that the organization be called "International Philosophers for the Prevention of Nuclear Omnicide."

Given that any major nuclear war will kill and critically injure such an enormous number of people—and so extremely degrade the conditions of life for the survivors—that it would have a negative value close to infinity, and given that there is a significant, non-null likelihood of a major nuclear war occurring, it follows that the disutility of any such war would be enormously high.

IPPNO's First International Conference was held at St. Louis, Missouri, United States, in May 1986; its second was held as part of the World Congress of Philosophy at Brighton, England, in August 1988. Then, in 1987, PPNO received Peace Messenger status at the United Nations.

Subsequently, the Soviet Secretariat was dissolved and Tom Daffern, an British philosopher of education, was elected IPPNO Coordinator at the IPPNO. Somerville selected Paul Allen to be Interim President. Ronald Santoni, previously Vice President, in turn succeeded Allen to become IPPNO President in the fall of 1991, and with Howard Friedman, who had been appointed Executive Secretary at the same time, the two reformed IPPNO's structure and began to radically expand its scope. Glen T. Martin was elected IPPNO President in the winter of 1995.

An article on IPPNO would be incomplete if it did not also refer to a kindred organization, Concerned Philosophers for Peace, earlier named "Pandora," but the history of their interrelationship goes beyond the scope of this paper. Suffice it to say that the two have played both overlapping and at the same time complementary and non-competing roles in their common dedication to bringing forth a just and peaceful world.

H. Friedman

International Relations is the study of relationships among countries, sovereign states, inter-governmental organizations (IGO), international non-governmental organizations (INGO), non-governmental organizations (NGO), and multinational corporations (MNC). During the last decade of the twentieth century, the area of international relations has undergone considerable changes. Up to the end of the 1980s, its main feature was bipolarity—a split of the world community into two countering blocks that led to a tense and wearisome struggle with each other, escaping, however, any direct military conflict. This circumstance had a decisive influence on international relations as

286   *Global Studies Encyclopedic Dictionary*

a whole and on international atmosphere. Starting at the end of the 1980s, a process of overcoming this bipolarity (the Cold War) began. Nowadays, this process is approaching completion.

The economic basis of modern international relations is a new international division of labor, which is being formed on the basis of past and present changes in the material base of the modern world. Because of globalization, industrially developed countries strengthen their position as economic centers, whereas the less developed countries, taken together, remain in the position of the dependent periphery.

Among industrially developed countries the United States holds the position of the most powerful and wealthy country in the world. However, judging by several parameters, the European Union has nowadays equaled the United States and even surpassed it. Taking into consideration a further expansion of the Union, its position will get even stronger. Japan still occupies the third place in this group of countries.

However, the future promises important changes. In twenty to twenty-five years, China could occupy a position of a world economic leader. India is becoming stronger as well. Several Asian states ("the tigers") are almost approaching a transition from developing to developed countries. These countries nowadays are not merely suppliers of raw materials and profitable sale markets for industrially developed states, but also manufacturers and exporters on a wide scale of industrial goods, including hi-tech ones. Similar changes are expected in other countries of Asia, and in Latin America. All this as a whole entails a possibility of further shifts not only in world economics, but also in world politics.

Politically, the situation already differs strikingly even from recent times. An increased number of states act as the traditional subjects of world politics. Since the second half of the twentieth century, it has grown from fifty to almost 200. One hundred eight-nine of them became members of the United Nations Organization. At the same time, there is an intensive process of regionalization-formation of pancontinental or narrower conglomerations of states that are already visible in Europe (the European Union), Asia (ASEAN and other narrower conglomerations), Africa (Organization of African Unity), North America (NAFTA), Latin America (MERCOSUR, Andes group). These regional organizations will in due course become the most active and influential players in the arena of international relations.

Alongside increased numbers of states and state conglomerations, recent decades witnessed increased influence of international economic organizations on the course of international relations. Among these organizations are the World Bank, the International Monetary Fund, the World Trade Organization, transnational industrial companies, and banks that monitor an overwhelming part of world industry and world financial assets. Some kind of privatization of world politics is going on.

With increased numbers and variety of world subjects, the sphere of international relations in general becomes increasingly complex and varied. We quite often hear that the modern world, compared to recent times, is unipolar, as the United States is the most powerful and wealthy state. The role and responsibility of the United States is indeed great. However, the "post-Cold

Imperatives to Cooperation of North & South – *Issyk-Kul Forum* 287

War" world is in fact multipolar. The number of countries that do not wish to be subject to anybody grows; they try to play their own role, increasing their chances due to an active policy.

The modern world presumes many problems that were inherited from the past. At the same time, new problems, perils, and challenges appear.

The modern system of international relations is characterized by increasing complexity, instability, and an unusual dynamism. It demands that all participants in world dialogue be attentive to developing processes and to their causes. Certainly, it demands high responsibility for the present and the future of the world.

V. V. Zagladin

International Relations, Democratization of: The only way out of modern crises created by international conflict is the democratization of international relations. At an early stage in industrial civilization, there were no generally accepted and observed rules in international relations. Capitalists reigned, laborers and other workers were removed from public life, and society lacked a channel for dialogue and compromise between different classes, so that social justice was not present. As a result, class conflicts increased, revolution became a high ideal for many people, and society became unstable, providing the ground for Marxism and Leninism to appear and develop.

In the nineteenth and early twentieth century, class conflicts prevailed. In many countries, social conflict found different paths of development. The October revolution occurred in Russia, and the "dictatorship of the proletariat" generated problems that are topical even now. In Western Europe and the United States, the ruling classes made gradual concessions to avoid social conflict. That helped prevent revolution. Due to various compromises, dialogue among different social groups became regular, social contradictions found prompt solutions, and the rules of the game were gradually improved. Thus, democracy was established and strengthened. As a result the society became stable and developed rapidly

A. Qinian

International Relations, Political Philosophy (Theory) of: The political philosophy (theory) of international relations is a branch of political philosophy (or normative political theory) focused on the philosophical aspects of the foreign policy of nation-states; world, and international relations; international conflicts, war and peace; ethical aspects of international interactions; epistemology, and the philosophical foundations of international processes.

Since the end of the 1970s, there has been a shift to normative questions—to ethics and morals on the international scene.

Following the intellectual impulse provided by the famous religious thinker and moral philosopher Reinhold Niebuhr, theorists started to apply normative attitudes of political philosophy and ethics to foreign policy. Step by step, this previously marginal trend in political theory turned into a separate branch of social science, or "international political theory."

An important topic in international relations is the philosophical rethinking of the problems of nuclear "deterrence" during the Cold War.

288   *Global Studies Encyclopedic Dictionary*

Unlike political theory that mostly deals with domestic political problems, the normative attitude toward international affairs faces some difficulties. First, many theorists of international relations continue to think that an effective normative theory of international relations is impossible in principle, because moral judgments cannot be applied to interstate relations. Second, there are some important epistemological problems. Because a normative theory of international relations has neither a long tradition of philosophical argumentation, nor a universally recognized language of categories and concepts, it often turns into a simple illustrating of the existent political theories by means of international examples. Thus, it is not self-sufficient. Third, the application of moral intuition to international relations is less well grounded than its application to domestic politics. This engenders tensions related to the foundations of international principles. Fourth, in the area of normative international studies, a theorist often lacks empirical data or correct information. As a result, a systematic normative theory of international relations has not been worked out up to the present, and normative considerations are still applied only to specific problems and questions.

Philosophical and political studies are concentrated on three central problems: (1) national interest and the role of power in international relations, (2) ethical aspects of world politics, and (3) the problem of globalization in the context of international interactions.

Political philosophers often appeal to the modern tradition of "natural law," sometimes called "the morality of states," as an alternative to the Hobbesian model of international relations (or moral skepticism).

To close the gap between domestic and foreign policy we need to rethink the principles of distributive justice in international politics. In recent discussions on international order after the end of the Cold War, many thinkers used the same type of justification that was used earlier by John Rawls in his famous *Theory of Justice*. In particular, they apply the comprehensive idea of "common contract" to the international sphere.

An important role in the development of international political theory is played by the so-called English School, whose positions are guided by political philosophy and social theory much more than their American colleagues. After the Cold War the revival of constructivist thinking took place because empirical international relations (IR) theorists met difficulties trying to explain the end of the Cold War and the systemic changes in world politics. According to some scholars, this was connected with the materialist and individualist orientation of International Relations theorists.

Nevertheless, international relations theorists belonging to alternative schools of political thought questioned many traditional postulates and methodological principles in the last decades of the twentieth century. Post-positivism (especially its postmodernist trend with its irrationalism and despise for theoretical knowledge) challenged the classical political theory questioning the possibility of knowing the world and the human ability to organize life on the foundation of reason and justice. Post-positivism interprets interrelations between cognition and society, science and power quite differently than the other trends in political philosophy. Neufeld distinguishes three main trends in post-positivism: neo-Marxism, based on the thoughts of Antonio Gramsci: postmodernism and feminism, each influenced International Relations theory.

Confrontation between post-positivism and positivism took the form of the struggle between rationalism and reflectivism. Instead of international relations theory, postmodernists wanted to describe world politics with the help of the genealogical approach proposed by Friedrich Nietzsche and further elaborated by Michel Foucault.

Reflection on "ungoverned society," "the decline of sovereign states," and the formation of the "global civil society" resulted in an overestimation of the importance of the category of "security" in the field of international studies. Under the influence of postmodernism a concept of "alternative," "non-offensive," "unprovocative" defense emerged, mainly worked out in the framework of European peace studies. In the process of its evolution, post-positivism became not just a trend challenging the public, but a respected direction of Western political and philosophical thought open to dialogue with its former opponents. It is possible to foresee that the process of integration of post-positivism into the mainstream of the traditional political philosophy will continue in the future.

T. A. Alekseeva

International Relations, Realism as an Approach to: Political realism is a dominant theory in the study of international relations, stemming from the thought of Thucydides, Machiavelli, Thomas Hobbes, and Jean-Jacques Rousseau among others, who argued that, in international affairs, leaders should protect the interests of their citizenry and provide security.

Realism is a state centric theory, holding that the state is the only actor in international relations. The state should maximize its power because the environment of international relations is anarchical and perilous. The first priority of state leaders is, therefore, the continued survival of the state. As there is no global government capable of providing world order in the manner that a government enforces law in domestic affairs, the state should bolster its own power to assure its continuation. Thus the concept of "self-help" is central to international political realism.

According to the realist position, peace is not a natural condition in international relations and is to be understood as the simple absence of war. Therefore, realists are much less optimistic than liberals in their expectations for world peace. This pessimistic perspective regarding world peace and co-operation among states has been confirmed by the failure of the "utopian" of establishing a world government. After World War I, liberal ideas were influential in establishing the League of Nations as an institution for achieving peace through cooperation. Lacking any real power, however, the League came to be little more than a debate club. Furthermore, the two big players in the international system, the United States and Soviet Union, did not join. The League of Nations failed to solve any major world problems and, critics maintain, participants acted only in self-interest, as realism would have anticipated.

Criticisms of realism include: (1) The contention that though realism may offer good explanations of world politics at any given time, it is unable to predict future changes. For example, realists failed to predict the end of the

290   *Global Studies Encyclopedic Dictionary*

Cold War. (2) By emphasizing national interest as the only role of the state in world affairs, realism promotes a narrow ethnocentric view on world. (3) Realism underestimates the potential power of NGOs (non-governmental organizations) and non-state organizations like the WTO, World Bank, and IMF. (4) Realism does not provide a clear definition of what constitutes national interest. What is national interest? Do states, in emphasizing national interest, really know what is in their national interest? Would a more cooperative stance better serve this interest?

D. Vrabic

International Security Governance: At the beginning of the 1990s, James N. Rosenau tried to theoretically grasp what he called the turbulence in world politics. These turbulences derive from the complex transformation process of basic interrelated parameters (structural, relational, orientational) in political systems. At the global level, the emerging new structure is a bifurcated system, consisting of the state-centric world and a multi-centered world. Power originates from diverse human and non-human capabilities instead of military ones. This inevitably leads to multiple rather than hierarchical relationships. At the national level, there is a tendency toward decentralized structures that have a pluralistic leadership. At the subnational level, we have relatively autonomous units loosely organized into flat hierarchies and networks. The emergence of the post-industrial era with its technological dynamics is at the heart of global turbulences.

The aforementioned developments have increased interactivity and interdependency. People change, positively (opportunities) and negatively (risks), the way in which international policy-making or international relations work. The outcome is what has been named "post-international politics."

All this has repercussions on our understanding of security. On the one hand, one should not forget that turbulence means uncertainty and does not correlate with violence. On the other hand, violent conflict or war might be one of the possible responses to turbulence.

The question is in how far the number of responses involving force will decrease or increase in importance as world politics becomes more turbulent. The answer partly depends on the changing nature of conflict. Although there is a clear trend toward a limitation in the use of coercion in the so-called OECD world, this does not mean that the world as a whole is heading for an era of perpetual peace. States will continue to maintain their ability to exert coercion and to wage war. However, these will be less viable and credible ways of exercising control over other actors, be they state, non-state, trans-state, or sub-state actors. The mode of coercion that has come to the fore is intrastate low-intensity conflict.

Terrorism fits into the picture of a low-intensity conflict. This kind of violence has often been mentioned in the context of failed states and societies. These are characterized by social fragmentation, violence, and deprivation. Such a context is likely to be the breeding ground or offers a favorable environment for terrorism. However, terrorism is not a new phenomenon as some people think since the events of 11 September 2001.

Imperatives to Cooperation of North & South – *Issyk-Kul Forum*   291

What is new, however, is the perception of the ongoing privatization of violence as a fundamental threat to international security and the intense international reaction to it. The UN Security Council has condemned acts of international terrorism as "one of the most serious threats to international peace and security in the twenty-first century," and recognized in this context "the inherent right of individual or collective self-defense in accordance with the Charter." NATO invoked the principle of Article 5 of the Washington Treaty for the first time in its history. The European Union adopted several Common Positions on combating terrorism. On 16 January 2002, the United Nations Security Council adopted as well for the first time in history a resolution with extraterritorial sanctions against al-Qaeda, a trans-national non-state actor.

What kind of international actor the European Union will really be is still to be determined. It is obvious that the European Union is much more than a nineteenth-century concert of powers balancing each other by power politics and shifting alliances, and less than a hierarchical state model. It can be described as an evolving multi-level decision-making system in which the member states predominate in CFSP matters but are increasingly tied by legal acts such as common strategies, joint actions, and common positions, and by the legitimacy criteria of efficiency that nourishes a trend toward federative structures. As to the external dimension, the European Union incorporates elements of civilian, military, and normative power. It remains to be seen, to what extent they contribute to the emergence of a European Union that resembles the model of a cooperative security provider.

H. G. Ehrhart

Internationalism: (Latin *inter*, "among" and Old French, *nation*, "nation") Internationalism is a movement that advocates greater economic and political cooperation among nations for the theoretical benefit of all. Proponents, e.g., the World Federalist Movement, claim that nations should cooperate because their long-term mutual interests are of greater value than their individual short term needs. As such, internationalism was central to the ideology of communist and socialist parties (proletarian internationalism). Communist internationalism was closely connected with class struggle: international solidarity was declared a characteristic of the oppressed classes; international unions of ruling classes were regarded extremely negatively. Internationalism here was juxtaposed to cosmopolitanism. These concepts can still be traced in the modern antiglobalist movement that considers globalization to be the result of a plot by rich countries against poor ones or of the upper class against the democratic masses.

The principles of internationalism also included statements (often demagogic) about the independence, freedom, and equality of all nations, about cooperation among nations, about antiracism, antichauvinism, and antinationalism, about the struggle against aggression and national oppression, and about cultural dialogue.

A. V. Katsura

Internationalism and Global Studies is a detailed study of the history of internationalism in various philosophical, religious, political, intellectual, and social contexts, from ancient times to the present day. One part of this method

292 *Global Studies Encyclopedic Dictionary*

is to analyze ancient philosophical systems (such as stoicism) that advocated universal harmony and internationalism as the path to world order, as well as movements within various religious traditions that have promoted a global perspective and whether they have encouraged peaceful coexistence with other cultural systems.

Global studies also considers the history and activities of international political organizations that have attempted to create a peaceful world order, such as the Congress of Europe (following the Napoleonic wars), the League of Nations, and the United Nations. Accordingly, Global studies seeks to understand the extent to which these organizations have succeeded or failed and the underlying reasons for their results. Additionally taken into account are kinds of deviant internationalism in which sectarian or biased approaches to internationalism seek to dominate and hegemonize the global agenda. These approaches often attempt to shape the world order according to a predetermined plan of cultural hegemony based on narrow cultural parameters that are usually enforced by violence rather than achieved through genuine inter-cultural and international negotiation and discourse. Such thinking underlies all forms of deviant international terrorist activity as well as most forms of inter-state violent conflict.

Global studies further analyzes manifestations of internationalism in intellectual life throughout different epochs of world history and various cultural situations, such as international scientific and humanistic discourse in Europe during Medieval Catholicism and the ease of movements across different universities. Another expression of internationalism is humanist enlightenment spreading from Italy in the early modern era, embracing both reformation humanism and counter-reformation scholarship. Internationalism also appears through the birth of the scientific enlightenment and the development of various national scientific academies and international scientific projects transcending national agendas and boundaries.

Also of interest to Global studies is the histories of social movements that have emerged as international phenomena, such as international working class solidarity movements and the trade union movement. It also considers, for example, the international peace movement and its various organizations, projects and campaigns, like the International Peace Bureau and the International Peace Research Association. Various international networks, structures and organizations of many social, commercial, and professional groups is of importance to this study, as are non-governmental organizations that specifically work to advance a sense of global identity and a community of common moral concerns. In addition to international enterprises promoting positive global identity, Global Studies also makes note of these organizations' contribution to the perceived devastation of localism and specific regional cultural creativity and self-sufficiency. Related to this, the more recent anti-globalization movement is an important aspect of this study, along with the way in which protest groups have attempted to challenge the hegemony of international corporate dominance.

Another aspect of this approach to internationalism is the study of global networks of international crime and criminal agencies, such as the Mafia, and the extent to which criminality has spread a "negative internationalism" prem-

Imperatives to Cooperation of North & South – *Issyk-Kul Forum*   293

ised on personal greed and the lust for power. These organizations undermine genuine internationalism with activities like international arms trade, narcotic trafficking, prostitution rackets, and child slavery. International criminal networks have also been interlinked with overtly terrorist movements. This leads Global Studies to include analyses of the history of attempts by legitimate international authorities to assert the rule of order and morality against international criminal rackets, through international police and anti-terrorist actions and through education, research, and media exposés.

T. C. Daffern

Internationalization: As a product of industrialism and capitalism, internationalization signifies progressive international interaction and the interdependence of economic and trading interests, connections, and relations at the end of the nineteenth and the beginning of the twentieth century. During the industrial revolution it led to the formation of a single world economy, to the massive integration of industrial and financial capital, and to intensification of the international division of labor.

Despite close relations between internationalization and globalization, they should not be confused. Internationalization can be defined as the intensification of international connections and exchanges; globalization means further development of this process where separate societies, countries, and regions look more and more like parts of a unity. In this context, internationalization acts as a launcher for globalization, gradually eliminating barriers in the way of the movement of goods, services, capital, and labor.

Globalization is a higher and qualitatively different stage of international interdependence; it begins with the appearance of transnational corporations, integrated markets, and the "globe without borders." While "empire" and "imperialism" did not associate with clearly defined historical realities, globalization could be treated as a claim for "world empire" or "global imperialism" that can be traced in such respectable terms as "global village," "mega-society," and "supersociety."

Among the real fruits of internationalization that became evident in the last decades of the twentieth century and that stipulated the transition to globalization are: (1) The universal character of economic and trading connections, accompanied by an adequate system of transport and communication. What is currently called the "information revolution" makes possible a planetary dialogue, free from time-space limits. (2) A general liberalization of markets, with restrictions in the sphere of trade and almost unobstructed in the sphere of capital movement. (3) An awareness of the ecological crisis as a universal problem that threatens the survival of the world and questions the predominant economic policy. It is clear that the named factors and tendencies "pull" behind themselves a lot of fundamental changes in the division of labor and the structure of production, technology, the social sphere, and the spiritual climate of society. Unlike internationalization, globalization has not yet become such a universal process and phenomenon. The majority of countries and regions have not yet achieved this stage, as they still need to pass the stage of local modernization. The proponents of globalization should take into consideration the increasing resistance of antiglobalist forces. But an

294 *Global Studies Encyclopedic Dictionary*

even more important obstacle to globalization is the heritage of the previous stage of internationalization reflected in the formula of "asymmetric interdependence" that still needs to be overcome.

Increasing (in geometrical progression) interdependence of countries and regions rests on the obviously unequal starting conditions and opportunities of the participants of globalization. Some countries have gone forward very far in their social and economic development; others have lagged behind noticeably and some even catastrophically. This situation enables the West as it enters the post-industrial stage to take leadership and govern the course of events. At the same time the overwhelming majority of countries and regions have found themselves in a position of outsiders and objects of globalization. This asymmetric global world, leaving unsolved all the "fatal" questions of former history discredits the very idea of globalization.

Arguably, symmetry is incompatible with globalization. As a reply to the Western concept of the global "electronic village," the developing countries discuss the issue of "informational imperialism," that is, the nonequivalence of information exchange between the North and the South. This is how they react to the increasing Western cultural influence that threatens their unique cultures and traditions. There are new financial and economic technologies, and political and military technologies, capable of undermining national sovereignty. Here it is necessary to take into account the sad experience of the internationalization epoch when there were attempts to solve global problems by means of compulsion (colonization and Westernization on the basis of the hegemonic principle of "divide and rule").

V. I. Tolstykh

Islamism is a religious and political current, the basic aim of which is the maximum implantation of religious standards in political sphere. Adherents of Islamism proceed from the dogma that it is possible to arrange internal and foreign policy of a country or activity of a political organization in accordance with their interpretation of Islamic regulations and principles. Throughout its history, Islamic political and legal thinking has accumulated a many contradictory ideas and conceptions. As a result of the non-uniformity of political culture in the Islamic community, the most comprehensible tendency in the evening-out the contradictions on basis of the radical political Islam has become more and more apparent since the late 1980s. The standards of permissible violence toward the representatives of other religions are arbitrarily extracted from the context of the Quran and prevail at the level of ordinary consciousness. They are used to justify extremist activities, such as international terrorism and to establish and to ensure functioning of Islamic parties and organizations, including the well-known al-Qaeda.

Along with extremist movements and organizations, the Islamic spectrum is comprised of political parties and organizations that adhere to moderate interpretation of Islam and reject radicalism. They are open to assimilation of other values consistent with their teaching. For instance, the documents and political practice of the Justice and Development Party that came to power in Turkey in 2002 combine respect for Islam as for a basis of national culture and psychology with realization of the fact that Turkey is a part of both western and the Islamic civilization, and the adherence to democratic principles of community development.

V. K. Egorov

Islam and Globalization: Conflict or Compromise? Globalization, despite its various sophisticated definitions and different dimensions such as economical, cultural, political, technological, and ideological ones, can be understood simply as a process of social change bringing distinct and different civilizations, cultures, communities or individuals into interaction with each other (Waters, 1995; Tomlinson, 1996). It is the intensification of worldwide social relations linking distant localities with instantaneous communications (Giddens, 1990). As for Islam, in spite of its ideological representations or political perceptions by some Western intellectuals, it is a religion based on a very simple monotheistic essence *Tawhid* "God's oneness," and on some universal moral principles seeking goodness for everybody regardless of their race, origin, background, belief, language, and color.

Globalization, beyond its literal meaning, is also defined as a creation of *a new global culture* with its attendant social structures (Beyer, 1994), as *an economic phenomenon* (Wallerstein, 1988), as *a world policy* creating value through collective authority (Meyer, 1980), as *an overall process* bringing about a single social world becoming increasingly interdependent as a "single Place" (Robertson), becoming a "global village" and "death of geography" (al-Roubaie). The theories of globalization are hotly debated and contested issues. These definitions will, therefore, continue to be reformulated and redefined depending on circumstances and perspectives. Unlike globalization, Islam, as a Semitic religion is beyond the definition discussions, but in spite of this fact it is subjected to some social or political speculations.

At the first glance, both globalization and Islam seem to have some polemical definitions from their opponents. They are both portrayed implicitly in negative terms and presented indirectly with some misleading understandings. As for Islam, instead of being taken as an Abrahamic religion based on a very simple monotheistic principle *Tawhid* and a great reaction against Arab paganism of past, it's been wrongly and unfairly associated, in globalization discussions, with the policies of some Middle Eastern societies and with their national traditions.

The understanding of globalization as an economic hegemony or cultural imperialism of the Western world was raised by the critics of globalization opponents. It also stemmed from an ideological stand based on some political assumptions due to the unequal economic, political, and cultural relations between rich and poor countries. Again, the understanding of Islam as a fundamentalist or dogmatic reaction against globalization and global values is another mistake of Western scholars. We can say that when people put globalization against Islam (or vice versa), what they have in mind is not the religion of Islam, but Muslims and their attitudes. Also, they can add that many Muslims are opponents of globalization. Even though this conviction is true to some extent, it does not change the reality of Islam free from some complicated practices of people.

296   *Global Studies Encyclopedic Dictionary*

Islam is not necessarily in conflict with the globalization as long as the theories or practices of globalization are ethical, humanitarian, egalitarian, just, and healthy. But, if they are in need to be discussed and questioned from a moral point of view, this is not only Islam's interest, but also that of all Semitic religions and commonsense. Both globalization and Islam, because of the misperception or misrepresentation, have some serious problems in being understood well. Islam greatly contributed to the development of democratic culture by ordering a consultancy system. With the principle of "no coercion in religion" (Koran 2:256) Islam taught people a lot of things about freedom of choice or expression even in religious beliefs. So Islam and humanistic values are not in conflict even with the effect of globalization.

The relation between globalization and Islam is ethical: The globalization process deeply affects all persons regardless of their identity, religion, faith, race, or country. It brings all of them closer to each other. With the instantaneous communication links, Muslims, like all others are being connected with others. Actually, they have been living for centuries with Jews and Christians everywhere. In its nature, Islam is already a religion beyond race, color, language, borders, and countries. It is a global faith and has global ethical aims. So as a global religion, Islam is not in a position to propose another alternative social process against globalization. It will only seek a moral globalization process.

Portraying Islam in a position that threatens any social phenomena like modernism or globalization or to present it as a barrier against contemporary social developments is equal to perceiving Islam as an ideology. Perhaps, economical, technological, or cultural results of the globalization process would be very risky and devastating to some. But ups and downs in social changes (in the globalization process) are still to be tested with the ethical principles (measures) of Religion (Islam). For example, Islam encourages business, property rights for everybody, working, producing, marketing, making money, and trade, but forbids limitless consumption, exploitation, and lying in selling and regaining borrowed money. These ethical rules are very crucial Islamic principles on economical life regardless of whether the globalization processes exist.

Globalization can be considered in both positive and negative ways. It can have advantages or disadvantages for all, but it cannot be presented as the hegemony of a specific culture over another one. It would be an utterly naïve project if one thinks that this process would help for some monopolizing cultural or economical life. Globalization in itself is against all kinds of hegemonies or monopolies. It breaks naturally all local prohibitions and threatens all local hegemonies or monopolies.

Cultures and civilizations are different parts of global structure of humankind. They are not counterparts of a chaotic organism, but constructive parts of a universal culture of all Humanity connected to each other by their own characteristics. So, trying to decompose them from each other and put them on opposite sides is not realistic. In addition, different cultures and civilizations are not the source of conflicts or wars, but they are uniting factors of human beings all together around humanistic, ethical, and artistic values. Political, ideological, and imperialistic confrontations of different nations are not valid reasons to erase their cultural or civilizational richness. As we know, the most devastating and tragic conflicts such World War I and II did not happen because of the diversities of civilization. Ethnicity and nationalism played an essential role in different ruthless conflicts that caused a lot of death and causalities like in the Balkans. Therefore, foreseeing a clash between cultures or civilizations in the future is only an imaginative description.

Jews, Christians, and Muslims have been living together all around the world for centuries. They will go on living together in different places with the other members of various beliefs. Maybe, they will be confronted again with many different reasons as they did in the past, but this would not be due to their religious identities or briefly their Abrahamic faith. In addition, there is no need to eradicate one of them for the sake of others.

The Earth is the best of all possible worlds for humans to live in. There is no other place yet to go for a better life. Therefore, humanity should share this tiny place (big house) equally and should take care of it all together without ignoring or harming other members of family. We (Jews, Christians, Muslims, Buddhists, or atheists) should aim at constituting a global ethic for a global serenity to find solutions for global problems. All religions should play a constructive role for humanity to set up this global ethic supporting humanistic, scientific, moral, cultural, and economical relations.

References: Meyer, John W. "The World Polity and the Authority of the Nation-State." In *Studies of the Modern World System*. Edited by A. Bergesen. New York: Academic Press, 1980. Wallerstein, I. "The Typology of Crises in the World-System," *Review (Fernand Braudel Center)* 11:4 (1988), 581–598. Robertson, R. "Internationalization and Globalization," *University Center for International Studies Newsletter*, University of Pittsburgh, (Spring 1989). Giddens, A. *The Consequences of Modernity*. Cambridge, UK: Polity, 1990. Al-Roubaie, A., and P. Beyer. *Religion and Globalization*. London: Sage, 1994. Waters, M. *Globalization*. New York: Routledge, 1995. Tomlinson, J. "Cultural Globalization," *The European Journal of Development* 8:2 (1996), 22–35. Selangor, D. E. *Globalization and the Muslim World*. Malaysia: Malita Jaya, 2002.

A. Topaloglu

Islam in the Modern World: One of the most distinct features of contemporary Islam is its readiness for dialogue. What we must speak about here is a dialogue among civilizations, between the East and the West, about tolerant communication of cultures in our modern times. Muslims who call for such dialogue are representatives of a powerful civilization and carriers of a rich culture. They believe that relations between people are to be based on a reasonable approach and upon dialogue, and not on force and duress. Dialogue between civilizations implies equality of nations and states.

One of the most serious problems in religious societies—and unfortunately Islam did not avoid this pitfall either—is a misconception that if you have religion you do not need reason. In the Middle Ages religion was opposed to reason and freedom; as a result all three suffered. In reality religion is a cradle and a major stronghold for the consolidation of reason and freedom. Although a human being is gifted with the divine spirit and experiences the beneficial effects of the dimensions lying beyond nature, time, and space, the Book of Nature is, nevertheless, comprehended by means of reason through

298   *Global Studies Encyclopedic Dictionary*

acquiring knowledge and studying science and philosophy. If we want a better future, we must be guided by the great and blessed gift of God: reason.

The world of religion and the world of arts are the ones addressing human beings. God appeals to the true individual beyond the historical essence of a human being, and that is why all God-given religions are intrinsically the same. The Eastern tradition explains "the eastern side" of a person's essence, and the West reveals the characteristics of the "western side." But in reality a human being is a place where the East of the spirit and the West of the reason meet. Denying any of these two sides would prevent us from grasping the significance of our existence. When striving to comprehend a personality, we should exercise caution to avoid either the pitfall of individualism or the pitfall of collectivism. It is not from a human being's individualist or collectivist substance that his or her uniqueness originates. Each person is unique because God's appeal is directed individually to each person. By responding to this appeal the human soul transcends its limits, and through this transcendence the human being's world becomes a world of justice and humanity.

The high estimation of reason is not to be interpreted as an appeal for a European-style of rationality and logo-centrism, as it happened during the period prior to postmodernism. As modern rationalism was born in Europe, Europe has to realize that it needs to give this teaching a critical appraisal and find solutions that would allow for the shunning of its destructive consequences. We know only too well that the irrepressible energy and vitality of European culture sprang from its critical approach to everything, including itself. But it is time for Europe to make another step forward and look at itself from the outside, to see itself as others see it. That is not to say that Europe should abandon its great cultural heritage or turn to a new type of obscurantism. Rather it is an appeal to European culture and civilization to see new vistas and to familiarize itself more intimately with global cultural geography.

It is important to understand that Islamic civilization (or, to be more precise, the civilization of Muslims and our current life) coexists with a so-called Western civilization. Whereas its achievements are many, its baneful consequences, especially for those living outside the Western world, are countless. Our era is a time when Western culture and civilization prevails; this has to be clear. But for this comprehension to be effective and useful, it is vital to look further than a superficial understanding of this civilization and to perceive its theoretical origins and basic values. A similar step forward on the part of the Western civilization is equally important.

We often find that Oriental studies treat the East merely as a subject matter and not as a "second participant in the dialogue." If a genuine dialogue between civilizations is to take place, the East must become a true participant in the discussion rather than remain an object of study. To make the project of a "Dialogue between civilizations" possible, Europe and the United States must take this crucial step forward. If today, the great Asian civilizations see themselves in the mirror of the West and recognize each other by means of this mirror, in the not so distant past Islam acted as such a mirror for the West. It was the mirror in which the West could see its own past and its philosophical and cultural heritage. It is well known that Islam, having experi-

Imperatives to Cooperation of North & South – *Issyk-Kul Forum*   299

enced the deep influence of Greek civilization, acted as an intermediary, introducing Europeans to the attainments of Greek thought and philosophy.

Civilizations exist while they are capable of finding answers to one new question after another, and of meeting the forever-changing needs of a human being. It can be contended that Western civilization is worn out and aged. Four centuries is a long time for a civilization, though some civilizations of the past were, perhaps, even more lasting. But science, technology, and electronic means of communication have drastically sped up the rate of change compared to the past. The life cycle of Western civilization, counting from the Renaissance to modern times, by no means can be called short, and it would not be an exaggeration to call Western civilization old. But the question is: has not Islamic civilization emerged once and then declined several ages before? Does the death of a civilization mean that we can no longer utilize its teachings as a foundation for our thoughts and actions? Does the birth and decline of Islam civilization mean that the time of Islam, a cornerstone for an Islam civilization, has passed? My answer to all these questions is negative.

The Islamic revolution was an event of grandiose importance in the history of the Iranian people and the Islamic community, and we can affirm that thanks to our revolution we freed ourselves from a host of alien values prevailing in our mentality. Acting in accordance with our true historical and cultural consciousness we laid a totally new foundation for administering our society. Our current crisis is a crisis of rebirth. A new Islamic civilization is being born. That is why a deep, thoughtful, and detailed dialogue with modern Islamic civilization could be exceptionally useful for finding just and practical solutions to a whole range of global problems that the international community currently faces. Crises in families; crisis in the relationship between persons and nature; crises in the ethics of scientific research; conflicts in conjunction with violence and terrorism–all of these and many other similar problems must be included in the agenda of a dialogue between the Islamic world and the Western civilization.

*Excerpted from Khatami, Seyyed Mohammad (President of the Islamic Republic of Iran). *Islam, Dialog i grazhdanskoe obshchestvo* (Islam: Dialogue and Civil Society). Moscow: Rosspen, 2001.

S. M. Khatami

Islamism's Participation in Globalization: Globalization, as an inevitable process of accelerated changes in human life, will influence all dimensions of human culture especially, religion. However, because globalization is a universal process that cannot have its frame without all of the other dimensions of human life, all of those dimensions have their role in globalization; hence religion must play a crucial role in constructing globalization: Globalization cannot be understood if we neglect the important role of religion.

I believe that all religions must have their active presence in the world today in spite of the impact of the global atmosphere that tries to give them a passive role. In order to maintain an active role, they must know their strengths, weaknesses, opportunities and threats. By this knowledge they will intelligently participate in globalization.

300   *Global Studies Encyclopedic Dictionary*

However, by focusing on the external opportunities and threats of globalization, I will address the manner of participation of Islam in Globalization. There are many views among contemporary thinkers regarding globalization. We can classify them in two branches of opportunities and threats for religions, especially for Islam. Opportunities include: (1) better mutual understanding of religions; (2) development of cultures and their religious dimension; (3) knowing the weaknesses and strength of every religion; (4) the need of the contemporary world for a moral solution for environmental pollution which makes religious activity necessary; (5) emphasis on regional culture instead of foreign culture as better culture; (6) the opportunity of attending in global activities by means of new technologies and possibilities and delivering the religious message to everybody; (7) aptitude of training interested talent, (8) developing of religions by coordinating a better world; (9) the possibility of active attendance in societies when the fashions of late modernity are removed by postmodern critics; (10) easier access to data in less developed countries; (11) renewal of religious life; (12) formation of pluralistic attitudes toward science, identity, and culture as the result of better understanding of other cultures and values.

Threats include: (1) cultural imperialism; (2) destruction of older traditions by the predominance of Western culture; (3) demolition of human values; (4) the destruction of cultural and mental security; (5) the cultural predominance of the most powerful countries on other cultures for a kind of cultural synchronization; (6) propagation of crimes and destruction of the religious resistance to decline; (7) destruction of a wide range of abilities and possibilities of all other peoples in the solution of political, environmental, and economic problems; (8) sovereignty of relativity and the loss of self reliance; (9) the undesired control and changing of general thoughts by the sources of power by means of information technology; (10) imposition of Western values and ideas, especially American ones; (11) imposition of humanism and secularism, which brought many tragedies during the last century, such as fascism, racism, and Nazism.

Some main approaches for the analysis of globalization include new liberalism, reformism, traditionalism, global socialism and postmodernism. All of these views have their own perspective on globalization and treat it in terms of their fundamental beliefs. However, how may the position of Islamic thought be brought to this universal process?

Islam, like Judaism, has religious jurisprudence, while this is not the case in the Christian emphasizes on morality. From this specification, religious laws flow into all dimensions of the life of a Muslim believer. Furthermore, the many social religious laws in Islam make this religion more involved in the social activities of a society. Islam is not only concerned with the personal relation between an individual and God. Many Islamic doctrines focus on the relation between the individual and society, emphasizing that this relationship might not be complete unless the social duties of a believer are fulfilled. The social duties of a Muslim believer contain all dimensions of life, including economic, political, cultural and social development. Therefore, if one wants to be a believer, one may not ignore social duties.

Because the social law in Islam is an essential specification for it, Islamic laws must examine every dimension of Globalization. Islam may not accept every economical frame, such as ones based on human autonomy apart from God's order. Certainly, it rejects some political ideas (like racism, imperialism, and compulsion in acceptance of a political government), and accepts some others. Therefore, Islam needs to be involved in Globalization, especially in the economy, politics, security, the environmental, and culture. Islam will have some challenges with Globalization because of its different and contrary values. Therefore, Islam will present more challenges to Globalization than other religions.

Another important difference between Islam and other religions involves its attention to mundane matters of human existence. While many activities may not seem to have religious implications, in Islam every human activity is seen as a religious one. The meaning of worship differs in Islam. Mundane activities are spiritual activities. We may not separate endeavors in this world from those pertaining to our lives in the next world. This orientation makes secularism more foreign to Islamic society than others.

However, this does not entail a rigid reaction of Islam against Globalization. The element of *Ijtehaad* in Islam makes the presence of Islam in the process possible and safe. *Ijtehaad* (exegesis of divine law on matters of theology and law) seeks to find a solution for making divine law compatible with spatial situations without abandoning Islamic foundations. This fluidity not only preserves the religious attitude toward construction of a better world based on spiritual values, but also makes the facts of the world today relevant to the a solution which should be compatible with the real problems.

Global relations have evoked the inescapable question: If our religion is the true one and other religions false, why do we see believers in other religions with strong beliefs that are like or better than we expect in religious persons?
There are three main replies to this question: pluralism, inclusivism, and exclusivism. Some philosophers of religion maintain pluralism; that is, all the religions are true and each is a way to the truth. Although the pluralists offer this solution to the question of religious diversity, is pluralism compatible with their religious doctrines and their religious tradition and culture? I think not. That so few Christian and Muslim thinkers accept religious pluralism shows that this attitude may continue to be primarily speculative.

H. Ayatollahy

Issyk-Kul Forum is a general name for the 1996–1997 Kyrgyzstan meetings of representatives of literary and arts circles, scientists, and intellectuals from various countries; the meetings aimed at connecting world cultures and reaching mutual understanding among them. The first Issyk-Kul forum took place at the time of Détente (Russian, *razryadka*, loosely meaning "relaxation of tension") when the Cold War was gradually receding into history. That is why the atmosphere of optimistic hopes for peace and prosperity in the near future for all humankind prevailed during the meeting.

The organizers had been nurturing the idea of gatherings like these for a long time: to get people who are close in spirit and thought together in a friendly circle, on a personal basis, and talk about the destiny of the world. In

302 *Global Studies Encyclopedic Dictionary*

olden times Kyrgyz elders used to hold "shernes" (meetings in the autumn, after the summer harvest works) which were gatherings of venerable people capable of holding leisurely conversations about the deepest issues of life. These were the conversations among equals in spirit and life experience; these talks were accompanied by a feast and tea.

In the autumn of 1986, humanists and creative individuals who are well known throughout the world were invited to the shores of Issyk-Kul Lake, a gem of Kyrgyzstan, to a Kyrgyz hearth, as guests. Soulmates—people in literature, the arts, and public thought who had similar ideas—gathered in one of the corners of the Earth just to have earnest talks. Participants included author and actor Peter Ustinoff (Great Britain); scientists-futurologists Alvin and Heidy Toffler (the United States); UNESCO Secretariat member Augusto Forti (Italy); painter Afewerk Tekle (Ethiopia); President of Indian Academy of Music and Drama Narayana Menon; Turkish composer Omer Livaneli; Club of Rome President Alexander King; Spanish public figure and litterateur Federico Major, who soon after became UNESCO General Director; Nobel prize laureate Claude Simon (France); Lisandro Otero (Cuba); Jashar Kemal (Turkey); and other remarkable people.

The motto of the meeting was "Survival through Creativity." The gathering of talented and courageous people put forward a topical and urgent idea that we a need for a new planetary way of thinking as an instrument for resolving imminent global problems. We need a purification and enhancement of the creative powers of any person who is capable and is called upon to pave the way toward a peaceful future on a prosperous planet.

The original aim did not include an intention of creating a new organization. The aim was simply to talk freely. But the mutual understanding attained led to the conclusion that the group should spread panhuman ideas and values in the world's artistic and intellectual milieu. All participants wanted to overcome the dissociation of our world, full of conflicts and split into military coalitions; of a world that was becoming more and more distressing and threatening for everybody. Humankind, having become tangled in the web of its own contradictions and biases, faced self-destruction; whereas technically human beings functioned like a demiurge, morally they were still savages. The question that naturally arose was: what does the aim of survival have to do with creativity?

We trust in reason in its global sense as in a responsible supreme substance that safeguards us from calamity, though in fact we are ourselves this "reason." Reason is capable of self-perfection, of eternal perfection, of eternal evolution. This intellectual evolution, spinning faster and faster in the vortex of grave global contradictions is now experiencing a new qualitative ascent, a new phase in its development. It could be exactly the moment when a *Homo sapiens* rises to become a *Homo sophos*. If reason is a rational logicality, wisdom is reason spiritualized by morality. In our artistic creativity, we first turn to the word, to spirit, to consciousness as such, which are all primary impulses of the vital functions of a thinking creature, in the hope that the never-ceasing study of our own substance could help humankind to survive and reinvent itself in the midst of irrepressible scientific and technological advance, in the situation when mass culture is governed by consumer interests and

Imperatives to Cooperation of North & South – *Issyk-Kul Forum* 303

tastes. It is here that we come across the global mission of cultivating a new consciousness, a new planetary creative thinking as a means of survival for human society at the turning point of the centuries and millennia.

All forum participants talked freely and without restraint; their speeches were penetrating and concerned. This was a meaningful conversation among like-minded people from various corners of the world. Fifteen years before the third millennium the following "Statement of the Issyl-Kul Forum" was adopted. This document says that for centuries, creativity has been helping the humankind's survival in this world, burdened with disasters, wars, and catastrophes. That is why it is vital that the new century be not only "The Century of the Planet," but also "The Century of Creativity." Survival and happiness through creativity-that is our answer to the dangers and threats in our trouble-ridden world. New ideas should penetrate every sphere of our life, including politics, creating, as a result, a new mentality in every country. The future should not depend exclusively on the decisions made by the politicians and on confrontation among world powers. Human genius; the force of the imagination of the talented; initiatives and discoveries made by the scientists; the dreams of poets; the aspirations of common people-they all are to have their say. Only all of these taken together could help to sow the seeds of a new mentality-both general and political.

It is quite possible that at the time the group somewhat overestimated the significance and influence that the voice of representatives of the culture might have. But none of the forum participants renounced the noble and principled convictions, and the group still considers creativity to be one of the major values in human society.

In 1997, the second Issyk-Kul Forum took place. Beyond representatives of the world of science, the arts, and culture, it included Kyrgyzstan President Askar Akaev, Turkish President Suleiman Demirel, ex-President of the Soviet Union Mikhail Gorbachev. The idea, running through the forum discussions, was the UNESCO idea about a "culture of the world," that is, new values, based on nonviolence and freedom, which could become common grounds for all humankind. The forum discussed the problems of the unity and diversity of cultures; of national and global aspects of culture; of the future model for the cultural development of Central Asia. The participants called on world leaders to develop a policy, which could make the movement "For Intellectual Convergence of Civilizations in the Twenty-First Century" a reality. The forum heard the address by the Club of Rome President Alexander King.

For the time being, Issyk-Kul Forum has fulfilled its historical mission. It will become a foundation for a new international movement called "Crossroads of World Cultures," the objective of the movement is to promote cultural contacts and a nonviolent resolution of world problems.

C. Aitmatov

⚛ J ⚛

Japanese Post-War Economic Miracle refers to the historical phenomenon of Japan's record period of economic growth after World War II. Japan's Gross National Product increased 4.6 times between 1955 and 1970. Its annual growth rate exceeded 10 percent twelve times and fell below 5 percent only twice between 1955 and 1973, whereas those of other Western nations never exceeded 10 percent except for once in West Germany.

(1) International Circumstances: The sophistication of the armaments developed during the Cold War. The demands of the Vietnam War so militarized the United States economy that United States citizens' demands for some mass-produced industrial products, such as fiber, home electric appliances, cars had to be satisfied with imported goods. Therefore, Japanese products flooded the markets of North America, especially during the latter half of the 1960s. Additionally, despite the constant United States fiscal deficit and currency crises, the fixed exchange rate of currency was favorable to Japan.

(2) Domestic Factors: As for the basic structure of the Gross National Expenditure, expansion of fixed capital formation in the private sector exceeded by a great degree that of the personal consumption expenditure and governmental purchase. In a word, we sacrificed personal consumption and welfare for the sake of capital accumulation.

An excessive supply of currency and credits were caused by public investment, financial investment and loans, credits of the Bank of Japan, and public debt. High-rate investment sometimes exceeded the real accumulation by private enterprises. The preference of the Japanese for saving money, which was caused by the low level of public welfare, strengthened this tendency.

Owing to the lack of a job system, a qualification system, or a job grading system, employees can be re-shuffled freely. Such an employment system facilitated flexibility in relation to technological innovations.

Policies on science and technology, which mainly consisted of a taxation system for the promotion of the importance of technology and specialized education, stimulated the technological development mainly composed of imported technologies.

The loss of colonies and a sphere of influence made it possible to build up new giant plants without cost for scrapping old facilities and equipment.

(3) Features of Technological Development and Their Consequences: By building large-scale apparatuses, "economy of scale" was pursued throughout. Building such giant plants, however, needed a long "pregnancy period" for capital. In addition to this, reduction of operation at the large-scale apparatuses sometimes leads to catastrophe in the rate of profit. So, maintenance of operation became the "Supreme Proposition." Thus, Japan suffered chronic inflation and the potential of excessive supply. Large-scale investment was often accompanied with "saving" on anti-pollution or safety facilities. Japan also suffered a variety of environmental pollutions.

Automation (Detroit Automation) secured speed-up in the mechanical processes. Construction of the industrial complex (Kombinat) secured succession of production. In some cases, an imbalance between the products (e.g., in

Japanese Post-War Economic Miracle – Just War Doctrine   305

an oil-refinement and petrochemical complex) called for a balancing measure in the balance of consumption (in this case, the balance between oil products and chemical products; thus, the "Motorization Drive" was set about).

H. Ichikawa

Jihad from a Nonviolent Perspective: The moral and ethical aptitudes and spiritual ideals of Islam operate by the inner willing of conscience and are expressed in praiseworthy character and admirable models of behavior. Among the principles that empower the individual in the Islamic tradition is jihad, which means to strive or struggle one's utmost in all of life's conditions. Islamic scholars have identified two kinds of jihad, the lesser jihad (*jihad-al-asghar*) and the greater jihad (*jihad al-akbar*). The lesser jihad is the war a Muslim wages to protect himself or herself against aggression; the greater jihad is a war a Muslim wages to purify himself or herself.

The serious misunderstanding of jihad among both Muslims and non-Muslims "as a war of aggression" (Engineer, 1992) has rendered jihad an external holy war that is completely disconnected from the greater holy war inside oneself (*jihad an-nafs*). The resulting separation of inward awareness from outward activity has fixated the attention of many Muslim activists, especially members of radical and fundamentalist groups, on the victory over others. This, in turn, subordinates Islamic spirituality-complete surrender of the self to the Divine—to militant politics. As Amin M. Kazak, scholar of Islam, observes, many radical Muslims replace the spiritual idea of jihad with programs of revolutionary struggle, arguing "that the idea of jihad as a struggle for self-control against evil temptation is not authentically Islamic" (Kazak, 1992).

From a nonviolent perspective, Islam is, at its core, a message of unity, peace, and reconciliation, and jihad is a nonviolent means for preserving this message. This message applies to the inner person, to society, and to the world. This understanding of jihad is contingent on tawhid, the Islamic principle of unity that provides a fertile soil for Islamic practice, faith, and spirituality. Unity is essential to Islam. Islam underscores the unity of God, the unity of the many streams of revelation, the unity of humanity, and ultimately the unity of existence (*wahdat-al-wajud*). Unity embraces and sustains diversity; the Whole is reflected in the parts. As the Qur'an affirms, "To Allah belong the East and the West: whithersoever ye turn, there is Allah's face; for Allah is All-Embracing, All-Knowing" (2:115).

Analyzing jihad, nonviolent activists posit a direct relationship between peace and jihad through the means employed and the ends achieved. As Karim Crow, former Project Coordinator for the Islam and Peace Project at Nonviolence International in Washington, DC, has argued, this relationship gives priority to "peaceful action": one must act justly and ethically when confronting injustice and demand just and ethical action from the oppressors. Crow connects this imperative to a central premise of nonviolence by stating, "The strategy of peaceful action, jihad, is based on the belief that the exercise of power depends on the consent of the ruled who, by withdrawing their consent, can neutralize or even control the power of their opponent" (1999).

While armed struggle can be likened to a branch that dies if torn from its root, unarmed struggle teaches reliance on God. As depicted by Jawdat Said,

306   *Global Studies Encyclopedic Dictionary*

contemporary Syrian Islamic reformist thinker, violence by its nature breeds egotism and can become a form of idolatry that saps the source of strength from struggle. Said's body of work represents an original attempt to re-think some of the most difficult issues facing Islam today: the role of force and violence in achieving social and political ends, the best paths for transformation of society, the problem of determinism and human liberty; the connection between human liberty and change, and the purposes disclosed by historical processes and by Providence.

In Islam, actions, which are disproportionate and improper, deviate from truth (*haqq*) and are ultimately self-destructive: they turn on themselves, incur God's disfavor, and obscure unity and interconnectedness. The Qur'an frequently cautions people against going to excess when attempting to pursue rights or correct injustice. This attitude itself is only an extension of the Islamic precept of adopting the middle way in all activities, of moderation and even-handedness. The Qur'an heaps utter condemnation on those who, by selfishly pursuing their own limited goals, seek infinite satisfaction from a finite world and bring destruction, oppression and violence (*fitnah*) down upon the rest of their fellows, "committing excesses on earth" (Qur'an 5:33). Islam also holds that greater harm is done to the world by indifference and unconcern-by the failure of good persons to rally in support of true causes-than by active deliberate mischief.

Compassion and mercy are integral and oft repeated attributes of God, to whom the Qur'an refers as al-Rahman, al-Rahim, the Merciful and the Compassionate. Although Islam does not limit God to any particular quality, attributes such as rahman and rahim suggest the nearness and the forbearance of God. God forgives and transforms those who turn toward him (tawba) with infinite mercy and graciousness. God's mercy extends to all worlds; the Qur'an describes the prophethood of Muhammad by stating, "We sent thee not, but as a Mercy for all creatures" (21:107).

The principal concern of shari'ah law, or law of Islam, is the maintenance of proper, harmonious relationships on and across all levels—between the individual and God, within the individual, within the family and community, among Muslims, between religions, and with all of humanity and creation. As has been the case with Christianity, Judaism, and other religious traditions, Muslim practices have not always reflected Islamic precepts, but the historical record of Muslim communities on matters of cultural and religious pluralism excelled that of pre-modern Europe. Religious tolerance is built into Islamic precepts, which designate the "People of the Book" as protected peoples.

At the most essential level, Islamic teaching on jihad as a form of spiritual striving affirms that a person must cultivate in the self the character traits of God (*takhalluq bi-akhlaq Allah*). In the daily life of the veritable practitioners of Islam, there is a practical demonstration of how to cherish the social and ethical values that lead men and women to the good life. Islam offers the stimulus and strength for performing deeds which are distinctively human in the deepest sense, to bring the human being nearer to God and to respecting the sanctity of human relationships, in which must be mirrored a glimmer of Divine attributes. The impassioned mind and the informed heart can together call forth the energy to move the planet toward realization. The Islamic tradition of jihad has always been the inner struggle to purify the self and behave in a manner that furthers rather than disrupts the divine harmony.

References: Engineer, Asghar 'Ali. "Sources of Nonviolence in Islam," *Gandhi Marg: Journal of Gandhi Peace Foundation* 14:1 (1992); Kazak, Amin M. "Belief Systems and Justice Without Violence in the Middle East." In *Justice without Violence*. Edited by P. Wehr, H. Burgess, and G. Burgess Boulder, CO: Lynne Rienner,1994; Crow, Karim. "Islamic Peaceful Action (al-Jihad al-Silmi): Nonviolent Approach to Justice and Peace in Islamic Societies," *Nonviolence International* (1999); Said, J. "Law, Religion, and the Prophetic Method of Social Change," *Journal of Law and Religion* 15: 1–2 (2000–2001), 83–150.

M. Sharify-Funk

Jonas, Hans (1903–1993) Hans Jonas studied with Edmund Husserl, Martin Heidegger, and Rudolf Bultmann. After emigrating from Germany in 1933, he was a professor of philosophy in Canada and then at the New School for Social Research in New York. He elaborated a philosophy of biology (*The Phenomenon of Life*, 1966), and his philosophy of nature was translated into several languages and influenced ecological debates in Germany (*Das Prinzip Verantwortung*, 1979; *The Imperative of Responsibility*, 1984). His contributions to Global Studies also include his aim to replace the ethics of loving one's near neighbor with a global ethics that includes people far away in space and time and his views on the possible intrinsic value of non-human organisms. In bioethics, he criticized cloning and rejected brain death as death of the whole person.

References: Jonas, H. *Das Prinzip Verantwortung*. Frankfurt am Main: Insel-Verlag, 1979; reprinted in English as *The Imperative of Responsibility*. Chicago: University of Chicago Press, 1984.

V. Hösle

Judaism and Nonviolence: Teachings about peace in the Jewish tradition overwhelmingly stress what peace theorists call "nonviolence" and "positive peace." The ideal of nonviolence aims at the radical transformation of individuals into proactive peacemakers who understand peace as the highest human value and respond even to violence with kindness. The ideal of positive peace aims not merely at the cessation of war, but the radical transformation of human society into a just order that precludes striving over resources.

(1) Ancestral Stories: offer popular models for self-development and are a significant part of the religious teaching to young children. Jewish ancestral stories are based on the Hebrew Bible, and are elaborated in *Midrash,* interpretive retellings of Biblical stories from the early Middle Ages.

(2) Prophecy: Judaic prophesy promises positive peace and repudiation of war, such as articulated by the prophet Isaiah, who prophesies that at the "end of days," after a series of cataclysmic battles, humans will repudiate war. God will institute a new order of righteousness, under which the rich and powerful will cease to take advantage of the poor and powerless.

(3) Kabbalah: Kabbalah, the mystical tradition in Judaism, includes instructions for ethical development. Rabbi Moses Cordovero, influential sixteenth-century writer on ethical development, praises God's restraint toward evildoers. God refrains from meting out punishment, Cordovero says, because

308 *Global Studies Encyclopedic Dictionary*

the essence of divinity is compassion. God actively practices tolerance, extending opportunities for repentance even to recalcitrant evildoers. Human beings, too, should be developing their powers of tolerance, restraint, and compassion. Moral power is tangible and gives individuals the tools to change reality as they reach out to others.

(4) Jewish Enlightenment: This nineteenth-century movement was an application of European Enlightenment philosophy. Influenced by the rise of democratic political orders, some Jewish thinkers taught that peace would come about when citizens of the world came together for collaborative self-governance. Moritz Lazarus argued that peace cannot merely be an inner state, but must be expressed through the social virtues of tolerance, open-mindedness, lack of self-centeredness, and cooperation. The universal public practice of these virtues in support of the common good would usher in the Messianic Age or, as the ancient prophets put it, "the end of days." These ideas continue to find expression in the Reform Movement in Judaism.

(5) Contemporary Application: Jewish religious teachings unequivocally emphasize nonviolence, while making some allowances for defensive just wars that are the last resort after an attempt to create peace. Jews actively debate the relevance of nonviolence and of just war in Israel today. With the World War II Holocaust of European Jewry in mind, proponents of just war argue that a strong nation-state with a strong proactive army is the best defense against the threat of genocide. They view all of Israel's military activities as justified actions of territorial security and self-defense. Proponents of nonviolence oppose what they call "holocaust theology," arguing that the memory of the holocaust should not overshadow nonviolent Jewish religious teaching. Hundreds of soldiers have refused to fight against the Palestinians, calling themselves "conscientious objectors." Nonviolent demonstrators place their bodies in the paths of Israeli bulldozers attempting to destroy Palestinian properties. Peace groups sponsor cooperative Palestinian-Jewish projects in the areas of education, citizen diplomacy, and dialogue. Pacifist rabbis and religious educators use the classroom, the sermon, the printed word, and the Internet to remind students that the ideal of nonviolence is central to Jewish religious thought.

L. D. Kaplan

Justice, Global: (Latin, *jus*, "right" "law") Global justice transcends national boundaries. It is universal in that it applies not only to all countries but also to some affinities and affiliations (Amartya Sen) hence making it more than international justice (John Rawls). It is foundational in that it exudes and demands what is important and necessary as viewed from a moral point of view. It presupposes fundamental values, beliefs, and principles underlying the relations among sovereign countries, organizations, corporations, and institutions. It is anchored in human dignity, moral equality, international law, shared humanity and global morality that serve as foundational prerequisites indispensably needed or aspired, if not postulated, for it to be meaningfully to talked about, valued and pursued.

Global justice is viewed as problematic due to the lack of any global sovereign power to enforce laws and corresponding sanctions (retributive justice) and established principles in distributing the benefits and burdens of international cooperative ventures (distributive justice). Of concern is whether it is based on reciprocity of mutual advantage considering that the global order is defined by rational self-interests and power-relations. While these are valid issues, a single global power with over-arching authority is too much to demand presently considering that current sovereign countries cannot rule among themselves cooperatively and rationally. Moreover, no overriding established principles for distributive justice currently exist, and the presence of international covenants and treaties insufficient to settle disputes and problems. Still, global justice, understood as international conflict-resolution aptly describes what the global system as it presently *is*, and what it *ought to be* can described as cooperative, collaborative, dialogical, and reciprocal interdependency; for example, on the order of Kai Nielsen's moral reciprocity as inspired by Immanuel Kant's precept on moral equality and autonomy.

Global justice concerns maintenance of global order, sustainability of the global environment, and ideal relations among peoples regardless of gender, race, history, or location, advancement of transcultural sensitivity and respect, moral tolerance and integrity, political autonomy and strengths, economic dependencies and growth. Equally important is the presence of global pluralism constituted by variegated ideas, faiths, valuations, worldviews, and the requirement for interdependence beyond the confines of domestic posturing and privileging. However, some dispute is noted about the grounds on which—interests, needs, rights, capabilities, or entitlements—it should be embodied.

As an ethical response, global justice seeks to resolve inequalities in dignity and liberties. This requires egalitarian seriousness, moral courage, relational ontology, human solidarity, and political will against decadent exploitative world systems. It is about shunning the moral blindness, complacency, and indifference that have perpetuated unjust global disparities between the North and South, First and Third World, rich and poor the powerful and the vulnerable. It fuels the moral outrage that should be felt by everyone to boycott, persecute, and lobby against the perpetrators and conduits of this crime against humanity. As a moral requirement, it entails global obligations to seek sustainable, long-term, and meaningful appropriations conducive to creating a global moral order. Since injustice and inequalities are iniquities that are structural and systemic in nature, deeply entrenched in the international political and economic order, the proper response should also be systemic and global in essence. It requires structural transformation of power-relations, economic, cultural, and political.

Global justice advances an end to absolute poverty, human misery, hunger, malnutrition, domination, subordination, exploitation, deprivation and ignorance that have long-stalled and emasculated the development of poor countries, and retarded the growth of their people who lack the capacity to lead the life they have reason to value. This has mainly resulted from the sugar-coating, glossing over and deceptive niceties of irresponsible rich countries, leaders and institutions owing to their unjust foreign policies, trade-offs, structural adjustments lacking even a simulacra of justice, fairness, impartiality, that is, owing to their lack of imaginative empathy. That is why the idea of reciprocity is suspect because of the undeniable unequal global positions of

310   *Global Studies Encyclopedic Dictionary*

countries and peoples translates into disequilibrium in their political, social, economic, cultural coordinates. Yet this hegemonic status quo should be transvalued as a measure of the efficacy of global justice not by all means possible but only by reasonable and life-affirming ways and means.

A. S. Layug

Just War Doctrine: To say, "just war" would be an oxymoron. In *The Leviathan*, Thomas Hobbes contends that human beings are naturally at war with one another, and where there is war, nothing can be unjust. However, little empirical data supports Hobbes's contention.

A more promising starting point assumes that human nature is pliable, having the potentiality for either caring or belligerence. At core, we are capable of uniting as family or community, or dividing as enemies. At some level, this involves choice, and moral criteria can be invoked to evaluate those decisions. As Aristotle held, we are naturally political and moral human beings.

Under this framework, wars do not occur void of moral content, but are judged either by their intentions (the deontological approach) or by their actual or anticipated ends (consequentialist approach). Just war doctrine is closer to the deontological approach.

Deontology, or, the study of duties goes back at least to Immanuel Kant, who tells us that we should never act as individuals in any way that could not also serve as a universal law. Preserving life could be a universal law, but war entails killing. Are there times when killing might be justified? Perhaps killing in self-defense, but that does not preserve the life of the assailant.

One way around this perplexity is to differentiate between killing and murder. Murder is defined as killing the innocent; hence murder, not killing, becomes the universally prohibited act. Absolute pacifists would disagree with this shift in meaning, but it is in keeping with just war doctrine. Wars fought in self-defense against aggressive enemies would then be morally justified, since soldiers forcefully invading a country are not innocent.

Based on the moral prohibition against murdering the innocent, the justification for war is still not obvious. Kant's position that never allows an exception to a law-like moral imperative would disqualify as immoral almost all wars. Most wars include soldiers on both sides killing the innocent. So a strict Kantian analysis would not justify war if innocent human beings are killed.

Another approach allows for some killing of the innocent within the framework of morality. The just war doctrine, going back to the Middle Ages, has its foundation in the principle of "double effect," under which an action that has two contrasting effects, one good and the other evil, may be permissible under some conditions provided only the good end is intended. The evil effect understandably cannot be intended although it may be foreseen.

The principle of double effect allows acts having double effects to be moral as long as the following four conditions are met:

    (1) The act, exclusive of the evil effect, must be good;

    (2) The one acting—individual or nation—must have a right intention;

    (3) The evil effect may not be intended as the means to the good effect;

    (4) The action must be in proportion to the good sought.

The requirements of just war, namely, just cause, right intention, discrimination, and proportionality, mirror these double effect conditions. A fifth condition, legitimate authority, demands that the defending state or nation is acting under proper public authority. Other frequently cited just war conditions can be reduced to these five conditions. All five conditions must be met for a war to be just.

Just cause is the good end for which a war is fought. A just cause is the defense of one's nation or the preservation of innocent life. It is defensive of life, not aggressive after a nation's self-interests. Some examples are defending against Iraq's invasion of Kuwait in 1990, Germany's invasion of Russia in 1941, and Japan's attack on Pearl Harbor in 1941.

Right intention runs deeper than reiterating the rightness of just cause, its purpose being to seek a peaceful resolution of the conflict before authorizing a military option. With right intention, war becomes a last resort; e.g., President Bill Clinton, before invading Haiti in 1994, sent Jimmy Carter, Sam Nunn, and Colin Powell to negotiate a settlement with the Haitian generals.

Discrimination prohibits the deliberate targeting of noncombatants who are not part of the military operation and have no war-making role. The discrimination clause distinguishes between intended and only foreseen actions. The indiscriminate atomic bombing of Hiroshima and Nagasaki was intentional and immoral, contrary to this condition. Today's more accurate technology, depending upon the targets chosen, should assist in fulfilling this condition.

Proportionality requires an overall assessment of the probability and cost of success. This is not a body count, but an honest analysis of whether the anticipated deaths are in proportion to the reasons for going to war. If war will only make the situation worse or slightly improved, then, as in the 2003 United States invasion of Iraq, proportionality has not been met.

Legitimate authority concerns procedure, not substance, necessitating that the proper public body or official authorize the war. This condition checks against autocratic behavior. In democratic nations, for instance, constitutions often require more than one person to authorize a war. For humanitarian interventions, a regional or international governing body must authorize the appropriate action.

Since just war doctrine is an application of the principle of double effect, all five conditions must be met for a defensive war to be considered just. Most situations readily satisfy one or more of these conditions, but adequately satisfying all five is rare. Some distinguish some conditions as *ad bellum* and others as *in bello*. In short wars, however, the means for fighting the war are known in advance, and are thus part of the *ad bellum* deliberations.

J. C. Kunkel

❦ K ❧

Khozin, Grigorii Sergeevitsh: (1933–2001) Soviet and Russian scholar, Professor of History, a member of the Russian Academy of Astronautics. From 1994, Grigorii Khozin worked at the Diplomatic Academy of the Russian Foreign Affairs Ministry where he created a research center for Global Studies (now, the World Economy and Global Studies Center). Khozin was one of the scholars who initiated theoretical studies of global issues in the Soviet Union; his attention was mainly focused on human (philosophical and political) aspects of space explorations and space activity. For him, equal cooperation of the whole of humankind was the solution for the global problem of space exploration as opposed to military confrontation of states in the outer space. The works by G. S. Khozin have been published both in Russia and abroad. His last book, *The Great Confrontation in Outer Space* (in Russian), was published posthumously in 2001.

Reference: Khozin, G. S. *Velikoe protivostoianie v kosmose* (The Great Confrontation in Space). Moscow: Veche, 2001.

A. V. Mitrofanova

Knowledge Economy occurs when most of the gross national product (GNP) is generated by producing new knowledge, informational goods and services, and equipment for transmitting and handling knowledge and information. For example, in the European Union and the United States, goods and services produced by high technologies constitute more than 70 percent of the GNP. The formation of a knowledge economy is supported by the exponential growth of demand in services provided via informational and communicational technologies (mobile and satellite phones, digital television and radio, telemedicine, Internet, e-government, and e-democracy). Transition to the stage of an informational society expands the horizons of economic development. New technological solutions permit a more economical use of non-renewable natural resources, even gradually decreasing their usage.

E. V. Mazour

ଓ L ଚ

Labor Market, World: The world labor market is the international field of supply and demand in hired labor, a constituent of the world market economy, along with commodity, service, capital, and other markets. The world labor market has no territorial limits, or state or regional borders. Its formation and development is the result of the migration of international labor, such as the export and import of labor resources.

Historically, the formation of the world labor market was connected with transformation of labor into a product or commodity. The world labor market became the natural result of the development of the capitalistic mode of production. The world labor market transfers labor from the countries with excessive labor to countries that demand additional labor. Thus, the development of the conditions for providing the mechanism of labor transfer, in particular, modern modes of transport and development of mass media, is of particular importance. State and private intermediary agencies involved in overseas employment of labor serve an important tool in the modern world labor market.

The world labor market features a large-scale sector of illegal employment related to migration control in the majority of developed countries and the high migration potential in developing and transitional countries, and the extreme profitability of disfranchised illegal migrant labor. Intermediary institutes in this sphere primarily consist of criminal structures involved in human trafficking.

The world labor market is considered to have been formed at the end of nineteenth century. However, later, under the pressure of globalization trends, the market was considerably changed, most of all in the aspect of the demand for foreign labor on the part of developed countries. In the modern world labor market the demand is usually focused on the two extremes of qualification. On the one hand, there always exists a stable demand for highly qualified professionals (scientists, senior managers, specialists in medicine, technological innovations, and information technologies).

Immigrant workers often occupy low-rated workplaces with difficult working conditions and low wages; for example, ancillary work in building, agriculture and industry, and occupations in public and domestic services. In fact, many workplaces in national markets of developed countries are assigned to *Gastarbeiter* (guest workers), so migrants have become a structural element of the global economy, and the world labor market has become a prerequisite for further globalization of world economy.

The situation in the world labor market is depends directly on cyclical nature of economic development within developed countries. At periods of economic upturn it provides additional labor for receiving countries, while at crisis periods it serves an "excretory valve" providing removal from the economic scene the part of population having become needless for the interests of capital development. The epoch of development of transnational corporations influenced world labor market formation. Since then, one of its rules has been not the movement of labor to production, but movement of production to the places where cheap and available labor is concentrated.

I. Ivakhnyuk

314   *Global Studies Encyclopedic Dictionary*

Land Ethics: Coined by the American ecologist Aldo Leopold, the term "land ethics" refers to a holistic normative theory, which is a departure from twenty-five hundred years of Western moral philosophy. Rather than the traditional focus on individual ethics, entire ecological systems, Leopold asserted, have moral considerability. Human beings are not merely members of human communities but also of biotic communities. This expansion of moral considerability fundamentally changes the relationship of human beings to the land from *Homo sapiens* as conqueror of the land community to *Homo sapiens* as a member and citizen of it."

The economic worldview is a major obstacle to achieving a land ethic. Currently our relations with the land are guided only by financial considerations. The problem with the economic value system is that it is incapable of recognizing non-economic (ecological) types of value. Leopold argues that this schism between economic and ecological paradigms is noticeable throughout disciplines that deal with the land—forestry, wildlife biology, agriculture. On the economic model, the value of the land is its resource, or instrumental, value. As outlined by John Locke in his theory of the creation of private property, nature itself has no inherent value; human beings, through labor, can transform the latent resource value of land into useful products. We should "release" as much value from the land as possible through development. On the ecological model, the land is a living thing with value above and beyond economic value. Leopold says that the land has intrinsic ("philosophical") value. Thus, the triumph of achieving an ethical relationship with the land requires the recognition of its intrinsic worth.

Leopold summarizes the land ethic, saying: "A thing is right when it tends to preserve the integrity, stability, and beauty of the biotic community. It is wrong when it tends otherwise" ([1948] 2001, p. 189).

American philosopher J. Baird Callicott furthers Leopold's insights, arguing that the entire enterprise of mainstream Occidental moral philosophy, which has been based on the individual, must be abandoned. Peter Singer's *Animal Liberation* (1975) and Tom Regan's *The Case for Animal Rights* (1983) shift the loci of moral considerability from human individuals to select non-human animals.

Individual organisms should not be thought of as having intrinsic value or rights. Individuals, taken in themselves, do not really impact ecosystems. What impacts ecosystems are species. An organism has value only insofar as it contributes to the overall integrity and stability of the larger biotic community in which it lives by virtue of being a member of a specific species, and this value differs. The upshot is that the "land ethic manifestly does not accord equal moral worth to each and every member of the biotic community." Therefore, according to Callicott, the axiology of land ethics is *nonegalitarian holism*. The center of value is the organic whole; individuals have no value in and of themselves independent of the biocommunity.

Callicott argues that "moral" behavior is animal instinct. Individuals who lack this instinct die off without reproducing at rates higher than their sociable counterparts do. In human beings, the proclivity to live together manifests itself in "sympathy," "beneficence," or "altruism." Selection for such behavior operates naturally at the level of communities.

Labor Market, World – The Limits to Growth   315

Callicott traces this biosocial moral theory from Adam Smith and Hume through Darwin to Leopold and the contemporary sociobiology of E. O. Wilson to elaborate his theory of bioempathy—to value nonhuman animals for their inherent worth—which ultimately boils down to enlightened self-interest. Our ability to value nonhuman species beyond our short-term instrumental needs enhances our own species' chance for survival. Adopting an ethic of the environment is prudential.

Bioempathy and understanding life science are connected. Although action springs from sentiment rather than reason, as David Hume pointed out, reason can help achieve what we desire. Knowledge of ecology can help us achieve survival. Because ecological science has taught us that the limits of the human community is not civilization but the biosphere at large, ecology is the wake-up call for bioempathy. This is the reason Callicott says, "the key to the emergence of a land ethic is, simply, universal ecological literacy."

Ecological literacy and bioempathy are the twin pillars of Callicott's formulation of the land ethic:

(1) Science has now discovered that the natural environment is a community or society to which we belong, no less than to the human global village (ecological literacy).

(2) We all generally have a positive attitude toward the community or society to which we belong (bioempathy).

(3) We ought to preserve the integrity, stability, and beauty of the biotic community (the land ethic).

Since having a positive attitude toward life and survival is constitutive of humanity, there is no is/ought or fact/value gap. Descriptions of life and health directly imply prescriptions. What ought to be affirmed—the integrity, stability, and beauty of biotic communities—flows directly from descriptions of healthy ecosystemic function

References: Singer, Peter. *Animal Liberation*. New York: New York Review, 1975. Regan, T. *The Case for Animal Rights*. Berkeley: University of California Press, 1983. Leopold, A. *A Sand County Almanac*. New York: Oxford University Press, 2001.

D. R. Keller

Land Resources, World: World land resources are lands of the world systematically used and suitable for some agricultural purposes, and which differ according to natural and historical characteristics. The term is understood as: reserves of arable lands; reserves of all agricultural lands (ploughed fields, pastures, hayfields, and so on); or territorial resources.

The area of the Earth's land with glaciers and continental waters is 14,900 million hectares; excluding continental waters and areas covered with ice, it is 13,383.5 million hectares. We can predict the reduction of total land area in the future caused by the rising of the level of the World's oceans.

The unique feature of the land as a natural resource is that only its upper layer, soil, is capable of being fertile and producing biomass.

The world dynamics of land reserves shows that the global area of arable lands (about 1.5 billion hectares) has not changed substantially in the past 20–30 years, but the Earth's population continues to increase. So, the area of arable land per capita constantly becomes smaller. In 1960, the area of arable

316 *Global Studies Encyclopedic Dictionary*

land per capita was 0.50 hectare; in 1975, 0.38 hectare; in 1985, 0.30. In 1995, it was only 0.24 hectare. This process continues and characterizes the relative decrease of productive land reserves of the planet; in different countries different ranges of this ratio are observed.

Currently a relatively stable character of the total area of arable lands is observed. This is explained by the reduction of areas of pastures, forests, and bushes and a corresponding increase of the area of "other lands." Forests are the first to suffer in this process.

About 4.5 billion hectares of non-productive ("other") lands now exist in the world, of which about 2.0 billion hectares are the result of anthropogenic influence and 2.5 billion hectares are naturally non-productive lands (deserts, rocky terrains, and so on).

According to the FAO estimates, land reserves of the world are not very favorable for agricultural development: 10 percent of the surface is covered with glaciers or ice, 15 percent is situated in very cold climate zones, 17 percent is situated in very dry climate, 18 percent is situated on steep slopes, 9 percent have non-fertile soils, 4 percent are marshy lands, and 5 percent are very poor soils.

In total, 78 percent of the land surface is unsuitable for cultivation, and only 22 percent are suitable for cultivation.

M. N. Stroganova

Languages: Estimates of the number of extant languages in the world today vary widely from 3,000 to 8,000, which has to do with the Western concept of "language" currently accepted throughout the globe. Before the rise of modernity (the concept and pattern of which originated in the West), the question of which language one spoke (or, more rarely, wrote) was of no significance to the vast majority, who were socially and spatially immobile and overwhelmingly illiterate. What mattered was successful communication. The rise of popular literacy in the nineteenth and twentieth century elevated the written form of an idiom (usually used by the elite) to the rank of a language. Other idioms were disparaged as "dialects." Subsequently, such dialects within the boundaries of a single state were anachronistically subordinated to the recognized (state) language as *its* dialects. Later, these dialects were slated for gradual disappearance when the educational system and mass media replaced them with the "real" (recognized written) language.

Presenting the basic axioms of linguistics, Leonard Bloomfield proposed that dialects are those language forms that are mutually comprehensible, while languages are not mutually comprehensible (1926). As comprehensibility is a quite subjective criterion, his argument did not hold. It became obvious that politics determines a language and which dialect belongs to a given language. Therefore, Chinese dialects, as mutually incomprehensible as German and French, are classified as part of the Chinese language. Speakers of low German in northern Germany and of Alemannic spoken in Bavaria and western Austria cannot successfully communicate with each other but, nevertheless, their dialects are subsumed in the umbrella of German. Although a low German speaker has no problem understanding Dutch, neither of them believe that they speak the same language. On the other hand, though Moldo-

Labor Market, World – The Limits to Growth

van and Romanian are nearly the same (much more similar than British and English), they remain two separate languages. Only script makes the Tajik language (written in Cyrillic) different from Iran's Persian (Farsi) noted in Arabic characters. As in the Moldovan-Romanian case, there is not even difference in script between Persian and Afghanistan's Dari. These examples amply prove that linguistics is not an exact science.

The 2000 edition of *Ethnologue*, the most renowned catalog of the world's languages, recorded 6,809 languages including 41,791 alternate names and dialect names. But when a language does not enjoy a specific written form, there is hardly any check on linguists and local users who may classify and re-classify such a language as a dialect, part of a different language or even as two different languages. Imagination, political expedience and social needs rule supreme in this process of language creation and re-creation.

In the field of written language, according to Wycliffe Bible Translators International, in 2003, "adequate translations" of the Bible were available in 405 languages, of the New Testament in 1,034 languages, and fragments of the Scriptures into 883 languages others. The Christian Holy Writ is available in approximately 2,200 written languages. This sounds optimistic when one remembers that the Unicode international character set for computers supports only 650 languages. In 2003, the Library of Congress in Washington DC contained holdings in about 500 languages, but so far the United Nations has endorsed translations of the Universal Declaration of Human Rights into only slightly over 300 languages.

In 2002, there were 191 member states of the United Nations, and in the same year, James Minahan's *Encyclopedia of the Stateless Nations* (2002) recorded over 300 such nations; as well, the same language sometimes enjoys official status in several states. Modernity introduced and conflated the thresholds of writing and nationhood. This number corresponds well with the extant 400 translations of the Bible into different languages and with 650 languages supported by the Unicode international character set, or with around 500 languages of the holdings in the Library of Congress. The Universal Declaration of Human Rights has been translated into approximately 300 languages. Apart from the extant states, approximately 100–150 stateless national movements are determined to develop their languages enough that they could serve the needs of modern politics and administration. All told, the number of significant (and, invariably, *written*) languages is around 300.

Solely oral language forms are considered to be languages, which amounts to an emulation of the above-presented system of written and politically endowed languages. While writing and political decisions reify a language, making it into a countable and discrete entity, outside this pale classification of languages is necessarily fluid. Anthropologists estimate that 5–6,000 ethnic groups inhabit the Earth, which closely corresponds to *Ethnologue*'s estimate of 6,809 languages (the surfeit of 800 languages over the number of ethnic groups is caused by the preservation of extinct or ritual languages, and invention of artificial languages, while many ethnic groups disappeared during the last two centuries). By definition, an ethnic group is a world to itself with its own language and customs. From the sociolinguistic perspective each ethnic group is a distinctive speech community complete

318 *Global Studies Encyclopedic Dictionary*

with its own language. In practice, this language is an ethnolect because it constitutes one of the markers that the group uses to produce and maintain its specific ethnicity.

The term "language," not unlike that of "nation," is arbitrary and ascriptive. In the world of nation-states, politicians and intellectual elites decide what a (written) language is or should be. They also impose such a decision, though in a less systematic manner, in relation to (oral) languages spoken by ethnic groups. These groups neither enjoy their own nation-states nor aspire to transform themselves into nations, but reside on the territory of a given nation-state and by default have to partake in the national, the world's landmasses (except inhabitable Antarctica) neatly divided among the extant nation-states. The most famous cases of such socio-linguistic engineering among oral-culture ethnic groups include creation and standardization of languages by missionaries for the sake of spreading the Christian Holy Word, and by the Soviet authorities who, similarly, wished to spread the knowledge of Marxism-Leninism to each inhabitant of the Soviet Union. Nowadays, the West continues to decide what counts as language and what does not. The ISO-639 registration authorities, namely: the Library of Congress, Washington DC, USA, and the International Information Center for Terminology, Vienna, Austria are responsible for conferring internationally recognized alphanumerical codes on languages. Such a code is the utmost recognition a language can receive today. If the code is denied to an aspiring language, it remains a dialect or, even worse, its existence is not acknowledged.

References: Bloomfield, Leonard. "A Set of Postulates for the Science of Language," *Language* 2 (1926), 153–164. Minahan, J. *Encyclopedia of the Stateless Nations.* Westport, CT: Greenwood Press, 2002.

T. Kamusella

Law and Legal Systems, International: Global studies considers international legal mechanisms such as the Universal Declaration of Human Rights and its adopted covenants, the law of war and peace, the Geneva Protocols governing the use of armed force to settle international disputes, the legal protocols of the International Court of Justice in The Hague, the work, history, background, future, and politics of the new International Criminal Court established by the Rome Convention; and the work of international United Nations agencies and their legal fields of operation, such as environmental and human rights law and cultural issues. It compares and contrasts the ways different legal systems train their officers and judges, and rules of evidence, trials, and sentencing procedures. It further involves analysis of attitudes toward crime in different cultures, the work of legal professions, and their adjunct agencies (police, intelligence, politicians) in pursing domestic and international criminals, issues of political corruption, and the extent to which different legal systems pursue or fail to pursue those who are suspected or known to be active in corrupt criminal practices. Philosophy of law is comparatively studied on a global basis, considering which cultures and nations advocate or base their legal systems on which philosophical or spiritual principles and traditions with which results in practice. It compares the legal bases of political constitutions, the legality of different political parties in different

Labor Market, World – The Limits to Growth 319

legal systems, the laws governing various aspects (e.g., financing) of political parties, and whether such mechanisms lead to corruption or transparency. The status of women in different legal systems, the rules governing women's rights, and the roles of women in the legal profession are compared as well. Both globally and comparatively, global studies considers questions of the (il)legality of nuclear weapons and weapons of mass destruction in general and the prospects for genuine multilateral and universal nuclear disarmament long-term. Another topic of analysis is academic work underway to reconcile legal systems derived from religious law and secular or philosophical traditions, including Chinese, Japanese, Christian, Jewish, Islamic, Hindu, Buddhist, Roman, Russian, Marxist, Napoleonic, Common law, and other legal traditions.

T. C. Daffern

Learning, Constructivist Theories of: Constructivist theories of learning are based upon the premise that learners construct meanings in their minds and integrate new knowledge into their mental constructs. In the United States and elsewhere, constructivism is taught on a large scale to student teachers in schools of education as the preferred method for teaching. Constructivists speak of "learning by doing," encouraging children to "be the authors of their own knowing." Opposition to constructivism comes from the more traditional or conservative policy makers, parents, and some teachers and researchers who advocate instead a method of more direct instruction where the teacher conveys information to the students more directly.

Constructivism is a broad term, covering many theories that all view students as the active learners or constructors of their knowledge. There are many theoretical perspectives that researchers use to understand how this mental construction takes place, but those perspectives can usefully be divided into two categories: cognitive and social. Cognitive constructivism approaches learning from the perspective of the individual. Social constructivism approaches learning from the perspective of the social environment and the learners' participation in it.

(1) Cognitive Constructivism: Cognitive psychology generally studies mental processes, including learning, thinking, forgetting, and language. The major foundations for cognitive theories of learning and teaching are developmental readiness and active, experiential learning. According to cognitive learning psychologists, cognition and learning are dependent on the learner's stage of development and the experiences provided for instruction. Children reach cognitive stages at particular ages, and, therefore, teachers should be able to predict or test when children are ready to learn new concepts or skills. In this view, instruction should be based upon the child's developmental readiness. Instruction should also be active and "authentic"—a process of exploration and integration of new material—rather than consisting of learning the material, fed to the student by the teacher or the textbook, by rote.

Jean Piaget is the seminal cognitive learning psychologist. In a widely influential set of works, he argued that children construct their own knowledge as they interact with the world around them. These interactions enable students to create schemas or mental models; the models are changed, enlarged, and made more complex as children continue to learn. Direct, re-

320 *Global Studies Encyclopedic Dictionary*

peated experience is the key for assimilating new information into the child's existing mental constructs. If the experience itself is different or new, the child will accommodate or modify his or her existing constructs to reach cognitive stability or equilibrium. In a Piagetian classroom, new facts are presented as aids to understanding the experience or as tools for problem solving in the context of the experience, not as isolated facts to be learned. Abstract concepts are assimilated, when children are developmentally ready, into the child's mental models, if the abstraction is provided in context.

(2) Social Constructivism: The teacher and the social context of the classroom play limited roles in Piaget's cognitive theories of development. According to social constructivists, by contrast, learning and development are social, collaborative activities. Important adults, elements of culture, language, and other aspects of the learner's environment all greatly affect the making of meaning and the development of the learner's mental abilities.

Lev Vygotsky, a key social constructivist in the field of learning and teaching, puts forth a social cognition learning model, in which the teacher and the classroom provide a social context for both cognitive development and learning. He describes a "zone of proximal development" as the primary vehicle for learning. This zone is the developmental space between those tasks children can accomplish on their own with no help (or those they already know) and those they cannot perform even with help from others. Students are able to learn something that falls within their zone of proximal development from adults or more advanced peers, mastering concepts and skills they cannot learn on their own. Thus, learning is a social, collaborative activity.

If a child uses new cognitive processes with the help of others such as teachers, parents, and fellow students, they will become skills that can be independently practiced. The social constructive theory of learning requires an involved teacher who is an active participant and guide for students.

D. M. Steiner

Limit, Philosophy of: (Latin *limen-liminis*; Greek *horos*) We can find an incipient formulation of limit in Anaximander's *apeiron*, which means that which does not have limits or boundaries, in the ontological sense.

Limit becomes a focus during modernity, from René Descartes to Immanuel Kant. By this time it is understood epistemologically; for this reason, limit is conceived in negative terms, separating knowledge (*Kennen*) from speculative thinking (*Denken*) and logical and conceptual intellection from reflexive knowledge. This restrictive sense of limit was dominant until Ludwig Wittgenstein, discussed limit within language ([1921] 1961). Thus, language encounters an insuperable wall of sense, beyond which lies the unnamed, mystical dimension of language.

During the 1970s and 1980s, in contemporary Spanish philosophy, Eugenio Trias formulated his "philosophy of limit." Typically the term "limit" has been used in mathematical or geographical domains, and his concept of limit was first comprehended in geometrical terms. Until Trias's time, considerations about limit were restrictive, but in his philosophy, limit received a positive sense, giving it a transcendental and even ontological dimension. Limit is the ontological principle of any reality. All being is designated by its

Labor Market, World – The Limits to Growth

limit: the only real being is a finite being, a being with limit. Trias has explored this ontological sense in the four principal domains of human experience: the symbolic use of intelligence (religion and aesthetics), the practical use of intelligence (ethics and politics) and, finally, its theoretical use (epistemology). These four domains compose four *quartiers* (neighborhoods) of an ideal city where the human condition is metaphorically expressed.

The philosophy of limit is the most important and recent Spanish contribution to philosophical discourse. Its heritage includes the most important currents in French and German thought of the twentieth century (structuralism, culture anthropology, semiotics or hermeneutics). On the other hand, Trias also builds on the metaphysical tradition of the nineteenth century, from Idealism to Post-Nietzscheanism, The philosophy of limit has been well received in Spain, South America, France, and Italy. Now, this philosophy waits what may be a very illuminating application to the limits of globalization.

Reference: *Tractatus logico-philosophicus*. New York, Humanities Press, [1921] 1961.

F. P.-B. Álvarez

The Limits to Growth (1972) was the first report to the Club of Rome—the largest center of alarmist ideology—and became a genuine manifesto of alarmism. Alarmism-related theories of crises and catastrophes rapidly developed, producing a whole branch of pessimistic social prognostication known as "catastrophism": a theory of general crisis of civilization, economy, and culture, up to forecasting the death of humanity in the fire of a thermonuclear conflict. In opposition to the pessimistic prognostication were theories in the post-war period that developed optimistic visions of the future: new industrial and post-industrial society and a new world order (economic, political, and moral). In the 1980s and 1990s, alarmism lost its acuteness and former influence. Nevertheless, alarmism has not disappeared; it was capable of being revived by what actually happened in the wake of the outbreak of international terrorism in September 2001.

Reference: Meadows, D. H., et al. *The Limits to Growth*. New York: Universe Books, 1972.

I. I. Kravchenko

☙ M ❧

Macroshift is a civilizational transformation where technology is the main moving force. According to Ervin Laslo, who coined the term, this transformation occurs in an era of overwhelming, quick, and irreversible shifts reaching the most remote corners of the Earth and touching all aspects of human life. It occurs when a critical mass of people understands the need to renovate the universal value system. Phases include:

(1) Emergence (1860–1960): "the phase of launching bifurcation" when innovations in "hard" technologies make possible more effective manufacturing and use of resources.

(2) Globalization (1960–2001): transformation nearly reaches its culmination, accompanied by increasing complexity of social structures and growing pressure on the natural environment.

(3) The decisive phase (seen in the twenty-first century): present critical (or chaotic) phase. New conditions question established values, worldviews, ethics, aspirations, and societies enter a period of fluctuation and instability.

(4) The future: When we will see either collapse, a stage of destruction if we do not change, or breakthrough, when a universal system emerges with the establishment of a new culture promoting values and behavioral schemes based on new morality and new consciousness.

A. N. Chumakov

Malthus, Thomas Robert (1766–1834): English economist, cleric, population theorist, after whom Malthusianism came to be named based on ideas put forth in *An Essay on the Principle of Population, as It Affects the Future Improvement of Society, with Remarks on the Speculations of Mr. Godwin, M. Condorcet and Other Writer*s (1798), wherein he put forth his theory about the relationship between population and availability of resources. He posited that population growth was increasing beyond the ability of the land to support it. He disapproved of mercy for the poor because it leads to exacerbation of the overpopulation problem. He mentioned that poverty only slightly depended on political rule or unequal property distribution because the rich are unable to provide the poor with labor and food. Thus, the poor have no right to demand labor and food from the rich.

Malthus's assumptions that hunger, epidemics, and wars are natural regulators of human "overproduction" engendered enthusiastic supporters (e.g., John Stuart Mill, Charles Darwin, Alfred Wallace, and John Maynard Keynes) as well as active opponents. In Marxist circles, his ideas were regarded as misanthropic, anti-scientific, and reactionary; he was labeled an epigone, a "cassock," a reptile servant of the reactionary bourgeoisie, and a plagiarist. Lenin wrote, "We are unequivocal enemies of neo-Malthusianism. Class-conscious workers will always conduct merciless struggle against any attempts to impose this reactionary and cowardly teaching on the strongest . . . class of modern society," and, "the most open declaration of war of the bourgeoisie against proletariat" (1965).

The achievements of the scientific revolution from the 1930s through the Cold War diminished interest in Malthus's ideas. They enjoyed renewed

Macroshift – Myth   323

interest after publication of *The Limits to Growth* (1972), which was congruent with what he had written150 years earlier.

References: Lenin, V. I. "The Working Class and NeoMalthusianism." In *On the Emancipation of Women*. Translated by K. Zetkin. Moscow: Progress Publishers, 1965. Meadows, D. H., et al. *The Limits to Growth*. New York: Universe Books, 1972.Malthus, T. *The Works of Thomas Robert Malthus*. Edited by E. A. Wrigley and D. Souden. London, 1986.

A. N. Chumakov

Management, Holistic: Holistic management concerns communities, cooperative corporations, and other organizations that follow the example of living organisms in coordinating interdependent groups and individuals without competition for dominance of one component over others. These "holistic" or "organic" structures usually provide for separating decision-making groups into more specialized and smaller units when they become too large or complex for intensive deliberation and adequate opportunity for all stakeholders—management, employees, clients, and watchdog or public interest groups—or their democratically chosen representatives to take a part in decision making. This may lead to networks of consultation with liaison by participants taking part in more than one constituent group, linked thus with related activities or with other levels of administration, including in the most complex cases networks of networks, or "nested" networks.

D. N. Everingham

Management, Strategic: Strategic Management is a method by which decision-makers structure their ideas on the condition of the managed system with its variability limitations and on practical approaches to developing and implementing a strategy. This method is a matrix or complex of variables. The deeper and more thoroughly these variables cover the subject, the more practical they are as navigation devices in the ocean of business and other forms of activity.

The process of strategic management can be described through concepts that include "system conditions" and "transition paths." System conditions reflect its natural and acquired properties, changed in the course of its history by multiple—successful or not—attempts to adapt to the requirements of the environment and to its own goals. The paths of systemic transitions are regulated by external and internal restraint factors (e.g., friction hampering the change of position). External factors descend from the environment of an organization (its political, economic, social, and technological variables); the internal ones reflect the tangible and intangible properties of the organization (production capacities, resource potential, product palette, trademark, behavior norms, and style, methods). The restraints are stable, but can be consciously controlled.

We can distinguish three paths: (1) those within the bounds of the current systemic condition (such as regulation, operative and tactic management, and functioning maintenance); (2) paths describing any transitions feasible within the given systemic condition (such as manufacturing of new products, entering new markets, elaboration of a new organizational architecture by means of confluence and absorption, and implementation of technological

324 *Global Studies Encyclopedic Dictionary*

innovations); and (3) paths representing attempts to reach objectively unattainable conditions.

The difference between strategic management and the other managerial activities lays in its orientation toward successful transition of the managed system to the new condition, that is, the second-type paths.

However, the third-type paths may also be successful in situations of bifurcation, when unthinkable results become possible. Nontrivial solutions, transforming a "miracle" and the logic of the impossible as a factor of strategy are of primary importance here.

Choice of path is critical to strategic management. It is difficult due to permanent changes inside and outside organizations; informational lack or overload; the multitude of purposes and interest groups and their mutual expectations and reciprocal reactions; cognitive, psychological, and social limitations of a manager's behavior, and the degree of risk acceptance. As a result, successful strategy is determined not only by scientific substantiation of a series of strategic solutions, but also by the art of management, intuition, and the ability to pick up hardly noticeable signals of reality and to coordinate the participants' motivations with the multitude of other circumstances, often external ones. The most promising direction in strategic management theory is related to global problems and the assessment of their possible solutions.

The increasing popularity of the term "strategic management" demonstrates the trend to more complex evaluation of the managerial goals, methods, results, and environment. In business practice strategic management relates mostly to the highest management level. In educational practice, strategic management claims to play the role of a meta-theory synthesizing such special disciplines as marketing, information systems, operational management, finance, and personnel management.

One of the newest features in the theoretical evolution of strategic management is the massive borrowing of ideas from methodological recommendations for the military and special services from 1950 to 1970. In the 1990s, many creative ideas were imported into strategic management theory from the natural sciences, for instance, from theories such as non-equilibrium thermodynamics, spins, catastrophes, games, dissipate structures, reflexivity, or neurolinguistic programming. Operational use of the mentioned models in the managerial context has significantly enriched the palette of management concepts and crystallized specific strategic and private managerial disciplines that had been separated from neoclassical, institutional, evolutionary, and entrepreneurial theories of corporation. Reconsidering the role and function of enterprise, offering various efficiency criteria, and setting up priorities for companies, these theories still serve as a source of more practically oriented managerial models.

Diverse interpretations of strategic management highlight its separate aspects: either the process of aims identification (vision, planning, guidelines, direction, and intentions) or the resources, including managerial principles and tools and efficiency evaluation.

Growing attention to strategic management theory evinces a serious gap between practical needs and the available management models as well as a transition state of the global economy and policy in general. It leads to the

Macroshift – Myth 325

devaluation of applied models and prognostic failures and to intensive search for new conceptual solutions. New strategic management models should take into account the following: (1) the logic of the behavior of economic actors cannot be reduced to the standard microeconomic clichés of rationality and presumes the handling of a wide range of permitted and non-permitted (legally and morally) financial, economic, military, political, diplomatic, informational, and reflective methods; (2) the practical foundations and patterns of competition evolve toward market-undermining methods (hypercompetition); and (3) managerial opportunities and limitations are influenced by globalization, while the actors of globalization (governmental, interstate, and international companies and banks) influence the world economy, international relations, and national sovereignties. As a result, the theory of strategic management shifts more and more from uni- and bi-dimensional to multi-dimensional, dynamic models.

A. I. Ageev

Mandela, Nelson: (1918–2013) (aka Madiba) Nelson Mendela was a South African nationalist, anti-apartheid revolutionary, and democratic socialist.

The 1948 South African National Party victory was followed by more stringent apartheid laws and stricter application of existing discriminatory laws. In 1951, the Separate Representation of Voters Bill removed persons of color from the common roll. This process sought to deny political representation and participation of blacks at all levels of government. This provided the context for the mounting tide of popular democratic resistance to the apartheid state, and during this period Mandela rose to prominence in the 1952 Defiance Campaign against Unjust Laws, which organized mass action tactics of boycotts, strikes, and civil disobedience (Lodge, 1983, p. 39). This was the largest scale nonviolent resistance ever seen in South Africa and the first campaign pursued jointly by all racial groups under the leadership of the African National Congress (ANC) and the South African Indian Congress (SAIC).

Although the campaign did not achieve the desired aim of overturning the apartheid laws, the resistance won UN recognition that the South African racial policy was an international issue; a UN Commission was established to investigate the situation. These years were crucial as the Defiance Campaign marked the ANC's move from moderation to militant non-racial cooperation in the resistance to apartheid.

Influenced by Marxism, Mandela attempted to equalize wealth distribution through his support for the 1955 Freedom Charter, which called for the nationalization of banks, gold mines, and land (it ultimately failed).

Working as a lawyer, Mandela was repeatedly arrested for seditious activities, and was tried for treason in 1956 (found not guilty). Historian Stephen Ellis found evidence that Mandela had been an active member of the South African Communist Party (SACP) despite his public denials (2011, pp. 667–668). Although initially committed to nonviolence, after the Sharpeville massacre in 1960 ("The Sharpeville Massacre," *Time Magazine*, 4 April 1960), in association with the SACP, Mandela co-founded the militant Umkhonto we Sizwe

326 *Global Studies Encyclopedic Dictionary*

("Spear of the Nation") in 1961, which waged a campaign of sabotage against the apartheid government ("Manifesto of the Umkhonto we Sizwe," 1961).

After a guerrilla attack against government installations in June 1961, South Africa and the United States classified Umkhonto we Sizwe as a terrorist organization, and banned it. Thereafter, Mandela was prosecuted for conspiracy to overthrow the state in the Rivonia Trial. In his speech from the dock at the beginning of that trial (1964), Mandela, while admitting to violent actions, denied that his actions constituted terrorism. In that, he can be characterized as a just warist, since he rationalized the violence as a necessary last resort.

After being convicted and sentenced to life in prison, Mandela wrote, "I was the symbol of justice in the court of the oppressor, the representative of the great ideals of freedom, fairness, and democracy in a society that dishonored those virtues. I realized then and there that I could carry on the fight even within the fortress of the enemy" (1994).

In 1980, at the urging of several top ANC officials, a Johannesburg newspaper launched a campaign to free Mandela by printing a petition that readers could sign to demand that he and other political prisoners be released. Although it met with strong resistance from the government, the campaign further established him as the embodiment of black South Africans' fight for freedom (Bio Ref Bank, 2009).

After mounting international attention to apartheid, Mandela was released from prison in 1990, during a time of escalating civil strife. He joined negotiations with President F. W. de Klerk to abolish apartheid and establish multiracial elections in 1994, in which he led the ANC to victory and became South Africa's first black president. He served as President of the ANC 1991–1997; internationally, he was Secretary General of the Non-Aligned Movement 1998–1999.

During Mandela's tenure in the Government of National Unity, he invited several other political parties to join the cabinet. As agreed during the negotiations to end apartheid, he promulgated a new constitution. He also created the Truth and Reconciliation Commission to investigate past human rights abuses. While continuing the former government's liberal economic policy, his administration also introduced measures to encourage land reform, combat poverty, and expand healthcare services.

Internationally, Mandela acted as mediator between Libya and the United Kingdom in the Pan Am Flight 103 bombing trial, and oversaw military intervention in Lesotho. After his term as President, he became an elder statesperson, focusing on charitable work in combating poverty and HIV/AIDS through the Nelson Mandela Foundation.

Along with the ANC, the Democratic National Front, a nonviolent political association, also played a significant role in ending apartheid in South Africa. Of even more decisive importance than the militant activities of the ANC in the fall of apartheid in South Africa, however, was international pressure in the form of financial sanctions. If not for the efficacy of movements such as the disinvestment campaign, which began in the 1960s and continued through the 1980s, South Africa arguably might have continued its program

Macroshift – Myth 327

of silencing anti-apartheid revolutionaries, with international support, since the ANC was often characterized as a terrorist organization.

References: Manifesto of Umkhonto we Sizwe. African National Congress. 16 December 1961. Accessed 12/23/2013. http://www.anc.org.za/show.php?id=77. Mandela, N. "Statement from the Dock, 20 April 20 1964." Accessed 22 December 2013. http://www.theguardian.com/world/2007/apr/23/nelsonmandela. Lodge, T. *Black Politics in South Africa since 1945.* New York: Longman, 1983. Mandela, N. *Long Walk to Freedom.* London: Little, Brown,1994. Biography Reference Bank. "Nelson Mandela (1918–2013)." H. W. Wilson, 2009, accession # 203044194.

E. D. Boepple

Mankind at the Turning Point (1974) was the second report to the Club of Rome made under the direction of Mihajlo D Mesarovic and Eduard Pestel on the basic purposes and a method of research. It was an apparent continuation of the first report, but with many critical remarks and authors' assumptions concerning the complexity of the global system. Considering differences among world regions, the authors instead of using the widely criticized "concept of zero growth" offered the concept of "organic growth," according to which each region of the world should carry out a special function, similarly to a cell of a living organism. Authors suggested that the world should be divided into different regions, believing that thus humankind can solve many problems, carrying out balanced development of the highly developed and developing states and territories. Their research described not only one system as in the first report but also ten subsystems-regions of the world—and more information and more complex techniques were used. In particular, the Meadows computer model had about a thousand mathematical equations, and the Mesarovic-Pestel model contained more than two thousand. Authors of the second report aimed not only at a more adequate description of the global situation but also at attempt to offer effective tools that can be used while making concrete decisions. The concept of "organic growth" turned out to be attractive enough and, despite criticism, received a wide resonance, and the report itself, and the previous one, was a powerful spur to research in the field of the global modeling developed in the 1970s.

Reference: Mesarovic, M. D., and E. Pestel. *Mankind at the Turning Point.* NewYork: Dutton, 1974.

A. N. Chumakov

Marxism and Global Values: The dissolution of the Soviet Union initiated important questions concerning the nature and future of Marxism. This essay examines the future of Marxism in relation to global values, specifically in relation to "Western" Marxism (non-Soviet or non-Orthodox Marxism).

Traditionally, Marxism is closely associated with liberation movements for women, racial and ethnic minorities, and developing countries. With the disintegration of the Soviet Union, the relevance of Marxism and socialism to these and other political movements is receiving increased attention. Marxism has also influenced capitalism, from some state involvement in the economy (rejection of laissez-faire capitalism) to some involvement of workers in the economic enterprises in which they are employed. Within philosophy, debates still occur over whether to distinguish an early and a late Marx and over the

328   *Global Studies Encyclopedic Dictionary*

proper relation of Marxism to phenomenology, existentialism, and structuralism, but they are abetting.

In the West, among contemporary advocates of a new world order, the World Policy Institute (formerly Institute for World Order) in particular seeks to advance global values rather than US economic and political hegemony. Robert Johansen's *The National Interest and the Human Interest* (1980)is especially relevant for illustrating global values. The four key world order or global values are: (1) peace without national military arsenals, (2) economic well being for all inhabitants on the earth, (3) universal human rights and social justice, and (4) ecological balance. Some thought the end of the bipolar global system would foster moving from East-West conflict to North-South cooperation. Instead, the post-Cold War era ushered in a unipolar world of US economic, political, and military hegemony.

Given the present geopolitical reality, global studies needs to address ways in which philosophical Marxism, in contrast to orthodox or state Marxism, may prove relevant. For advancing global values neither current state capitalism nor prior state socialism is adequate. State capitalism has focused on the production of great wealth and has supported, to a lesser extent, the advancement of human rights. However, state capitalism is also associated with frequent wars and significant levels of environmental damage. By contrast, the state socialism of the former Soviet Union focused on economic growth, sometimes at the cost of environmental damage and low priority to human rights. In the West, the impact of government has been primarily internal in relation to regulation, and the state has functioned poorly in controlling external markets, with the important exception of those instances, mostly but not entirely in the past, of colonial aggression.

The importance of the re-appropriation of Marxism is significant. This version of Marxism can detail and critique: (1) the degree of devastation that state capitalism has wrought on the environment (at least in the developing world, where the United States, in particular, has aggravated these problems); (2) the implications of stressing economic growth over human growth (the dysfunctional consequences of pursuing ever more profit); (3) the ways in which the forms of bureaucracy in the United States continue to interfere with advancing social justice; (4) the manner in which the current unipolar system of United States hegemony makes the unchecked use of military force more likely. Philosophical Marxism can expose how unipolar globalization is undercutting pursuit of humanitarian global values.

On a constructive level, philosophical Marxism can foster global values that facilitate pursuing policymaking from a global, rather than a narrowly national, perspective. *The German Ideology* and *Capital* show how Marx's concept of history pretty closely articulates most of the needed global values.

In the first place, from a global perspective, an orientation to the human race should include horizontal (trans-national) and vertical (trans-class) considerations. In his *Economic and Philosophical Manuscripts*, Marx's concept of *Gattungswesen* (species-being) is quite compatible with such an inclusive concept of the human race. Marx's concept is both descriptive and prescriptive. Marx uses the potentiality of any member of the species as a critical index against which to measure the actuality of any member or class of the spe-

Macroshift – Myth   329

cies. This concern continues throughout Marx's writings. In *Capital*, for example, he notes the human interest in the labor-time (and its distribution) to attain subsistence. Marx also states the obvious fundamental importance of "species preservation" in his first premise of history. In *The German Ideology*, he states that persons "must be in a position to live in order to be able to 'make' history." As Marx states in *Capital* ([1867] 1978), conditions of necessity (*Reich der Notwendigkeit*) must be met before the possibility of freedom (*Reich der Freiheit*) can begin.

At economic and political levels, the focus should be on service to "human needs." Marx noted in his second premise of history the centrality of satisfying human needs. Throughout his writings Marx forcibly argues that socioeconomic development should bring satisfaction to all people and not depravation to any specific class. Advocates of global values can learn from Marxism that consideration of human needs should be disengaged from elitist theories and, instead, should affirm a "politics of liberation" at national and individual levels.

Finally, the planetary eco-system as a whole needs attention. Marx was aware, even in the nineteenth century, of the global ecological effects of human productive activity. Criticizing the Romantic dichotomy between persons and nature, Marx observes in *The German Ideology* ([1932] 1978) that pure nature no longer exists "except perhaps on a few Australian coral islands of recent origin." This global, ecological concern should serve to check socioeconomic development and to insure planetary habitability.

To be avoided, however, is a trade-off in which political exploitation is exchanged for economic exploitation. Advocates of global values can profit from a return to philosophical Marxism for a critique of exploitation. Marxism can also begin to champion global values of ecological balance, economic justice, human rights, and world peace.

References: Johansen, Robert C. *The National Interest and the Human Interest.* Princeton, 1980. *Capital.* In *The Marx-Engels Reader.* Translated by R. Tucker. New York, 1978. *The German Ideology*, also in *The Marx-Engels Reader.*

W. C. Gay

Maximum Concentration Limit (MCL) of contaminant is the maximum concentration of contaminant in landscape components that, in their daily influence over a long period of time, do not provoke negative consequences. Three general types are distinguished: MCL in the atmosphere (averaged over a specific time), in water, or in soil, that does not harmfully influence people or the environment (including hygienic conditions of water usage) over the course of their periodic influence or their lifetime influence (including remote effects). Usually MCL lists also specify a class of danger of contaminants and a limiting parameter of danger (for the first case, it is also necessary to specify the time interval to which the standard is referred).

In addition, there are several other classes of MCL:

(1) Maximum one-time MCL in the air of populated areas: concentration of contaminants in the air that should not invoke any reflex reactions in human beings after twenty minutes of aspiration.

330 *Global Studies Encyclopedic Dictionary*

(2) Daily average MCL in the air of populated areas: concentration of contaminants in the air that should not harm a person directly or indirectly within unlimited time (years) of aspiration.

(3) MCL in the air of a work area: concentration of contaminants in work zone air that—referring to daily (except week-ends) work for eight hours or other work schedules, but not more than forty-one hours per week during the whole length of service—should not cause any disease or deflection in a person's health that can be revealed by modern research methods during the work process or affect life of the present and succeeding generations (the work area is considered as the space of up to two meters above the floor or a platform where workers stay constantly or permanently).

(4) MCL in reservoirs for household: potable water for public use, including factors that influence hygienic conditions of water.

(5) MCL in reservoirs of fish: farming water concentration of contaminants in water the constant influence of which does not cause fish extinction or extinction of organisms that serve as fish forage; gradual disappearance of those or other kinds of fish to which the reservoir used to be applicable, or the substitute of organisms, valuable as forage, by ones less valuable or unusable as forage; damage to the commercial quality of the fish dwelling in a reservoir, for example, the appearance of off-flavors and odors; changes that during some seasons or in the nearest future can lead to fish extinction, substitution of valuable kinds of fish by less valuable, or to a total loss of the fish-farming value of a reservoir or its part.

(6) MCL (Allowable Residual Amount, ARA; or Maximum Residual Amount Limit, MRAL) in food stuffs: maximum concentration of contaminants in food stuffs that, over an unlimited time (and in daily influence), does not cause any diseases or deflections in persons' state of health.

(7) MCL in the arable layer of soil: concentration of contaminants in the upper, arable layer of soil which should not influence negatively in a direct or indirect way either a person's health or the self-cleaning ability of the soil.

V. V. Snakin

Media and Global Studies: is the detailed study of the way in which different forms of media impact world affairs and the role the media plays in covering global events and emergencies. Key issues include media neutrality, crisis reporting, and security. Issues include neutrality and objectivity of reporting, questions of ownership, political orientation, state versus private control, and the power of advertising. Other issues include the accuracy and veracity of media reports and the potential dangers of image, spin, and short-term advantage versus ethical reporting. Similarly, gender and media bias and coverage of religious affairs are issues of concern. Additionally, global studies is concerned with the extent to which media support or fail to support the work of the United Nations and its agencies. Issues of security include the insecurity of journalists from danger and violence against media outlets. Finally, global studies analyses the morality of media, its long-term vision in politics and governance, and the efforts of international nongovernmental organizations working on different aspects of the global problematic.

T. C. Daffern

Macroshift – Myth   331

Media Participation (Participatory Communication): Mass communication can be understood not as a series of practices to be restrictively controlled by media professionals, but as a human right to people to participate in media that cuts across entire societies. Originally proposed in 1969, by the French civil servant Jean d'Arcy, who aimed to broaden the right to be informed, this principle is embedded in article nineteen of the Universal Declaration of Human Rights. Although the definition of the right to communicate was highly debated, Jim Richstad and Michael Anderson wrote that the right to communicate included active participation in the communication process (1981).

Jan Servaes posits that the field of participatory communication is characterised by two interrelated points of view: Paulo Freire's dialogical pedagogy and the UNESCO debates about access, participation, and self-management (1970s). Participation is situated in a context of the reduction of power imbalances, both on the broad social, political and economic level (the relations between oppressors and repressed) and on the level of the educational system, where students and teachers strive for knowledge in a non-authoritative collaboration that fosters partnership.

The second point of view is situated within the context of the UNESCO debates about a New World Information and Communication Order (NWICO) and a New International Economic Order (NIEO). These debates (see Belgrade 1977 debates in Servaes, 1999) also tried to define the concepts of access, participation, and self-management. The resulting definition of access stresses the available opportunities to choose varied and relevant programs and to have a means of feedback to transmit the public's reactions and demands to the media organizations. Media participation and self-management are seen as processes that imply a higher levels of public involvement, ranging from mere representation and consultation of the public in decision-making processes to more advanced forms of participation such as decision-making within communication enterprises and where the public is fully involved in the formulation of communication policies and plans.

Within communication studies, attempts have been made to introduce the notion of interaction as an intermediary layer between access and participation, to limit the wide scope of meanings attributed to these concepts. From a policy studies perspective, complex typologies have been developed to tackle all variations in meaning. An important and early example is Sherry Arnstein's ladder of citizen participation (1969). More recent is the OECD's three-stage model, which distinguishes information dissemination and consultation from active participation. In this model, consultation implies a limited two-way relationship, where governments consider citizens' feedback on policy making. Although downplaying the more radical meanings of participation, the OECD-model emphasizes the more advanced nature of the two-way relationship between citizens and government by defining participation as a situation where citizens are actively engaged in decision-making and policy-making.

Participation is often categorized by the degree to which power is equally distributed among the participants. Pateman (1972) for instance, introduced a difference between full and partial participation, partial being a process where the parties involved can all exert an influence on the outcome, but the

332 *Global Studies Encyclopedic Dictionary*

ultimate decision-making rests with one party. In the case of full participation, all participants hold equal powers to determine the outcome. Other terms have been used to construct a hierarchically ordered system within the definitions of participation on the basis of a real-unreal dichotomy. In the field of political participation, for example, Sidney Verba (1961) used the concept of "pseudo-participation" to construct this difference. In discussions on organizational participation, George Strauss (1998) refers to the notion of "manipulative participation," in order to create a similar distinction.

Shirley A. White (1994) writes that there can only be genuine participation when the people are empowered to control the action to be taken. She considers power and control to be pivotal concepts that contribute to understanding the diversity of expectations and anticipated outcomes of participation. Jan Servaes (1999) mentions that this "real" participation has to be seen as participation that directly addresses power and its distribution in society. For him too, participation touches the very core of power relationships.

References: Arnstein, S. R. "A Ladder of Citizen Participation," *JAIP*, 35:4 (July 1969), 216–224. Verba, S. *Small Groups and Political Behaviour*. Princeton: Princeton University Press, 1961. Pateman, C. *Participation and Democratic Theory*. Cambridge, UK: University Press, 1972. Richstad, J., and M. Anderson *Crisis in International News*. New York: Columbia University Press, 1981. White, S. *Participatory Communication*. Beverly Hills: Sage, 1994. Strauss, G. "An Overview." In *Organizational Participation*. Edited by F. Heller, E. Pusic, G. Strauss, 8–39. New York: Oxford University Press, 1998. Servaes, J. *Communication for Development*. Cresskill, NJ: Hampton Press,1999. Freire, P. *Pedagogy of the Oppressed*. New York: Continuum, 2000.

N. Carpentier

Mergers and Acquisitions (M&A): Financial transactions aimed at the integration of companies or banks into one economic entity to obtain a competitive advantage and maximize financial value, achieved by voluntary agreement among participants. They are asymmetric. Thus, asset and transaction control is completely carried out by one of the parties to the merger. Types are horizontal integration, which joins competitors or manufacturers of basic goods (product extension) or companies working on different markets (market extension). and vertical integration, which involves joining up with consumers (forward integration) or suppliers (backward integration). Conglomeration is the integration of companies not related by their kind of activity for the sake of diversification to reduce risk. In all countries, governments regulate M&A. As M&As increase monopolization within a market, antimonopoly organs control many such transactions, such as the Celler/Kefauver Act in the USA and the Monopolies and Mergers Acts and Competitions Acts in Great Britain, which prohibit M&As when a company owns a liberal share of the market.

The last M&A wave started at the end of the 1990s, and was related to increased trends toward globalization, the growth of international competition, and the transition of developed economies to innovation development based on the constant updating of production and management methods. Most M&As occur in the sphere of telecommunications, financial services, the aerospace industry, oil and gas, and the pharmaceutical industries. The common

M&A market (market for corporate control) was valued at approximately $3 trillion at the beginning of the twenty-first century.

R. A. Kurbanov

Microelectronics and Society: For Better or for Worse (1982) was the twelfth report to the Club of Rome, which represented the conclusion of active academic investigations over ten years under the Club of Rome patronage. It revealed a new way to the solution of global problems: the way of fundamental reforms of the economic and social realm and of the whole way of an individual life and living conditions. The report mentioned how the research cohered with the traditional range of problems of the Club of Rome and it directed attention to new technologies based on microelectronics, concluding, "microelectronics will be global, and hardly any kind of society can escape it." By means of miniaturization, automation, and computerization, microelectronics will totally transfigure our world, and allow solving the apparently insuperable problems of a global character. The world's menaces include population explosion, violation growth, and individuals' alienation from society. The main question is whether it allows this new power to aggravate the situation create new forms of the society that are more egalitarian and which would provide all people with a worthy life and moderate prosperity. The report is optimistic, but admonishes that such transformations will require sacrifices for the privileged minority.

Reference: Friedrichs, G., and A. Schaff, eds. *Microelectronics and Society, for Better or for Worse*. New York: Pergamon, 1982.

A. N. Chumakov

Migration (from Latin *migratio*, "resettlement," "moving") refers to people (migrants) moving across boundaries of any territories to temporarily or permanently change the place of constant residence.

Internal migration does not influence the aggregate population of the country, whereas international migration results in changes of population of countries and is frequently attended by changing of citizenship. or bet According to temporal criteria, there can be specified nonreturnable (internal resettlements, emigration, immigration) or returnable migration (includes long-term, seasonal, pendular, occasional migration).

Forced Migration is migration compelled by conditions beyond immigrants' control (political and national persecutions, natural disasters, technogenic disasters, ecological catastrophes, and military operations). Those affected include refugees, displaced persons (forced settlements), and persons seeking asylum.

The term "refugee" refers to persons forced to leave their place of constant residence as a result of various emergency situations and to migrate to another state; "displaced person" is applied to those forced to migrate within the frontiers of one state. The term "ecological refugee" appeared in the beginning of 1990s, after the report of the United Nation's Environmental Program and refers to persons forced to leave their habitual residence because of noticeable deterioration of the environment. In the beginning of the twenty-first century, the number of ecological refugees in the world had already ex-

334   *Global Studies Encyclopedic Dictionary*

ceeded twenty-five million. Under current estimates, this number will increase to more than 150 million by 2050.

Globalization of the processes of migration is a phenomenon closely related to the globalization of the world economy. This refers to the increase of international migration, the increasing number of countries participating in world migration, and the shift in the qualitative structure of migration flows according to the demands of the globalizing labor market.

Without globalization of population, the flows of the modern economic system characterized by openness and mutual supplementation would not exist. On the other hand, the globalization of the world economic system has stimulated migration of highly skilled professionals representing a human element of worldwide activity of the multinationals. Globalization of industry and the spread of information and computer technologies result in unification of qualifications, skills, and standards of management. For many occupations, nationality is no longer of its previous importance and national boundaries are not a barrier to employment. As a result, recipient countries encourage the inflow of professional migrants: senior managers and executives, engineers and technicians, scientists, entrepreneurs, and students, and restrict the inflow of non- or low-skilled migrants.

Nearly all nations of the world are involved in global migration. They are traditionally divided into immigration countries or recipient countries (mainly industrially advanced states) and emigration countries or sending countries (mainly developing and transition states that supply world labor market with a cheap non-skilled or a highly skilled labor force). At the same time, the complex structure of migration flows tends to erode the distinction between receiving countries and sending countries.

Another significant feature of globalization in the field of migration is the impossibility of solving many problems related to international migration on the national level. Currently, the problems concerning the growing number of refugees or the illegal migration boom have become global issues not only due to their scale but also because of the fact that their management and regulation can be effective only by means of joint efforts of all countries involved. This is also true for temporary labor migration that is now becoming the major form of international migration.

The salient feature of international migration is the emergence of so-called global cities (such as New York, London, or Paris) acting as the centers of global migration systems. The economies and labor markets of these cities are strongly dependent on migrant workers, and migrant communities provide the necessary numbers of migrants despite official restrictions.

International Migration is the territorial (spatial) moving of people through state boundaries, connected with changing of place of constant residence and citizenship conditioned by various factors (family, national, political and others) or with stay in the country of entrance having long-term (more than one year) seasonal or pendulous nature and with circular (occasional) trips for job, rest, or treatment.

We can enumerate the following basic kinds of international migration: (1) irrevocable; (2) enforced; (3) pendulous (daily or weekly moving from one country to another caused usually by labor or study activity); (4) illegal;

Macroshift – Myth 335

(5) seasonal; (6) labor (moving of able-bodied population having usually returnable nature after completion of labor activity in the country of entrance).

International Migration of Labor is a complex of territorial intergovernmental movements of population that seek to establish legal working activity in the country of entrance during a definite time, after which the migrant returns to the country of departure. Illegal immigration, on the other hand, is interpreted as an entrance of citizens of one country into another for indefinite time, usually with the purpose of illegal employment.

Among centers that attract labor internationally, the largest are North America, Australia and New Zealand, countries of the European Union (Great Britain, Germany, France), oil-exporting countries of the Middle East (Kuwait, United Arabian Emirates, and Saudi Arabia), and the more developed countries of the Asian-Pacific region (Brunei, Malaysia, Singapore, Siangan, Taiwan, South Korea, Japan) and, more recently, Russia.

I. A. Aleshkovsky, V. A. Iontsev, I. Ivakhnyuk

Modernity, Global Challenges of: This term refers to a set of ecological, demographic, economic, technological, cultural, and other critical problems of world civilization at the beginning of the twenty-first century.

History has destroyed the illusion that after the defeat of the communist system at the end of 1980s, the evolution of world civilization had come to its final stage. The Western world began to neglect international institutions and norms of the former bipolar world. This situation led to a new approach to international relations, "global egoism"—the presumed right to interfere into any event in any part of the world based on "humanitarian" considerations and adherence to progressive values.

The new egotistic development strategy aims to "privatize the future." The modern Western world has started a new stage of its development that is frequently called post-industrial. This term refers, first, to the economic aspect of modern civilization that is connected with the development of a new global economy based on high technologies. Some global monetary economy states and private groups extract super-profits out of financial speculations and debts. At the same time, many Western leaders are interested in preserving the under-development of the rest of the world. One key result is growing polarization, inequality, and dependence of the rest of the world on Western knowledge.

A new idea has emerged, namely, the self-restriction of the development of civilization as a means to survive and perhaps even advance global social justice. When the West claims that its intellectual and technological leadership ensures the progress of humankind, opponents object, saying that to survive the world, we should "return to nature" and "fear God." Thus, the fundamentalism of "the chosen Western people" confronts the fundamentalism of its opponents. This "clash of civilizations," however, could lead to mutual destruction.

At present, the United States and the West as a whole control the monetary, technological, and post-industrial sectors of the global economy, whereas the Asian Pacific region is gradually turning into a world factory, the new industrial center. The world of raw resources is beneath, and Russia, unfortunately, is located quite close to it. Nevertheless, in the next fifty years, the West will become the "world minority" despite its political and economic

336 *Global Studies Encyclopedic Dictionary*

expansion and human history will cease to be a history of the Western civilization. The real New World will be something about which we know little.

These problems should make people think about envisioning some project of the future for the population of the whole planet, not just for "the golden billion." But awareness of human unity engenders many new problems. To some extent, the political history of humankind was always a consecutive transition to greater integration and consent, democracy and cooperation, smoothing and partial overcoming of social contradictions. Accordingly, in the course of transition to another-planetary-level of perception and comprehension of social unity, the world may well face new social splits, the formation of a world elite and a world proletariat, and preconditions for a world civil war that could use means as diverse as cyber war and terror. At the same time, setting limits to the development of the West may provoke internal class conflicts that had been settled earlier. The West may have to choose between domestic and global tensions.

The modern world is also threatened by the so-called revenge of values. World civilization is on a threshold of a new global revolution of values, a new "revolution of consciousness" that could let humankind meet the challenges of the present and overcome the limited opportunities of extensive development. To survive, humankind needs first to find the foundations for moral solidarity and principles and values of global social consensus beyond national, confessional, or civilizational borders. Such moral and ethical progress needs to be placed above scientific and technical progress. Still, in relation to our global future, we can choose. This is undoubtedly humanity's main hope.

Y. M. Luzhkov

Modernity, Industrial: A set of relatively stable, integrated sociological/cultural and technological/economic features of modern industrial capitalist societies. Any social organism is a combination of multiple aspects of traditional, industrial, and post-industrial societies forming a relatively stable integrity. However, the nature of their interactions in any social organism is just as unique as fingerprints. If we compare, for instance, Swedish and Spanish "capitalism" we can find more differences than similarities by many civilizational parameters even though these differences are not so significant from the standpoint of rigid economic centrism. In the capitalism of the Atlantic there are relatively few so-called remnants of the past, civilizational traits peculiar to a traditional society. In the countries "undertaking modernization" where transition to the path of industrial development is far from being complete, the representation of traits of traditional societies in the general civilizational fabric is much more prominent. The practice of the modernization provides plenty of evidence that efforts to forcibly eradicate these values in a bid "for whipping up the horse of history" can provoke a powerful potential for a "back to the past" movement-neo-archaization. The experience of social modernization in Asia and the Pacific region testifies to the fact that traditional values deeply ingrained in the public consciousness (group loyalty, labor ethics, ability for selfless behavior, and self-sacrifice) are frequently found to be well-suited for tackling modernization problems per se.

Macroshift – Myth 337

Industrial modernity is the product of the transformation of traditional culture, the Reformation, and further dissemination of Protestant ethics in Europe with its peculiar cult of productive labor. In industrial society, direct personified social relationships and relations of personal dependence are replaced by those mediated with things (commodity), anonymous, social relations with personally unfamiliar characters (actors) of the social/economic process. Emancipation as liberation from the fetters of personal dependency is a prerequisite for the formation of the value of personal freedom. Non-economic coercion peculiar to traditional society is replaced with economic dependency in consequence of private property for the means of production.

Entry to industrial contemporaneity brought about hitherto unprecedented improvements in the quality of life. The powerful development of commodity production led not only to the saturation of the market with essential products, but also formed new demands unknown to traditional society. The quality of medical service improved markedly and life expectancy grew sharply. But there was a price to pay for progress. From a cultural and anthropological point of view "modern" society can be characterized by depersonalization of social relationships, perception of the Other as a set of social functions, identification of person with his or her social role. The anonymity of social relationships is fraught with human's alienation. A customer of a modern medical institution will sometimes recall with nostalgia Chekhov's description of the warm and caring hands of a provincial physician. For modern doctors as operators of modern diagnostic equipment, a customer, though a unique and original personality, is just playing a social role—that of a patient. Alienation as a cultural product of "modern" society is an inevitable consequence of direct personal social relationships being replaced with indirect functionally depersonalized ones. Social communications, the invisible threads of a social fabric, in the industrial society assume the form of commodity exchange (products, labor, services). For the mass consciousness social dependence appears in a transformed form as the power of money and things. Commodity-money fetishism becomes the most evident manifestation of human's alienation. Notions about "money ruling the world" become the faith symbol of the modernist consciousness. Thus, the "religious disenchantment of the world" (Weber)—secularization—comes back as a new "enchantment" of the industrial society with things and symbols as mediators of social communications.

Compared with the rate of development of traditional society, "modern" society is characterized by the acceleration of social and economic development. The "time-arrow" becomes a temporal metaphor of industrial modernity symbolizing the irreversibility of historical time, unlike the traditional metaphor of a "time-circle" ("wheel of time"). As distinct from the syncretic social space of traditional society, the cultural product of "modern" society is the division of private and public spheres: urban space is expressly divided into living, production and social and cultural quarters and areas, that is, the concept of the value of the private-cultural product of "modern" society.

In contrast to traditional values, things that are artificial, human-made, are, in the hierarchy of values of the industrial society, incomparably more valuable than something natural or created by nature. Industrial modernity is

338   *Global Studies Encyclopedic Dictionary*

a triumph of technology regarded not only as labor saving, but also as a means to enslave persons. The biological rhythms of human beings are sacrificed in favor of technology. Night shifts, which are an attribute of continuous production cycle, have long been taken for granted. One of the greatest revolutions of the New Time was started by French bakers who protested against night work just to satisfy the whims of the rich to have fresh bread by morning! "Mechanization" in terms of notions about human beings gave way to "the new parlance" of the Russian poetic language of the post-revolutionary modernization period: "Intellect has given us steel wings-hands and the fiery motor in place of the heart." Writers become "engineers of human' souls," and a metaphor for a society as an organic whole which is peculiar to traditional mentality is replaced by a notion about social system designed by means of social engineering. Social designing of reality becomes one of the highlights of modern industrial society.

The modern ecological and anthropological crisis is a challenge to the entire system of the universal worldview of "modern" society. Unlimited industrial expansion legitimized by meta-narration of progress created such significant global ecological problems that pose a threat to the very survival of human beings.

Attempts at social legitimization of "modern" society developed into conceptions that attribute the negative characteristics of "modern" society to the fact that the modernist project is yet to be completed. Therefore, it is assumed that the process of alienation and disintegration of structures of communicative rationality should be overcome in the process of further evolution, completion of the modernist project through conscious moral regulation of economic and technological development. Such is Habermas's interpretation of modern society as an incomplete project and Latura's conviction that we have never been modern.

N. M. Smirnova

Moiseyev, Nikita: (1917–2000) Russian scientist, mathematician, expert on systems theory, philosopher, and sociopolitical thinker. Among his most notable scientific achievements is the mathematical model of probable biospheric consequences of nuclear war ("nuclear winter" and "nuclear night"), which has served as scientific confirmation of the catastrophic consequences of nuclear weapons use. It has greatly influenced world politics, having stimulated negotiations between leading states on reduction of nuclear arms.

Moiseyev developed the concept of "universal evolutionism," which brings together the evolution of abiocoen (nonliving components of an environment), life, human beings, and society. He characterizes modern societal development as leading to a global civilizational crisis, making humankind face fundamental ecological, social, and political threats. According to him, to escape the global crisis we must first appreciate the seriousness of the threat to human existence, and then gradually move toward civilization based on co-evolution of society and nature. A key element for such a transition is education based on a new world outlook that originates from understanding fundamental interrelations among nature, human beings, and the society.

Macroshift – Myth 339

The issue of Russia's niche in the globalizing world takes a special place in Moiseyev's works. He thought the events of the 1990s in Russia were catastrophic and threatened its very existence. To restore Russia's strength in these conditions, he suggested increasing Russia's human and intellectual potential, developing high technologies, promoting interaction among civilizations, and spreading ideas that could inspire people to act purposefully and vigorously.

A. G. Sytin

Mondialisation is the French word for "globalization," but having a somewhat different meaning. Jacques Derrida introduced the concept. In his works dedicated to social philosophy (1990–2000), Derrida compares and contrasts the terms with the German "*Globalisierung*." The concept of globalization relies on a "spatial" (Greek) understanding of the concept of the universe, which depicts the world as a cosmos, and it differs in many respects from the concept of *mondialisation*. Derrida reveals that globalization might lead to a concealed geopolitical aggression or direct cultural expansionism bringing about the unification of values and legislation. He thinks that globalization is synonymous with "neocolonialism" and "neoimperialism." Globalization is a sort of generalization, a widespread equalizing and even otherworldly process, which falls apart from the world in which we live. The phenomenon of globalization does resonate with our everyday life because of their different scales. Globalization tends to embody and universalize all kinds of various social organizations. In some sense, besides affecting life on the Earth, globalization affects the whole universe.

When Derrida explains the concept of *mondialisation*, he stresses the function of creation of a common world space rather than unification of different worlds leading to disappearance of differences. One necessary stage of *mondialisation* is the development of common legal, economic and political principles. The end goal of *mondialisation* is a new system of differences at the horizon of a common world rather than any kind of world identity. The process of *mondialisation* enables and provides generalization and liberation from historical roots and geographic boundaries.

D. A. Olshansky.

Monopolar World: A geopolitical reality that has taken shape since the end of the Cold War and the disintegration of the Soviet Union, wherein the United States became recognized in some circles as the sole pole of power. This viewpoint is not generally accepted and the model of a monopolar world is sometimes severely criticized.

A. D. Korolev

Multiculturalism: Descriptively, multiculturalism refers to multiple cultures existing within a single social space. Normatively, the term refers to promotion of conflict-free cultural diversity. The modern concept of multiculturalism was formed as a result of discussions between the communitarians (Alisdair MacIntyre, Charles Taylor, Michael Walzer) and adherents of the theory of justice (John Rawls). The central topic of the discussions was the question of whether liberal values will prove sufficiently "neutral" with respect to the multitude of cultural differences (Taylor, 1994). Claims that liberalism em-

340 *Global Studies Encyclopedic Dictionary*

bodied the concept of justice did not correspond with cultural pluralism; communitarians focused on the morality of small local communities and realized the necessity of a transition from the policy of "incorporation" of individual and group differences into larger structures to "the politics of recognition" of their right to exist as "the others." Multiculturalism is an attempt at such a transition.

The main theoretical and practical problem related to multiculturalism is not how justified are the arguments about the multitude of life practices, but how deep are the differences underlying this multitude. Pluralization of European social and cultural life on the basis of Western cultural values and rationality is often considered as a possible universally recognized standard and a general value on a global scale. Max Weber saw the roots of Western rationality in the "idea of salvation" that is present in all cultures of the world and that is interpreted as people's strife to escape suffering. Having compared the economic ethics of world religions, Weber demonstrated a close link between rationality and cultural type. Rationality serves as a means for explicating cultural values in the world, and the image of this world depends on the nature of these values rather than on the structure of rationality. Apparently, this is the reason for sociocultural pluralism that cannot be reduced to general foundations at the level of the ultimate foundation of culture. It is the difference upon which cultural identification is based. Thus, culture, at the same time, unites people into stable communities and separates these communities.

Cultural pluralism now has a status of ontological pluralism, which means the impossibility of developing an integrating form, that is, some averaged cultural value. Cultural identity expresses the essence of human freedom, its formation, and realization. Individuals and social groups differ in many features: sex, skin color, language, religion, history, and moral values. The higher the level of individual and collective consciousness, the more sensitive to differences people are, because identification is based on difference. The quest for difference leads to finding additional identifying features, and to the "deepening" of the boundaries between various individuals and groups.

The sociocultural specifics of identification can be explained through one feature of modern pluralism. On the one hand, there is increasing sensitivity to cultural and value differences; on the other hand, there is a tendency toward social unity in the form of globalization. The correct understanding of cultural identity allows the diversity and pluralism to become a basis for power rather than weakness, and a means for cooperation rather than a pretext for conflict. Moreover, respect for the Other (to difference) could be considered as a wish or demand for justice, while orientation toward unity reflects the idea of social solidarity. Solidarity implies the absence or overcoming of social inconsistency on the individual, groups, and state level, that is, in solidarity societies overcome the dichotomy of private interest and public welfare. As for justice, its meaning is modified in the conditions of the pluralism of values. In this case, the just demand would not be that of equality of opportunities, but the attention to differences in history, religion, and ways of life of individuals and social groups forming common social space.

Multiculturalism, or multiculturality, could be considered as the "golden mean" between postmodernism and "discourse ethics" or a compromise in intercultural dialogue. Multiculturalism implies an effort to remove the ten-

Macroshift – Myth 341

sion between the local and the universal, between the desire of local cultural groups to preserve their identity and the necessity to provide integrity of a large society. It suggests a more constructive and promising solution to the problem of order in the circumstances of cultural pluralism. It asserts respect for differences, but, at the same time, does not give up the search for universality. Considering history from the multiculturalist point of view, one could believe that globalization theory with its tendency toward a linear-progress interpretation of world development needs corrections and modifications. Development could be considered as not only a linear progressive one, but also as oscillatory or spiral-like one, and the concept of historical asynchronism seems to make sense as well.

Reference: Taylor, C. *Multiculturalism*. Princeton, NJ: Princeton University Press, 1994.

N. S. Kirabaev

Myth as a practice and human phenomenon is universal, occurring within the historical development of all human communities. Like language, art, and tool-making, it reflects some basic aspect of the human condition—the hunger for explanation and for story. Myth is a form of "animating" experience, by making the physical world and its properties (time, death, change, and luck,) have a teleology.

Myths give some of our earliest evidence of human explanation and justification of the world and human roles within it. Myth as such is the forerunner of philosophy and science, echoing the same will to understand existence.

Philosophy considers myth as a human practice and a component of culture. "Metanarrative" (grand narrative) is Jean-Francois Lyotard's term for the central, undergirding story or stories that give a general structure to a culture and its citizens. He cites two metanarratives that undergird and guide culture: Marxism and Christianity. Lyotard regards the "postmodern" age, the current age, as delegitimating metanarratives, by which he means metanarratives are being re-understood as human constructs, fallible, and contingent accounts of the world. But he notes as well that a real hunger for metanarratives continues. Finally, Lyotard refers to "the terror of a metanarrative," meaning the status of the metanarrative that creates in its believers a license to kill for the sake of supporting the goal or message of the metanarrative.

Today, myth is a double-edged sword: the richness of imagery and imagination (testament to mind's power of explanation) is weighed against the ill of teleology (the danger that meaningless be described as meaningful). This aspect is the tension between the human hunger that existence have purpose or goal versus the scientific/rational dismissal of final causes.

Myth is thus both an ability and a danger. We may be well advised to regard grand narratives in global studies in light of both of these prospects

D. L. Stegall

⋈ N ⋈

Nation(alism): (Latin *natio*, derived from *nascor*, "I am born") Presently, the term "nation" is the most accepted unit of social-political organization. Ideally, it should correspond to the nation-state, the basic unit of the political organization of the world from the turn of the twentieth century.

For a stateless nation to win its own nation-state in the world already tightly divided among the extant nation-states, it would have to seize a piece of a territory in the possession of one or more established nation-states. After the break-up of the Soviet Union and of the multinational federation of Yugoslavia, it has been next to impossible for aspiring stateless nations to achieve fully independent nation-state status. Such a feat came true only in the case of the Eritreans after the long and bloody war with Ethiopia (1993) and of the Slovaks and the Czechs (1993) because the mono-national Czechoslovakia of the postulated Czechoslovaks had never materialized.

Within the broader framework of nation building, the homogeneity of the nation can be achieved through employing either the civic or the ethnic. In the first case, it is the state that makes its population into a nation, granting the inhabitants with citizenship. In ethnic nationalism, one's nationality is identified with a specific ethnicity defined through culture, language, religion, history, traditional way of life, tradition, and mythology.

Usually, the extant nation-states profess nationalisms that are mixtures of the ethnic and the civic to a varying degree. However, broadly speaking, ethnic nation-states predominate in Eurasia while civic ones elsewhere. Comparing the civic and ethnic modes of nation building, we can say that in the absence of their own nation-states, stateless nations have no choice but to ground their claims to status as a nation in ethnicity. Conversely, there are no civic stateless nations unless in the wartime conditions when their nation-states find themselves under a foreign occupation. Occupation, however, rarely results in internationally recognized liquidation of a state, let alone of the nation associated with it.

Ernest Gellner (1983) defines nationalism as a political principle that holds that the political and the national unit should be congruent. These "units" are the nation-state and the nation, respectively. This ideal may be reached in two ways: (1) The state may turn its population into citizens. In this model of civic nationalism, citizenship equals nationality and citizenry equals the nation. Thus, the state is transformed into a nation-state. (2) In the scope of ethnic nationalism, a group of activists may establish a national movement grounded in ethnicity delimited by faith, language, tradition, or way of life. The nation-builders claim those sharing (or perceived to share) a specific ethnicity to be a nation. Should the targeted population espouse this view they do become such a nation. The problem with Gellner's formulation is that it refuses to treat nationalism as an ideology.

The popular conclusion is that it is impossible to label nationalism as an ideology because representatives of the whole political spectrum (leftists, centrists, and conservatives) all espouse national tenets. This approach reifies ideology as something unique and pristine that cannot be merged with anything else, let alone another ideology. But some present-day ideologies are such mixtures containing two or more ideologies and principles (for example, social democracy, liberal conservatism, Christian democracy or National Socialism).

Nationalism has attained the status of the sole "infrastructural" ideology of the modern world thanks to its simplicity, the malleability of its principles, and the popular appeal connected to the biologically determined human desire to live in a group. The envisaged objective of spreading nationalism to every corner of the globe took two centuries and now seems to have been completed. At least civic nationalists perceive the situation in this manner. From the ethnic point of view of making nations there is no end.

The roots of nationalism are European—Western. But this ideology proved potent and malleable enough to become the first global ideology. Its universality is attested to by the use of its tenets for the political and social organization of the whole world at the turn of the twentieth and twenty-first century. For the time being, nationalism seems to be the first and only form of global order accepted by virtually all the actors of international relations and most political groupings.

References: Gellner, E. *Nations and Nationalism*. Ithaca, NY: Cornell University Press, 1983.

T. Kamusella

National Security involves the safety of citizens, society, vital state interests, national treasures, and modes of life against a wide range of different external and internal threats (political, economic, military, ecological, or psychological). National security consists of three interrelated levels: personal, social, and state security, the interrelation of which is dynamic and is determined by the nature of social relations, political and economic structure, and the degree of constitutional state and civil society development. National security is fully realized when state and social safety is not an end in itself, but is means for personal safety. National security can be broken both by military force and in other ways; for example, by terrorism or the illegal import and export of raw materials, resources, products, currency, works of art and other treasures, in addition to industrial and other kinds of espionage, cultural expansion, and the distribution of subversive information. Internal national security concerns the safety of individuals, society, and the state against internal threats to their steady, stable functioning and development. A nations' external national security is based on the recognition and observance of basic international law principles and is ensured by active foreign policy and other activities, including military assertion, and if necessary, activities to protect national interests on the international scene.

V. A. Barishpolets

NATO Response Force (NRF) refers to the international joint forces of the member countries of the North Atlantic Treaty Organization (NATO), whose mission it is to respond swiftly to various types of crises anywhere in the world, take part in military operations, aid in the separation of the armies of belligerent powers, and temporarily stabilize hostile situations before the arrival of larger follow-up NATO forces. It was created at the Prague summit in

344 *Global Studies Encyclopedic Dictionary*

November of 2002, and is committed to overcoming terrorism and the dissemination of weapons of mass destruction. Training is done by NATO during peace-making military actions. The inauguration of the two first multinational task forces numbering 6,300 troops was held in October 2003; then these groups were placed under the command of the NATO northern regional headquarters in Brunssum, The Netherlands.

The force reached full operational capability in 2006, combining land, air, sea and special forces. The forces rotate every six months. The NRF can be tailored to any specific operation. The fully operational NRF consists of a brigade-size land component with forced entry capability, a naval task force composed of one carrier battle group, an amphibious task group and a surface action group, and an air component capable of 200 combat sorties per day. It has approximately 21,000 troops. Forces will be ready to deploy within five to thirty days of notice, and will be self-sustainable for thirty days thereafter.

The NATO Response Force may be used for exercising force or countering armed aggression. It may be deployed in any region of the world as a stand-alone force in response to Article 5 (collective defense) or non-Article 5 crisis operations. The capabilities of the NRF will be developed in conformity with the Prague Capabilities Commitments, and the European NATO member countries will focus their efforts on the development of programs for creation of weapons of accuracy, pilotless drones and transport aircraft, defense in case of chemical, bacteriological, radiological events and measures against the anti-air capabilities of an enemy.

Command and control of the NRF is accomplished through a small, deployable joint task force (DJTF) headquarters capable of planning, coordinating, and conducting military operations by multinational task groups. Command will rotate yearly among static parent headquarters of the Joint Force Command (JFC) in Brunssum, The Netherlands, and Naples, Italy. Its multinational task groups afford NATO its main crisis response capability.

The NRF will become more important post-2014, after the NATO-led International Security Assistance Force (ISAF) complete its mission in Afghanistan. It will demonstrate operational readiness and act as a "testbed" for Alliance Transformation. It can be used in the implementation of the Connected Forces Initiative (CFI) as a vehicle for greater cooperation in education and training, increased exercises and better use of technology.

P. A. Perebeinos

Neutrality: (Latin, *neuter*, "neither") In international law, neutrality refers to the stance of a nation that refrains from participation in a war with other states or refrains from participation in military blocs in peacetime. In a broader sense, neutrality can mean refraining from participation in any interstate coalition. Usually neutral duties and rights are incorporated in treaties reached with other countries. The rights and duties of neutral powers in a war on land and the position of belligerent powers in respect to neutral powers are regulated by two conventions: Laws of War: Rights and Duties of Neutral Powers in Naval War (Hague XIII), 18 October 1907; Laws of War: Rights and Duties of Neutral Powers and Persons in Case of War on Land (Hague V), 18 October 1907"; and other international acts and agreements, which forbid use of territory held by neutral powers for military actions. The policy of neutrality is based on a balance of forces, that is, an impartial attitude toward the belligerents and military blocs. Such a policy was widespread in Europe in modern times and currently serves as one of the pillars of international policy.

Specific preconditions for neutrality include the geopolitical situation of the country and national traditions. Thus, Belgium followed a policy of neutrality for a long time, as this country was virtually a point of the intersection of the interests of Great Britain, France, and Germany. Switzerland strove to follow the same policy being situated in the central part of Europe to such an extent that it did not join the UN when it was created, regarding UN membership as incompatible with neutrality, and not wanting to participate in collective security or economic measures of the UN against other countries. In a sense, we can draw a parallel between the policy of neutrality and the position of modern European skeptics, for example, those in Denmark, who exercise great caution in respect to the continuance of the integration into the European Community (EC). The policy of neutrality has played a useful role by mitigating the policy of the balance of forces and serving as a buffer between belligerent powers or groups of countries. In addition, neutral states may act as intermediaries. Such a policy was beneficial for neutral states as the result of their non-participation in the measures taken by allied countries. As interdependence and globalization progress, neutrality has begun to wane, so that in the early 1990s, Switzerland was admitted to the UN, and in 1995, Sweden and Austria were admitted to membership in the European Union.

G. A. Drobot, V. N. Drobot

No Limits to Learning (1979) was the seventh report to the Club of Rome. Three independent groups in the United States, Morocco, and Romania worked on this project. of the study deals with issues connected to changes of human behavior in our constantly changing world—especially education as a social institution. It was noted that education studies were pushed by student unrest in the West between 1964 and 1968.

By the example of this report, it can be noted that by then the orientation of research by the Club of Rome had undergone significant transformation, from a "purely" scientific approach to pragmatic resolution of global problems, through acknowledging the role and importance of social factors to "internal potentialities of a person." The report called for global humanitarian issues to be put into the center of discussion, suggesting that the emphasis should be moved from issues of physical character to purely human issues, to a "life support system." It opined that the future is frequently and wrongly represented as an extrapolation of the present, without taking into account the occurrence of qualitatively new phenomena. When making decisions people should predict the results of their actions considering their future consequences.

The report indicated that education is dependent on the arms race, which absorbs huge financial and intellectual resources of society, and military expenditures exceeded that spent on education, public health services, power, and food supplies combined. UNESCO that in 1980, the number of illiterate people worldwide was 820 million, one fifth the world population. Out of this

346 *Global Studies Encyclopedic Dictionary*

number, the major lived in developing countries, which considerably complicated the solution of many of the global problems that existed there.

Reference: Botkin, J. W., M. Elmandjra, and M. Malita. *No Limits to Learning.* New York: Pergamon, 1979.

A. N. Chumakov

Nomadic Empires (Steppe Empires or Central/Inner Asian Empires) (from Greek *nomades*): Cattle breeders who inhabit the Eurasian steppe regions and move seasonally have been called nomads. Their nomadic style, in which organization of production is dispersed, limited their societal development. The biological means of production (cattle breeding) cannot be mechanized; therefore the life of nomads is rather static and conservative. The style of nomadic existence is more mobile and free in comparison with the life of farmers. Two circumstances determine the specification of nomadic states. First, the need to create a permanent state is absent. Second, the frequent recurrence of political processes is also absent. Most economic processes are within the scope of private isolated households. Social life is strictly regulated by traditions. The cycle of state development typically involves moving from separate clans, to a prince's armed forces, to the formation of Nomadic Empires. The specification of an economic-cultural type and a mode of production also determine the distinguishing features of social life. These features involve a clan structure, private-family property in cattle and social property in pastures, a lack of the development of exploitation within clans, and a warlike character. Scientists give various definitions of nomadic existence. For example, they are defined as pre-class societies, a specific form of feudalism, a specific form of production, or a specific civilization. In the modern multi-linear approach to historical processes, nomads or nomadic civilizations are a necessary original part the global process of human development.

S. Enchtuya

Nonaligned Movement (NAM) means an informal association of countries, mainly developing nations, which seek to avoid political or ideological affiliations with major power blocs. India, Indonesia, Egypt, Nigeria, Columbia, Chili are among the leaders of the bloc of non-aligned countries. The official nonaligned position of the member nations was declared at the First Conference of Nonaligned Nations held in September of 1961, in Belgrade, on the initiative of Nehru, Nkruma, Nasser, Tito, and Sukarno; twenty-five developing countries participated. The nonalignment doctrine was developed in the late 1940s. It relied on the general democratic principles of the national liberation movement, ideas of sovereignty and independence of nations, and the principle of political and economic equality. NAM allowed the developing countries to avoid strict affiliation with one of the two social-and-political systems, communism and capitalism, and protected them against attempts of external powers to draw them into the ideological, military and political confrontation between the two opposed blocks. In the 1960s and 1970s, as dozens of young countries gained or regained sovereignty and joined the NAM, willing to coordinate their actions and take joint measures, NAM achieved interstate association status, first at a sub-regional, then regional, and finally a global level.

Nation(alism) – Nuclear Winter　　　　347

Since the collapse of the communist system in the early 1990s, the NAM has been confronted with perhaps the worst crisis in its history. The main idea of nonalignment, which permitted the avoidance of affiliation with opposed political-and-ideological power blocs, no longer seemed important. This crisis was exacerbated by an internal crisis: the collapse and disintegration of the Socialist Federal Republic of Yugoslavia, and the withdrawal from the NAM membership of several influential member countries. Presently, the problems of the nonaligned movement are also exacerbated by the continuing social and economic differentiation of the member countries. Another serious problem undermining the movement's internal equilibrium is the rapid military build-up (both, in nuclear arms and conventional arms) and militarization of the economy in several nonaligned countries.

However, the NAM survived and proved it was not just a product of the confrontation of two superpowers despite its emergence during the Cold War. The main goals of the NAM are related to globalization and the interdependence of the countries in the modern world: (1) a commitment to overcome poverty and to achieve sustainable development of third world countries. To that end, NAM countries seek to avoid confrontation with developed countries and to reach a political compromise with them to attract investments, benefit from protectionism, and ease debts. (2) promotion of disarmament; (3) reform of the UN to increase its efficiency and promote its democratization.

G. A. Drobot, V. N. Drobot

Noosphere (Greek *nous*, "mind" "reason"; and *sphaira*, "sphere") is a hypothetical future state of human society and its interaction with nature, wherein mind would occupy the place of priority. French scientist and philosopher E. Le Roy and, later, French philosopher Pierre Teilhard de Chardin and Russian academician Vladimir I. Vernadsky began using the term during the 1930s.

Vernadsky put forward the idea of the formation of the noosphere as the main direction of the further development of humanity. He believed that the establishment of the noosphere is a natural historical process and that it would emerge spontaneously. However, now, in the age of global problems, it has become clear that spontaneous formation of the noosphere is impossible and it can be achieved only through a sociotechnological designing of the future with the help of the human mind and, first, noosphere-oriented science.

The problem of noospherogenesis is studied by noospherology, that is, science about the sphere of mind. Noospherology is a part of global studies oriented toward the future, and, vice versa, global studies can be considered as a component of noospherology that deals with the present-day problems of noospheric development. Noospherology is an interdisciplinary and integrative field of scientific research that deals with the laws and processes of the establishment of the sphere of mind and the prospects for survival and sustainable development of our civilization with the most complete implementation of humanistic principles and ideals. Such system-evolutionary approach to the noosphere and the stress on global problems gradually replaces the traditional vision of the future as found in concepts such as post-industrial society, and unstable development.

A. D. Ursul

348   *Global Studies Encyclopedic Dictionary*

North-South (The Rich North and the Poor South): Coined during the 1990s, after the end of the Cold War and the collapse of the Soviet Union, these terms refer to a categorization of the world's states. These momentous events changed the shape of global politics overnight. The bipolar world of the Cold War became non-polar or multi-polar before it became obvious that what had emerged was a monopolar world with the United States as the sole remaining superpower.

The Cold War was fought with the weapon of economy. The Soviet bloc collapsed because it was outdone by the West in the race of providing for armaments and for population. In the 1980s, the Soviet Union was spending over 40 percent of its budget on the military, which left pitifully little not only for its own citizens, but for planned projection of Soviet/communist power worldwide. At the same time, the United States army received less than eight per cent from the federal budget, which left Washington with ample elbow-room to project its power over the globe and to effectively contain the Soviet bloc. When the Soviet bloc broke up the customarily inefficient economies of post-Soviet and post-communist states declined even more, being so much geared to the military, isolated from the world market and traditionally full of disregard for the needs of the average consumer.

The former categorization of the globe's states into the "First World" (or the West), "Second World" (or the Soviet bloc together with other communist states in Europe and Asia) and "Third World" (or the postcolonial and usually underdeveloped states) ceased to make sense. From the political point of view the Second World disappeared with the breakup of the Soviet Union—the superpower that headed and dominated the Soviet bloc. When one analyzes the fate of economy in the post-Soviet and post-communist states, it is obvious that the Second World or the Soviet bloc as an economic entity disappeared too. Poor economic performance relegated all the post-Soviet states but Estonia, Latvia and Lithuania (together with the rich urban enclaves of Moscow and St. Petersburg) to the group of the developing or underdeveloped states. In a singularly unique and unprecedented systemic transition almost all the former European member states of the Soviet bloc have made it into the club of the rich Western states grouped in the Organization for Economic Cooperation and Development (OECD).

Recognizing the dramatically changed political, economic and social realities in the post-Cold War world, the UN proposed that the overall quality of life should become the yardstick with which one should categorize the extant states. The main contention is that one-fifth of the earth's over six billion inhabitants live in the prosperous states, which garner four-fifths of the globe's total income. This leaves the less fortunate four-fifths of the planet's population living in the developing states with one-fifth of the world's income. The astounding disparity in prosperity forms the global divide between rich and poor.

In the language of journalism these findings were translated into the economic chasm between the rich North and the poor South. In this manner the old pre-nineteenth century Western European categorization of the European states was evoked. In this old scheme the militarily and economically astute states where agricultural and industrial revolutions commenced (France, the

Netherlands, Northern Italian states, Prussia, the United Kingdom, the western sections of the Holy Roman Empire) were identified as "the North." Opposed to it was the poor and under-performing "South" that consisted of the eastern half of the Habsburg realms, the Ottoman Empire, Portugal, the southern half of the Apennine Peninsula and Spain. During the nineteenth and twentieth century, the novel East-West division replaced the erstwhile North-South one.

Obviously, it is not geography that determines which state belongs to the North or South nowadays. The litmus test is economic performance underwritten by a state's membership in the North's or North-controlled organizations such as the exclusive clubs of G-7 (today, G-8 after the Russian Federation was allowed to join the political part of the deliberations at this group's meetings. The economic leg of the talks is still reserved for the former G-7) and OECD, or the more welcoming WTO (World Trade Organization). However, the vast majority of the states belonging to the rich North are located in the Northern Hemisphere. The exceptions located in the Southern Hemisphere are few including, among others: Argentina, Australia, Chile, New Zealand and South Africa. States categorized as belonging to the poor South much more many, paradoxically, the majority of them are located in the Northern Hemisphere. Let us enumerate the most notable ones: China, Egypt, India, Iran, Kazakhstan, Mexico, Nigeria, Pakistan, Russia or Ukraine. The paradox is explained by the fact that the vast majority of the globe's landmasses are located in the Northern Hemisphere.

T. Kamusella

Nuclear Deterrence Theory is a rationale of the political function of nuclear weapons, intended to prevent the first use of nuclear weapons by any country based on the fear of unacceptable damage that might be caused by nuclear retaliation. Nuclear deterrence theory (or conception), when made a policy principle in the area of the nuclear armament of a nuclear power, becomes the doctrine of nuclear deterrence.

Nuclear deterrence is neither a military-technical, nor a legal mechanism, but a political psychological one. US President John F. Kennedy compared nuclear weapons to "the sword of Damocles hanging by a thin thread above our heads." When nuclear weapons are available to one country, it will prevent another one from being the first to use them. But in order to make this principle work the former country has to be sure that: (1) the latter one possesses nuclear weapons; (2) the latter one is technically capable, and both psychologically and politically prepared to use nuclear weapons in response; (3) the consequences of a potential nuclear retaliation will be unacceptable for the state and society. During the 1960s, US Secretary of Defense Robert McNamara formulated the threshold criteria of damage inadmissibility ("McNamara's criteria"), assuming that no state would be first to start a nuclear attack if it runs the risk of losing at least one third of the population and 70 percent of the industrial potential. However, such criteria essentially depend on the specific historical situation and on national psychology. Currently, many politicians and experts consider the consequences of even a single nuclear explosion to be unbearable for an inhabited industrial region (especially

350   *Global Studies Encyclopedic Dictionary*

regarding contemporary nuclear explosive yields that have constantly increased since the twenty-kiloton bombs dropped on Hiroshima and Nagasaki).

Nuclear deterrence does not necessarily require nuclear parity between the potential enemies. The Cuban missile crisis of 1963 occurred when the United States threatened to use nuclear weapons in its demand that Soviet nuclear missiles be withdrawn from Cuba, but abstained from doing so in fear of a retaliatory attack. This crisis is a classic example of nuclear deterrence. Although during the crisis the Soviet Union possessed about seventeen times fewer nuclear missiles than the United States, this proportion was sufficient for the nuclear deterrence psychology to act.

Mutually assured destruction (MAD) may be used as a strategy by two or more nuclear powers or potential enemies. However, nuclear deterrence may be turned against a non-nuclear enemy as well; in this case, a state's nuclear weapons are used to repulse an attack of superior conventional forces (chemical and biological weapons) in cases critical for national survival.

An attempt to overcome the limits of nuclear deterrence is reflected in the "limited nuclear war" concept of the 1970s, which assumed an efficient high-precision application of nuclear weapons upon military targets, without an exchange of unacceptable nuclear strikes. At the same time, nuclear deterrence mentality was supported in the 1970s and 1980s by the study of the potential effect of massive nuclear weapons use on the ecology, climate, and planetary biosphere (including the mathematical model of so-called nuclear winter—a complex of deep, nearly irreversible, planetary scale consequences of a nuclear catastrophe that could lead to the annihilation of humanity and to the destruction of the many forms of biological life on the planet).

Nuclear deterrence theory was the theoretical basis for the US-Soviet Anti-Ballistic Missile Treaty (1972), which prevented the development of the missile and nuclear attack defense technologies for thirty years. It aimed to (1) make the retaliatory attack unacceptable in case of the escalation of war; (2) keep the parties from the first use of nuclear weapons; (3) slow down the nuclear arms race. The US withdrawal from the ABM Treaty in 2002 was accompanied by attempts to switch nuclear deterrence doctrine to the doctrine of assured survival of one of the nuclear powers and to base nuclear policy on the balance of defensive and nuclear offensive arming.

Russia's Military Doctrine (approved 21 April 2000) designates Russian nuclear weapons to be a means for the deterrence of aggression against Russia's and its allies' military security, and to assure international stability and peace. Russia reserves the right to use nuclear weapons in response to the application of nuclear or any other kind of weapons of mass destruction against itself or its allies and in response to a high-scale aggression with the use of conventional arms in a situation critical for national security.

Nuclear deterrence is not necessarily a global military or political phenomenon, although it emerged as a theory and a political doctrine in the period of the United States-Soviet global confrontation and the Cold War. Considering the nuclear arsenals of the five largest nuclear powers, and those of India, Pakistan, and Israel, we may speak of regional nuclear deterrence or nuclear deterrence as a component of the general military strategy of a state. At the same time, nuclear deterrence works only in the situation of a confrontation between states, not working or working inadequately with non-state actors (such as terrorist organizations and radical movements) or with possibly insufficiently rational nuclear states (such as North Korea, the most recent new member of the "nuclear club").

The contemporary evolution of the nuclear deterrence doctrine presumes a transfer from "hard" nuclear deterrence (when each element of the nuclear triad can cause maximum damage during a retaliatory nuclear attack) to a "soft" or "limited" one (when only specific elements of the triad survive assuredly; the aim of a nuclear attack is a maximum damage not to the population and the industrial potential but to the enemy's military targets). Nevertheless, nuclear deterrence remains an inhumane political and psychological mechanism for trying to prevent war by means of the intimidation posed by the mass annihilation of the weapons' disastrous power and the inability to be protected from it. A shift from nuclear deterrence as a "negative mechanism" to positive approaches to international peace, confidence, interaction, and cooperation, together with a total nuclear arms' prohibition and liquidation, remains a prospective goal of the reconstruction of the international security system in the twenty-first century.

A. I. Nikitin

Nuclear Power Engineering is an industry employing nuclear power to produce electricity and heat; it is also a field of science and technology concentrating on the development of methods and facilities of nuclear energy transformation into electric and thermal energy. Problems caused by nuclear power engineering (NPE) development are of an acute political, ecological, and global character. The problems induced by humankind's technical achievements became global when the Soviet Union developed the atomic bomb in 1949; then the world entered an epoch of attempting to balance nuclear forces among countries each able to annihilate the planet.

The first atomic power station was started in 1954, in Obninsk, Soviet Union. Soon the United States set up the "Atoms for Peace" program supported by United States President Dwight D. Eisenhower and the United States Nuclear Power Committee. Its head, Lewis Strauss, presented an atomic power station in 1954, as an example of an energy source "too cheap to meter." The first industrial reactor in the United States was built in 1956, in the Argon National Laboratory (Illinois); by 1957, the Shippingport reactor produced sixty Mw. In 1963, several atomic power stations produced already 200 Mw; soon, new stations with an output of 600 Mw were constructed. The 1960s were marked as a period of the rapid growth of atomic power stations (APS). By the mid-1970s, 240 stations had been built worldwide, most of them with an output of more than 1000 Mw. Besides nuclear reactors, stationary APS contain "transport" nuclear power devices for nuclear power submarines and other vessels with nuclear power engines.

However, as early as in the 1970s, active discussions had started on problems of atomic power engineering safety and on permissible levels of radiation. By that time, the basic dangers of nuclear stations were revealed: water, air, and soil pollution caused by radioactive waste release in cases of emergency and serious problems concerning the storage of nuclear fuel wastes.

352 *Global Studies Encyclopedic Dictionary*

Despite the 1973 oil crisis, specialists gave their preference to energy saving and searching for alternative energy sources. In 1975, the International Atomic Energy Agency worked out the Norms of Nuclear Safety (non-obligatory for the government).

B. G. Rezhabek

Nuclear Strategy, American: As yet, no one has experienced a nuclear exchange between belligerents. Strategizing must begin with conjectures based on applied psychology and conclude with speculation regarding the outcomes of nuclear war. Still, strategic conjectures regarding nuclear weapons cannot be confirmed by empirical testing.

However, let us assume that uncovering the psychological, bureaucratic, and economic origins of strategizing merely generates irrelevant, ad hominem arguments: strategists may be doing the right thing for the wrong reasons. Or perhaps strategists may accurately represent themselves as theorists motivated primarily by exigent circumstances minimizing the danger of nuclear war. Further, let us assume that decision-makers ordinarily embrace such a high-minded, prudent strategy. Unfortunately, it does not follow that prudence will necessarily guide decision-making at extraordinary critical junctures-unanticipated crises when decision-makers decide between peace and war.

In sum, it is inappropriate to hazard breathtaking generalization about the origin and nature of nuclear strategy. Strategizing cannot be reduced to impassioned religious devotion or icy-cold economic calculation. By the same token, strategizing cannot be reduced to a system akin to the disinterested claims found in textbook science. Finally, even the most carefully conceived, prudent strategy may be disregarded in crisis situations.

R. Hirschbein

Nuclear Warfare and Morality: In each decade of the nuclear age, philosophers have provided critical reflections on the nature, use, and consequences of nuclear weapons. These reflections have passed though five phases and have addressed the morality of producing, testing, deploying, and using nuclear weapons. The first phase stretches from the atomic bombing of Hiroshima to above ground nuclear tests at Bikini Atoll. During the 1950s and 1960s, the second phase focuses on above ground testing of the hydrogen bomb. The third phase addresses increasing shifts during the 1970s and 1980s, to counterforce weapons and nuclear war fighting strategies. The fourth phase responds to the disintegration of the Soviet Union in 1991, and the problems of nuclear proliferation in the post-Cold War world. The fifth stage responds to the U.S. "war against terrorism."

General Background: Nuclear weapons undercut the traditional distinction between military combatants and non-combatants. The strategy of nuclear deterrence is based on the claim that the prohibitively high cost of nuclear war makes it unlikely. The doctrine of Mutual Assured Destruction (MAD) is the primary symbol of this dangerous calculus. If deterrence fails, non-combatants will be the primary victims. Because of radioactive fall-out, the use of nuclear weapons entails the precipitation of ecological warfare.

The Nature of Nuclear Weapons: Conventional and nuclear weapons are alike in that they explode by rapidly releasing large amounts of energy, producing heat and blast. Their distinction hinges on the manner in which energy is released. Conventional weapons rely on chemical reactions in which the atoms in the explosive material are simply rearranged; nuclear weapons rely on the formation of different nuclei by means of subatomic reactions in which protons and neutrons are redistributed.

The two types of nuclear weapons are based on the two ways in which chain reactions can be obtained, that is, the "fission" (splitting) of the heaviest atomic nuclei (uranium-235 and plutonium-239) and "fusion" (joining) of the lightest atomic nuclei (hydrogen isotopes). Fission of one pound of uranium or plutonium releases the explosive energy of the explosion of 8,000 tons of TNT. The complete fusion of one pound of the hydrogen isotope deuterium would release the same explosive energy as the explosion of 26,000 tons of TNT.

The other qualitatively distinct characteristic of nuclear weapons is radiation. Beyond the initial phenomenon of radiation sickness, radiation causes long-range carcinogenic and mutagenic damage. Moreover, these radiological effects, because of the worldwide distribution of fall-out, spill over into parts of the world totally non-involved in the conflict.

First Phase of Moral Response: From Hiroshima to Bikini: In the first phase, several prominent philosophers proclaimed the theme of social responsibility. On 8 August 1945, two days after the atomic bombing of Hiroshima, Albert Camus was the first philosopher to voice ethical concern with an essay in the resistance newspaper *Combat*. On 18 August 1945, Bertrand Russell began his prolonged responses. Also in 1945, Jean-Paul Sartre and John Dewey responded critically."

Second Phase of Moral Response, Above ground Tests of hydrogen bomb: In the second phase, debate on the extinction thesis received increased attention. During the 1950s, hope for international control of atomic weapons was displaced by the harsh realities of the Cold War: the Baruch Plan had been rejected, the hydrogen bomb had been developed, the Chinese Revolution had succeeded, and the Korean War had begun. Against this backdrop, in 1958, Bertrand Russell and Sidney Hook carried on a heated exchange with each arguing from opposite extreme positions. Russell argued nuclear war would destroy all humanity, and Hook argued Soviet communism would destroy all freedom. Russell lost sight of the fact that not all of humanity would perish in a nuclear war; while Hook lost sight of the fact that no society was completely devoid of freedom. Russell was the philosopher who spoke most extensively about the nuclear war throughout this period. He made a dramatic broadcast against the hydrogen bomb for the BBC, initiated the anti-nuclear Pugwash movement, contributed to the Campaign for Nuclear Disarmament, and in 1959, published his classic *Common Sense and Nuclear Warfare*.

Third Phase of Moral Response: The Emergence of Counterforce Strategy: The third phase swelled into prominence because of public concern over the nuclear threat during the 1970s and 1980s. The American Academy of Sciences warned of ozone depletion from nuclear detonations, Physicians for Social Responsibility declared the unmanageability of medical problems, Jonathan Schell in his famous antinuclear manifesto *The Fate of the Earth* used the term "second death" to refer to the meaning of annihilating humanity in nuclear war, and Carl Sagan popularized the notion of nuclear winter.

354 *Global Studies Encyclopedic Dictionary*

Fourth Phase of Moral Response, End of the Cold War: During the fourth phase, philosophers turned to the disintegration of the Soviet Union and the continued proliferation of nation-states with nuclear arsenals.

Fifth Phase of Moral Response, The Future of Violence, Terrorism, and War: A fifth phase response began to emerge following the attacks of 11 September 2001. Philosophers are critically assessing connections among violence, terrorism, and war. Increasingly, these differences are seen more as ones of degree than kind. Given the persistence of their moral critiques, philosophers likely will continue to side with victims of violence and injustice and seek to advance a world that renounces nuclear weapons and other weapons of mass destruction.

References: Russell, B. *Common Sense and Nuclear Warfare.* New York: Simon and Schuster, 1959. Schell, J. *The Fate of the Earth.* New York: Knopf, 1982.

W. C. Gay

Nuclear Weapons Testing refers to experiments carried out to determine the effectiveness, yield, and explosive capability of nuclear weapons.

The first recorded nuclear test was conducted on 16 July 1945, in Alamogordo, New Mexico, United States. On 6 and 9 August 1945, the United States dropped atomic bombs on the Japanese cities Hiroshima and Nagasaki. The first Soviet nuclear explosive test was conducted on 29 August 1949, at the Semipalatinsk nuclear polygon under the direction of Igor V. Kurchatov. The nuclear device was placed on a thirty-meter high tower, and its ability equaled twenty-two kilotons in trinitrotoluol equivalent. In the same location, the first Soviet airplane-dropped nuclear bomb testing was conducted on 18 October 1951. The last atmospheric testing at the Semipalatinsk polygon was conducted on 24 December 1962.

By 21 September 1955, when the Soviet Union conducted the fist deep water nuclear explosion at Novaya Zemlya nuclear polygon, nuclear tests had been held in all the spheres: on the ground, in the air, under water, and underground. The deep water nuclear explosion at Novaya Zemlya, which started this polygon functioning, was the ninety-third to occur; all the tests conducted by that time included: United States on the ground (seventeen explosions), in the air (forty-one), under water (two), and underground (two); Great Britain (three explosions); Soviet Union at the Semipalatinsk polygon (twenty-two). Later, China, France, India, and Pakistan started conducting nuclear tests.

All in all, the United States conducted 143 atmospheric nuclear tests, and Great Britain conducted ten tests. Joining the nuclear marathon in 1960, France conducted (till 1963) four land surface explosions and five underground explosions and continued atmospheric nuclear testing until 1975, in a medium forbidden since 1963. The last (forty-fifth) French nuclear explosion in the atmosphere was conducted in 1974.

China adopted the nuclear weapons development program in 1955 and in 1964; the first nuclear weapon testing occurred at the polygon near Lobnor Lake in northwest China, in the Sinkiang Uighur Autonomous Region. In 1980, China conducted its last atmospheric nuclear test; subsequent ones occurred underground. In sum, forty-five tests were conducted at the polygon (twenty-three atmospheric and underground explosions) until 1996, when the Chinese Peoples Republic signed the Comprehensive Nuclear Test Ban Treaty.

On 5 August 1963, in Moscow, the Soviet Union, the United States and Great Britain signed the Treaty Banning Nuclear Weapon Tests in the atmosphere, in outer space, and under water. Nowadays, more than 110 countries are parties to the treaty. In June 1974, in Moscow, the USSR and the United States signed the Underground Nuclear Testing Restraints Treaty, which set a capacity threshold of 150 kg that has been observed since April 1976. In May 1976, the United States and the Soviet Union signed the Peaceful Nuclear Explosions Treaty that stipulated specific restraints on their capacity. Both the treaties and their relevant protocols were ratified in 1990.

A. A. Belikov, V. V. Gapchukov, N. F. Yerin

Nuclear Winter: The biospheric consequences of a large-scale nuclear war have been studied since the beginning of the 1960s. These studies have resulted in treaties banning nuclear tests in the atmosphere, ocean, and underground. However, the continuing nuclear arms race is still a cause for deep concern. In 1982, *AMBIO*, a well-known Swedish journal, published a special issue devoted to the climatic consequences of a nuclear war, wherein potential effects of a hypothetical nuclear conflict on the biosphere and the human population were discussed. Targets in North America, Europe, and Asia (mostly over the territory of the former Soviet Union) were be attacked by nuclear explosions of a total explosive power equivalent of 5,000 MT of TNT in the course of such a conflict. Contamination of the atmosphere was seen to be caused by fine dust particles resulting from nuclear explosions at the ground, and by sooty aerosols produced in massive fires. Crutzen and Birks (1982) concluded that sooty aerosol clouds could attenuate solar radiation for an extended period of time causing a cooling of the Earth's surface and producing a large-scale biospheric catastrophe.

The term "Nuclear Winter" was coined by the so-called TTAPS team (Richard P. Turco, Owen Toon, Thomas P. Ackerman, James B. Pollack and Carl Sagan) who undertook a computational modeling study of the atmospheric consequences of nuclear war, publishing their results in *Science* in December 1983. TTAPS conducted the first systematic study of factors responsible for the environmental and climatic effects of a nuclear war. The milestone of the "Nuclear Winter" theory is the calculation of the amount of smoke produced by fires, and a lifetime of the aerosol clouds that intercept solar radiation on its way to the Earth's surface.

Smoke absorbs solar radiation more effectively than a thermal one, because an absorption coefficient of sooty particles for visible wavelengths is about one order of magnitude larger than for the infrared wavelengths. A sudden change of the optical properties of atmosphere causes deep changes in the energy balance of the climate system. In order to estimate these changes one has to know the spatial (horizontal and vertical) distribution of aerosol particles, their size distribution, optical properties, and chemical composition. The initial pollution of atmosphere depends on the scenario of a nuclear war. The most important factors are the distribution of targets, the type of nuclear explosions (ground or atmospheric), and the effectiveness of smoke genera-

356   *Global Studies Encyclopedic Dictionary*

tion in massive fires. Lifetime, magnitude, and duration of the radiative effect of aerosol particles depend on their horizontal and vertical transport and microphysical transformations, which, in turn, depend on the season, chemical composition of aerosols, and their optical properties. To evaluate these factors, TTAPS considered different scenarios of nuclear war with total power of explosions from 100 to 25,000 MT. A series of scenarios has been also considered by the National Research Council (NRC) at the National Academy of Sciences (NRC, 1985) and during the international research project SCOPE-ENUWAR (Scientific Committee on Problem of Environment-Environmental Consequences of Nuclear War) (Ackerman et al., 1985).

Many important industrial centers are located in highly urbanized regions. Therefore, according to the scenarios of a nuclear war, 15–30 percent of the strategic nuclear potential will be directly or indirectly used to destroy the largest urban centers. Ground explosions of 0.1 to 1 MT will be used to attack the industrial and defense targets. A ground explosion with a power equivalent of 1 MT TNT produces up to 0.6 MT of fine dust. A significant portion of this dust could be injected in the lower stratosphere where it can stay more than a year defining long-term residual climatic effects. Nuclear explosions in the atmosphere will cause fires over wide areas. For example, the explosion of 1 MT can burn up to 500 km^2 of forests or a city with a population of one million. Cities are oversaturated with burnable materials. From 2–8 percent of the material burned in the fires of forests, cities, industrial centers, and fuel storages will be converted to smoke. Urban fires can cause the strongest atmospheric pollution because of the tremendous amount of the organic and petroleum-based burnable materials concentrated there. During an attack specifically against cities (Turco et al, 1983) only 100 MT of total explosive power is enough to burn the largest urban centers and emit into the atmosphere about 200 MT of soot. This is as much as could be produced in a full-scale nuclear war. According to the basic scenario of 6,500 MT developed by the National Research Council, 150 MT of soot will be produced in urban fires. Forest fires will add an additional 30 MT. An average aerosol atmospheric loading will reach 1 g/m^2 per hemisphere. It corresponds to the optical depth of extinction фe = 4, and optical depth of absorption фa = 1.4. At such a level of pollution clouds of soot almost cease the solar heating of the Earth's surface in the Northern Hemisphere where the war will take place. The energy regime of the climate system will change drastically. The continental air temperature will drop by tens of degrees Celsius. This will characterize the onset of a "Nuclear Winter." The upper troposphere will be overheated because of aerosol absorption of solar and infrared radiation. The nonlinear response of the climate system to strong aerosol radiative forcing will tend to prolong the "Nuclear Winter." For example, the stable vertical stratification will cause stabilization of pollutants in the atmosphere for a long time.

The studies on the theory of Nuclear Winter continued up to the mid-1990s, and were devoted mostly to decreasing uncertainties. The theory of Nuclear Winter played an important role in a peaceful ending of the Cold War. The results of this work were further used in the on-going research on the effects of natural and anthropogenic aerosols on climate. One of the final publications on the Nuclear Winter theory was the last paper of TTAPS, who concluded that nuclear explosions will result in devastating effects on the biosphere and will cause more human casualties worldwide than the direct effects of nuclear explosions.

References: National Research Council. *Long-Term Worldwide Effects of Multiple Nuclear Weapons Detonations.* Washington, DC: National Academy of Sciences. 1975. Crutzen, P. J., and Birks, J. W. "The Atmosphere after a Nuclear War," *Ambio* (Allen Press) 11: 2/3 (1982), 114–125. TTAPS. "Nuclear Winter," *Science* 222: 4630, 1283–1292.

G. Stenchikov

O

Opinion, World Public is the opinion of the world public on issues of international, world, and universal significance.

Ongoing worldwide social changes generate conditions that shape world public opinion as a specific social phenomenon. Among these are the emergence and aggravation of global problems which concern vital interests not only of individual nations and peoples, but also of humanity as a whole; growing worldwide communications, interweaving of people's interests, greater mutual contacts; increasing intellectualization of the lifestyle of modern humankind; the rise of popular self-awareness and political activism, the increasing activity and role of mass media and communications; the emergence of a qualitatively new type of consciousness: universal, planetary consciousness, which exists and manifests itself as world public opinion.

To clarify the mechanism of the functioning of world public opinion, it is important to keep in mind the distinction between its subject (bearer) and its mouthpiece. The subject of world public opinion is the planetary population engendering universal, planetary consciousness. Smaller sections of society can act as the mouthpiece of such an opinion: individual states and nations, international organizations and unions, social groups, individual political and public figures, private citizens-in short, all those who not only share a given position of world public opinion, but also publicly convey it. A special role in the mechanism of expressing world public opinion belongs to those international institutions that by the very nature of their activities are called upon to safeguard universal moral values and to search for the optimal balance between national and universal interests. The United Nations is especially responsible for strengthening peace and international cooperation.

World public opinion is formed toward the international reality in its geopolitical aspect. Global problems of modernity are at the core of world public opinion, and the progress of humankind depends tremendously and increasingly on whether these problems have been solved. The problems constantly attracting the attention of world public include: prevention of global nuclear war and securing the peace needed for the development of all peoples on the globe; peaceful exploration of space and of the world oceanic resources; saving necessary natural resources, food, and industrial raw materials and energy sources for the further economic development of humankind; prevention of ecological crisis engendered by the pollution of the environment with industrial and agricultural waste; overcoming economic underdevelopment, eradication of poverty and illiteracy in the developing world; fighting the "demographic explosion" that hampers its socioeconomic progress; early prediction and prevention of different negative consequences of the scientific and technological revolution; elimination of infectious diseases; and the struggle against international terrorism.

The global character of these problems lies not in their being "equally" important for any social system in the present-day world, but in the fact that each problem concerns the interests of all humankind even if found at the level of a separate region. Their solution requires the concerted efforts of all

Opinion, World Public – Ozone Layer Depletion　　359

states and can be substantially facilitated, if the squandering of power and resources in the arms race stops. A contradiction that lies at the base of each global problem creates the conditions for the specific reaction on the part of world public opinion. Conversely, the coming of this or that contradiction to the fore and its aggravation makes humankind sensible about the specific need for solving the global problem in question and provokes increased and sustained worldwide concerns that are the engine helping an opinion to be formed and expressed.

The contents of the world public opinion are not equal to the sum of national public opinions. World public opinion comprises those features of the national public opinion that have a unifying, integrative nature and that are conditioned by the universal elements of morality, and by some common psychological forms of moral experience and the manifestation of passion and social feelings.

The spiritual and practical potential of world public opinion is most visibly revealed in its reactions toward the problems of preventing world nuclear war, cessation of the arms race, and elimination of world terrorism.

All social functions of world public opinion (evaluation, control, protection, communication, and mobilization) are closely interrelated; they mutually complement and reinforce each other. Together, they form the regulative function of world public opinion, expressing its ability to influence the moral and political climate in the world and to contribute to international governance. Thus, world public opinion reflects, in an indirect and generalized manner, the attitude of humankind toward global problems, events, and processes affecting its vital interests and needs embodied in value judgments or practical actions.

V. G. Gorshkov

Our Common Future (1987) was the report of the World Commission on Environment and Development, widely known as the "G. H. Brundtland Report," to the General Assembly of the United Nations. The Commission was formed in coordination with Resolution 38/161 of the UN General Assembly, passed in the autumn of 1983, with a condition that at least half of commission members should represent developing countries. Gro Harlem Brundtland from Norway, then leader of the Norwegian working party, became the chair, and Mansur Halid, former minister for foreign affairs of Sudan, deputy chair.

The Commission was formed during a period of unprecedented menacing forecasts concerning the destiny of humankind. Its major goal to make recommendations for a more just and more secure future, based on politics and practices that would serve expansion while maintaining the stability of the ecological basis of social development. Several key directions were singled out for analysis: prospects concerning population, the environment, and steady development; a system of decision-making for support of environmental management; food security, agriculture, and forestry; international cooperation, and power, industry, and international economic relations connected with the environment and development. These problems were examined in connection with economic and social policies at that time an into the future.

The Commission worked as an independent body and published preliminary results in 1987, under the following titles: *Energy 2000* and *Food 2000*.

360   *Global Studies Encyclopedic Dictionary*

In total, the Commission, together with many scientists and experts working under its patronage in many countries of the world, completed more than seventy-five research projects and reports, which created a basis for the final (summary) report that presented in autumn of 1987, at the forty-second session of the UN General Assembly and resonated worldwide.

References: World Commission on Environment and Development. *Our Common Future.* New York: Oxford University Press, 1987. —. *Energy 2000.* London: Zed Books, 1987. —. *Food 2000.* London: Zed Books, 1987.

A. N. Chumakov

Our Global Neighborhood (1995) was the report of the independent Commission on Global Governance made for the fiftieth anniversary of the United Nations. The commission's research analyzed the world transformation that occurred after 1945, justifying the necessity of changes in governing world processes. The report stated that we live in times that demand an original and innovative approach to issues of global governance and cooperation, but it cautioned that global governance is not a global government. The authors stressed that they were not recommending any move in the direction of global government, as that would be even less democratic than the world in which we live now—a world more obedient to authority, more favorable to hegemonic ambitions, and giving a role to states and governments that would be to the detriment of the rights of ordinary people. At the same time, the report emphasized that the point was not about the creation of a world with no principles and norms.

A chaotic, disorganized world would be a bigger danger. That is why we must reach a balance in which international affairs would consider the interests of all people of the planet, and global organizations would be based on universal values and correspond to the realities of global variety.

The report concluded with several recommendations, including rather radical ones on security in the broadest sense of the word for both nations and the planet. They gave recommendations on problems connected to economic interdependence and on reforming the UN, which would ensure an increase of ordinary people's role through the organization of world civil society.

Reference: Commission on Global Governance. *Our Global Neighborhood.* New York: Oxford University Press, 1995.

A. N. Chumakov

Ozone Layer Depletion occurs due to the emission of ozone-depleting chemicals into the atmosphere. The main mass of ozone (90 percent) is in the stratosphere in a layer at altitudes from 10–50 km with maximum concentration at altitudes from 19–23 km from the Earth's surface. The ozone layer performs the important function of protecting living organisms from the harmful influence of short-wave ultraviolet solar radiation. A continuous natural process of the formation and depletion of ozone occurs in the stratosphere. Ozone is formed as a result of the impact of solar radiation on oxygen in the course of photosynthesis and is depleted due to a series of catalytic cycles that involve natural nitrogen, chlorine, bromine, and hydrogen. The observed ozone depletion was explained by natural and anthropogenic causes. The first explanation names possible sources of influence on the concentration of ozone (first volcanic activity and changes of solar activity), but it does not specify particular mechanisms of such influence. The second explanation connects ozone depletion to wide industrial application of chlorofluorocarbons (CFCs), hydrochlorofluorocarbons (HCFCs), and halons: substances artificially synthesized in the twentieth century. Their mass production began in the 1950s, and in the 1980s, worldwide production exceeded 1.1 million tons annually. The most probable cause of ozone depletion is the combined effect of natural and anthropogenic factors, but the important role of CFC in this process is fully proved. The effect of increased radiation on living organisms is very negative: it is pernicious to many kinds of oceanic plankton and various land microorganisms; it causes skin cancer, cataract, and other diseases in human beings and can damage DNA genetic material.

Emission of anthropogenic chlorofluorocarbons (CFC) and other compounds based on halogen (such as chlorine, bromine, or fluorine) and nitrogen oxides, accelerates natural depletion of ozone and disturbs the chemical balance in the stratosphere, which leads to diminished ozone concentration. Due to a wide application of substances containing halogen as the drive gases in aerosol sprays, in the production of foamy substances, in refrigerators, and cleaning solvents and disinfectants (ethyl bromide), their concentration in the atmosphere has grown rapidly causing the depletion of the ozone layer. Some CFC and halogenated carbohydrates can remain in the atmosphere over 100 years, and the emitted amount is sufficient to continue to deplete the ozone layer for decades. Jets emitting significant amounts of nitrogen oxides into the lower stratosphere also pose a danger to the ozone layer. Within the last fifteen to twenty years a stable tendency of ozone decrease was discovered. Global loss of ozone is estimated at about 5 percent (6.5 percent over the Northern Hemisphere and 9.5 percent over the Southern Hemisphere). The maximum decrease of the amount of ozone occurs in the lower stratosphere, approximately 10 percent per decade. The maximum ozone loss was registered between 1992 and 1995 over the Southern Hemisphere. The "ozone hole" over the Antarctic was ten million km^2 in 1995; the maximum decrease of ozone in the lower stratosphere (15–18 km) was over 50 percent.

A strong depletion of the ozone layer over the Northern Hemisphere was observed for the first time in 1995. Ozone loss by 1 percent causes an increase of ultraviolet radiation by approximately 1.3 percent.

To prevent the depletion of the ozone layer, a Vienna Convention on Protection of the Ozone Layer was adopted in 1985. The Montreal Protocol on substances destroying the ozone layer was adopted in 1987, followed by additional documents issued in London (1990) and Copenhagen (1992) that envisage measures aimed at step-by-step reduction of production and use of ozone-depleting substances. On the condition of full observance of these documents, the rehabilitation of the ozone layer may be completed by the middle of the twenty-first century.

The necessity of reducing CFC production and application and finally discontinuing of their venting in the air (though they may be used in CFC closed technological cycles) was realized at the end of the 1980s. In 1987, the Montreal Protocol on substances that destroy the ozonosphere was opened for signing, according to which industrially developed countries, including the former Soviet Union, took on the obligation to reduce and then discontinue production of CFCs. The periods for fulfilling these obligations (1995, 2000, and 2005) were established according to the effect of CFCs. The Montreal Protocol (together with its supplements), signed and ratified by more than eighty countries, is being fulfilled now. However, CFCs that are already in the atmosphere will affect the ozonosphere for dozens of years.

Different time estimations concerning the stability of the conditions of the ozonosphere have been calculated, but the problem is complicated by the fact that the apparent warming of the climate negatively affects this process since a rise in the temperature rise increases CFC catalytic activity.

M. K. Vermishev; V. I. Danilov-Danilyan

ଓ P ଠ

Pacifism: (Latin, *pax* for "peace" or *pacis* "compact" and *facere*, to make) Pacifism is moral opposition to war and other forms of violence. Literally, pacifism means peacemaking or agreement making. It is often incorrectly used interchangeably with passivism (Latin *passivus*, "suffering"), which means to be inert or inactive, suffering acceptance. While pacifists may be passivists, more often, pacifists are activists, using nonviolent methods to resolve conflicts and as the primary means in their struggles for personal and political goals.

Conceptually, pacifism includes a wide range of moral positions, from an absolute prohibition of all use of force against sentient beings at one end of the spectrum, to a pragmatic restraint of selected forms of force under various conditions at the other. Disagreements among pacifists are common, about their reasons for their position, over the extent of their commitment to nonviolence, and the grounds they offer to support their views. Understanding that most people are pacifists to some degree is important in setting aside the common inclination to caricature pacifism as a fanatical or extreme position.

Pacifism consists of two parts, one negative and critical—the moral opposition, the other positive and constructive—the commitment to cooperative social and political conduct based on agreement. This dual nature reflects the recognition that peace is not merely the absence of war, but a condition of order that arises from within a group through willing cooperation of its members rather than being imposed from the outside by the domination of others.

The purest and most extreme form of pacifism, "absolute pacifism," is the view that any violence against living creatures is morally under any circumstances. Although many aspire to this position, few actually practice it.

Because war is the extreme form of violence between persons, most pacifist literature discusses pacifism as it relates to war. Moral opposition to war is discussed across the history of Western philosophy. Early considerations of the morality of war can be found in ancient Greek texts; for example, in Plato's *Republic* (469b–471c). Nonetheless, pacifism as the moral opposition to war per se, as distinct from moral opposition to particular wars, seems to emerge in Western culture among first-century Christians. Although Imperial Rome was virtually always at war, devoted early Christians carried their pacifism to the extreme of non-resistance to evil despite their own persecution by Romans. Whether pacifism is an essential feature of Christianity remains controversial; Jesus himself is quoted both for and against pacifist views. Regardless, the idea that peace and cooperative social order are superior to war and violence-enforced order emerged among Jesus' early followers.

The Emperor Constantine's declaration of tolerance to Christianity (313 CE) and his own conversion to Christianity led the Church away from pacifist values. Once the Church and the Empire were united, soldiers could be Christians and vice versa. Early in the fifth century, Augustine combined the classical code of war with Christian doctrine to create a just war position. Jesus's expression, "resist not evil," was interpreted to apply not to actions but to attitudes. Killing and love were seen to be compatible if hate were avoided because salvation of the soul, not life of the body, was considered the crucial

364 *Global Studies Encyclopedic Dictionary*

goal. Injury and death could even benefit a sinner if justice were vindicated and peace restored. As long as massacres, looting, and atrocities were avoided, war could be just.

With the rise of Augustine's just war view, pacifism remained largely dormant until the sixteenth-century reforming sects such as the Mennonites and Anabaptists. Erasmus advocated pacifism in the sixteenth century, and, in the late eighteenth century, Immanuel Kant discussed the elimination of war and developed a moral view that some interpreters have understood to entail pacifism. Adin Ballou articulated a pragmatic form of pacifism in the nineteenth century, and William James explored pacifist philosophy in the early twentieth century.

The volume of pacifist literature grew dramatically in the twentieth century, largely due to the work of notable pacifists such as Leo Tolstoy, Jane Addams, Albert Schweitzer, Mohandas Gandhi, Dorothy Day, and Martin Luther King Jr, whose thoughts and actions have inspired serious consideration of alternatives to war and violence. The rapid growth in pacifist literature is perhaps due also to the immense scale on which war was waged throughout the twentieth century, the bloodiest century in history, and to the ever-advancing technology of war, weapons of war, and their proliferation. Large-scale killing is increasingly possible for more and more nations, and one of the lessons of history is that nations tend to use their weapons to do what their weapons make possible.

Arguments favoring pacifism focus on a variety of concerns regarding the evils of war, which include human suffering (especially among innocents), the uncontrollability of warfare, and the morally degrading effects of war on its participants.

Techniques of nonviolent direct action vary with the degree of coercion they employ. Genuine consensus requires unforced mutual agreement. Where agreement is genuine, no one is coerced. Negotiation involves give-and-take in a context of constrained choices. As coercion increases by degree, nonviolent techniques vary incrementally until we reach the other end of the spectrum, but they stop short of physical violence. Thus, in practice, pacifists vary over the degree of coercion they will employ or endorse, making choices for less rather than more violence if they cannot reach pure nonviolence.

D. L. Cady

***Paideia* (Education) and Globalism**. In the culture of ancient Greece, the term *paideia* referred to the rearing and education of ideal citizens of the polis, articulated by Plato in *The Laws*. It included both subject matter and training in proper attitudes and values.

Plato's dialogues regarding *paideia*, have long been the topics of debate in the history of education. Critics have argued, variously and vigorously, that Platonic notions of *paideia* are culture-bound, fascistic, racist, and sexist.

Nowhere have these debates been more evident than in the discussion regarding the nature and purpose of higher education. One of the more influential discussions on this debate can be found in Jean-Francois Lyotard's now famous monograph *The Postmodern Condition* (1979), in which he argues that the two narratives traditionally guiding higher education are now defunct.

On the one hand, according to Lyotard, there was the narrative of genius and speculative science, related in many ways to Plato's idea, with the university designed to be the place where the intellectual elite might cultivate knowledge freely and independently, unimpeded by the day-to-day concerns of conventional life. This notion has its most influential model in what today is called the Humboldtean conception of the modern research university, concerned not only with research but also with the character development (*Bildungsformierung*). The other, much at odds with the Humboldtean conception, is the narrative of emancipation that comes to us by way of the French Revolution. In this narrative, "humanity is the hero of liberty" with the "rights of man" providing the legitimacy for "equal access" to the education once reserved only for the privileged few, even if this access comes at the expense of quality. What today is called "nation building" might be viewed as the extension of the narrative of emancipation, especially when universities are principal instruments or agents of this process through grants and various research programs.

The problem, according to Lyotard, is that the ultimate aim of both narratives has been altogether subsumed by the language of performance. The goals of higher education in contemporary mass societies have to do with neither the well-being of individuals nor with the well-being of community. The lofty values of speculative science and social emancipation have been altogether subordinated to the instrumental or cyborgian value of performance. What counts is not the actual production of educational systems measured in terms of social and cultural consequence, but making certain that the system works at optimal levels of economic performance and efficiency.

This is why Lyotard argued famously that the principal identifying characteristic of the postmodern age is the loss of a meta-narrative that might guide today's educators. Whether the "business of education" has degraded education to the extent Lyotard implies, the reduction of the meaning of education to categories of performance is value-free in the sense of being altogether free from the kind of value inquiry traditionally associated with the meaning of *paideia*. Obviously, if the objective of global education is to achieve higher levels of efficiency and commerce, this is a great danger.

Reference: Lyotard, J.-F. *La condition postmoderne*. Paris: Éditions de Minuit, 1979.

A. M. Olson

Pan-Islamism is the movement and ideology that emerged at the end of the nineteenth century. The idea of creating a political alliance of Muslim peoples was put forward for the first time by Jamal al-Din al-Afghani in the 1860s. Having visited Iran, Iraq, India, Afghanistan, and Turkey, al-Afghani realized the existence of common problems shared by Muslims in different countries. Economic and social underdevelopment, spiritual apathy, local tyrannical regimes, and colonialism concerned him. He believed that political aims could be achieved only through religion: Islam is a common ideological platform capable of uniting peoples in their struggle against colonialism and making them believe in a future revival. Al-Afghani thought that nationalism prevented Muslims from consolidating their efforts against their common enemy, colonialism. Hence, he preached religious solidarity.

366 *Global Studies Encyclopedic Dictionary*

Initially, Al-Afghani envisioned the alliance of Muslim countries as an agreement between Afghanistan, Belukhistan, Kashgar, Yarkand, Bukhara, and Kokand, with the approval of the Sultan of Turkey and with financial support from Indian Muslims. A newspaper *Al-'Urwa al-Wuthqa* (The Inseparable Link), which he started to publish in March 1884, became the mouthpiece of Pan-Islamic propaganda. Though published for only eight months, its impact on developments in Muslim countries was immense. Some historians suggest that *Al-'Urwa al-Wuthqa* was the paper of a secret organization with the same name, founded by al-Afghani with a membership of Muslims coming from India, Egypt, North Africa, and Syria.

Time and again, al-Afghani made new plans for an Islamic alliance, appealing to the Khedive of Egypt, the Mahdi of Sudan, the Shah of Iran, and the Sultan of Turkey. These activities made his contemporaries and historians suspect him of being a secret agent of various governments. In fact, al-Afghani did receive aid, including financial, from several monarchs who sought to use Pan-Islamism to preserve and strengthen their own power and prestige. However, for al-Afghani, their support was only a means for achieving his main purpose—political independence of Muslim peoples.

Al-Afghani's activity toward the establishment of a parliamentary democracy in Egypt and his conviction that overthrowing an unjust monarch lends adequate evidence that monarchy was not his political ideal. However, believing that elimination of foreign rule was of primary importance, al-Afghani temporarily set aside the problem of the future political system in Muslim countries. He hoped to use Muslims' devotion to the heads of the their community for their consolidation in the struggle against imperialism. Al-Afghani tried to find support not only from Muslim monarchs but also from the ruling elite of Europe, playing on contradictions between the colonial powers. But neither Asian monarchs nor Western governments were interested in the Pan-Islamic project.

Al-Afghani's ideas corresponded with peoples' interests because he promoted their liberation from colonial dominance. Nevertheless, al-Afghani can be considered an ideologist of the national bourgeoisie. His interpretation of Islam in a spirit of human activism, and his criticism of religious dogmatism and adherence to rationalism contributed to the elimination of feudalism. Al-Afghani propagated capitalist principles and opposed socialist ideas, considering communism to be "the most disgusting" doctrine, since it rejects private property and natural social inequality. However, by the end of his life, his views had changed. He started to claim that socialism does not contradict Islam and referred to the first followers of Mohammed as great socialists.

Pan-Islamism has played different roles in the history of Eastern nations. In India, the Caliphate movement awakened Muslims to the struggle against British rule. (It is significant that Mahatma Gandhi and the Indian National Congress supported the Caliphatists.) In other countries, it also acquired anti-imperialist coloring. At the same time, the development and spread of nationalistic ideologies was making Pan-Islamism more and more an obstacle in the way to political liberation of a particular country, the means by which the Ottoman Empire tried to maintain its hegemony over the Muslim world.

In the 1930s, when the caliphate was abolished, it became clear that hopes to consolidate Muslim peoples into one Islamic state were unrealistic. While still recognizing it as an ideal, many Muslims realized the impossibility of its fulfillment under the existing social and historical conditions. However, the idea was not completely abandoned. As an alternative, the idea of Islamic unity in the form of different international alliances or organizations emerged.

The Pan-Islamists constantly inspired a universal spirit of a Muslim "brotherhood," *millat*, against the "regional narrowness" of nationalism. The millat slogan very often served as a cover for hegemonic aspirations of the ruling circles of some countries. For instance, during the early period of the Pan-Islamic movement, Turkey claimed its dominance in the world of Islam, and later, after Pakistan had emerged, it was its turn to claim a special role in the future of the Muslim world. Also, hegemonic aspirations were typical for Saudi Arabia that, considering itself to be the guardian of the purity of the Islamic faith, made several attempts to create the Islamic Pact under its own aegis.

A specific way of realizing the idea of a political Muslim alliance is the Organization of the Islamic Conference (OIC) founded in Rabat (1969). Its establishment was caused mainly by the strivings of its member-countries to oppose economic, political, and ideological colonialism, to strengthen national sovereignty and economic independence, and to play a more active role on the international arena. OIC received a status of an observer at the United Nations General Assembly thirtieth session (1975). The highest political body of OIC involves a meeting of the leaders of its member-countries. Meetings of foreign affairs ministers take place annually. OIC has established the Islamic Bank of Trade and Development, the Islamic Solidarity Fund, the Islamic Institute of Defense Technology, and the Islamic Committee on Economic, Social and Cultural Affairs. The Activity of OIC demonstrates a striving by its members for a new international economic order.

Recently, Pan-Islamism has manifested itself in a form of "Islamic solidarity," the theoretical foundations of which were formulated at Muslim universities such as al-Azhar (Egypt), az-Zaitun (Tunis), al-Karaviin (Morocco), and by the World Muslim Organization (an international organization of Muslim theologians) founded in 1970, and the League of Muslim World.

The concept of Islamic solidarity demonstrates the plurality of social forces that have joined the movement; consequently, a variety of aims is pursued, which may include consolidated struggle against imperialism, strengthening political and economic independence, and preserving cultural identity. Sometimes, economic solidarity competes with foreign multinationals or struggle against democratization for establishing "Islamic order." Pan-Islamism is also propagated by fundamentalist organizations such as "Muslim Brethren" and "Jamaat-i islami," some of which are inclined toward terrorism.

References: al-Afghani, J. "La dernierce lettre de Djamal-ed-Din envoyee de Constantinople a l'adresse d'un ami persan" (Letter of Djamal-ed-Din, Envoy to Constantinople, Addressed to a Friend). In *Djamal-ed-Din Assad Abadi dit Afghani, par Homa Pakdaman*. Edited by H. Pakdaman. Paris: Maisonneuve et Larose, 1969.

M. Stepanyants

368   *Global Studies Encyclopedic Dictionary*

Pantheism (Greek *pan*, or "all"; *theos*, "God") as a philosophical system is based on the recognition of different forms of identity and unity of God the Creator and the created world, which distinguishes it from creationist ideas of the Creator and creation in monotheistic religions. The English philosopher John Toland introduced the term "pantheist" in 1705.

The phenomenon of pantheism has existed since antiquity. Since people became aware of the existence of the material and the spiritual, they have been trying to understand their interrelationship and to overcome their split, at least in human consciousness. Thus came into being animism, hylozoism, and, with the rise of the idea of God, pantheism. The concepts of pantheism included elements of both materialism and idealism. That is why a popular vision of pantheism with its principle that "all is God" as an "idealist system" is conditional, since the essential pantheist principle that "all is God" logically presupposes the assertion that "God is all." There has rarely been a pantheist in the history of philosophy who would not have recognized both principles.

In the Western tradition, pantheism is represented in the teachings of John Scotus Erigena, David Dinant, Meister Eckhart, Nicholas of Cusa, Giordano Bruno, Girolamo Cardano, Tommaso Campanella, and Benedict Spinoza. Asian pantheism is reflected in the heritage of Mansur al-Hallaj, Ibn Al-Kuzata Miyaneji, Jalal ad-Din Rumi, Mahmud Shabestari, Yunus Emre, Shams Tabrizi, Fazlullah Naimi, Imad ad-Din Masimi, Jalal ad-Din Davvani and many other thinkers. The commonality between Eastern and Western pantheism lies in the recognition of unity and identity of the Creator and creation, being and non-being, parts and whole, and micro- and microcosm.

Z. A. K. Kuli-Zade

Peace Culture refers to a social orientation that avoids or rejects war and pursues social justice and the nonviolent resolution of conflict. It is needed to maintain sustainability in our "global village." After the horrible terrorist attacks on the World Trade Center in New York and the Pentagon in Washington, DC (11 September 2001), the growing global culture of violence and terror has become one of the greatest risk factors for the sustainability and future development of human civilization. Thus, a whole new philosophical conception and ethical "peace culture system" must be established.

In our era of high technology and globalization, societies are in a constant state of dynamic transformation. We need a world at peace so that it can function and flourish securely, and one of the basic requirements for attaining sustainability and security involves the remaking of modern culture, which, at all levels, contains much violence. The creation of an innovative global and regional "Culture of Peace System," should be accompanied and promoted by an objective, balanced, and ethical media. This new peace culture and network should powerfully counteract and gradually replace the often-prevalent culture of violence. Its major aims should be: (1) to address cultural and ethical root-causes of violence, conflicts, terror, and war; (2) to build harmonious bridges of culture, understanding, and respect among people, ethnic entities, and nations; (3) to create, develop, and spread a culture of a peace climate nationally and globally; (4) to nurture, promote, and develop the concept and identity of the "global citizen" and global citizenship, through a pluralistic

Pacifism –Political Science and Global Studies 369

global culture; (5) to promote and render "peace news" and "peace culture developments" as primarily newsworthy, instead of mainly violent news.

A powerful and influential new peace culture system, with the aid of a responsible media, could inculcate and diffuse the required constructive and harmonious philosophical concepts, values, ideas, and ethics, through arts, literature, education, and entertainment, which would reach the various sectors of society and the masses. What people watch, hear, and read, and the kind of culture, films, literature, and art to which they are exposed influence their thoughts, feelings, and ethics throughout their lives. Therefore, the communications and stories we watch on television and in films, and to which we listen on the radio should include programs that create and build and do not destroy. A popular phrase states, "We are the product of the stories we are told." If we think of religions at a general philosophical level, they are indeed mainly built on stories and parables.

Culture has the capability of a mutual transferring of these aspects between civilizations. By doing so, it can create bridges of understanding and mutual recognition of basic ingredients of a common humanity. Culture is both a preserving and a productive force, by transmitting the cultural patterns of the past and the present to the future, and an important innovative influence, by its power to inculcate new attitudes, thoughts, values, and norms, which could be organized and promoted by the new required system.

This innovative system could also propound the philosophical concept that culture is a key factor in promoting global peace. By instilling the recognition of the "other," respect for his or her identity and culture, and a commitment to solving conflicts and differences by peaceful means, the chances for peace can be greatly enhanced. By contrast, if the cultural system instills, as it does now, philosophical and psychological attitudes of self-centeredness, rejection, mistrust, and hatred of the "other," and calls for and justifies resorting to violence to solve conflicts, then sustainability and the very existence of humanity may be endangered.

A reformed culture, communications, and educational system can facilitate the creative voices of the many lands around the globe yearning for peace to be heard and to be listened to. Writers, filmmakers, television producers, dramatists, and poets, encouraged by the spreading of a culture of peace, would give voice to the craving of the preponderant majority of the global community to build a world beyond war, terror, and violence. The need for both philosophical and actual reform on an international scale, concerning culture, literature, and the arts, has become crucial, and can indeed undermine and replace the culture that breeds violence, terror, and crime.

A. Aharoni, I. M. Matskevich

Peacemaking Operations (PMOs) are systematic and organized activities of states and intergovernmental unions to intervene in conflicts with the aim to stop and settle the disputes. This activity is legitimate and regulated by the United Nations political mandate or by regional organizations. During the fifty years of UN existence, more than eighty states contributed troops and other personnel to international peacemaking operations. More than 900 thousand military and civil specialists participated in PMO. Currently, more than

370 *Global Studies Encyclopedic Dictionary*

fourteen states have agreements with the UN specifying their readiness to send troops for current and future operations. During the "peak" of peacemaking activity in 1993, the annual costs of the operations amounted to $4 billion, with more than 80,000 people deployed in conflict regions. The number of participants had been reduced to between 12,000 and 15,000 by the late 1990s, with an annual budget of about $1 billion. However, in 2000, only operations in the former Yugoslavia involved more than 65,000 people.

By legal status, one should distinguish the following types of PMO:

(1) Peacekeeping operations (PKO) carried out by the international community by approbation of the conflicting parties and, in case of a non-international conflict, by approbation of political leaders of the state where the conflict takes place. As a rule, PKO are carried out for keeping not yet broken or already restored peace, or, before or after the violent stage of a conflict.

(2) Peace-enforcement operations (PEO), which include elements of coercive actions and are carried out under the UN mandate but without any consent either of the conflicting parties or of the legitimate political government of the country where the conflict takes place.

Functional types of PMO include:

(1) Observers' PMO (OPMO): International observers' missions (civil, military, or mixed) or fact-finding missions;

(2) Military PMO (MPMO) International peacemaking operations employing military forces and with elements of military activities;

(3) Police PMO (PPMO) International peacemaking police operations;

(4) Humanitarian PMO (HPMO) International peacemaking operations to render humanitarian assistance in conflict regions (to be distinguished from rendering assistance to victims of natural disasters and catastrophes).

In practice, military, police, economic, or diplomatic measures can be applied in every type of operations. Political and diplomatic measures promoting negotiations, political settlement, and assisting in working out agreements on the post-conflict status can be regarded as a particular "nonmilitary," political-and-diplomatic type of PMO (DPMO), but which can also be considered as a stage or component of all types of PMOs. These measures are usually taken both before and after the conflict mediation.

The types of PMO differ also by their actors:

(1) UN peacemaking operations (on behalf of the world community under UN mandate);

(2) Regional PMO, of regional intergovernmental organizations under mandates of regional organizations complying with the provisions of Chapter VIII of the UN Charter (e.g., operations of the OSCE).

Finally, in political-diplomatic practice, debates are under way on the concept of domestic PMO, which, from the point of view of international law, are a police operation of a state on its own territory, with the aim to suppress and settle non-international conflicts between intrastate subjects (e.g., ethnic groups or regions of a state). For example, in Russia, military forces were employed to separate and partially disarm the parties during the armed conflict between Ingushetia and North Ossetia, in 1992. The status of PMO only arises in cases where the government acts as a "third party," equidistant from the conflicting parties and "pulling them apart." In the case of struggle of the cen-

Pacifism –Political Science and Global Studies 371

tral government with regional separatists or political insurgents (government acting as one of the conflicting parties), domestic PMO status is not applicable.

Delegation of authority to intervene into conflicts has become a new international norm. Authority is delegated individually to the newly formed coalitions of states providing contingents, means, armament, or equipment. Meanwhile, the mechanism of such a delegation of authority is not yet worked out, and sometimes it takes the form of an illegitimate acquisition of authority.

The concept of PMO does not exist in the UN Charter. Initially, another task was envisaged: to apply compulsory measures by the Security Council, including the use of the armed forces, against states committing acts of aggression, breaking, or threatening the peace. In 1946–1949, the idea was discussed of establishing permanent military UN forces under the command of the Military-Headquarters Committee (MHC) and under the political guidance of the Security Council. However, in practice, UN activity aimed at inhibiting conflicts was developing in a different direction. First, the institution of military observers was organized, sent on behalf of the international society with missions into the regions of already existing potential conflicts. Accordingly, two of the first military observer missions were sent to Palestine and to the region of conflicts between India and Pakistan in 1948 and 1949. Later on, such UN missions were sent to Lebanon and Yemen (1963), Afghanistan and Pakistan (1988–1990), Angola (from 1991), Iran and Iraq (1988–1991), Iraq and Kuwait (from 1991), Georgia, and Haiti (from 1993), Tajikistan (from 1994). Next, in circumvention of the MHC, which has never started to operate because of the contradictions between the West and the Soviet Union, the UN military operations took different shape than the one envisaged by chapter VII of the UN Charter.

With all its successes and importance of its proper peacemaking efforts, the UN has not become a universal agent of peacemaking activity that might be efficiently used by the world community in all conflicts. In practice, the UN seems to be not always a necessary and almost always an insufficient part of the peacemaking process. It appears to be selective with respect to the conflicts in which it intervenes, the scale of measures, and the functions to be applied: a great number of blood-shedding conflicts of the last decades remained beyond UN peacemaking activity.

Doctrinal UN provisions do not sufficiently motivate this organization to interfere in conflicts, especially if these are not international, but domestic, since the UN Charter is basically oriented at a priority of national sovereignty and non-interference in domestic affairs. However, the role and the scale of non-international conflicts constantly grow, thus pushing the UN to gradually modify the rules and principles of interference. The UN Charter (primarily chapter VIII) and other provisions and principles of international law acknowledge the possibility and necessity to use regional bodies and agreements to settle conflicts and keep peace and stability in the region.

A. I. Nikitin

Perestroika is a Russian term meaning "restructuring," which was first used to describe the radical restructuring of the economy of the Soviet Union initiated by Mikhail Gorbachev after he was elected General Secretary of the

Central Committee of the Communist Party of the Soviet Union in March of 1985. Afterwards, this term has been incorporated into political lexicon in many countries. The historical meaning of perestroika (1985–1991) entails the dismantling Stalinist heritage, the democratization of political and economic life, the replacement of the administrative-and-command system with a market society, and the radical changing of the federal structure, as formerly the Soviet Union was virtually a unitary state. It can be regarded as a process of modernization, which was invoked to some extent by challenges of the globalization, although to a greater extent it was driven by internal factors: the resources of the Soviet system, which ensured the rapid development of the country at a previous time, were exhausted, leading to economic stagnation. People saw no bright spots on the economic horizon. The challenges of globalization brought the necessity for a transition from a closed or semiclosed society (and the Soviet system could be characterized like this) to openness to the external world, glasnost and political pluralism. Perestroika was of great importance and significance for the entire world as these changes in foreign policy put an end to both the Cold War and the nuclear arms race. Thanks to the perestroika, the two parts of Germany could reunite and the Eastern European countries became truly independent, ending the ideological schism of Europe. As a result, the international climate drastically changed. On a global scale, perestroika was Russia's largest step toward integration into the modern global structure.

A. S. Chernyaev

Personalism is the philosophical theory that acknowledges the personality as a fundamental creative reality. The basic concern of society must be the intellectual perfection of human beings as persons. Personalism was first developed in the end of nineteenth century and has three basic versions: (1) *Theistic*, (2) *Atheistic*, and (3) *Idealistic* (the person is the ultimate reality). The first appearance the term "personalism" in a dictionary appeared in 1915, by Ralph Flewelling, then editor of the journal *Personalism*. Traces of personalism can be found in the history of ideas, especially in the non-materialistic schools, as far back as Anaxagoras, Heraclitus, and Protagoras.

G. Mouladoudis

Philosophy (Greek *philo*, "love" and *sofia*, "wisdom") is a specific form of social consciousness and inquiry developing a system of knowledge about the fundamental principles and foundations of human life; about the most general essential characteristics of the human attitude toward nature, society and spiritual life in all its basic manifestations. Contrary to mythological and religious outlooks leaning upon faith and fantastic worldviews, philosophy is based on theoretical methods of understanding reality using specific logical and epistemological criteria to validate its provisions.

A social function of philosophy is to find new theoretical landmarks through rational understanding of cultural universals, their critical analysis, and subsequent building of new theories. Rational explication of the meanings of the universals of culture in philosophy begins with what might be called catching the general in qualitatively different spheres of human culture, understanding their unity and wholeness. Therefore, not so much concepts

Pacifism –Political Science and Global Studies 373

but images, metaphors, and analogies become the initial forms of philosophical categories. This peculiarity can be clearly seen as we look at the origins of philosophy. Even in the relatively elaborated philosophical systems of Antiquity many fundamental categories still contain some elements of symbolic and mythological world outlooks (such as Heraclitus' Firelogos, or Anaxagoras's Nous).

For the Eastern philosophical schools, traditionalism and a tendency to justify the established social values were typical. The rational and logical components, and association with science, were relatively weak here, but the ideas about the cosmological nature of consciousness, principles, and techniques of everyday wisdom, moral education, and spiritual self-control were being worked out and validated in detail. All these theoretical orientations naturally became a part of the culture of the traditional agricultural civilizations with their typical orientation toward reproducing the extant way of life, hierarchy of clans and casts, and fixing individuals within a system of rigidly defined corporate relations.

Ancient culture engendered a different style of doing philosophy. It was modeled after social life of a polis, based on trade, industrial arts and democracy, and characterized by the greater dynamism compared to other traditional societies. Here a philosophy emerged that was oriented toward connections with science and constructing logical and rational systems of knowledge. Ancient philosophy contained in embryo the basic research directions, which were to define future philosophical development. In the course of Western philosophical development from the Renaissance and the Modern period to the Age of Enlightenment, the basic theoretical ideas were enunciated and validated that defined the transition from traditional civilizations to a totally new type of civilizational development—a technogenic civilization that was born with capitalism. During this historical period, a great philosophical revolution occurred. A new vision was formed of human beings as active subjects determined to reshape the world and of nature as a consistently ordered field where human activity should be applied was developed. The value of scientific rationality as the regulative foundation of human activity was established, and the ideas of social contract, individual sovereignty, and natural human rights were validated.

Crises of the twentieth century (ecological and anthropological) endangered human survival itself. The demand emerged for finding strategies for the attitude of human beings to nature and for interpersonal communications, thereby making the issue of new theoretical guidelines more acute. This is the main task of the modern philosophical inquiry. In this process the dialogue between Western and Eastern philosophical traditions that is a part of a broader intercultural dialogue plays an increasingly important role.

V. S. Stiopin

Policy (Greek *apódeixis*, "a showing or setting forth") refers to ideas determining purposeful activity aimed at forming vitally important relationships among states, peoples, nations, classes, and social groups within a society.

The core of policy is winning, retaining, and consolidating state power, putting it to use in the interest of the development of the society's well being

374 *Global Studies Encyclopedic Dictionary*

and its security, for solving other internal and external political, socioeconomic, and defense tasks. The primary political tasks include the quest for ways of realization of national interests through the development of the political system, supervising principal social processes, governing the state, and supervising all kinds of public activities, class, ideological, and party struggle.

Foreign and domestic policies are two instruments of the state allowing for realization of national interests. A domestic policy preoccupied with solving internal social tasks is determinant, consisting of economic, social, technological, cultural, and other policies of a state. Foreign policy covers the activities of a state (the ruling party) in the international arena, dealing with interstate relations, first, among neighboring states and aims at promoting the state's vital interests and security, at strengthening its international positions and increasing its impact on global socioeconomic and military-political development.

The level of economic development and political order of a society determine state policy. The most important factor responsible for the viability and effectiveness of a policy is its correlation with the material and spiritual demands of a society and a correct assessment of the real economic capabilities of a state, its ethnic peculiarities, political, and geographical situation.

D. O. Rogozin

Politics, World: World politics is a branch of science and a university discipline that appeared in the United States in the last quarter of the twentieth century. It considers the new political system that is being formed within the framework of globalization, which differs from the Westphalian one that has existed for more than 350 years.

The Westphalian system (model) appeared as a result of the Treaty of Westphalia, the peace treaty signed in 1648, that ended Europe's Thirty Years' War, which recognized national sovereignty as a key principle and thus gave rise to the political system that subsequently spread over the European boundaries and was named the "state-centric world system." The principle of national sovereignty presumed that each state had absolute power within its borders and developed its own foreign policy. This right had to be respected by other states. A system of intra- and inter-national relations with corresponding mechanisms and management and political and legal norms began to shape. Since that time, the state was the "cell"—the building block-of the world political system. Within the classical Westphalian system, the only actors of international interactions were the states that presented themselves or formed coalitions to reach specific goals. Corresponding academic branches reflecting political relations on different levels were developing. The analysis of state activities became the focus of political science, and the analysis of intergovernmental relations became the focus of international relations.

At the end of the twentieth century, the role of new—so-called nontraditional—actors on the world arena has increased. They include intergovernmental organizations (such as the United Nations (UN), the Organization for Security and Cooperation in Europe (OSCE), and the World Trade Organization (WTO), transnational corporations (Ford, Coca-Cola, Microsoft), and international nongovernmental organizations (such as Green Peace, the International Red Cross, and Medicines Sans Frontiers). According to such criteria as quantity, political impact, and financial resources they began to compete with the state, though the state remained the principal actor. In addition to it the development of new information and communication technologies helped to greatly intensify international interactions and make state borders transparent. As a result, discussions began on the vanishing of national sovereignty, the erosion of the Westphalian world system, and the transparency between domestic and foreign policy and accordingly, between political science and international relations. The new emerging branch of science, world politics, combined to some extent both political science and international relations.

Meanwhile, economic factors were becoming increasingly significant. The Japanese economy and several European economies destroyed during the World War II were on the rise. The problem of the North-South economic gap became more pressing. The energy crisis of 1973–1974 provided some of the most irresistible evidence that the role of economic factors in defining the position of the state in the world structure had been growing. At that time the Organization of Petroleum Exporting Countries (OPEC) limited oil production. The United States, with its enormous military-political potential, could do nothing about it. It had to carry on negotiations with OPEC countries and thus make mutual concessions to solve the problem. The growing role of the economic factor stimulated the formation of a separate academic branch under the name "International political economy" (IPE).

Another important trend in the development of world politics aimed at defining the interconnection between the domestic policy of different countries and the international political economy. Peter Katzenstein, Robert O. Keohane, and Stephen Krasner (1998) tried to define the determinants of the foreign economic policy of states and of the strategies of different corporations. Most of these works were empirically grounded case studies. They showed that the internal political organization of the state has a significant influence on economic behavior on both national and supranational levels. Scientific works connecting international factors with internal ones played an important role in the formation of world politics as a discipline.

Later on, researchers focused their attention on how domestic policy influenced the foreign policy and, vice versa, what affect the international system had on domestic political structures and processes. Boundaries between foreign policy and domestic policy and correspondingly between the two kinds of studies were getting increasingly transparent.

Important patterns emerged in the discipline of political science; e.g., public policy came to be considered the result of the conflict of different group interests. It was shown that groups formed coalitions to realize their goals. These coalitions were not steady; they changed depending on the circumstances. The work of the American scholar Graham T. Allison (1971) was of great importance for the development of research in this area. Allison showed that the state was not as much unitary as it previously had been perceived. State policy can be comprehended if we interpret it as the interaction and competition of different structures within the state. On this view, national interest is formed through compromises made by different groups.

Another trend of studies that greatly influenced the formation of world politics as a discipline was the study of international institutions. After World

376 *Global Studies Encyclopedic Dictionary*

War II, there was a sudden increase in the number of international institutions. They became a topic of research for political scientists, economists, and specialists in international relations. In the 1960s, works on European integration appeared. These works showed the tendency of scholars not only to describe the activities of international organizations but also to give a comparative analysis. Special emphasis was placed on UN activities and its peace operations. Thus, a subject for discussion became the necessary adaptation of UN activities to the realities of the Cold War and superpower competition.

Finally, the most significant contribution to the development of world politics was made by research studies devoted to describing transnational relations, the assertion of the plurality of actors on the world arena, and the analysis of interdependence in the modern world. They exposed a fundamental shift in world political architecture that was followed by a fundamental reformulation of the whole international agenda-security issues, conflicts and their settlement, ethnic relations, and environmental protection.

Spheres of activities of all contemporary actors in the world arena are strongly intertwined. Earlier, for instance, intrastate regions tended to influence only domestic political processes, while international organizations tended to influence issues of foreign policy; that seemed to be quite logical. Now the situation has changed. International organizations (NATO, OSCE, and the United Nations, in particular) increasingly interfere in such internal affairs as settlement of domestic conflicts, observance of human rights, and developing state financial policy (the International Monetary Fund). Intrastate regions tend to play an equal role in comparison with the state role in foreign policy. It often causes anxiety and confusion for the central authorities.

Contemporary world politics is a fast developing branch of science that embraces the full spectrum of international problems for the globalizing world. The latter means that non-state players became actors in the world arena and nation-state borders became transparent.

References: Allison, G. *Essence of Decision.* Boston: Little, Brown, 1971. Katzenstein, P., R. O. Keohane, and S. Krasner. *Exploration and Contestation in the Study of World Politics.* Cambridge, MA: MIT Press, 1998.

M. M. Lebedeva

Political Concensus (Greek *ta politika*, "affairs of state" and Latin *consensus*, "agreement" "accord") The stability of a political system depends on the degree of consensus among the governed. By means of political consensus the real, actual legitimatization of political authority is carried out.

The character of consensus and methods of consensus formation and maintenance vary in stable and unstable public systems. Three levels of a consensus include: (1) Consensus at a level of community (basic consensus), (2) consensus at a level of a mode (procedural consensus), and (3) consensus at a level of governmental policy.

The first level of consensus involves fundamental values; viz. whether society members share the same values and purposes. Procedural consensus is necessary to establish rule of law and administrative procedures, including specification of methods to settle conflict and disputes; these are usually codified in constitutions, legislative acts, and contractual documents. This is the

Pacifism –Political Science and Global Studies 377

necessary condition and actual precondition for democracy. The third level of consensus involves consent of governing representatives necessary to establish policies; viz. the process of making and implementing political decisions. This is the weakest, situational type of consensus. Strong political stability can be achieved only through the presence of all three types of consensus, building on basic and procedural consensus.

A. V. Glukhova

Political Culture is a collection of attitudes, beliefs, and standards that comprise the behavior of the political system on the whole and of individuals, formed as the result of a lengthy historical development.

Political culture is the result of lengthy historical development, during which varied forms of political organization and political conscience of the people are formed. It is made of political attitudes, morality, and traditions. This term is used in: (1) comparative cross-cultural studies to characterize the distinctive features of national political systems, and (2) analysis of people's political behavior, including the specificities of their political participation, electoral behavior, level of political activism and political commitments.

American political analysts Gabriel Almond and Sidney Verba categorize political culture into "parochial," "subordinate," and "participational" types, based on political orientation. Modern liberal democracies are characterized by a combination of participational orientations and political obedience. This underpins the stability and efficiency of the modern "civil culture"—the culture of citizens who are loyal toward the political system, positively disposed and competent. On the contrary, if the political culture includes ideological or social hegemony (class or ethnos superiority) it becomes a conservative force impeding social transformation processes.

Political culture sets the parameters of the interpretation of political events. Correspondingly, the motivations of people's political behavior, the reasons for conflicts, and so on, are not always caused by patent factors. Sometimes they are the consequence of a stable system of standards and symbols inherent in the political culture. Affiliation with a political culture creates the sense of common political identity.

I. V. Mitina

Political Science and Global Studies is a subdivision of modern global studies, which focuses on political aspects of globalization. Beyond merely describing global problems, it attempts to offer recommendations for means and methods of problem resolution. The methodology involves civilizational and sociocultural approaches that consider globalization as possible and desirable.

This field endeavors to develop anticipatory knowledge and acceptable alternatives for the future. Within the civilizational approach, the subject of global studies in political science is globalization in its historical dynamics, viz., the formation of a single interconnected world through creative dialogue of civilizations.

Dialogue is considered to be constitutive of a new logic of the global world, which promotes humane global thinking as a chance for coexistence and interaction on a global scale. It views dialogue among civilizations as an adequate answer to the contemporary global ethno-confessional, geopolitical, ecological, moral, and cultural challenges. Aspects include:

(1) comparative analysis of the basic sociocultural values of world civilizations, determining the value dimension of political dialogue in the global world; development of a new universalism through comprehension of "the blooming complexity" of planetary existence;

(2) comparative analysis of political strategies of civilizational interactions; promoting humanitarian peacemaking as an adequate answer to the recurrences of the "new barbarism" and brutality;

(3) development of humanistic globalism as an alternative to the uniformity and hegemonism in the global world;

(4) enunciation of a new concept of humanitarian consensus as a process of value coordination among different civilizations;

(5) study of the emerging problems of global political time and space;

(6) study of the deformations caused by globalization (exchange of mass culture "anti-values," nonequivalent information exchange);

(7) comparative analysis of the alternative paths of globalization from the position of the post-classical concept of progress.

I. A. Vasilenko

○Q ○

Al-Qaeda: (Arabic *al-Qāʿidah* "the base") broad-based militant Islamist organization founded by Osama bin Laden in the late 1980s. Al-Qaeda began as a logistical network to support Muslims fighting against the Soviet Union during the Afghan War. Members were recruited throughout the Islamic world. When the Soviets withdrew from Afghanistan in 1989, the organization dispersed but continued to oppose what it considers corrupt Islamic regimes and foreign presence in Islamic lands. Based in Sudan during the early 1990s, the group eventually reestablished its headquarters in Afghanistan (circa 1996) under the patronage of the Taliban militia.

Al-Qaeda has merged with other militant Islamist organizations, including Egypt's Islamic Jihad and the Islamic Group, and on several occasions its leaders declared jihad against the United States. Camps for Muslim militants train tens of thousands in paramilitary skills, and its agents are engaged in terrorist attacks around the world, including the destruction of the US embassies in Nairobi, Kenya, and Dar es Salaam, Tanzania (1998). In 2000, a suicide bomber attacked the *USS Cole* in Aden, Yemen. Numerous other suicide attacks have been conducted around the world, with the most damaging being that against the World Trade Center in New York and the Pentagon on 11 September 2001, after which the United States led an invasion of Afghanistan, which developed into a protracted US presence in that country. This can be seen as a vicious circle of events, of course, since the US presence there and elsewhere in the Middle East has only served to intensify al-Qaeda's resistance against the United States and others they consider complicit.

While the war in Afghanistan has allegedly compromised the "headquarters" of al-Qaeda, its members and philosophy continue from other locations. Its guerrilla attacks continue to be impossible to completely eradicate.

Reference: *Encyclopædia Britannica Online*, s. v. "al-Qaeda." Accessed 17 January 2014, http://www.britannica.com/EBchecked/topic/734613/al-Qaeda.

E. D. Boepple

Quarter, Brook No: (brook, "to endure"; Old English, *brucan* "use" "enjoy" cohabit with"; give quarter, "spare from immediate death") To brook no quarter is a military term meaning to offer no clemency or mercy to the enemy, also often called a "take no prisoners" policy. Under Article 23d of the 1907 Hague Convention IV, The Laws and Customs of War on Land, it is especially forbidden to give orders to brook no quarter. Even if orders to brook no quarter are prohibited, the potential damage of modern weapons can cause the same result. Because such weapons can fail to discriminate between combatants and noncombatants, some guerrilla wars, the saturation bombings of World War II, and the use of nuclear weapons can make such military actions ones that brook no quarter in their consequences quite apart from any declaratory policy rejecting such a practice.

W. C. Gay

Quixote Center is a multi-issue, grassroots organization founded in the Catholic social justice tradition, which organizes campaigns for systemic

380 *Global Studies Encyclopedic Dictionary*

change. Operating independently of church and government, it strives toward just, peaceful, and equitable policies and practices. (1) Quest for Peace seeks peace and friendship with Nicaragua by advocating just US policies, such as opposing US funding of the Contra War against the Sandinista government in the 1980s, and supporting Nicaraguan human development organizations. (2) Haiti Reborn builds grassroots support in the United States for improved policies toward the people of Haiti and supports Haitian development organizations, in particular ones that promote reforestation and sustainable agriculture. (3) Catholics Speak Out and Priests for Equality encourage Roman Catholic reformers pursuing the direction of the Second Vatican Council and equality for all men and women in both church and society, including advocacy of ordination of women, ending celibacy, and promoting LGBT rights.

Reference: http://quixote.org.

W. C. Gay

∽ R ∾

Racism, Environmental: Inequitable exposure to pollution based on socio-economic status. In the United States, sources of pollution such as toxic waste dumps, incinerators, manufacturing plants, and other facilities are typically located in lower income neighborhoods—often African American and Latino—because poorer communities do not have the political and legal clout that wealthy communities do. Conversely, environmental justice is the equitable distribution of pollution regardless of social class or race. In the United States, the legal foundation of environmental justice is the Civil Rights Act of 1964, specifically Title VI that prohibits exclusion of any person or group from participating in any program receiving federal financial assistance.

Common attempted justifications for the inequitable distribution of pollution point sources are the following. (1) According to a Utilitarian viewpoint, the greatest good may be effectuated by subjecting select communities to inordinate amounts of pollution. The problem is public policy based solely on Utilitarian criteria opens the door for the discrimination of select individuals and/or groups. (2) From a Libertarian perspective, individuals and corporations should be able to do with their private property as they wish, including the opening of toxic waste facilities. The problem is ecological systems do not conform to property lines. (3) Pundits of lassie-faire Capitalism argue that the market should operate unfettered by governmental intrusion, especially laws protecting communities from pollution. The problem is unmitigated market forces result in unethical social practices, such as human and drug trafficking, bribery, collusion, and the formation of monopolies.

An alternative conception of justice that lays the ethical foundation for concluding that environmental inequity is unjust is the work of American philosopher John Rawls, who argues that a just society is one in which the least well off still live in the most favorable conditions possible, and that discrepancies between the best off and worst off are only justified if the least well off benefit from those social inequities. On this model, environmental injustice occurs when some persons are exposed to pollution and industrial toxins from which they do not benefit. Environmental justice is the distribution of pollution and industrial toxins in such a way that the living conditions of those most exposed are better off than they would be without the presence of industry.

Increasingly environmentalists and civil rights activists in the United States agree that some sort of environmental injustice or environmental racism exists. In 1982, the largely African American community of Warren County, North Carolina, fought a polychlorinated biphenyl (PCB) disposal site. In 1984, a leak in a Bhopal, India Union Carbide plant killed 4,000 people, and consequently residents of Kanawha Valley, West Virginia, questioned Union Carbide officials about the safety of the plant located there and discovered that Environmental Protection Agency (EPA) officials were stonewalling efforts to investigate community health concerns. This incident demonstrates that, since this area is predominantly Caucasian, environmental injustice is more accurately characterized by socioeconomic status rather than race. In 1990, the EPA began to consider the issue of environmental injustice.

382 *Global Studies Encyclopedic Dictionary*

In 1993, building on the foundation of the Civil Rights Act of 1964, Congress passed the Environmental Justice Act, the Environmental Equal Rights Act, and the Environmental Health Equity Information Act. In 1994, President Clinton signed executive order "Federal Actions to Address Environmental Justice in Minority Populations and Low-Income Populations."

Why has the mainstream environmental movement waited until recently to address the issue of environmental racism? In part in the United States, academic environmental ethics has tended to be holistic (focused on whole ecological systems rather than individuals) and non-anthropocentric (not human-centered). Moreover, environmentalism as a political force has its roots in affluent white citizens aiming to preserve wilderness. Mainstream environmentalism has also been institutional (focused on legislation) rather than grass roots (focused on activism). Interestingly, the Civil Rights Movement in many ways has had opposite emphases: it has focused in individual rights and liberties, been human-centered, and found its political strength in the grass roots.

Since the 1990s, for a growing number of civil rights and environmental activists, in order to take the issue of civil rights seriously, the problem of environmental racism and injustice cannot be overlooked.

References: Rawls, J. *A Theory of Justice*. Oxford: Carendon Press, 1972.

D. R. Keller

Radioactivity, Pollution of the Environment by: by origin all the radioactive substances found in the environment can be divided into natural (potassium-40, elements of uranium and thorium families, and cosmic-origin radionuclides) and artificial substances (fragments of atomic nuclei fission, activation products). The radiation situation is formed by the emanation of these radionuclides and by cosmic radiation. Natural radionuclides are actually ubiquitous: potassium, as one of the rock-forming elements, and uranium and thorium, as scattered elements, are constituents of soils, rocks, waters (in small amounts), plants, and animals. Air contains radon and the products of its decay belonging to the uranium and thorium families. Natural radioactivity (together with cosmic radiation) has existed throughout the geological history of the Earth and has always been one of the factors of the evolution of life.

Human interference with natural processes can change the radioactive situation even without carrying in artificial radionuclides. In this case, anthropogenic changing of the radiation background is caused by accumulation of natural radionuclides, employment of increased-activity materials in the building industry, and retaining of radon released from the ground by buildings.

Artificial radionuclides are mainly generated in the course of the fission of heavy nuclei in nuclear reactors and fission nuclear explosions, and by the activation of nuclei typical for fusion nuclear explosions (thermonuclear bombs), although they can emerge in the course of nuclear reactor operation.

The problem of radioactive pollution has appeared recently, in the late 1940s, with the beginning of the production of plutonium for weapons and the conducting of tests of nuclear weapons. The 1950s and 1960s were a time when this kind of anthropogenic effect on the environment turned into a global ecological problem.

Pollution ubiquity is caused by radioactive products of middle and high power nuclear explosions getting into stratosphere, from where, after mixing and staying in this pool, they precipitate onto the Earth surface for many months. The periods of the highest emission of radionuclides into stratosphere were 1957–1958 and 1961–1962.

Environmental pollution by mass-scale fall-out of products of tests of nuclear weapons had been forming from the end of the 1950s until the mid-1960s. In 1963, an agreement banning nuclear tests in the air, space, and waters was signed in Moscow between the Soviet Union and the United States. France and China refused to join it and continued conducting nuclear tests in the air until 1974 and 1980, respectively.

Radioactive fall-out can be classified into the following groups: (1) near or local (dozens of km from the source, particles <50 mkm), (2) distant (first hundreds of km from the source, particles size 10–50 mkm), (3) semi-global (tropospheric, fall-out can continue for one or two weeks, traces can be monitored to thousands of km from the source, particle sizes up to the order of mkm), and (4) global (stratospheric, continuing for many weeks, months, and even years, spreading over the entire surface of the globe, particle sizes a fraction of mkm).

The main ingress of plutonium-238 on the surface of the Earth is due not to nuclear explosions, but to burning in the atmosphere of the American space satellite SNAP-9A in 1964, where this radionuclide was used as a nuclear energy source. The conducting of surface nuclear explosions resulted in the formation of local radioactive traces. In the Soviet Union, surface nuclear explosions were mainly conducted on nuclear proving grounds. Currently, aero-gamma surveys register only two traces of all the surface and underground excavation nuclear explosions, conducted on the proving ground in Semipalatinsk; one of them is caused by the nuclear explosion of 24 September 1951, and the other by the nuclear explosion of 12 August 1953. The radionuclide composition of fall-out in the traces of surface explosions differs from that of global fall-out by the presence of long-lived induced-activity radionuclides, such as europium-152 and some other rare-earth radionuclides.

Currently, the era of mass test of nuclear weapons in the air and other environments seems to have come to an end. However, this epoch left a tangible legacy, global pollution in the form of long-living radionuclides, ubiquitous all over the globe.

Currently, radionuclide pollution of the environment caused by severe accidents with nuclear reactors has become an issue of special concern. In the course of such severe reactor accidents, the high-temperature melting of the whole active zone or its part occurs. Moreover, unless the reactor is covered by a strong and durable hood, as happened in the accident at the "Three-Mile-Island" nuclear power station (1979, United States), most of the radionuclides escape into the atmosphere. A similar case was the accident at the Windscale reactor (1957, United Kingdom). But should a severe accident be accompanied by the explosion of the active zone (even if thermal), as happened in the block 4 of the Chernobyl nuclear power station, the total range of radionuclides accumulated in reactor at the moment of explosion escapes into the atmosphere (although the percentages of radionuclides can vary). The Fuku-

384 *Global Studies Encyclopedic Dictionary*

shima Dai-ichi plant Japan has been leaking hundreds of tons of contaminated underground water into the sea since shortly after a massive 2011 earthquake and tsunami damaged the complex. Several leaks from tanks storing radioactive water were present as late as Autumn 2013.

The consequences of an accident related to artificial radioactivity escaping into the environment are pollution of the atmosphere, the hydrosphere, and the landscape components (soil, natural and anthropogenic flora and artificial objects). The main landscape component is the soil, where artificial radionuclides are accumulated and redistributed. Radioactive pollution of the geological environment can be caused by underground nuclear explosions conducted in the adits of the proving grounds and by the burial of high-activity nuclear wastes. The main hazard of geological environment pollution is a possible pollution of ground waters.

Pollution of natural environments leads to the exposure of people and other biological objects to various radiation doses, added up from the direct influence of radionuclide radiation (external radiation) and from the consumption of food by living organisms (internal radiation).

Y. A. Israel

Rawls, John (1921–2002): American political philosopher, Rawls is best known for a single book, *A Theory of Justice* (1972), which played a major role in determining the direction of political theory discussions in the Anglo-Saxon world and beyond for several decades. Its premise is that imaginary parties placed in a hypothetical "original position," in which they know some general facts but not the social positions that they would occupy in the real world, and charged with deciding basic principles of justice to govern that world, would choose, first, a principle of equal liberty for all, and then a principle of fair distribution of goods. The latter, which Rawls called "the difference principle," mandates that inequalities in distribution are permissible only to the extent to which they will benefit the society's least advantaged members.

In the middle years of his career, Rawls devoted increasing attention to the problem of trying to achieve consensus about political institutions in modern pluralistic societies in which there is an extreme diversity of religious and other comprehensive worldviews. In *Political Liberalism* (1993), a collection of lectures and essays from these years, he defends the possibility of persuading virtually everyone to agree on basic political arrangements without having to agree on philosophical principles, thus retreating from some of the stronger ethical claims that he had made in *A Theory of Justice*.

Whereas his earlier studies had deliberately confined themselves to the question of justice in a single, closed society, in *The Law of Peoples* (1999), John Rawls tried to consider the global society. A disappointment to many readers, this work distinguishes between liberal and "decent" peoples, the latter meaning primarily societies that are hierarchical rather than democratic but are still characterized by some principles of fairness, and it also takes for granted the existence of "rogue" or "outlaw" states, the existence of which is said to justify the maintenance of nuclear weapons by "liberal" ones. Moreover, in this book Rawls denies that it is necessary to apply his own earlier principles of distributive justice within single societies to the global scene,

Racism, Environmental – Russian Philosophical Society 385

instead going so far as to claim that impoverished, "burdened" societies are such because of, at least in most cases, deficiencies in their own cultures, work ethics, and reproductive practices. While some philosophers continue to attempt to derive from Rawlsian principles a more generous global conception of justice than his own, others claim to perceive the beginning of an end of liberal democracy as it was known in the twentieth century, coincident with the death of one of its leading late-twentieth century theorists.

Rawls, J. *A Theory of Justice.* Oxford: Carendon Press, 1972. —. *Political Liberalism.* New York: Columbia University Press,1993. —. *The Law of Peoples.* Cambridge, MA: Harvard University Press,1999.

W. McBride

Red Data Book of the Russian Federation, initiated in 1982, by the decree of Council of Ministers of Russian Soviet Federative Socialist Republic, is a list of rare organisms and organisms on the brink of extinction. It contains an annotated index of species and subspecies, indicating their former and current spread, reproduction features, and measures that have been taken and need to be taken for the species' conservation. Information collected for the red books was initiated by the International Union for the Conservation of Nature and Natural Resources (IUCN) in 1949. The first edition of the R.S.F.S.R. red book was released in the 1980s—the volume *Animals* in 1983, and the volume *Plants* in 1988. International, national, and local editions of the red books have subsequently been published on a flora and fauna comprising and a variety of systematical groups. Red books are central to the conservation of biological diversity.

As of 1 November 1997, the *Red Data Book of the Russian Federation* has registered 487 plant species (440 phanerogams, or flowering plants, 11 gymnospermous plants, 10 filicinae, 22 anophytes, 4 mosses), 29 lichens, 17 mushrooms, 415 animal species (155 invertebrates, including 34 insects; 4 cyclostomata, 39 fishes, 8 amphibia, 21 reptiles, 123 birds, 65 mammals). At the same time, 38 animal species have been removed from the red book of the Russian Federation because their condition and number are no longer under threat, thanks to the conservation measures taken.

Many regions of such country as large as Russia surpass the largest European countries in not only area, but also in biological diversity, and are characterized by the high concentration of rare and endemic species. Therefore, apart from the federal red book, there is a need for a regional red book for every region, comprising a very limited list of animals, plants, and mushrooms, which are to be conserved.

The basis for the publication of the regional *Red Books* was the respective classifications of IUCN and the *Red Books* of the Soviet Union and R.S.F.S.R., and the legislation devised for them. From the middle of the 1980s until the present time thirty-seven regional red books have been published, a little more than a third of all the Russian Federation subjects. The significance of the regional red books increased greatly in the 1990s, when local government of the Federation gained more power. Besides, more than 10 years have passed since the publication of the first edition of the Russian red book; the question of the next, enlarged edition is still up in the air, while

386 *Global Studies Encyclopedic Dictionary*

there is a threat of extinction related to many rare species, especially in the regions with intense development of new areas. Apart from the red books, many regions have devised and mostly approved regional lists of animal and plant species in need for conservation. Usually these lists consist of Russian and Latin names of the species, conserved in the region, without an indication of their spread, and number. Many official lists have departmental status, are not widely published, and are still virtually inaccessible for a wide circle.

V. V. Snakin

Refugee is a person forced to leave their place of constant residence because of any emergency and to migrate to another state, receiving the status of a "refugee" in that state in accordance with international agreements and national law. The main reasons for leaving one's own country include military and political conflicts, catastrophes, and natural disasters and the persecution of people with specific allegiances or affiliations. In accordance with the 1951 United Nations Geneva Convention, refuges are defined as persons who are outside their country of nationality or habitual residence, have a well-founded fear of persecution because of race, religion, nationality, membership in a particular social group or political opinion, and are unable or unwilling to avail themselves of the protection of that country, or to return there, for fear of persecution. A person committing a war crime, a crime against humanity, or any grave deliberate crime may not be granted refugee status. As well, within the framework of modern international law, states limit the definition of "refugee," excluding people forced to migrate because of economic reasons (for instance, because of poverty).

Especially large streams of refugees appeared as the result of the World Wars. After World War II, a network of institutions, systems, and legislative acts attempting to resolve refugees' problems evolved, the base of which is the Office of the UN High Commissioner for Refugees (UNHCR) and the 1951 United Nations Convention on the Status of Refugees.

The number of refugees worldwide reached its peak at 17.8 million in 1992. Since then, it steadily decreased, and by the beginning of 2004, according to UNHCR information, there were about 9.7 million refugees worldwide (this number does not include about 4 million Palestine refugees in the Middle East falling under the mandate of the UN Relief and Works Agency for Palestine Refugees (UNRWA). The main cause of this reduction has been the resolution of long standing conflicts (first in Angola, Afghanistan, Bosnia and Herzegovina, Liberia, Mozambique, and Rwanda). The largest country of origin for refugees is Afghanistan (2.1 million); Sudan occupies the second place (606 thousand), and Burundi the third (531 thousand). The total number of refugees forced to leave their countries in 2003 was 310 thousand. The most significant overall movement of refugees was recorded in the countries of tropical Africa. The greatest numbers of refugees are now in the countries of Asia and Africa (about 60 percent) and Europe (about 25 percent). The largest countries by number of the received refugees are Pakistan (1.1 million.), Iran (1 million.) and Germany (960 thousand).

I. A. Aleshkovsky, V. A. Iontsev

Racism, Environmental – Russian Philosophical Society 387

Regionalism/Regionalization are concepts that have been used in the fields of political science, global studies, and international economics with growing frequency in the post-1991 world. The end of the bi-polar world brought about by the collapse of communism in Eastern Europe led to an increased focus on different regions of the world and their development as political and economic units. As such, references to regionalization and regionalism appeared with greater frequency in the academic literature of the 1990s. A decade later, the terms have become virtually synonymous with one another and are frequently understood to be interchangeable. However, the use of these two concepts can be confusing, as they do not always refer to the same set of issues.

In the simplest sense, both regionalization and regionalism refer to the process of examining, understanding, identifying, and analyzing the features of regional development. While each concept makes reference to the exploration of regional issues, be they economic, political, or social, and while both terms are commonly used interchangeably, they have much different origins and ultimately different definitions.

The term regionalization has been in use for several decades, primarily as a concept in the fields of geography and taxonomy. Geographers and taxonomists in identifying different features of regional importance, either organic or physical, understand regionalization, in this context, as a system of classification for use. However, in the post-Cold War world, use of the concept of regionalization has grown beyond the language of geography and taxonomy. The term has been co-opted by political scientists, international relations experts, and political economists seeking to understand post-Cold War developmental trends. Utilization of the concept of regionalization has proven to be useful in opening research perspectives into regional development, as it has been taken as a process-oriented approach to the phenomenon of regional integration. As scholars have examined the increasingly important processes of regional development, they have adapted their language of inquiry to current themes in interstate affairs. In this respect, the attention given in the last decade to the so-called process of globalization has played an important role in the development of the concept of regionalization as an analytic tool for understanding interstate and international relations.

The concept of regionalization lends itself to debate as it is often used to focus on different perspectives of the same phenomenon. In contemporary academic literature, regionalization has been used to explain both the move toward regional political and economic development and the move away from the process of globalization, the shared component being the development of regional patterns of behavior. It has been understood to be both a response to the forces of globalization (strengthening regional interests over global influences) and as an intermediary step toward full globalization (formation of regional blocs prior to the implementation of a universal political and economic system). Those scholars who support the latter see regionalization as simply an evolutionary step toward the realization of a unified global and economic system. Those scholars favoring the former definition see regionalization as the realization that global processes are not inevitable. This debate regarding the characteristics of regionalization, of whether it is a natural developmental process or a response to globalization pressures, is part of the

388 *Global Studies Encyclopedic Dictionary*

difficuly in settling upon an exact, or exacting, definition of the concept. Yet it is precisely this debate that opens up new areas of inquiry as more data is collected, more observations are made, and more opinions are explored.

The term regionalism has a longer history of use than regionalization in the vernacular of political science and international relations, international economics, and global studies, but it has only gained momentum as a theoretical approach after the collapse of the League of Nations in the 1930s, and the establishment of the bi-polar political world in the 1940s (Calleya). In much of the scholarly literature, the debate leading to an exploration of the characteristics of regionalism has been a result of the formation of European regional organizations in the 1950s, such as the European Coal and Steal Community (ECSC) and European Atomic Energy Community (EURATOM). The emergence of such new integration schemes, existing inside a global political structure that was ideologically, politically, and economically constrained, was cause for a greater examination of the nature of regional organization and the structure of its development.

As a concept, regionalism offers a rich diversity of approaches in understanding regional development. It is a concept that is of particular importance for political scientists and international relations theorists as it is part of the debate within these fields of the "level of analysis" question, an ongoing argument about what is the most appropriate level in which to analyze interstate relations. Early post-World War II scholars focused much of their energy on defining the field through normative speculation regarding the proper ordering of the world. From this grew a plethora of approaches that were based on the more seemingly sound theories and continuously collected descriptive data.

Ultimately, regionalism and regionalization are both concepts that seek to explore the nature of regional development. While regionalization focuses on the process aspect of regional integration, regionalism examines the theoretical framework of this same phenomenon. As a concept, regionalism is the more developed due primarily to the application of standard political theory in its development, and the tendency of political scientists and international economists, among others, to focus on theory in creating greater understanding of observed phenomena. In addition to following many of the theoretical debates within these fields, theories of regionalism to some extent mirror regionalization theories by also following the process aspects of regional development, that is, insofar as they contribute to the greater theoretical model that is being explored. Regionalization, due to its focus on process itself, tends to be more limited in its exploration. Yet, as the newer of the two concepts in the examination of regional integration, it appears to be more promising and less static than regionalism theories. In the post-Cold War world, new forces have been unleashed that no scholars have had sufficient time to understand or the proper tools with which to understand them. In this respect, regionalization theories may prove to be the key in creating a greater understanding of the dynamic phenomenon of regional integration.

References: Calleya, Stephen C. *Navigating Regional Dynamics in the Post-Cold War World.* Brookfield, VT: Dartmouth, 1997.

S. M. Cox

Regional Industrial System: Regional industrial systems, which cultivate regional industrial competence, can be understood by virtue of two typical theoretical patterns: (1) regional manufacturing synthesis theory and (2) industrial cluster theory.

Regional manufacturing synthesis theory was developed in the former Soviet Union and is often based on the strategic resource of developing special regions. It sets up specialized sectors and service parts for developing and using resources. While its scope is not restricted administratively, it has some regional administrative organization. The structural pattern and mechanism of this kind of regional industry, used for comprehensive development and use of resources, has worked efficiently in the planned economic system. On the other hand, it has the disadvantages of a planned economy. For instance, enterprise is the object of "arrangement," and higher authority, which means that the enterprises are totally passive in their positions, decides the relationships among enterprises. Consequently, there is neither competition of any form nor any motive force for enterprises to innovate. On the other hand, given the regional division of labor based on static advantages, enterprises have no prompt industrial adjusting mechanism. In addition, the central administrative organization, whose job is to manage completely and comprehensively, keeps a kind of rigid relationship with local government, and its functions are not divided organically. A lack of competence and vigor lies in the regional manufacturing synthesis because it has the following characteristics: it is resource-directed; it is government-oriented in form, and it must obey the direction of planning from beginning to end.

The regional industrial cluster theory, though a product of the market economic system, has something in common with regional synthesis theory. But they have distinctive meaning. The most obvious distinction is that enterprises in competition have been made the core of the regional organization. Therefore, regional competence is closely related to enterprises, and this theory becomes an effective micro-organizing form and theoretical reference that assures that the regional economy will develop effectively and coordinately in the market economic system. Regional industrial cluster theory emphasizes a concentration of enterprises of related industry in the same region, and they are reliable, cooperative, and competitive in relation to each other. This concentration stems from competition, which serves as both a productive organizing form and a business managing means. It is the production of a market economy. Industrial cluster includes various kinds of corporations, but the core of it belongs to related industries. They promote each other to higher and higher level through free cooperation and competition. Two characteristics of the industrial cluster are pointed out in this theory. The first is the active motive power of enterprises and the second is quick economic foundation. The active motivation of enterprises ensures close cooperation among them. Meanwhile, technology communication speeds up so that each corporation obtains its advantage in competition. However, without the support of a reasonable industrial structure and quick economic foundation, no corporation can obtain or give full play to its final competitive advantage.

Therefore, to form an industrial cluster as soon as possible, it is vital to produce a more active cultivation of enterprises and quick economic founda-

390   *Global Studies Encyclopedic Dictionary*

tion. The following are functions of the regional industrial cluster theory: (1) Relate the regional industrial structural problems of the medium-level to the corporations and enterprise organizations of the micro-world; emphasis is put on setting up the regional competitive structure, of which the core is enterprises in a market economic system. (2) Emphasis is also put on the importance of an abundant economic foundation in the forming of a regional industrial cluster, which gives clear direction to government in encouraging and helping to develop an industrial cluster. (3) The concept of "enterprise motivation" is put forward. It means that through competition and cooperation enterprises of related industries may obtain their competitive advantages, which motivate the industrial cluster to form and become consolidated.

It is worthy of special attention that an industrial cluster does not simply refer to the gathering of the enterprises in a specific region and space. The key of the industrial cluster-organizing pattern is to obtain a clustering innovation advantage to increase the regional industrial competence through clustering innovation, thus realizing the directed strategy of development facilitated by clustering innovation.

The so-called clustering innovation refers to gains in technology innovation achieved by the application of clustering advantage. So, it can be defined as a kind of innovative organization pattern, with enterprises of the same industry or related industries that are based on a specialized division of labor and cooperation as its main body. By means of gathering the enterprises in a specific region or in nearby areas, the effect of clustering innovation can be obtained, thus gaining innovation advantage. The structure of this organization lies between the market and its levels. Being steadier and more flexible, it helps the enterprises maintain a kind of long and steady relationship of innovative cooperation. In practice, this definition includes the following contents: (1) The premise is the specialized vision of labor and cooperation with an emphasis on communication among enterprises. (2) As the main innovative body, the enterprises must belong to the same or related industry, and they should keep in industrial contact with each other often. (3) It is done by means of gathering together the enterprises themselves or by a specific kind of organic concentration of enterprises by government that is to say, by drawing near geographically.(4) The functional purpose is to obtain innovation advantage. More emphasis needs to be placed on the nature of the concentration of the enterprises.

L. You-Jin, G. Qing-Lin

Religion: (Latin *religio*, "respect for what is sacred" "reverence for the gods" conscientiousness," "sense of right" "moral obligation" "divine service" "mode of worship" "cult"; Late Latin, "monastic life") Religion is a form of worldview; a social phenomenon that includes religious consciousness, religious cults, and religious organizations. Religion is often related to believing in the supernatural and that it influences human life. Religion positions morality and provides one of the foundations of spiritual culture.

Religions are divided into primeval, local, and world ones. Totemism, magic, fetishism and animism, veneration of spirits, ancestors, chieftains, and shamanism are considered primeval religious beliefs. Religion is based on tradition, which may be oral (stories, legends passed down by word of mouth) and written (Holy Scriptures, often specially guarded against any changes, like the Bible or the Koran). With the spread of writing, the influence of oral tradition declines, while the role of Holy Scriptures increases. Judaism, Christianity, and Islam are under the common name of "religions of the Book."

In primeval religions, material objects and natural phenomena were objects of religious worship (deities). Later on deities personified social authority. The local forms of religion initially emerged within one people or group of peoples united into a state (such as religions of ancient Egypt, Mesopotamia, Persia, Ancient Greece and Rome, Hinduism, Shintoism, and Judaism). Then religious systems developed that spread among the peoples of different countries and continents. Buddhism, Christianity, and Islam are world religions.

Early forms of religion can still be found in tropical Africa, among ethnic minorities of Asia and Indians from the basin of the Amazon River, aboriginals of Australia, and others areas. Their followers number around 103 million people with the majority coming from Africa (70 million) and Asia (30 million). The followers of the most ancient monotheistic religion of Judaism are mostly ethnic Jews, most of whom live in the United States (5.8 million) and Israel (4.6 million). Christianity is the largest world religion in terms of the number of believers. Christians account for around 1955 million-approximately 34 percent of the global population, that is, every third inhabitant of the Earth is a Christian. Followers of Islam make around 1126 million (19 percent of the global population). The Buddhist population is estimated at around 500 million.

Modern interfaith relations are often characterized by conflicts caused by intolerance and xenophobia. The adherents of other religions are perceived as "enemies of the faith," who are subject to isolation, public denunciation, quite often-violent treatment, and even physical extermination. Nevertheless, several international religious organizations and movements unite people of different religions (such as the Ecumenical Movement, the World Council of Churches, and the International Association of Religious Freedom (IARF). They advocate the idea of creating a world community in which believers of different religions could cooperate constructively respecting the opinion of others. They recognize religious freedom as one of the most important elements of freedom of consciousness, which is an integral characteristic of democracy and civil society.

A. G. Ganzha, G. S. Senatskaya, S. F. Khribar

Religions and Global Studies: The study of religions concerned with how religions impact societies, define religious doctrines, relate to other religions and philosophies, and promote peace and justice. Global Studies begins by analyzing the histories of different religions and the ways they have contributed to the formation of different civilizations. Related questions involve the influences of religions on educational history and the links between religions and economics. How have different religious organizations existed as economic institutions? What impact have they had on wider economic history? Similarly, Global Studies considers how religions are linked with global politics and international relations, exploring the extent to which religions have

392 *Global Studies Encyclopedic Dictionary*

influenced and inspired different persons of state in past and present world affairs. Secondly, this study analyzes how religions have addressed different contradictory impulses, such as ruling elites, on one hand, who have imposed dogmas and orthodoxy on others, and heretical factions, on the other hand, who have opposed such fixed interpretations of truth. Of related concern is how religions emphasize the importance of different teachers, leaders, founders, or gurus as compared to their emphasis on the innate and natural unfolding of wisdom in each human person.

A third aspect of this study involves how religions relate to other religions and philosophies. Global Studies considers how diverse religions have interconnected their ideas and borrowed from one another, as well as how different religions have interacted with philosophy, at different times viewing philosophers as either central to the work of a given religion or as heretics to be outcast. Global Studies considers how religions promote peace and justice, including the extent to which they support or oppose full gender equality, granting to men and women equal capacities, rights, and aptitudes for attaining the highest states of being capable of humans within their respective faiths. Like attention is given to how fully different religions advocate, endorse, accept, or oppose and resist acts of violence, including war, murder, terrorism, hostage taking, genocide, assassination, and gender-violence. Additionally important is whether and how religions consider peace attainable in a global perspective and how religions describe the relation of peace to justice. Related to this aspect of analysis are the ways in which religions view the future shape of global society according to their own eschatology, visions of global judgment, last judgment, or supernatural intervention in human affairs. Alternatively, religions might understand justice as the innate natural way in which global society comes to order its own affairs according to increasingly higher human norms of morality. Ultimately, Global Studies is concerned with the links between religions and global future studies, the ways in which each perceives and envisions its own future and the future evolution of society. Global Studies seeks to discover the extent to which reconciliation is possible and achievable between religiously inspired perceptions of reality and humanistically oriented perceptions of existence, and how both positions can coexist in peace in today's complex world.

T. C. Daffern

Religious Movements, New (NRM) (Non-Traditional Religions, Non-Traditional Cults) are new confessions based on the idea of syncretism among the main principles of world religions. Institutionally, these new confessions are oppositional relative to the traditional (world and national) religions. They appeared in the second half of the twentieth century.

The formation of NRM is tightly connected with the modernization of traditional Christian ideas (as with the Mormons, or the Church of Jesus Christ of Latter-day Saints and Jehovah's Witnesses). On the other hand, NRM represent a fusion of Western and Eastern religious traditions (as in the teaching of the Unification Church; the International Society for Krishna Consciousness; Shinrikyo and so on). Furthermore, NRM tend to unify the religious and scientific mentality, giving rise to a new type of religious mythology (e.g., International Scientist Church, Russian neo-Pagan Movements).

Characteristic of NRM is the desacralization of traditional religious values and the birth of new sacral systems that deal with global issues: the erasure of traditional boundaries between East and West; the uniting of world religions into a whole; the synthesis of scientific and religious conceptions aimed to reform the human nature. The ideology and ritual practice of NRM is marked by laconism that grades into primitivism, which corresponds to the demands of some modern human beings, who seek to find simplified means of comprehending the world around them.

L. V. Denisova

Religion in Primeval Societies and Modern Religious Movements: In primeval societies universal phenomena of spiritual culture are manifested in local, ethnically tinted forms and connected with specific ethnic communities. They help people to realize their unity and, at the same time, differences between them and the representatives of other communities, tribes, nations. Supra-ethnic religions are manifested in their highest form as world religions-Christianity, Islam, and Buddhism. They try to overcome local limitations of ethnic religions, their ethnocentricity. These religions go beyond communal or tribal boundaries, becoming the religions of greater tribal aggregations and then nations (for instance, China and Japan prior to penetration of Buddhism), or vast historical and cultural areas (like the Indian subcontinent). This tendency finds its ultimate expression in a religious appeal to all of humanity irrespective of ethnic, geographic, historical, cultural, or racial boundaries. Losing connection to a specific society, religion loses connection to a specific locus, limited territory.

Revolutionary change occurred in the middle of first millennium BCE on vast expanses of the globe from ancient Greece to China and marked the transition from primeval locally limited mythological consciousness to universal historical consciousness, when people became aware of their place in history. They also became aware of the tragedy of being, the imminence of catastrophe, and the yearning for salvation. It was during that epoch that the foundations of world religions were laid-religions full of ethical pathos.

In the nineteenth and twentieth century, the epoch of crisis-ridden world empires, many traditional societies were absorbed in prophetic and messianic, eschatological and apocalyptic movements. It was as through people sensed the imminent end of history, the entry of the world into a new era, an epoch of hitherto unknown cataclysms. Millions of people took part in these movements-from Indians of North America to Papuans of New Guinea, from primeval hunters of Australia to shepherds of Mountain Altai. Indigenous beliefs were curiously interwoven in their ideology with alien elements of world religions, mainly Christianity.

One of the most important features of primeval societies was that social processes were manifested largely through religion. Therefore, social protest movements often resembled religious movements. The belief that God shall send to the Earth a national leader, who will rally the people and drive out strangers from their native land, influenced the ideology of early Christianity

394   *Global Studies Encyclopedic Dictionary*

(apart from the idea of victory over death, resurrection, and the premonition of a forthcoming universal catastrophe). In the same way, present-day religious mass movements emerge amid the conflagration of social and national crisis and conflict between indigenous population and alien conquerors.

Eschatological and prophetic principles are ingrained in the depths of mythological consciousness. However, historical consciousness is making its way through mythological consciousness, which is characterized by the idea of the irreversibility of time. At the same time, a new supra-ethnic consciousness comes to replace ethnic religious consciousness. The new religious consciousness becomes increasingly more universal; it addresses more and more the view that human personality looks for support not in a society, but in an individual.

V. R. Kabo

Religious Politization is the use of political means to achieve religious goals that has become a global phenomenon. Contrary to popular opinion, the use of religion to justify some political action is not a sign of religious politicization. For the supporters of a politicized religion, political power is only the means to achieve the real goal, such as, for example, the Islamic state, or the earthly Kingdom of God. Along similar lines, Mark Juergensmeyer (1992) makes a distinction between wars justified by religion and religious wars. In the latter case the political battles themselves are seen first as religious events.

Politicization is typical for all religions in some stages of their development. Moreover, in the same period, while some of the believers can view a religion politically, others may not (there is "political Islam" and "mere Islam"). Modern Western societies are characterized by "personalization" of religion, as one's private business. But this does not prevent religion from interference into politics. For example, in the West we can find "Christian parties" and some politically engaged clergymen (Archbishop Macarios III was the first president of Cyprus). In the United States, Protestant fundamentalists demonstrate political activity especially in the anti-abortion movement. But these developments do not signify genuine politicization of religion. To define them, some scholars suggest using the term "public religions" because in this case religious people do politics not to impose their beliefs on the others but to take part in public debate. Even the most radical American Protestants who are very politically engaged, mostly wish to become nothing but a successful "lobbying group" or (in case of Mennonites, for instance) to build their own isolated communities where they could live according to their principles but not make the whole society live like them.

Contrary to this, political religions want to dominate the whole given society including non-religious people or members of the other religious groups (examples: the Taliban regime in Afghanistan, the Islamic republic in Iran). Moreover, a German scholar Bassam Tibi stresses that there is a difference between how political religion is understood by tribal and world religions (Hinduism and Islam, for instance): while radical Hindu activists only seek a political territorialization of Hinduism within the boundaries of India, Islamic fundamentalism is an absolute worldview, "a vision of a worldwide order based on Islam."

The causes of politicization are very well researched, especially with regard to the Muslim world. The majority of scholars link a "sudden" rise of political Islam at the end of the 1970s, with a modernization failure that has led, not to the growth of people's wealth and economic development, but to a severe crisis or several crises among which we can list national identity crisis, political democracy crisis, and economic development crisis. By the end of the 1970s, these crises had led both to politicization of Islam in Iran and to politicization of Catholicism in Latin America in the form of Liberation Theology.

So-called official, institutionalized religion never became a moving force of politicization. For example, in Iran, it was not the traditional ulama (spiritual leaders) that committed the Islamic revolution, but, as John Esposito puts it, "a religiously minded lay intelligentsia." Official religious leaders mostly associate themselves with the ruling elite and, even when criticizing it, normally profess religious quietism or non-involvement of religious people in politics. Lay people who are not traditionalists or a part of the religious establishment mostly support politicization. This does not mean that there can be no representatives of official religion among the politicization supporters, but they would be private persons and not a "church" (or its analog in the non-Christian world) as an institution.

The known cases of politicization demonstrate that, as a rule, politicization occurs not in the traditional and archaic societies where the majority of the population is practicing religion (participate in rituals, observe traditions, know the basic doctrines), but in the modernized and secularized ones. The reason for this obvious contradiction between fundamentalism (accurate performing of religious duties and theological "literacy") and the religious establishment, on the one hand, and politicized religion, on the other hand, is that the purpose of politicization is not a return to the past or a conservation of the archaic present. Politicization is, in fact, an attempt to give a religious answer to the most acute questions of modernity. Religious politicization demarcates not a total rejection of modernization but a desire to have a successful modernization. If a society has not been modernized on the Western liberal grounds, attempts will be made to do it on religious grounds. This does not prevent the backers of politicized religion from calling for the return to "traditional" values. In spite of such calls, politicized religion is a totally modern phenomenon that emerged as a response to the challenges of our civilization.

Since politicization occurs through a considerable break with the initial religious tradition, it cannot happen in the form of a direct application of religion to politics (if such a thing is possible at all). Normally politicization takes place through a mediatory ideology that can differ significantly from the original religion given that that "lay intelligentsia" that backs politicization mostly forgets the "traditional" tenets. In the case of Islam differences between the religion and the ideology (so-called Islamism) are not so evident because Islam is as much a religion as a system of government and a way of life. While there is such thing as Islamic governance and Islamic law (shari'a), many religions have no such concepts or have forgotten them long ago. That is why the differences between Islam as a religion and Islamism as an ideology are hidden. In the other cases a mediatory ideology can differ from the initial religion significantly. For example, so-called Progressive Catholicism or Liberation Theology, being a

396 *Global Studies Encyclopedic Dictionary*

mixture of Latin American "people's" Catholicism and Marxism, is considered the mediatory ideology for political Catholicism.

References: Esposito, J. L. *The Islamic Threat.* New York: Oxford University Press, 1992. Juergensmeyer, M. "Sacrifice and Cosmic War." In *Violence and the Sacred in the Modern World.* London: Frank Cass, 1992.—. *The New Cold War?.* Berkeley: University of California Press, 1993. Tibi, B. *The Challenge of Fundamentalism.* Berkeley: University of California Press, 1998.

A. V. Mitrofanova

Reshaping the International Order (1976) was the third report to the Club of Rome and was directed by the Dutch scientist and Nobel Prize winner in economy J. Tinbergen. The oil crisis, which erupted in 1973, highlighted the negative features of the existing international economic order and growing grievance of developing countries and became a subject of special discussion in the United Nations. In February 1974, the Club of Rome held a conference in Salzburg, where it was recognized that economic-mathematical patterns were not enough for coping with the problems that had arisen and that socio-political and ideological aspects that affect a crisis situation should be studied. At this conference Tinbergen offered to head the working up of the new report, which would take into account the conclusions made and give possible variations of reorganization of the current international relations. In attempts to solve this problem, the authors of the draft made a significant step forward compared with previous research in understanding the complex interdependence of global problems. They noted that no important problem of the modern world can be solved separately and that attempts to act in this way almost inevitably lead to an aggravation of other problems, seemingly not related to them. The report showed that the average income of the wealthiest strata of the world population steadily grows, and in 1970, it had exceeded thirteen times the average income of the poorest strata; a gap between the highest incomes of the most developed countries and the lowest incomes of the least developed ones was even wider. Finally, the report recommended changing rates of income growth per capita so that in developing countries it would be 5 percent higher than in developed ones, and focused on reorganization of power structures all over the world, including economic, financial, political, and military relations.

By the middle of the 1970s, global problems became a subject of wide discussion in the United Nations, and the issues cited in the third report were the focus of interest of other reports, in particular of Pearson's Commission organized under the initiative of the World Bank, former Secretary of Defense of the United States, Robert McNamara, and Brandt's Commission. As a result, the third report was not as tremendously successful as the two previous ones and led the Club of Rome to reconsider the strategy of its activity toward transition from the quantitative analysis of global problems to the analysis of qualitative characteristics of human life, the purposes of development and system of values.

A. N. Chumakov

Responsibility, Global: Global responsibility denotes an intellectual position presumes a moral duty to contribute to the advancement of collective human well-being is taken as paramount and self-evident. It is connected with human rights, arguing that each human right is based on a shadow human responsibility, and that we cannot advance one without the notion; so global human rights brings with it global responsibilities, and not merely to other human beings, but to the wider web of life of which we are part.

T. C. Daffern

***Responsibility to Protect* (2001)** is a report on Intervention and State Sovereignty that was presented by the International Commission to the United Nations Secretary General. The Commission was created by a proposal of Canada in September 2000, for the purpose of developing a new approach to "humanitarian intervention," which caused serious disputes after United States intervention in Somali, Bosnia, and Kosovo. In the report, the Commission opted not to use the term "humanitarian intervention," so that humanitarianism would not be associated with military actions. Still, it allowed for intervention intended to protect people, in cases when the state is not willing or capable of protecting its citizens from suffering or hardships resulting from internecine wars, mutinies, and repressions. The principle of "the responsibility to protect" set out in the report presumes that it is necessary to: (1) lay down more specific rules, procedures and liminal criteria of intervention; (2) evaluate the validity of military intervention, if it becomes necessary after other approaches proved ineffective; (3) ensure that military intervention is carried out only for the declared purpose; efficiently and with minimal harm to the people and the state; (4) help eliminate the cause of the conflict, simultaneously consolidating the prospects of establishing a secure and stable world. The UN Security Council must approve all military intervention. Permanent members of the Security Council should abstain from using veto in such cases unless their crucial national interests are concerned. If the Council denies the request or does not consider it within a reasonable term, the issue shall be considered at an extraordinary session of the UN General Assembly. Alternatively, military action may be carried out within the jurisdiction of regional or subregional organizations under Chapter VIII of the UN Charter, on the condition that they subsequently apply to United Nations for the corresponding sanction.

A. B. Veber

Rio-92 refers to the UN conference on environment and development held in 1992, in Rio de Janeiro, Brazil. The conference was timed to commemorate the twentieth anniversary of the UN environment conference held in Stockholm in 1972. It was intended to evaluate the results of the past period and to set new tasks.

During the first decade after Stockholm-72, a nature conservation infrastructure was being developed in the most of the countries of the world. Various environment, energy, and resource conservation technologies have been developed and applied. In developed countries, some negative environment effects have exhibited a tendency to decrease. However, the global ecological indices were becoming worse, a larger number of people suffered from ecological degradation and it was becoming more and more evident that the ac-

398 *Global Studies Encyclopedic Dictionary*

tions hitherto taken were absolutely insufficient to solve the problems. To analyze the present situation, the United Nations established a special commission, known as "Brundtland Commission" (named after its head, Norwegian political figure, Gru Harlem Brundtland). A report of the commission was published in 1987. Its main idea was that ecological problems may only be solved under the condition of realization of a wide range of social and economical measures and they require revision of the whole strategy of human progress and change to sustainable development. This idea held as a guiding principle in the course of preparation for RIO-92.

To prepare for the conference, the United Nations established a special organizing committee. Morris Strong, a well-known Canadian politician and businessman noted for his nature conservation activity was appointed General Secretary. Working out the documents to be presented at the conference required coordinated efforts of politicians and diplomats, but especially of scientists from all over the world. Though the maximum program concerning the RIO-92 documentation package has not been implemented (there was not time enough to work out the Declaration of Earth, and a convention on Forest Conservation and Exploitation also has not been agreed upon), RIO-92 turned to be unprecedented in its scale and profundity. The conference was also unrivaled by the number of participating countries and level of representation: 178 countries took part in the conference; the heads of 114 delegations (including all the delegations of developed countries) were heads of states and governments. Simultaneously a Global Forum on environmental problems with nine thousand accredited organizations and twenty-nine thousand participants was conducted in Rio de Janeiro. The forum program included over a thousand sessions and other events, which were visited by almost half a million people.

The qualitative difference between the RIO-92 and Stockholm-72 conferences consisted in changing the focus from the problems of environmental conservation as such to social, economic, and political problems, to which specific solution could ease the ecological crisis and prevent ecological catastrophe. The Declaration of Environment and Development (the so-called Rio Declaration) became the main political document of RIO-92. An attempt to accept a new approach to development and to arrange world policy priorities in a different way emerged. In particular, it was declared that environmental protection should become an imperative component of development and should not be considered apart from it. However, the definition of problems was not accompanied by an estimate of their complexity or description of actions adequate to ensure considerable progress in their solution or at least to initiate significant changes in social, economical, and political spheres leading to the improvement of the biosphere and contributing to the breaking of the destructive tendencies of human development. In spite of some correct ascertainments, the following two decades have shown that the document has remained nothing more than just a "declaration."

The Twenty-First Century Agenda is the most voluminous of all the documents approved at RIO-92 (over 1000 pages of close print). This document is an attempt to unfold the Rio Declaration provisions into a detailed comprehensive long-term action program. The following points are considered in the twenty-first Century Agenda in the following sequence: (1) social, demographic and economic problems; (2) protection and rational conservation of natural resources for development purposes; (3) strengthening of main population groups; (4) instruments of realization, including finance, science, education and international cooperation. However, analysis of various problems of sustainable development and proposals on their solution are distinguished by extreme unevenness.

Thus, the problem of poverty is accentuated, while demographic problems are considered very superficially. Causes of overpopulation in most developing countries remained uncovered. Influence of the measures on demographic processes has not been estimated. The problem of debt repayment of developing countries to the developed ones is actually concealed, though the debt itself is mentioned in the document more than once. The problem of the role of global natural resources (developing countries' natural resources) in the formation of the economic wealth of developed countries is completely left aside. At the same time, the principle of "differentiated responsibility" is proclaimed, though it is not clear how such a differentiation can be executed. No reduction of military expenditures has been proposed. No opportunities of the at least partial transfer to the needs of sustainable development are considered. Discourse about advanced technologies transferred from developed countries to the developing ones that contribute to the reduction of power consumption in industry is completely non-constructive. The threat of dangerous waste removal from developed countries to developing ones is not estimated; no actions are foreseen to eliminate the threat.

During the conference preparation and conduct, many attempts were made to make the Rio Declaration and the Twenty-First Century Agenda more concrete, to use more definite, clear formulations, and to include in the documents some commitments to be assumed by the countries. However, the consensus rule in force in the UN system stipulates that any proposal infringing upon interests of a country (as it seems to the country) for the sake of the goals of civilization as a whole turns out to be the "no-go" one. The inadmissibility of such procedures in a world in deep social and ecological crisis could be heard more than once at RIO-92 and other forums, nevertheless UN did not even any proposals regarding the procedures alteration.

In addition to the mentioned general documents, the Statement on Principles of Forest Management, Conservation and Sustainable Development was approved at RIO-92. The Framework Convention on Problems of Climate Change and the Convention for Biological Diversity were prepared for signing. The first of the documents did not include any commitments. It was assumed that the document just forestalls the more concrete and harsh forest convention to be approved shortly after RIO-92. However, even after ten years, by the occurrence of the World Summit on Sustainable Development Johannesburg-2002, such a convention has not been developed.

The summing-up of the decade after RIO-92 at the summit in Johannesburg revealed that just a slight progress, if any, has been reached in all the directions marked out in the Twenty-First Century Agenda. Serious challenges were faced in course of realization of the Framework Convention on Problems of Climate Change and the Convention on Biological Diversity.

V. I. Danilov-Danilyan

400   *Global Studies Encyclopedic Dictionary*

Risk: in the concept of risk, basic experiences and problems of a highly industrialized and, to a great extent, scientist societies are crystallized. We can therefore designate "risk" with full justification as a social-theoretical term that denotes a characteristic feature of modern societies. Our society seems to be paradoxically organized: we can describe it as a "Risk and Catastrophe Society," as the Munich sociologist, Ulrich Beck, has done with vivid pathos and with convincing evidence; with equal justification one could speak of an "Insurance Society," which has raised security to the rank of a central value. Modern societies obviously increase security and insecurity at the same time. This contradictory development is reflected in the concept of "risk," and this is the reason for its importance for social theory.

The origins of deliberations and activities concerning the modern conception of risk in Western society stem from the end of the Middle Ages and the dawn of the Renaissance: awakening awareness of risks, comprehension, and intellectual elaboration of risky situations, and the development of strategies for acting at risk, are inseparably bound to the ideological and lifeworld changes taking place during the Reformation and the Enlightenment, with the rise of civil, "worldly" society. "Risk" is therefore a "topos" of the entire modern era, and not just of the immediate present.

On one hand, risk resides in processes external to and separate from the observer, but which only become effective through action; on the other hand, risk refers to daring as a characteristic of human behavior. In both cases, the consequences of human actions are addressed-consequences, on the one hand, as an attribute of a process set in motion, or an event initiated by human activity, and, on the other hand, as the result of an option for action.

The declared goal of risk research is to subject it to rational calculation. Some disciplines (e.g., math and engineering) have concerned themselves for some time with risk/uncertainty; others (e.g., sociology and political science) have only recently become a subject of dedicated research to risk.

Risk research had its technical point of departure in questions of the manageability, security, and dependability of technical systems and technological processes, and in the causes and effects of technical failures. At the outset, it was primarily limited to cost-benefit calculations of accidents, and to problems of risk acceptance. In the technical area, ensuring dependability and quality was held to be amenable to planning and testing to a great extent, methods of objectivization and systematization of the search for and elimination of flaws and weaknesses were developed. The goal of quality and reliability control was to reduce the residual risk (whatever was and is understood under it) to an "acceptable" level at justifiable cost in time and money. Normatively, technical risk analysis is based on the principle of cost efficiency. Characteristic for this approach was the fact that it applied primarily to "traditional" technical risks, that is, to such risks for which an individual accountability for the risk's consequences, an assessment of the practical damage, and the delimitation of causes and effects of damage in space and time could, to a great extent, be assumed. This would mean that practical experience and comparative data were available, and that there was the possibility of testing theoretical approaches in practice.

The structural characteristic of the "new" technical risks consists in the fact that they are inseparably bound to non-rational and unintentional effects of rationally planned decisions and actions. For such risks, limiting the consequences in space and time is almost impossible, individual responsibility for the cause and for the subsequent effects is not ascribable, "exact" knowledge about the probability of event and the extent of damage is hardly ascertainable. Therefore, it is justified to speak of hypothetical risks, because the following applies: scientifically-elaborated long-term planning and probabilistic calculations of risk replace the successive adaptation of technical systems to the requirements of the situation (e.g., stricter safety norms due to accidents); practical experience is increasingly being replaced by hypothetical assumptions (models, idealizations, complexity reduction); empirical knowledge is being displaced by calculations of probability; tests cannot be carried out in sufficient number; experiments or observations cannot be repeated at will; the potentials and likelihood of danger can (and should), as a result, no longer be ascertained and reduced according to traditional "trial-and-error."

In recent years, our understanding of and manner of dealing with risks in society has become more differentiated. This changed attitude has been brought about by the risk debate in many sciences. Among the results to the present, the following three differentiations should be singled out: (1) the distinction between risk (as the risk of decision), danger, and chance, (2) the difference between one's own estimation of the risks and the estimation of the same risks by other actors, and (3) the difference between the so-called objective scientific risk assessment, and subjective individual risk perception and assessment. Above and beyond these insights, knowledge about questions of the construction and communication of risks in modern society remains to a great extent controversial, as is knowledge about dealing with risks, that is, risk management.

To the extent to which it became apparent that modern technologies and their risks were complex and less comprehensible than originally assumed, it became obvious that managing and coping with risks should not be understood only or primarily as a question of safety engineering (and therefore as the exclusive responsibility of experts), or as an administrative problem (and therefore as a matter of priority for bureaucracy), but that the early identification of the risks connected with technical and technological development and their containment, control, or even avoidance is a complex matter of societal concern largely in a human-made (or at least human-influenced) world. In the final analysis, it is primarily a question of principles of public supervision and co-determination, of the multidimensionality of criteria for decision-making, and of fairness-principles that even the best technical risk analysis cannot take into consideration (which points up its possibilities and its limitations).

G. Banse

Risk, Ecological: Ecological risk refers to the probability of environment degradation or shifting into the state of instability caused by current or planned economic activity, and emergency situations of a natural and technogenic nature; a potential loss of control over ecological events—the probability of the occurrence of an unfavorable event (e.g., hazardous natural phenomenon or ecological accident), the amount of potential damage, the uncertainty of the time of occurrence, the intensity and consequences of the potential unfavorable event. Ecological risk can be evaluated quantitatively by the product of the negative impact probability of the source of the risk (for instance, pollution of a natural object) and the amount of potential damage resulting from this impact. In the area of radiation safety, risk depends on the probability of death of an individual caused by the ascertained effect or on the ability to identify this effect within a time interval upon receiving a radiation dose. The risk can be lowered by introducing a protection system, but not totally eliminated. The purpose of risk management is to assure security.

Global technological development can be seen almost everywhere on the globe in the form of direct or indirect anthropogenic transformations of the natural landscape. First this is based on the unified interrelationships among the natural components in the form of general energy and mass exchange occurring in the geo- and biospheres.

The current situation is such that the total volume of anthropogenic emission in many cases is comparable to and often surpasses the volume of emission from natural sources. Thus, natural and anthropogenic sources of nitric oxide and nitrogen dioxide account for the emission of thirty and thirty-five to forty million tons per year respectively. Anthropogenic emission sources of lead exceed natural ones by almost ten times.

Anthropogenic changes occurring in the hydrosphere significantly deteriorate the quality of water causing increasing shortages of the water supply needed for daily use. The hydrosphere contains around 1.6 billion km^3 of free water of which ocean water accounts for 1.37 billion km^3. 90 million km^3 is available on the continents of which 60 million km^3 is subterranean (almost all this water is salty); 27 million km^3 of water is preserved in the Antarctic, Arctic, and mountain glaciers. Useful reserves of accessible fresh water in rivers, lakes, and underground to a depth of 1 km amount to 3 million km^3.

This relatively small volume of fresh water, thanks to its constant renewal based on natural water circulation is enough, according to scientific estimates, to resolve the problem of the water supply for the needs of the global population, but only on the condition of the full preservation of its quality. Unfortunately, precisely this condition is not being duly met at present. All non-organic dissolved substances and up to 10 percent of organic pollutants remain in purified sewage water even after the most efficient purification, including biological ones. Melted water can become safe for consumption only after repeated dilution with clean natural water.

The quoted data indicate a real ecological risk, existing both on local and global scales. The danger of disturbing the natural balance at various levels (regional, national, continental, global) can serve as an integral criterion of such a risk. This danger is a potential characteristic of irreparable losses (or irreversible degradation processes) quantitatively related to anthropogenic factors of industrial production. The general principle of environmental protection can be formulated in this case as the minimization of losses.

I. I. Mazour

Risk Management Technologies are modern technologies aimed at acquiring information, decision-making, and risk management. For example, in seismology there is a new branch: seismology of emergency situations. A new approach to human protection against earthquakes includes the global geoinformation system (GIS), technologies for estimating individual seismic risk, forecasting the consequences of earthquakes, and working out effective response scenarios. In the case of a destructive earthquake, the GIS "Extremum" needs only two hours to estimate possible human losses, the number of people in blockages, the number of necessary rescuers, technical equipment, and the amount of life-support means for the injured population. GIS "Extremum" has provided effective rescue works in Neftegorsk (Soviet Union, 1985), Turkey, Greece, Taiwan (1999), and India (2001) when half of the injured were rescued.

An important research direction is working out and implementing technology that allows defining stability and earthquake resistance of buildings and constructions with the help of mobile diagnostic systems. This technology defines frequencies of a building's internal oscillation and other parameters and helps to issue a certificate of building's safety describing hidden defects, building's earthquake resistance, residual life of the building and recommendations concerning its strengthening. With the help of this certificate the whole set of buildings' safety problems can be solved, and real estate dealership (sale, insurance) can be carried out at higher scientific and technical levels. Further development of this technology implied developing mathematical models of buildings and constructions subject to natural strains (such as vibration, wind, snow, and temperature influences).

Another trend in the perspective of scientific research is the remote sounding of the Earth from the outer space. It can help to reveal wilderness fires, smoke generation in built-up area, flood flows and inundations, and droughts in real time; there is an opportunity to discover a fire in a territory with less than 400-square-meters area even without a smoke cloud, just by the radiation of heat. This system allows not only for discovering old and new fires, but also for finding deserted and fresh cutting areas and estimating soil conditions and the degree of pollution.

M. A. Shakhramanyan

Risk and Postmodern Society: A historically unique level of private wealth and social security that is supported by a highly effective social safety net characterizes Western societies. We can ask, therefore, how the contentious language of risk has been able to gain so much ground in the public arena in modern society. In response, we can identify at least three ongoing debates that deal with the issue of society's danger to itself:

First, we confront the consequences of using complex advanced technologies. Whether rooted in physics, chemistry, or biology, these technologies have a high potential for devastating catastrophes. In the event of accidents or total failures, the damage is out of all proportion to the purposes of the technologies. What is more, existing facilities for compensation based on operator liability also fail, because the scale of damage is so huge that it is uninsurable. A characteristic of advanced technologies is that total control is

404   *Global Studies Encyclopedic Dictionary*

not possible. Accidents can only be made more improbable, not ruled out altogether. If the vulnerability to catastrophe can only be contained but not eliminated, the technical problem of safety measures becomes a social problem of the acceptance of possible human-made catastrophes.

Second, In the course of the risk debate, it has emerged that a further dimension of uncertainty is generated socially. The discussion involves the growing discrepancy between the intentions and the consequences of technological actions. Through and with the help of genetic engineering, humanity can now try and manipulate the conditions of its own evolution. Precisely because it gives humanity access to the self-replicating mechanisms of the biological foundation of human life, genetic engineering sharply impacts humanity's cultural understanding of itself and its identity. It is impossible from our present vantage point to forecast the scale of the associated social and cultural changes and shifts in humanity's view of itself. We are seeing an intervention in evolution whose effects cannot be even remotely predicted.

A third type of uncertainty appears in the non-spectacular consequences of daily actions, that is, the long-term ecological changes due to everyday acts and decisions. Whether these involve road transport, CO_2 production, clearing the tropical rain forests, or the massive use of detergents, the consequences of our behavior are the destruction of forests, possible climatic change, and the pollution of our ground water. Typical of dangers that result from everyday conduct in modern life is the long interval and complex relationship between cause and effect. Moreover, the effects of this type of uncertainty can only be made evident by science, yet the gap between action, consequences, and causes is so great that it is impossible to establish a clear relationship between them.

The very ambiguity of ecological damage and the globality of the consequences make prevention difficult.

What is common to all areas of human-generated hazards is that we cannot predict with certainty how great the danger really is. The modern issue of risk involves an "irresolvable ambivalence." Not only is uncertainty produced on a previously unknown scale, but all attempts to solve the possible problems make us even more aware how fragile modern societies happen to be. This can be briefly illustrated in terms of the three cases cited above (advanced technology, genetic engineering, and ecological consequences).

The development of advanced technologies has led to complex and hard-to-control industrial structures where there is a risk that their actual purpose of producing energy and materials will be increasingly overshadowed by their side effects (impact on humanity and nature). Empirical studies show that adding safety installations increases the complexity of the system as a whole and so makes it more vulnerable to accidents. Society is aware that technically created risks are not being solved but at best transformed into a different kind of uncertainty. This situation, which results in widespread awareness of contingencies, combines the knowledge that other decisions could have been taken with the realization that nobody can rule out a disaster, however small the probability of its occurrence is calculated to be.

The example of genetic engineering shows that risk includes the opportunity to shape. Transforming the dangers of denaturing humanity into risks—starting to calculate the potential and drawbacks of intervening in evolution without regard to metasocial rules (religion, tradition)—is a prerequisite for scientific and experimental exploration of the biological mechanisms creating life. The more human action is involved in the shaping process, the faster social structures will change, that is, they become more dependent on decisions; at the same time, the unforeseen consequences of action increase and, perhaps of decisive importance, the future becomes less predictable, as it is based on decisions which could have been different. Absence of knowledge becomes central in its importance for decisions relating to the future.

The most difficult problem, however, is probably the problem of ecological consequences. Changes in nature and in human-made second and third "nature" can be fast or slow, sudden or gradual. They are the results of virtually invisible causes or actions by vast numbers of people. Changes in the ecosystem are not amenable to linear or causative interpretation; so, they lie outside all present classical models of analysis and notions of reality and action tied to ideas of causality. They make us aware to a great extent of the complexity and interdependence of the world, as the synergetic effects are caused by the acts of many individuals. Here, uncertainty is a matter of attribution: first, there is uncertainty about whether the causality yielded by research for the emerging problem is really valid (destruction of forests) or entirely different factors are playing a role. Second, more and more consequences and effects are attributed to humanity (climatic catastrophe) without knowing precisely whether conditions of natural evolution not yet amenable to our influence are responsible for the changes. This situation could be described as one of enhanced responsibility with growing uncertainty.

The language of risk reflects a new uncertainty in society, which takes the form of conscious perception of the future as contingent on the present. Seen in these sociological terms, risk already attributes possible damage to decision-making attitudes, though it is impossible to know the scale of the damage, the emergence of the damage, or if there will be damage at all. This ignorance (unpredictability of the consequences of a decision) becomes part of the decision. The only thing that is certain is that a decision must be made, as there is no social entity which future damage can be attributed to, leaving only decision-making under conditions of uncertainty. The expansion of the potential for decision and the disappearance of any metasocial rules with the resulting pressure to choose options have resulted in society increasingly viewing its future today in terms of risk.

We can point to many indications that modern society really perceives its future in the shape of a currently existing risk. One need only think of the possibility of insuring oneself against many accidents. Insurance does not create certainty that the accident will not take place. It merely guarantees that the property situation of the person affected does not change. Industry provides the possibility to insure oneself. One has to make a decision on this, however. In this way, dangers against which one could insure oneself are changed into risks. The risk lies in the decision between insuring and not insuring oneself. The switch from danger to risks is the counter-intuitive, unintended purpose of many institutions of modern society, which were originally conceived for completely different purposes.

G. Bechmann

406 *Global Studies Encyclopedic Dictionary*

Road Maps to the Future (1980) was the tenth report to the Club of Rome. Its author Bohdan Hawrylyshin, the International Management Institute director, submitted for a wide discussion the model of the future world order based on how different cultures, religions, and ways of life co-exist. Putting the emphasis on the range of social and human problems he pronounced the possibility, or even necessity of the various, perhaps opposite, social systems' convergence, propounding at the same time retention of the market economy as a required stipulation. As is stressed in the report, the team's working principles dwell in every society even if "dormant" or being out of sight. If they are aroused one day, the social order based on these principles, which might be incarnated differently in different countries, according to their historical experience, development level, scale, human and material resources will be set up in the world. The report not only designated ten years ahead the general outline of the events that have then followed (radical reformations in the socialist countries), but also pointed out the main "stumbling-block" on the way from a "planned" to a market economy: the return to private property. In this research the global problems were relegated to the background, and the social and political institutions' analysis was regarded as the most important prerequisite for understanding the global problems' kernel and the ways to surmount them.

Reference: Hawrylyshin, B. *Road Maps to the Future*. New York: Pergamon, 1980.

A. N. Chumakov

Roerich, Nikolai: (1874–1947) Nikolai Roerich demonstrated the originality of Russian culture and its links to the East and West. In the 1930s, he put forward an idea of a pact for the international protection of cultural monuments in wartime. In 1935, the Roerich Pact was signed in Washington, D. C. by twenty-one nations of the American continent. It served as the basis for the 1954 Hague Convention concerning the protection of cultural values during armed conflicts.

V. V. Frolov, F. T. Yanshina

Russian Philosophical Society (RPhS) is a social-scientific institution that unites citizens of Russia who work in the field of philosophy. The society is open to all, including people who are not citizens of Russia, regardless of residence. Legally constituted in 1992, as an assignee of the Soviet Philosophical Society created in 1971, and attached to the Soviet Academy of Sciences, it has an extensive network of primary organizations and branches, with over 3000 members. The supreme governing body of RPhS is the general meeting of its members conducted at least once every five years. The Third Russian Congress of Philosophy took place 16–20 September 2002, in Rostov-on-Don, where a new twenty-six person Presidium of the RPhS was elected.

The main directions of the activities of the RPhS include organization and realization of scientific events, publishing, establishment and development of scientific contacts both in Russia and abroad. The Russian Congress of Philosophy (RPhC) conducted under the aegis of the Russian Philosophical Society meets once every two or three years in different regions of Russia.

Since 1997, RPhS has published *Bulletin of the Russian Philosophical Society*. Also, the Society periodically publishes a collection of analytical works, including *Proceedings by Members of RPhS*, proceedings of conferences, and books dedicated to topical problems. Together with the Russian Ecological Academy, RPhS published the international interdisciplinary *Global Studies Encyclopedia* (2003).

RPhS is a full member of the International Federation of Philosophical Societies, and its members participate in all World Congresses of Philosophy.

A. N. Chumakov

❈ S ❈

Sartre, Jean-Paul: (1905–1980) was a French philosopher, essayist, novelist, and dramatist who was a dominant figure in the middle part of the twentieth century in France and worldwide.

During the Second World War, after his return from a German prisoner of war camp, Sartre clandestinely supported the Resistance. He took bold stances in favor of ending colonialism (especially French colonialism in Indochina and then in Algeria) and achieving peace between the East and West, and he attacked privilege. He traveled extensively. During the (American) war in Vietnam, he presided over an unofficial War Crimes Tribunal along with the British philosopher, Bertrand Russell. Until near-blindness toward the end of Sartre's life halted most of his political activities, he continued to be a strong, radical advocate for global justice.

W. McBride

Security Policy, Global: A branch of global studies in political science, Global Security Policy examines policy-making with respect to global security and the opportunities and means of the political regulation of global space.

The globalization of politics reflects the need to prioritize universal human values. Different approaches to the issues of ensuring global security have arisen. Epistemologically, global studies in political science dealing with security should disclose the distinctive features of politics under global threats; investigate political ways and means for individual societies and civilization to adapt to the imperative of survival; and search for mechanisms, methods, and directions to control global political interactions.

A global security policy would be complex and would necessarily be inseparably linked to political processes and social life. Its structure would depend on the level and sphere of activity: it can be focused on different spheres (economic, ecological, military, information, and sociocultural); it can manifest itself in different spaces (global, regional, national, and local). In a wider sense, global security policy is a policy of minimizing global risks.

To ensure global security through political means, it is important to understand the mechanisms of the functioning of global policy. The character of the current political culture and political consciousness is of vital importance. Internationalization of the political sphere, its globalization, causes the formation of the global political culture as a system of relations and processes of production and reproduction of constituent elements in the course of the interchange of generations. Global political culture is a new phenomenon, which is influenced significantly by the following factors: (1) scientific and technological progress; (2) development of communications and emergence of global informational space; (3) formation of transnational financial and political elites.

In the analysis of global processes, it is important to investigate different changes underway in the modern world and to forecast processes that could lead to radical qualitative changes. Global security policy should be based on scientific prognostication. However, it is difficult to meet this requirement for the following reasons: (1) For a long time, prognostication as a state-sponsored activity has been non-existent even in developed countries.

Sartre, Jean-Paul – Sustainable Development, World Summit on 409

(2) Scientific prognoses did not adequately stimulate state agencies to make the necessary decisions. (3) Prognostication models were ambiguous and mutually excluding, and that made decision-making difficult, especially at the international level. (4) Even a correct prognosis may not be useful if seen in the light of various political and economic interests.

Non-military aspects of global security are becoming extremely important in the twenty-first century (e.g., guarding against cyber attacks). Ensuring global security is linked to strengthening international law and the role of the UN, making negotiation more effective; reducing nuclear and conventional armaments; successful modeling and forecasting of the global development; and implementing the strategy of sustainable development.

A. L. Kostin

Self-Government, Local: Local self-government is an activity of a population undertaken independently and under its own responsibility to solve issues of local concerns directly or through specially elected bodies. In democratic countries, it is recognized as one of the foundations of the democratic system. John Stuart Mill contrasted local government (direct rule by those ruled) with centralized representative levels of government. Local self-government was to promote justice, equality, brotherhood and sisterhood, and municipal socialism. The central issues concerned the balance of centralization with decentralization. Cultural, ethnic, national, and socioeconomic problems can be most fully resolved within local communities and through local self-government. Therefore, against the background of globalization, the significance of different forms of local self-government will increase, and it will become a specific adaptation mechanism enabling local communities to enter smoothly into the system of supranational relations and universal standards. Alienation of populations from government, and further unification of the cultural-historical and economic foundations of human life, call for the creation of compensatory mechanisms in the form of strengthening local self-government, both as a school of governing and as a tool for psychological monitoring of a society.

Reference: Mill, J. S. "Of Local Representative Bodies." In *Considerations on Representative Government*. London: Parker, Son, and Bourn, 1861.

I. I. Rusin

Semiosphere (Greek *semeion*, "sign, feature" and *sphaira*, "sphere") is the spiritual coat of the planet. Yuri M. Lotman, a Russian specialist in semiotics, who offered it in addition to such notions as "noosphere" and "pneumatosphere," introduced the term. The semiosphere is a special semiotic space comprised of both the sum of individual languages and the sociocultural field in which they function. According to Lotman, the semioshere includes, first, natural languages as the broad system of signs and texts. This sphere of signs, connected to the pneumatosphere, the sphere of spiritual expression, is a part of the noosphere, the sphere of reason.

A. N. Chumakov

Social Change is one of the basic concepts of modern sociology, describing reality from the viewpoint of its mobility, fluidity, instability, changeability,

Global Studies Encyclopedic Dictionary

mutability, and dynamics; it is the principal unit of sociopolitical analysis in the study of social processes and events. The theory of social change, proposed in 1922 by American sociologist William Ogburn, currently dominates the field of social development studies.

One approach to social change denies any purposefulness, another holds that all changes are necessarily purposeful. Talcott Parsons asserts that the process of change implies increasing "universal adaptive capability" through the development of evolutionary universals in society. Conflict theories focus on redistribution of power and authority (a theory examined by Max Weber) as a purpose of change. Empirical studies have identified the purpose of change to be linked to demographic and ecological factors, and to spatial mobility. The patterns of sociocultural dynamics have been systematized by Pitirim Sorokin (1941) as "the principle of immanent changes" and "the border principle" and represented as phases repeating in the rhythm of the sociocultural process.

Causes and mechanisms of social change can be classified as economic, social, and cultural. Social change is constitutive of human nature; sociological study focuses on subjective perception of social relations. Ralf Dahrendorf speaks of "life chances" to emphasize the human context of social change. Within the structural-functional approach to social change, causes of are linked to social differentiation and stratification.

Contemporary interpretations of social change are characterized by a theoretical move away from linear schemes and flat images of the sociocultural realm to be replaced by three-dimensional images of reality, viz. by schemes and models implying plurality of alternatives and diverse combination of factors.

An innovative approach is the use of Illya Prigozhin's non-equilibrium thermodynamics to the processes of social change (Prigozhin, Stengers, and Priogine, 1984). Sociological theory has been radically changed by the acknowledgement that social system development is naturally spontaneous and stochastic, that instability and non-equilibrium are immanent to it. Hence, the notion of "social change" has been extended to include stochastic, fluctuating, and deviant processes.

The development of the concept of social change has stimulated the emergence of a whole range of new programs in the field of empirical sociopolitical research: analysis of innovative and stabilizing behavior, of human adaptive capacities, or of the mechanism of sociocultural transmission. It has also given rise to a new paradigm in understanding social processes and phenomena, which is a theoretical response to the challenge of civilization, caused by changing existential situation of humanity represented by a complex of threats, risk zones, insecurity of human existence, and globalization of local developments. This knowledge is methodologically based on: restricted rationalism; the absence of a global trend toward ascending progress; fluctuations; pluralism, nonlinearity, alternativeness, non-equilibrium, multidimensionality, abruptness, spontaneity, stochasticity, insecurity, and systemic self-organization. These methodological instruments enable us to understand the widest range of social developments and phenomena of different nature and scope (including the disintegration of the Soviet Union, global terrorism, the information revolution, and the ethnic problems in the Balkans).

Reference: Ogburn, W. F. *Social Change with Respect to Culture and Original Nature*. New York: B. W. Huebsch, 1922. Sorokin, P. A. *Social and Cultural Dynamics*. New York: American Book Company, 1937–1941. Prigozhin, I., I. Stengers, and I. Priogine. *Order Out of Chaos*. New York: Bantam Books, 1984.

I. M. Predborskaya

Social Contract is a model of the origin of society and legitimacy of the state as authority over the governed. Originating in Europe during the seventeenth and eighteenth century (espoused by Thomas Hobbes, John Locke, and Jean-Jacques Rousseau) social contract later became mainly associated with the Western European pattern of a welfare state. Nowadays, the extended version of social contract, brought about by globalization process, is being widely employed as a prerequisite of more equitable world form and ensuring global public benefit (such as social or ecological). UNO experts have developed a plan of the Contract among Nations, the subject of which would be mutual obligations of rich and poor countries, international financial institutions, UNO specialized institutions, private sector and civil community organizations, focused on Millennium Development Goals, stated by the world leaders in the Millennium Declaration, the resolution of the UN General Meeting of 8 September 2000. Social contract of a global scope presumes such a transformation of world political arrangement, which would open an opportunity to control transnational market powers and guide world development in the interests of all nations.

A. B. Veber

Social Doctrine of the Roman Catholic Church is "the social concern of the Church, directed toward an authentic development of man and society which would respect and promote all the dimensions of the human person" (*Sollicitudo*). The SDC is a doctrinal corpus which "builds up gradually, as the Church . . . reads events as they unfold in the course of history" (*Sollicitudo*) and is a reflection about reality that the Church brings into effect, "with the support also of rational reflection and of the human sciences," for the purpose of guiding people toward "their vocation as responsible builders of earthly society" (*Sollicitudo*). The term "Church's social discourse" refers teachings regarding ethics, economics, society, and politics first discussed in *Rerum Novarum* (1891). The doctrine is intended to "help in promoting both the correct definition of the problems being faced and the best solution to them" (*Sollicitudo*; cf. Catholic Church, 1992). It contains three divisions: reflection upon principles, criteria for judgment, and directives for action. A fundamental feature that it is not a rigid, dogmatic, a priori system to be applied to a provisional, changing, or contingent social reality. Instead, it is a doctrinal corpus in progress, which is developed throughout the course of history by the Church, based on elaboration provided by the human sciences, especially as they respond to external trends from various human societies.

The social teaching of the Church has progressed through three periods: (1) the publication of *Rerum Novarum* in 1891, when the Church recognized

412 *Global Studies Encyclopedic Dictionary*

the seriousness of the oppression of the working-class. The response was not derived from social phenomena, but was deducted from philosophical and theological principia, especially from natural-rights and revelation. (2) With Pius XI (1922–1939), the social thought of the Church entered a second phase, characterized by a longing for the re-birth of the "Christian social model" from inside the lay society. The Church describes this model as the "Perfect Society." Addressing economics, the encyclical *"Quadragesimo Anno* (1931) attempted to defend the catholic "third way" against the "socialistic way" (applied in Russia since 1917) and against the "capitalistic way" (called into question by the Wall Street Crash and the Great Depression in 1929). (3) Since Vatican II, the Church has demonstrated willingness to establish a dialogue with the world. This marked a turning point in the Church's social discourse; it no longer focused on conflicts among social classes (as it did in the nineteenth century), or confrontations among national economic systems (as in the first half of twentieth century). The entire economic order, worldwide, is polarized first by the political tensions between West and East and by the economic tensions between North and South, tensions which remain as irresolvable imbalances.

From a historical point of view, Roman Catholic social teaching is characterized by two important features: *change*, insofar as it is attentive to mutable historical and social conditions and to different methods chosen to interpret reality, and *continuity*, because the inspiration from the gospel and the pastoral contributions of the Pontiffs remains constant. Some of these constants include the view that the human person possesses an inherent dignity which must be protected in all cases; that the dignity of work is more valuable than the product work; that private property has a social function, but must be reconciled with the universal allocation of material goods which were created for everybody; that principles of justice, equality, and responsibility shall be applied to all the agents of economic activity; that the right to work is one of the fundamental rights of humankind; and that a worker's right to join a union must be protected and promoted as one of the means by which the justice of the relationships between enterprise and workers can be assured. The role of public powers as far as economics and society are concerned involve the determination of the juridical context of productive activities, protecting the classes which are the most exposed to economical exploitation and regulating energy resources and other major means of production.

The principles of reflection upon social doctrine, both natural and Christian, include the primacy of the human person, the principle of subsidiarity and solidarity (*Sollicitudo*). The first principle refers to humans, their superiority over other things and their unalienable dignity: humanities' superiority is affirmed as the foundation of human greatness, human nature is understood as being in God's "image and resemblance."

Following the subsidiarity principle, humans are to be placed in the appropriate conditions to fully realize their humanity, so that human potentialities develop in response to human needs, before wishing and asking for external intervention. "A community of a higher order should not interfere in the internal life of a community of a lower order, depriving the latter of its functions, but rather should support it in case of need and help to coordinate

Sartre, Jean-Paul – Sustainable Development, World Summit on   413

its activity with the activities of the rest of society, always with a view to the common good" (*Centesimus annus*).

References: Leo XIII. *Rerum Novarum* (On the Condition of Labor). Rome: Vatican, 1881. Pius XI. *On Social Reconstruction* (*Quadragesimo Anno*). Boston: Daughters of St. Paul, 1931. John Paul II, *Encyclical Letter Sollicitudo rei socialis* (Concern for Social Life). Vatican City: Libreria Editrice Vaticana, 1987. —. *On the Hundreth Anniversary of Rerum Novarum* (*Centesimus annus*). Washington, DC: Office for Pub. and Promotion Services, US Catholic Conference, 1991. Roman Catholic Church. *Catechism of the Catholic Church*. Vatican City: Liguori Publications, 1992.

A. Marocco

Social Entities: Those who hold that institutions exist accept the ontological status of Social Entities; epiphenomenologists deny this.

Among epiphenomenal positions are those that are motivated by general *naturalistic* or *physicalistic* assumptions. According to philosophers such as Willard V. O. Quine or David Lewis all basic structures of reality are physical structures. Another epiphenomenal stance simply denies the existence of everything that depends ontologically on the intentions of persons. Because all socialstructures are dependent on the intentions of at least one person, they do not exist. Peter Van Inwagen and Gary Rosenkrantz are philosophers who take this view. The main problem with this view is that "independence from intentions" is a problematic criterion for existence.

Keith Campbell and David Armstrong defend a qualified epiphenomenalist position regarding the existence of social structures, asserting that social structures are *supervenient* phenomena. "Supervenience" means dependence without reduction or elimination. Supervenient structures are not, despite their irreducibility, entities in their own right. The problem of supervenience is that it is a puzzle in itself rather than a solution.

Another position regarding the ontology of social structures is realism. Realists regard social structures as entities in a strict sense. John Searle, Jonathan Lowe, and Roderick Chisholm are well-known realists concerning social entities. The first main problem is to explain their ontological status, especially their dependence on individual persons. The second problem involves giving an account on the "conditions of identity" of social entities. A third difficult problem has to do with explaining the causality of social entities. Social entities have causes and effects.

Are there only individual human persons, or, in addition to individuals, is there such a thing as society? What is the relation between individuals and institutions? Can individual persons influence global structures? Whatever answers are offered, they will (at least implicitly) depend on the answerer's opinion about whether social entities exist.

C. Kanzian

Social Partnerships refer to a system of relations between wage laborers (and labor representatives), employers (and employer representatives), and governmental bodies, which target employee- employer interest reconciliation by regulating social and labor relations. The concept of social partnerships concept is widespread in economically developed countries, forming the

414 *Global Studies Encyclopedic Dictionary*

regulatory basis of relations between wage and salary earners and employers. Its essence consists in minimizing the role of official policy and if possible excluding social conflicts in the labor field. Preserving social peace is the main objective of social partnership. Focus is placed on determining working conditions and adjusting labor disputes.

Such social partnership issues in Bulgaria, Hungary, Germany, and Italy have been addressed in their Constitutions and in special acts. Several acts on the federal level have been established in the United States: for example the National Labor Relations Law of 1935 (Wagner's law), the Workmen Act of 1947, and the Fair Labor Standards Act of 1938.

Many means of implementing social partnership have been developed. Social partnership in Bulgaria is realized in labor contracts and in other forms. In Germany, basic forms are collective (tariff) contracts, employees' participation in production management, and social payment system. In the United States and Switzerland collective contract elaboration is one of the most important forms of carrying out social partnership.

Globalization inevitably requires the reformation of all social spheres. Much attention is paid to framework agreements that represent a new means of association and collective negotiations and freedom in the specific conditions of globalization. The world's leading companies and trade unions can collectively form new negotiations and partnerships. Transnational corporations exhibiting corporate social responsibility also put forward social, economic and ecological initiatives.

K. N. Gusov

Social Theory, Critical: Critical Social theory is a trend in contemporary social philosophy, which originates from an updated version of "critical theory" which describes the neo-Marxist philosophy of the Frankfurt School, developed in Germany in the 1930s. Critical theory maintained that the principle obstacle to human liberation is ideology. It was based on the theory of communicative action of Jürgen Habermas (1981), shaped in the course of debates between Habermas and Michel Foucault regarding the effective strategy for sociophilosophical criticism. The main elements of that strategy are: (1) the reinterpretation and immanent criticism of "modernity" based on disclosing its complex structure and demonstrating the play of freedom and dominance in modern life and thinking, (2) "hybrid discourse," combining philosophical education with multidisciplinary social studies, and (3) the generation of politically relevant knowledge.

Intensive development of the transnational dimension of social life and globalization as the thematic priorities of social sciences in the 1990s required a substantial revision of views on the public sphere traditionally understood in terms of a classical nation-state. Globalization raised the issue of the legitimization of supranational political structures. For Habermas this issue assumes the form of the supranational public sphere of the cosmopolitan public guided by the principle of "constitutional patriotism." In national consciousness, there was always a tension between the universal value orientations of a constitutional state and democracy (the "political nation") and the particularism of the "ethnic nation." After culture and politics become more differentiated in classical nation-states, conditions arise for "constitutional patriotism." As distinct from the patriotism that sprang up within national consciousness, constitutional patriotism no longer correlates with a particular nation, but with abstract methods and principles of democratic political culture. Gaining ground among the cosmopolitan public, it becomes a factor in shaping the supranational public sphere capable of performing the function of legitimization.

Contemporary critical theoreticians categorize themes of transnational public spheres according to several aspects: (1) the emergence of new public spheres, ensuring the democratic character of supranational political structures and transnational policy as a whole; (2) new social conflicts generated by the clash of legitimacy arising within transnational public spheres with the legitimacy of nation-states (the rise of ethno-nationalism and ethnic violence as the reverse side of globalization); (3) the building of an adequate political language for the potential postnational order; and (4) the problem of "cyber-democracy" in the context of the virtualization of social life.

Reference: Habermas, J. *The Theory of Communicative Action*. Boston: Beacon, 1981.

V. N. Fourse

Society is a sphere of human existence that includes the life and activity of people as bearers of various connections developing between them and their attitudes (public attitudes, communications) as founders and subjects of material and spiritual culture. Society exists in nature and at the same time differs from it as a historically arisen product and a result of the activity of people. A person is a living organism, a part of the Earth's biosphere, though the physical appearance and nervous system of human beings were structured in the process of the formation of society and under the influence of this process. Nevertheless, as a social creature and a subject of culture, the person is a product of society. The individual creates the society and is created by the society. Belonging to society is a distinguishing feature of a person.

The term "society" has different meaning and content used in social studies as compared with ordinary speech. Ancient philosophy identified society with the state. According to Plato, people unite in a state because they need each other. A state is their joint settlement and labor division among them, which allows them to satisfy their diverse needs. Aristotle called a person "a political animal"; he tightly connected the features of human beings to their belonging to a state. In the Middle Ages, a person existed in a system of strict submission to heavenly authority and authority given by God. This corresponded to the distinction between an ideal "Heavenly city" and a vicious "Earthly city," made by Augustine and to the assertion of the superiority of ecclesiastical authority above secular authority.

In the seventeenth and eighteenth century, the concept of society started to gradually lose its identification with the state. Contributing to that process was the theory of the public agreement (or social contract), accepted by many thinkers of those times, according to which the state is created on the basis of an agreement between people. Basic to this theory was a thesis that initially people existed in their natural state and had inalienable natural rights (such as individual freedom). The conclusion of a public agreement meant a transition

416 *Global Studies Encyclopedic Dictionary*

from a natural into a civil state and establishment of the state. The necessity of such a transition received different explanations.

Thomas Hobbes (seventeenth century) considered the war of everyone against everyone a natural thing, and a state was established to stop mutual destruction. This could be done only by a powerful state with loyal subjects. The legitimacy of power is stipulated by the fact that individuals transfer their natural rights to the state. In exchange for renouncing their natural rights they receive protection of their lives and peace. Jean-Jacques Rousseau (eighteenth century), opposing Hobbes, believed that (in a state of nature) people lived peacefully, freely, and happily, having their natural rights. The establishment of a public agreement (or social contract) was brought about by the desire to save some equality because of the advent of private property. People were sovereign. They transferred part of their natural rights to the state. Thus, the state does not absorb the whole society, but corresponds to civil society. The difference between the two (the state and civil society) is expressed in Hegel's philosophy, which presented civil society as a kingdom of economic relations opposed to the state as a center of World Spirit that was the basis of the historical process.

Unlike nature, society includes a subject. This subject is a person with a consciousness, free will, and capable of teleologism. Since society is a product of human activity, conscious activity plays a decisive role in its formation, change, and development. (The history of social thinking has been and still is considered the grounds for the conclusion that society does not contain objective determinants and regularities.) Concepts have also been developed, stipulating that a distinguishing feature of human existence is culture with its ideals, values, and norms, because social connections and relations exist in nature as well (as with gregarious animals or ant colonies). Indeed, the complexity and variety of a society's life give grounds to various methodological constructions (frequently opposing each other), and the establishment of a scientific approach in the sphere of social knowledge turns out to be an extraordinarily difficult task. At the same time, social philosophy has developed various "ways of thinking," selected major methodological directions, accumulated huge experience. It shows that a scientific methodology for comprehending society is impossible without recognizing the fact that it includes objective principles, independent of human consciousness; these principles serve as a basis for existence and the activity of the objective regularities of social life. Marxism posits that these principles are the material interaction of persons and nature in the process of social production; material, social relations give the framework interrelations between society and nature; sub-personal social structures are also preserved in the process. Generations starting their life act in conditions that are given to them objectively, being created by preceding generations. Besides, changes that take place in society are the consequence of interaction of multidirectional forces and different tendencies, where the end result in most cases does not coincide with the objectives that people set. Taking all this into consideration, it is possible to comprehend and scientifically explain human activity and the role of subjective-ideological, spiritual, and personal, factors in the life and development of society.

Society is originally part of nature and can exist on Earth only in constant interaction with nature. Nature, including the biological nature of per-

Sartre, Jean-Paul – Sustainable Development, World Summit on   417

sons themselves, is a natural basis of society, and people's health is a social problem, significant for society. Various relations between society and nature have always interested philosophy and social studies. Still, up to the second half of the twentieth century study of the influence of nature as a "geographical medium" on society was emphasized. There were polemics, concerning whether the environment defines the development of society (geographical determinism) or not, whether the biological nature of people determine their activity or not. There was a utilitarian attitude toward nature, which was taken only as a source of material and energy necessary for life. Since society cannot exist and develop without transforming nature (adapting it to the needs of society), it was judged that the purpose of this transformation is the establishment of the domination of society over nature.

During the last decades of the twentieth century, the picture was qualitatively changed; all former aims were replaced by their opposites. The growing volume of production has forced us to realize that there is a limit to this growth. Earth's resources are limited, and, in the near future, if the former way of development continues, many vital mineral and power resources will be exhausted. Together with local and regional consequences, global negative consequences of society's influence on nature also exist. Waste products and human activity as a whole pollute lands, the world's oceans and atmosphere; they start damaging nature, and nature cannot manage this anymore. Deterioration of "ecology" harms people's health. The variety of animal and plant species decreases. Anthropogenic pressure upon the nature becomes intolerable. There is a crisis in the interrelations of society and nature. If this tendency continues, society will destroy the complex self-replicating system of the Earth's biosphere and thus will completely undermine the natural conditions of its own existence. This ecological problem is one of the most acute global problems of modernity. The specificity of global problems is that they can be solved only by the combined efforts of all of humankind and they oblige people to work together. The process of economic, financial, and information globalization, as taking place currently, should promote this unification as fundamental.

V. J. Kelle

Society, Closed: A "closed" society is a type of social life, or a pattern of distinguished to an "open" society. Classical features of a closed society are authoritativeness and hierarchy and typically evinces communal thinking, or, a thinking of a "community of coreligionists." Communal thinking is understood to lie at the root of racial, ethnic, and religious prejudices. Such thinking gives rise to feelings of community participation, provides easy acceptance of the authority of power and the state. Rigidly programmable fetishes manipulating human behavior (such as "kulak," "people's enemy," "leader of the world proletariat," or "teacher") are prevalent.

In a closed society, communication is based on sub-personal norms. Thinking reacts to reality symptomatology only by means of modifications of dominant dogma. Ideology acquires a mystical power over people's outlook, and the influence of information incorporated in this ideology replaces efforts of individual comprehension and responsibility with a mechanism of irresponsibility. A fundamentally nonreflexive method of thinking is characteris-

418 *Global Studies Encyclopedic Dictionary*

tic of a closed society. Participants in real sociopolitical events realize little, or do not realize at all, why they pursue these aims and not those or what makes them act this way and not the other.

A closed society can be fairly compared to an organism subject to instinct, as opposed to reason. Sometimes thinking begins to resemble a special type of belief. A citizen of a closed community should have a specific conviction, belief in the accepted doctrine, which is represented as true, self-sufficient, and tolerating no objections. Revival of tribal spirit and a cult of collectivism constitute the atmosphere of a closed society. A citizen of a closed society is characterized by a constantly alert attitude and seizure of alien influences that gives rise to a stable type of reaction of assault or defense, directed either outside or inside. Independent thinking is impossible in a closed society, and is regarded as deviant.

Reality, having no real reviewer, is mystified, being characterized by features that do not exist. At the same time, a real empirical subject is obliged to know the technique of double consciousness. No doubt all this is reflected in inter-civilizational dialogue and gives rise to estimating it as belonging to the area of the incomprehensible, irrational, and absurd. A closed society loses in both the speed of acquiring information and in its quality. The whole structure of such a type of society, not adjusted to dynamic adaptation to constantly changing external conditions, appears vulnerable in many respects. At the same time, the hierarchical power structures of a closed society are capable of fast regeneration.

T. G. Leshkevich

Society, Network: Network society refers to a society that emerges in the information era and in which the dominant functions and processes are organized on the basis of computer network principles. New information technologies provide a material basis for multiple penetrations of networks into the structure of a society. Manuel Castells introduced the concept of "Network society," coined by Manuel Castells in 1996, posits that network logic entails the emergence of a new social determinant on a level higher than the specific interests giving rise to such networks: the power of structure turns out to be more powerful than the structure of power. Belonging or not belonging to this or that network, along with the dynamics of some networks with respect to other networks, is the most important sources of power and change in a modern society. This allows us to characterize our society as a network society, of which the distinctive feature is the domination of social morphology over social action.

The concept of "network" is linked most frequently to telecommunications, multimedia technologies, and the infrastructure of the "information society." Hierarchic management structures are being replaced with network ones. Network structures are especially useful for open socioeconomic systems oriented toward cooperation. Market activities, above all in the area of direct sales, are also carried out based on network principle. Effectiveness of networks is emphasized in the area of conflict prevention and resolution and human rights and in the system of civil control and participation. Remote interactive education is also based on the network principle. This principle increasingly becomes universal and modern organizational structures begin to absorb more and more information structures and information technologies.

Network structures possess high dynamic characteristics, related above all to information exchange. At the same time, the network matrix sets static parameters and models individual blocs and segments and the lines of tension between them. Network structure plays a fundamental role in building an effective communications system; it significantly reduces the volume and cost of the information exchanged. Such a structure is optimal for decision-making in an area of uncertainty; it possesses a high degree of adaptability, as it tends to give priority to its own values, not to signals coming from the external environment. Networks are open structures capable of infinite expansion by incorporating new units able to communicate within a given network (using similar communication codes). A social structure having a network basis is characterized by high dynamics and openness to innovations, without the risk of becoming unbalanced. Networks facilitate economic development; network morphology becomes a source for restructuring power relationships, for democratic control and participation, and for the formation of electronic government (e-government). Thus, new economic forms are being built around global networks of capital, management, and information, while network-based access to technological skills and knowledge becomes the foundation of productivity and competitiveness.

I. A. Malkovskaya

Society, Open: Open society is a type of social organization built upon the principle of "openness" that can be interpreted in different ways: openness as the fundamental infinity of a social project; openness of the foundations of individual and collective action to rational criticism; openness as an ability to assimilate diverse principles and values and as preparedness to sensible and equal dialogue with other cultures and societies. Andre Bergson and Karl Popper, in particular, advanced this concept.

In an open society, rationality is a precondition for democracy, and vice versa. Open society is, thus, an externalization of critical rationalism just like critical rationalism is a quintessence of the meaning of an open society.

The compelling need for an open society is especially evident with regard to global crises that threaten the future of humankind, some of which are caused by anthropogenic factors: overpopulation; expansion of natural resources needed to maintain the lives of rapidly growing human masses; ecological collapse; military catastrophe; globalization of terrorism; and extremism. Open society is not a guarantee against such threats, but it can set against them the conscious will of people united not by forceful regimentation of their behavior but by a rational action plan adopted democratically.

According to the American entrepreneur and philanthropist George Soros, global capitalism is an infinite and even distorted form of an open society. Community and private interests are imbalanced. Market mechanisms in themselves are unable to provide an optimal allocation of resources and world economic stability gives rise to destructive financial, economic, and social consequences and increasing tensions in international relations that are fraught with total crisis. This gives rise to the demand for a reconsideration of some of the characteristics of an open society that used to considered necessary (for example, the principle of state non-interference into market for-

420 *Global Studies Encyclopedic Dictionary*

mation and functioning). The very notion of "open society, thus, must be subject to rational criticism with regard to the reality of the modern world.

V. N. Porus

Sociogenesis (English, "society" and Greek *genesis*, "origin") is the generation, becoming, and development of a social structure through establishment of material, social, spiritual, emotional, moral, aesthetic, and other connections among people. Social genesis is a progressive process of "humanization" of humankind as a biological species, forestalling filling-in with social origin of his essence as that of a biosocial being. Humankind as a biosocial being is constantly evolving during thousands of years from an animal to a human state. Its biosocial content is progressively filled with social components: knowledge, intellection, speech, an ability to communicate, mutual activity, sense of beauty, and moral values. The process of sociogenesis is the starting point, the momentum; it sets the direction and character of globalization development in people life. Globalization is implemented through growth of social connections, intensification of communicating among people and their groups, through progressive movement from first steps of human life globalization to the global human world.

K. A. Barlybayev

Soil: (Latin *solum*, "base," "ground") Soil is that substance formed on the Earth's surface in that part of the biosphere that is the zone of contact and interpenetration of the atmosphere, hydrosphere, and the lithosphere and which has a maximum concentration of the living matter of our planet. Soil also forms a planetary envelope called pedosphere.

The upper boundary of soil is defined as the surface that contacts the atmosphere. Determining the lower boundary depends on how soil is understood as a natural-historical body, as a habitat for plants, or as an object of engineering-technological amelioration. Presently, the lower boundary of soil usually corresponds to the maximal depth of the penetration of biological objects. In practice, this requires determining the depth to which the root systems of local plant species penetrate and to what depth soil animals burrow. Soil is usually studied to a depth of 1.5–2.0 m, and for some purposes, up to 3.5 m.

Soil composition includes humus, which refers to any organic matter that has reached a point of stability, where it will break down no further and might, if conditions do not change, remain as it is for centuries, if not millennia. Humus significantly influences the texture of soil and contributes to moisture and nutrient retention. The essential part of soil—its live phase—is constituted by live organisms: the root systems of plants, soil animals of different sizes including unicellular protozoa, and a great diversity of microorganisms. Thus, soil is a multiphase system that has solid, liquid, gaseous, and live states as distinct from other natural bodies. It is impossible to separate soil microorganisms from soil humus even in laboratory conditions; hence, a total content of the organic matter in soil is analyzed.

T. V. Prokofieva

Soil Degradation refers generally to all processes that aggravate productivity of soils. In the narrow sense, soil degradation refers to processes of deterioration of soil structure, loss of humus and natural compounds, and sometimes washing away of silt in black earth.

The term "Soil degradation" is connected with the long history of development of agriculture and the cultivation of land resources. Although soil science was born more than a hundred years ago, the term "soil degradation" still has no precise definition due to its being many-sided, and being related both with natural and anthropogenic circumstances, including economic activity. Nevertheless, a summary of definitions of that can be found in modern literature includes the following common elements: (1) The notion of soil degradation is basically understood through the totality of soil-formation processes that lead to changes in soils and soil layers in comparison with the standard ones (according to both natural standards and productivity standards). (2) Soil degradation leads to decreasing fertility and productivity of soil or the quality of products. (3) Degradation of soils leads to increasing expenses for soil rehabilitation and raising the productivity level.

In connection with global losses of land resources, a new ecological aspect was added to the definition of soil degradation: it results in functional changes in soils as elements of the ecological system and in deviations from ecological standards and the aggravation of parameters that are important for biota and people. This aspect focuses more on the anthropocentric aspects and is connected with the decline of the biological productivity of degraded soils. This definition of soils degradation that takes into consideration human and environmental "well-being" is currently the most popular although it seems simple by comparison with the general systemic approach to soil degradation. Within the framework of the latter, the degradation of soils as complex bio-inert systems should be regarded as a process of gradual loss of elements and structural degradation of these systems.

Obviously, systemic vision of degradation corresponds to the term "soil degradation" in the case of such processes as erosion and deflation, but is not identical to them in such cases as formation of saline black earth in the course of irrigation. Perhaps, in the future soil scientists will be able to differentiate these two sides of the concept, but today the environmentalist point of view is predominant. The secondary changes of soils that are a result of human activity are called anthropogenic degradation of soils. These changes result in partial or total degradation of the fertile layer of soil or lead to its destruction.

Soil is a renewable resource that does not expire while being exploited, but persists and can improve. Partial loss of the fertile layer can be restored, but its total destruction is irreversible and finally leads to deformation, collapse, or deep degradation of the landscape. This idea is the foundation of modern "landscape adaptive" or "biological" agriculture.

Soil degradation is often caused by anthropogenic activities. Generally, it is very difficult to distinguish between natural and anthropogenic factors of soil degradation, while anthropogenic influence, as a rule, rapidly changes the intensity of the natural process within soils and causes new ones. Entirely anthropogenic processes are rare.

422   *Global Studies Encyclopedic Dictionary*

International and national law regulates measures on degradation prevention. The UN Food and Agricultural Organization (FAO) adopted the "World Soil Charter" in 1982, which called on governments of all countries to regard the soil layer as a universal heritage. Today the necessity of soil protection is proved by international documents, such as "Agenda 21" developed at the UN Convention on Struggle against Desertification (1994), and the UN Convention on Biodiversity (1993). In Russia, the necessity of soil protection is legally stated in the "Environmental Protection Act" (2002).

G. S. Kust

Somerville, John (1905–1994) was a tireless activist in behalf of world peace, human decency, and planetary survival. A scholar of social philosophy and ethics, and an internationally known expert on Marxism, he worked relentlessly against the Cold War and in behalf of dialogue between Soviet and American philosophers. His efforts led to the first bi-national conferences of American and Soviet philosophers-in Mexico City in 1963, and in New York City in 1964. Sometimes judged controversial by the paranoid standards of McCarthyism, Somerville persisted in placing personal integrity and global reconciliation above personal or professional advancement.

Committed to the cause of world peace and the prevention of nuclear extermination, Somerville was the first philosopher to mobilize North American philosophers professionally against the inhumanities of the nuclear threat and to provide us with a new vocabulary to discuss the threat of nuclear extinction ("omnicide"). With like-minded others, he founded International Philosophers for the Prevention of Nuclear Omnicide (IPPNO; now renamed International Philosophers for Peace) and the Union of American and Japanese Professionals against Nuclear Omnicide. His message was strong and unambiguous: "Those who take no action against [nuclear] weapons are, in effect, casting their votes for omnicide." Such literary figures and scientists as Thomas Mann, Albert Einstein, and Bernard Lown acknowledged his work.

R. E. Santoni

Sovereignty: In political science, sovereignty is usually defined as the state's most essential attribute in the form of its complete self-sufficiency; i.e., its supremacy in domestic policy and its independence in foreign policy. The notion of sovereignty, formed at the beginning of the modern age in works by Machiavelli, Jean Bodin, Thomas Hobbs, and others, germinated in the system of international relations after the Thirty Years War and 1648 Peace Treaties of Westphalia, and became more widespread in the nineteenth century. The UN Charter contains regulations on sovereign equality of states. However, the notion of sovereignty is quite difficult and ambiguous, and, in political science, one gradually becomes aware of the need to reconsider the concept in connection with globalization processes. In political science, the subject of the "diffusing" or "disappearing" of national sovereignty has been raised, beginning in the late twentieth and early twenty-first century.

The process of globalization greatly affects the change and reduction of the nomenclature and scope of contemporary states' sovereign authorities. We might even speak about the transition of most countries—and the system of international relations in general—into a new type of sovereignty. In the future, the change of the nature of sovereignty will influence all of the most important processes of global evolution. So, the transformation of sovereignty should be considered as one of the most significant contemporary trends.

During the first post-war decades, the modern nation-state became the leading type of government world-wide, and the principle of sovereignty reached its zenith. Simultaneously, sovereignty began to decline in the mid twentieth century. States' sovereignty has been considerably reduced in respect to war and peace; creating, preserving, and testing new types of weapons; many economic aspects, such as foreign trade duties and the size of budget deficits; rules concerning prisoners and using prisoners' labor; capital punishment; fundamental rules and election procedures; and other important areas.

The change of sovereign rights is a complex, nonlinear, bilateral process. Some factors fairly undermining the states' sovereignty are strengthening, such as technological and economic innovations, the world community's aspiration to avoid wars and settle common issues, a rapid increase of the number of contacts, and the growth of the number of democratic regimes in the world. Other threats to state sovereignty include global financial flows, activities of multinational corporations and global media empires, and the Internet.

Most countries voluntarily and deliberately limit the assertion of their sovereign rights. Such a reduction of rights becomes profitable because when joining supranational formations and international organizations the states hope to get in return quite real advantages and benefits (including strengthening their own positions or transferring some of their problems to the international community). Voluntariness in the process of reducing the scope of sovereign authorities for the sake of attaining additional prestige and benefits is one of the most important factors and is the factor that determines the process's irreversibility. This aspect of the process is highly underestimated.

Although the trend of sovereignty reduction refers to an overwhelming majority of countries, the speed and direction of relevant changes and the reasons for them (as well as the respective balance of benefits and implications) significantly vary in different countries and civilizations. The future of different cultures greatly depends on the way the processes of transformation of sovereignty will proceed.

The state will remain the leading subject of international relations for quite a long time. Its role may increase during some periods; nevertheless, in prospect, the tendency of reduction and transformation of national sovereignty will increase.

References: Berger, P. L. "Four Faces of Global Culture," *National Interest* 49 (Fall 1997). Buzan, B. "New Patterns of Global Security in the Twenty-First Century," *International Affairs* 67.3 (1991). Grinin, L. E. "The State in the Past and in the Future," *Herald of the Russian Academy of Sciences* 79.5 (2009); — "Globalization and Sovereignty: Why Do States Abandon Their Sovereign Prerogatives?," *Age of Globalization* 1 (2010). Held, D. "The Changing Structure of International Law." In *The Global Transformations Reader*. 2nd ed. Edited by D. Held and A. McGrew Cambridge: Polity Press, 2003. Maritain, J. "The Concept of Sovereignty," *The American Political Science Review* 44.2 (1950). Stankiewicz, W. J., ed. *In Defense of Sovereignty*. New York: Oxford University Press, 1969.

L. Grinin

424 *Global Studies Encyclopedic Dictionary*

Soviet Union, Global Studies in: Global studies in the Soviet Union is a system of knowledge that emerged between 1970 and 1990 at the meeting point of social, natural, and technical sciences and which is a complex of political activities oriented toward analyzing and solving socionatural contradictions of universal (planetary) scope and character. Beginning in the mid-1970s, the leading Soviet sociopolitical periodicals (*Pravda, Kommunist, Problemi mira i sotsializma*) were presenting several publications dedicated to global problems. The leading Soviet philosophical journal *Voprosy filosofii* published several round-table discussions on global (universal) issues. At the beginning of the 1980s the first monographs were issued. The central Soviet scientific institutes promoted global studies.

From the first half of the 1980s, global studies were conducted under the aegis of the Scientific Council on Philosophical and Social Problems of Science and Technology of the Soviet Academy of Sciences (headed by Izbrannye T. Frolov) and its Global Problems Panel (headed by Vadim V. Zagladin). In collaboration with other most significant research centers, they organized representative conferences and symposia in the Soviet Union and abroad, held discussions about the most important global problems and published fundamental monographs.

Foundations for further development of global studies were laid by studies of the system "persons-society-biosphere." Most of the work was done under the aegis of the Scientific Council on the Problems of Biosphere of the Soviet Academy of Sciences, headed by Alexander L. Yanshin. Especially important directions of global ecological studies include (1) philosophical and methodological direction; (2) geographical direction; (3) scientific and technical direction; (4) economic and legal direction; (5) educational direction; (6) ideological direction; and (7) international direction.

By the mid-1980s, the following thesis was formulated on the basis of studying global ecological simulators worked together by United States and Soviet scientists: if humankind did not overcome the arms race of the superpowers and avoid a thermonuclear conflict, not only would the material foundations of the modern civilization be destroyed, but the total extinction of life on the planet would result from an abrupt temperature lapse ("nuclear winter"). This thesis became the ground for the "new political thinking" that rejected confrontation of the "two worlds" (socialist and capitalist) and stressed the need for constructive interaction of the planetary community in order to solve universal tasks of the entire world. Despite that contradictions and fundamental differences among the members of the world community still existed, a trend to integration and non-confrontation gradually began to prevail. Human survival was recognized as a central problem, and the policy of peaceful coexistence of states with different social systems came to be seen as the only precondition for secure and constructive development of our civilization. This is how Soviet geologist Vladimir Vernadsky's idea came true that humankind would become a "global force" acting and thinking not only at the biospheric level but also at the "planetary level."

Given the "rules of the game" of that time, Soviet global studies were conceptually based on Marxist ideology. Nevertheless, despite their historical limitations, they achieved several innovations in evaluating civilizational development. First, they constructed a system of problems of global status; established their hierarchy and links; developed and put into practice global simulation techniques that enabled them to study problems of universal scope. Second, they overcame the class-based (formational) approach and analyzed civilizational perspectives in an ever-broader sociocultural context. Third, formal (and, much more often, informal) research groups emerged oriented toward studying the system of global problems of modernity.

The dissolution of the Soviet Union has directly influenced the development of global studies. There are several reasons why interest in global studies is diminishing, including the theoretical vacuum created by the rejection of a single (Marxist) ideology; the prevalence of a "Russian" approach to evaluating world processes, and organizational and financial hardships. However, by the end of the 1980s, and at the beginning of the 1990s, interest in studying processes of global scope is rising again in Russia. The experience of Soviet global studies is used to analyze the globalization process and the phenomenon of "sustainable development."

V. A. Los

Space Era is the stage of humankind's development that explores space beyond Earth's atmosphere and space objects in real-time, making them not only a subject of theoretical study but a sphere of practical activity. Human space explorations began on 4 October 1957, when the Soviet Union launched the world's first artificial Earth satellite, Sputnik. In 1959, human beings were able to view the Moon as a whole for the first time because of photos taken by the Soviet explorer Luna-3. The flight of Soviet cosmonaut Yuri Gagarin, the first human being to journey into outer space, when his Vostok spacecraft completed an orbit of the Earth on 12 April 1961, established a new pace of human history, "closing" the planet spatially in real-time mode. The United States became the second nation to achieve manned spaceflight with the suborbital flight of astronaut Alan Shepard aboard Freedom 7 on 5 May 1961. US Apolo 11 was the first space flight to land on the moon with human passengers, on 16 July 1969.

The space era promoted a unique, peaceful, international cooperation among many countries. Since the first human spaceflight by the Soviet Union, citizens of thirty-eight countries have flown in space (twenty-four "first flights" occurred on Soviet or Russian flights, while the United States carried thirteen). Human beings have been continually present in space since 2000, on the International Space Station. Currently, only Russia and China maintain human spaceflight capability independent of international cooperation. As of 2013, human spaceflights are only launched by the Soyuz program conducted by the Russian Federal Space Agency and the Shenzhou program conducted by the Chinese National Space Administration. The United States lost human spaceflight launch capability upon retirement of the space shuttle in 2011.

In recent years, there has been a gradual movement towards more commercial means of spaceflight. The first private human spaceflight took place on 21 June 2004, when SpaceShipOne conducted a suborbital flight. A number of non-governmental startup companies have sprung up, hoping to create a space tourism industry. In the United States, NASA has also tried to stimu-

426   *Global Studies Encyclopedic Dictionary*

late private spaceflight through programs such as Commercial Crew Development (CCDev) and Commercial Orbital Transportation Services (COTS).

Manned space flights, for the first time in human history, allowed us to see the entire planet in a single glance. Since that moment, the world where people live is not perceived from the inside only. Now it is possible to perceive and to explore it from the outside, and from every side. For the first time in human history, people were able to become external observers of what happens on the Earth. They are now able fully understand not only the uniqueness of their existence in the endless outer space but also the fragility of their "home" both in the face of natural forces and the technogenic impact on nature generated by people.

From the beginning of the space era, outer space became as much a source of human heritage as the natural earthly environment. This has, on the one hand, noticeably extended the sphere of human activity and enhanced human ability to influence biospheric processes on the Earth. On the other hand, this provided human beings with principally new opportunities to explore outer space and nature of the world. The era of exploring new worlds began from flights to the Moon, Mars, Venus, other planets and objects of the Solar System and was continued by sending space ships out of further into the solar system of planets.

Outer space explorations has had enormous impact on the development of communication and information technologies such as radio, television, and telephone. In many aspects, this contributed to the development of informational revolution and made electronic mass media the most important vehicle of policy-making and the influencing of mass consciousness.

A. N. Chumakov

Space Exploration and Global Development: Space exploration provides a unique opportunity to solve many practical tasks of global development. It opens the way to nontraditional means of providing socioeconomic stability and national security; facilitates the access to inexhaustible raw material and energy resources; and stimulates economic and political cooperation among nations. The cooperative realization of space projects could contribute to sustainable development and building of a global security system.

The important problems of the Earth and the Solar system could be solved in the process of fundamental space research. Joint scientific projects in outer space (namely, diversification of advanced technologies needed for the construction of space devices) can promote worldwide cooperation within the world community. For example, an idea has been proposed to evacuate radioactive waste, enclosed in special containers, from the Earth with the help of space rockets. This waste could be kept in distant orbits until humankind finds means of its processing and utilization.

Currently, almost all the advanced countries of the world have intensified their space programs. More than 120 countries are taking part in such programs, with more than twenty having their own satellites.

Sartre, Jean-Paul – Sustainable Development, World Summit on 427

At present, realization of large-scale space projects is only possible under UN aegis. The world community must assume responsibility for the achievements of space explorations, taking into account the interests of all countries.

V. K. Postanogov

Spheres of the Earth are various coats covering Earth. The idea of such spheres was first formed after people had come to the conclusion that our planet has a form of a globe and accumulated the material needed for the related generalizations and for understanding these spheres as holistic systems. It only happened in the nineteenth century, that is, more than three hundred years after the great geographic discoveries.

The emergence of the term "biology," coined by French naturalist Jean Baptiste Lamark in 1802, was an important step on the way to such generalizations. Lamark laid the foundation for the future broad generalization in the sphere of studying life as a planetary phenomenon. In 1875, Austrian geologist Eduard Suess introduced the notion of "biosphere" to designate the area where the life existed on the Earth. He also offered new terms—lithosphere and hydrosphere, respectively—to describe the land and the water coat of the Earth. Much earlier the gas coat of the Earth had been designated as "atmosphere" and from the second half of the nineteenth century people began to observe and study it deliberately on a regular basis. These were the first terms intended to describe the largest planetary systems, the number of which grew following scientific discoveries and generalizations about new facets of the holistic world.

French geographer and sociologist Jacques Élisée Reclus attempted to provide a common picture of human development and description of various countries (1876–1894). After this work, but especially after the publication of *The Proliferation of Organisms and Their Role in the Machinery of the Biosphere* by Vladimir I. Vernadsky (1926), the vision of the "many coats" of the Earth has become a part of scientific consciousness, which has added similar terms. For example, the sum of the geographic coats of the planet is the atmosphere (divided into the troposphere, stratosphere and mesosphere); the hydrosphere and lithosphere was renamed as geosphere, or the area of the non-living (the inert) matter, and the basis for life on the Earth is called the biosphere. In 1944, Viktor Borisovich Sochava distinguished the phytosphere and E. M. Lavrinenko, in 1949, the phytogeosphere.

The complex notion of "biogeosphere" was created as a synonym of "the membrane of life," the layer of life's "condensation" (Vernadsky). In 1940, Vladimir N. Sukachev offered the term "biogeocenosis" to characterize the homogenous areas of the Earth's surface with specific combinations of the living and inert components in dynamic interaction. Much earlier, in the 1920s, the term "noosphere" emerges in works by Jules Le Roi, Pierre Teiard de Chardin, and Vernadsky. Then, Pavel A. Florensky offered the term "pneumatosphere" and Yu. M. Lotman created the term "semiosphere." By the middle of the twentieth century, a whole sphere of scientific knowledge emerged where human beings and various forms of spiritual and social life engendered by them were considered as a planetary phenomenon. Such terms

428 *Global Studies Encyclopedic Dictionary*

as "anthroposphere," "technosphere" and "sociosphere" also entered the language of science.

The multi-coat vision of the contemporary state of the planet is now represented by various planetary spheres among which the geosphere, biosphere, and sociosphere are the most basic. Taken in their interdependence and dynamics, they represent a single, holistic geobiosociosystem.

References: Reclus, J. E. *La nouvelle géographie universelle* (A New Universal Geography). 19 vols. Paris: Hachette et cie., 1875–1894. Vernadsky, V. I. *La multiplication des organismes et son rôle dans le mécanisme de la biosphère* (The Proliferation of Organisms and Their Role in the Machinery of the Biosphere). 1926.

A. N. Chumakov

Stateless Nation is a nation without its own nation-state. State is much older than nation. The first states emerged five millennia ago in Mesopotamia and Egypt, having been made and unmade throughout history. Only in the twentieth century did all the inhabited territory of the Earth get tightly divided among states. Nation emerged in an evolutionary manner in England during the sixteenth century. This model entailed the state that made its population into a nation. The result of this process was the nation-state. The codification of this standard in nation- and nation-state-building came with the American and French Revolutions. They produced the model civic nation-state. Its ethnic counterpart came into being when the ethnically construed national movements founded the Italian and German nation-states in 1860 and 1871, respectively. Italian and German nationalists, however, legitimized the creation of their nation-states referring to the tradition of statehood conveniently provided by the medieval Kingdom of Italy and the Holy Roman Empire. Italian national activist Giuseppe Mazzini concluded this discourse on legitimate and illegitimate national claims in his 1857 map of the ideal Europe of nation-states. He established that there was place enough only for eleven "true nations," this is, for those with their states or some established tradition of statehood. Hence, his map included Poland but not Ireland or Slovakia. In the popular view, at that time, Ireland was seen to be as much an inalienable part of the United Kingdom as Slovakia (Upper Hungary) was of Hungary.

The utmost recognition for stateless nations came after World War I. The West European powers led by the United States recognized the right to self-determination as the principle of creating and maintaining international order (predominantly in Europe). On this basis, the non-national empires of Austria-Hungary and the Ottomans were replaced with a plethora of states granted to ethnic nations, this is, stateless ones and without history. This process was repeated in the wake of the 1991 break-up of the Soviet Union and Yugoslavia. Again, stateless nations gained their own nation-states, though in other parts of the globe a different model of nation-state-building has predominated to this day. In the wake of decolonization postcolonial states were burdened with the task of shaping their nations on a civic basis.

In the today's world, tightly divided into nation-states, is a clear tendency to deny legitimacy to stateless nations. Founding any new nation-state for such a nation means curtailing the territorial size and legitimacy of an already existing nation-state or even several of them. This would entail undermining the post-1945 global state system earmarked for nation-states only and grounded in the principle of the inviolability of borders, that is, borders that cannot be changed through a unilateral action.

If the treatment of ethnic minorities is upgraded to that of national minorities, by default, many ethnic groups would become de facto nations even without the necessity of applying the label of nation to itself. Thus, the dilemma of stateless nations is still around. What is the way out of it? Perhaps, it requires making the ethnic foundations of Eurasian nation-states more civic, espousing multiculturalism and some collective rights for minorities, and regionalization/federalization. This last method would allow granting the most aspiring stateless nations with their own ersatz nation-states without the necessity of destroying the existing nation-states. Last but not least, an overlaying supra-state structure (such as that of the European Union) may also make it possible to create new separate nation-states in a peaceful manner that would be beneficial to the stateless nation in the quest for its statehood and not harmful to nations already enjoying their nation-states.

T. Kamusella

State System, Global: At the beginning of the twenty-first century, all the inhabited landmass of the world is divided among the almost two hundred extant states. These states are organized according to the tenets of the nation-state model. The relations among the states are regulated by the set of agreed-upon principles. First, each state is sovereign. Second, to be such, a state must declare itself a nation-state and the vast majority of other states must recognize the nascent state's existence and its national declaration. Third, the recognized state's established borders are inviolable and cannot be altered through a unilateral action. Fourth, agreed-upon international law regulates the day-to-day relations among the states. Fifth, there is no authority higher than the state that would control the behavior of the states in their relations with one another.

Since all the extant states share the same model of political and social organization, the dynamics of their behavior is very similar. This homogeneity allows for a high degree of compatibility among the states construed as the only full actors of international relations. In turn, the maintenance of durable mutual contacts by each state with almost all the other ones is possible, thanks to the observance of the basic rules that regulate international relations. The resultant high degree of interrelatedness makes these states and the relations among them into a system. Despite the lack of any supra-state authority, this system is not chaotic but self-regulating because almost all the states are committed to maintaining the existence of the system. The violators are isolated or their misbehavior is corrected through a collective action undertaken by other states. This is a viable course of action because at any given time most states are intent on preserving this system rather than destroying it or replacing it with another model of interstate relations. At most, the states tend to wish to add some minor modification to the system in order to ensure its smooth running. This is usually executed through introducing a new agreed-on principle of international law.

430   *Global Studies Encyclopedic Dictionary*

The contemporary, largely homogeneous political organization of the world is a very recent phenomenon, though social scientists tend to view it in a stationary manner as timeless. The fundamental categories of analysis and the basic subjects of research in sociology and political science are society and state, respectively. But though many kinds of society can be distinguished, the default society on which the sociologist would focus is the nation. In a similar fashion (but less pronounced), the political scientist does not research any kind of polity. The field of political science is the nation-state because, virtually, no other models of state obtain nowadays. This situation leaves different kinds of statehood organization to the scrutiny of historians, while the sociologist's lack of attention to non-national modes of social organization places this problematic in the court of anthropology.

Historically, international relations has been marked by anarchy, this is, the absence of authoritative institutions and norms above independent and sovereign nation-states. However, during the last two centuries, there has emerged an increasing number of implicit principles that regulate the establishment of nation-states and their behavior within the framework of the global state system. This system has spread worldwide. Even uninhabited Antarctica was provisionally parceled among seven nation-states. Moreover, the UN Convention on the Law and the Sea (1982) allows coastal nation-states to exercise full sovereignty over a territorial sea up to twelve nautical miles and limited jurisdiction in an exclusive economic zone up to 200 nautical miles. The development of the unclaimed seabed is to be regulated by an international organization.

Hence, the anarchy of international relations has been gradually put in order. Even a quasi supra-state institution has emerged in the form of the Security Council and United Nations Assembly. The community of recognized nation-states that are not permanent members of this Council dominates the latter. Both these groups of states (with the seniority resting with the former) decide whether to include a state in the global state system.

The question remains how a global system is going to fare in future. It can get fortified, should all or the vast majority of the nation-states adopt democracy, market economies, and human rights observance as virtually universal elements of such a system. Alternatively, the development of continent-wide supra-state political-economic blocs may diversify the homogeneity of the global state system. In this scenario, the world's richest nation-states would band together in blocs, leaving other nation-states beyond the North-South poverty barrier unable to spawn such supra-state structures and to stand up to the rich. In a third scenario, US political scientist Samuel P. Huntington opines, in the wake of the bloody unmaking of Yugoslavia, a new world order "civilization" could replace the nation and the nation-state as the main unit of social and political organization of the globe.

T. Kamusella

Stockholm-72 refers to the UN Conference on the Environment held in Stockholm, Sweden, in 1972.

In the mid-1960s, most of the developed countries suddenly became aware of local and regional ecological problems caused by the massive destruction of natural ecosystems, environmental pollution, and other negative influences on the environment caused by economic growth based on extensive use of industrial technologies in manufacturing, agriculture, the transport system, and public services sector of the economy. These technologies require considerable consumption of natural resources per unit of the output product and generate dangerous amounts of wastes that pollute air, soil, surface and subsurface waters, and seas. Ecosystems in the regions of high industrial concentration and intensive agriculture lost their ability to fully process the growing flow of pollution, entailing the onset of environmental degradation with inevitable dreadful consequences for human health and economic losses.

Pollution of air and fresh water sources has caused widespread allergies, oncological diseases, genetic deviations, weakened resistance to "traditional" diseases, and poisonings; e.g., some new ecological diseases have been registered in Japan. Leading problems include air pollution in industrial cities and megalopolises; water ecosystem degradation in the developed regions and in the zones of their ecological influence; radiation threat stemming from nuclear weapons production, testing, and storage, nuclear power engineering; and the use of radioactive elements in manufacturing. Eutrophication turned to be a threat even for such immense water objects as the Great Lakes in North America. Soil corrosion endangered agriculture and forestry of many regions due to industrial and transport emissions of sulfur and nitrogen compounds resulting in acid rains falling thousands of kilometers away from the emitting sites. Growing deforestation, desertification, and biological diversity reduction was recognized as a symptom of global ecological imbalance.

Since the end of the 1960s, many nongovernmental ecological organizations in the US and Europe have started protest actions against polluting companies, and anti-ecological government policies. "Green" parties started the fight for positions in the local parliaments; some international environmental organizations were founded. Ecological factors became significant for both domestic and foreign policy and international relations. Environmental protection required coordination of the actions of all the world states and the elaboration of a concerted ecological policy.

The aim of the Stockholm-72 conference was to examine these tasks at the highest international level. It took two years to prepare this conference. The Soviet Union took an active part in the preparation, but refused to send its delegation to Stockholm in protest because East Germany had not been allowed to participate in the conference.

The main finding of the discussions of the conference was that the contemporary development pattern is incompatible with ecological sustainability and environmental protection. Development strategy had to be altered both with respect to the applied industrial technologies and consumption volume and structure, which should be brought into line with biospheric capabilities. This conclusion was reflected somewhat in the Stockholm Declaration and Stockholm Actions Plan adopted by the conference.

The Stockholm Declaration was based on five principles: the right of all to a friendly environment and personal responsibility for the state of this environment, which led to some political conclusions (such as the inadmissibility of apartheid, segregation, and various form of oppression); the necessity to preserve natural resources for the sake of the present and future generations;

432 *Global Studies Encyclopedic Dictionary*

the demand for the preservation of the ability of the Earth to reproduce renewable resources; the importance of wild life preservation; the necessity to carefully use nonrenewable resources and fairly distribute the profits gained from their exploitation.

The Stockholm Actions Plan contained 109 recommendations, categorized by: evaluation of the environmental condition; environmental control; global pollution exposure and monitoring; ecological education, culture, and information; development and the environment. Only the last of the five sections was devoted to the social and economic issues governing the condition of the environment, rather than to the ecological problems themselves and their reflection in the public consciousness. In the 1970s, most ecologists, not mentioning the other forum participants, still thought that ecological problems could be solved technologically, through the use of technical facilities, organizational solutions, financial and economic measures, and other traditional tools though reoriented toward a new goal.

The main practical result of Stockholm-72 was the decision to establish a permanent United Nations body—the UN Environmental Program (UNEP), located in Nairobi, Kenya, intended to coordinate the efforts of nations in preserving the environment and working out the solutions to global ecological problems. UNEP has developed and implemented dozens of projects (e.g., keeping the register of dangerous chemicals or ecological monitoring), published several reports on the world environmental condition (the first published in 1984), and prepared a significant number of international environmental conventions, agreements, and protocols.

According to the Stockholm Actions Plan, the majority of countries (including all the developed countries) have begun to create ecological legislation and regulation; in the 1970s, governmental agencies dealing with environmental protection emerged, and national reports on the environment began to be regularly published. Stockholm-72 served as an impulse to expand research projects on the preservation of nature and ecology including international projects. The most important of these are "Humans and the Biosphere," "World Climate Program," and "International Geosphere and Biosphere Program." Special attention was paid at the conference to ecological education. However, time has shown that the main goals of the conference have not been reached. The conference Rio-92 was to sum up twenty years after Stockholm-72 and to find new solutions to the old problems.

V. I. Danilov-Danilyan

Superpower: Coined in the 1960s, to designate the two world nuclear powers—the United States and the Soviet Union, the term "superpower" implied that the world was seen as bipolar. It was believed that each superpower was a separate military and political leader of a pole, embracing a specific group of nations (e.g., the socialist camp and the imperialistic camp). Many nations (first, in Asia, Africa, and Latin America) not belonging to one of the camps; were referred to as the "Third World." After the disintegration of the Soviet Union, the United States became the sole global power. Many question a widely used expression "monopolar world," and the United States (or, broadly speaking, G7 or NATO) pretensions to supremacy in governing globalization processes have given rise to such phenomena as the anti-globalist movement. However, some scholars suggest that concentration of power in the hands of one nation will no longer be characteristic of world politics. A new power configuration will arise which some theorists talk about in terms of a multi-polar world.

A. V. Katsura

Sustainable Development is development that meets the needs of present generations without compromising the ability of future generations to meet their needs. This definition was formulated at the United Nations Conference on the Environment and Development (Rio de Janeiro, 1992), but it cannot be regarded as constructive, because to put it to use, one would need to compare the vital needs of the current generation with the needs of future generations to which we have no access. The degree of uncertainty regarding the future is constantly growing due to the pace of science and technical advance and the rate of the accumulation of social changes

Potential access to the planet's resources of different generations can only be the same if non-renewable resources are not utilized at all (this condition is necessary but not sufficient). Any exclusion of a non-renewable resource from nature irreversibly reduces opportunities of future generations.

We can come out with a more precise definition if we apply it to the specific aim, namely, to analyze the possibilities of averting irreversible environmental changes (or changes in the biosphere, in general) occurring as a result of accumulated anthropogenic impact. But then another issue comes into the fore, namely, whether it is possible to assess the level of allowed impact on the biosphere. Unlike the task of comparing the urgent needs of different generations, this problem is a natural scientific one.

At first the issue of sustainable development was examined in the context of searching for an answer to the ecological challenge, but such an answer presumes integral resolving of a host of economic, social, demographic, scientific, technological, and other problems of modern civilization.

A system's development can be considered sustainable if a system retains any of its essential invariants, that is, does not alter or endanger its essential quality, relation, limitation, subsystem, or element—anything immanent to the central, critical aspect of the system's existence.

With this approach in mind, sustainability becomes virtually a synonym for survival of a civilization in general. However, any attempt to go one or several structural levels down and, correspondingly, to reduce these notions becomes fraught with the most complicated methodological problems.

The biosphere's ecological capacity represents a limit not to be exceeded in the course of the development of civilization. Specific estimates of the biosphere's ecological capacity are well known. What is beyond doubt is that there is a limit to the allowed human impact on the biosphere. The situation with other trends in sustainable development analysis is less clear.

We can single out three aspects crucial for ensuring the sustainability of the development of civilization: first, environmental protection (by guaranteeing that the anthropogenic impact does not exceed the biosphere's ecological

434   *Global Studies Encyclopedic Dictionary*

capacity); second, protection of human health to prevent biological degeneration; and third, formation, preservation, and maintenance of the mechanisms (such as social, economic, or political) that ensure achieving the objectives set by the first two factors and suppress socially destructive structures and mechanisms that civilization might create (apparently, this is its immanent property). The first aspect is ecological, the second is sociomedical, and the third is sociohumanitarian and fuses all the other factors and problems of sustainable development that are tightly interwoven.

The measures necessary for the transition to sustainable development are still unclear in many ways. Four aspects could be singled out: preservation of natural ecosystems; stabilization of the world's population size; ecologization of production; rationalization of consumption.

It is obvious that it is inherent in any known economic power to strive to use any measure, situation, event, and process for its own advantage. But it is transnational corporations, big businesses, and the "golden billion" in general who learned to do it better than the others; so, it is not unlikely that they could use the "new world antiglobalist revolution," which the ultra-leftists dream about, to their advantage as well. That is exactly the main issue: to ensure the priority of the objective of humankind's survival over any other aim that any participant of the process of world history might have.

V. I. Danilov-Danilyan

Sustainable Life, Permanently: "Permanently sustainable life" is a term coined by Joseph Varvoushek, the first and last minister of Federal Czechoslovakia. He proposed to use this term instead of "sustainable development," which is quite often interpreted as an economic concept, and which, in Varvoushek's opinion, hampers the search for solutions to environmental challenges and other problems relating to social development. Varvoushek has realized that besides protecting the environment one should actively develop and introduce the concept of permanently sustainable life. He based his opinion on the understanding that the modern environmental crisis is caused by the human impact on our planet and the imperfection of the means and technologies utilized. The concept of permanently sustainable life focuses on the search for harmony with nature.

Varvoushek considered two possible scenarios of the future development of humankind. The pessimistic scenario suggests that humankind is under threat of disasters and humans will eventually die off. The second scenario provides for a solution, which must be chosen. It is oriented toward a search for solutions in the form of a well thought-out evolution suggesting approaches to be taken to overcome the existing problems and measures for prevention of such kind of problems. This approach is based on the general protection of the environment on our planet.

V. Diurchik

Sustainable Development, World Summit on: The World Summit on Sustainable Development was held in Johannesburg, South Africa, 25 August–4 September 2002. The 21,000 attendees included 9,101 individual delegates, 8,227 NGO representatives, and 4,012 accredited mass media representatives. It was aimed at adopting an action plan to implement Agenda 21, adopted by the UN Conference on Environment and Development (Rio de Janeiro, 1992)., Priorities included: poverty, water supply, energy, health protection, agriculture, and preservation of biological diversity. Among the obligations and goals adopted by the summit are:

(1) Halve the number of people having no free access to water and living in inappropriate sanitary conditions by 2015.
(2) Decrease the rate of biological diversity decline by 2010.
(3) Extend access of developing countries to ecologically friendly chemicals and diminish production of chemicals destroying the ozone layer.
(4) Ten-year programs on sustainable (balanced) consumption/production.
(5) Develop a system of response to natural disasters.

Participants of the Science and Technology Forum within the summit proposed to decrease the unreasonable consumption of natural resources in rich countries. They pointed out a contradiction between sustainable development values and the consumerism of modern society, the vulnerability of the neo-liberal globalization model.

Based on the principles of precaution, participants called upon the world community to hinder the propagation of genetic engineering and genetically modified organisms. Failure to solve the problem of bio-security would endanger life itself. Jacques Chirac, the President of France, mentioned that we should not let the twenty-first century become for our children and grandchildren the century of humanity's crime against life itself.

A great danger for future human generations is the global informationization of society. While genetic engineering undermines the natural basis of life, interactive information destroys the sociocultural core of nations. It is important nowadays to protect not only biological diversity but also the diversity of cultures. Unfortunately, new information technologies are used as a means of cultural colonialism to create a global technoculture, which assimilates traditional values.

The World Civil Forum participants agreed that there could be only one answer to culture-oppressing globalization, namely, cultural diversity. They reminded us about the responsibility of politicians, scientists, and civil society to civilize globalization and to bring to the forefront the interests and perspectives of human beings.

The most important question is to what extent world society is able to change the basic direction of global development and to ensure a safe and sustainable future. Heads of states and governments unanimously adopted the Political Declaration, according to which they undertook responsibility and obligation to create a society which will be humane, just, and considerate toward people's needs and will acknowledge the necessity to preserve the human dignity of all members of this society. According to this Declaration, the top priority of sustainable development is human well-being and security. World leaders proved their political will, determination, and adherence to Agenda 21 and to the Rio Declaration. At the same time, they admitted that global changes have occurred after Rio, and we should look for new models of sustainable development (considering globalization and the liberalization of capital markets and the wide spread of informational and telecommunication technologies). The benefits of globalization benefits were evident: during

436 *Global Studies Encyclopedic Dictionary*

the 1990s, when the economies of developing countries were integrated into the world economy and were growing twice as quickly as the economies of the developed countries. However, not all people and all countries could have profited from globalization. In the 1990s, there was a sharp increase in the gap between the rich and the poor. The income of 1 percent of the richest people is equal to the total income of 57 percent of the poorest people. The UN Secretary General Kofi Annan characterized the current situation of humanity as "super dangerous" and proposed that the Summit participants examine several recommendations aimed at such a control over globalization, which will contribute to sustainable development of humanity:

(1) Promote coordinated management of macroeconomic politics at the national and international levels.
(2) Eliminate agriculture subsidizing and inefficient energy use in the rich countries; to facilitate the access of goods from the developing countries into the markets of the developed countries.
(3) Strengthen the WTO so that it could provide institutional frames and rules for a just and non-discriminating world trade system.
(4) Encourage partner initiatives of the rich countries to finance sustainable development of the poor countries or countries in transition.
(5) Assist in all possible ways the introduction of sustainable (balanced) models of consumption and production.
(6) Direct international investments into the production of ecologically safe products and promoting eco-efficiency in all countries.
(7) Develop sustainable consumption and production through state regulations, including the system of ecologically oriented national accounts, and ecological reform of taxation.

A peculiar feature of this summit, as distinct from the Rio summit, is an active participation of the respected representatives of big business, the private sector, and international financial structures, such as the WTO, the World Business League, and the World Bank. Upon China's suggestion, the summit acknowledged the necessity of reforming the international financial architecture to provide openness, equality, and sustainable development of all countries. The summit acknowledged financial mobilization as the first step toward sustainable development of all peoples in the twenty-first century.

Sufficient financial resources exist in the world for a transition to sustainable development. Taxes alone on international currency operations of only 0.05 percent would have given more than 150 billion dollars per year. Redundant financial speculations are now up to 95 percent of the total volume of currency operations. Operations in financial markets are seventy times as large as trade of goods and services. Only 2–3 percent of the daily monetary turnover is used to pay bills; the rest are financial transactions. How to divert the capital flows toward sustainable development is the problem to be solved by UN agencies.

Sustainable development should lead to a better quality of life all over the world without any violation of acceptable bounds of the anthropogenic load over the Earth's ecosystems. In its scale, the Johannesburg Summit adopted an unprecedented specific plan of actions aimed at realization of goals and objectives to this end. The most important distinction of the Johan-

Sartre, Jean-Paul – Sustainable Development, World Summit on   437

nesburg Summit from the previous meetings is the emergence of "partner initiatives": an innovating mechanism of coordination of actions of governments, large corporations, and nongovernmental organizations aimed at the realization of specific plans and projects on sustainable development.

The summit participants agreed on the necessity of stronger responsibility and accountability of the private sector, especially large corporations, for sustainable development. The private sector is obliged to contribute more to the creation of more just and sustainable, humane communities than it does at present. Heads of states and governments undertook a collective responsibility for the three crucial problems of sustainable development: social progress, environmental protection, and economic prosperity. The problem is whether the political signals of the summit will be able to modify the direction of world development and to ensure a safe and sustainable future for humanity.

V. V. Mantatov

ॐ T ॐ

Teaching, Ethical Aspects of: More than teaching static knowledge, schools should be learning centers that transmit the skills of learning and where students learn to seek of knowledge. What sorts of skills in ethics does a good life call for in our modern, or perhaps, in our postmodern society?

One of the aims of education should be to prepare students to be good persons. The ideal is to nurture socially constructive citizens who have found their place in the community and who conduct themselves appropriately. On the one hand, we must give people a general education that permits them to fill the roles the community finds desirable and acceptable. It is not too much to ask that citizens should be able to survive in their own community without needing to cross its borders. General education contributing to this modest goal should also include civil ethics.

Ethics does not seek objectivity in the way philosophy "proper" and the sciences do. This does not mean that ethics would reject the demand of reliability. Whatever demands we can set for objective knowledge in general, we can set for the background knowledge ethics is using. However, the goal of ethics is to challenge students to make their own valuations and choices as to what significance they give to these facts in their own life. Similarly, ethics challenges them to think about the place of values and choices in their world. For these reasons, ethics is the search of a synoptic view, where the requirements of objectivity are reconciled with the necessity of living one's life from the perspective of subjectivity.

P. Elo, A. Haapanen, M. Kabata, H. Lämsä, J. Savolainen

Technocracy: (Greek *techne*, "crafting skill, craft"; and *kratos*, "power, ule"; literally, "rule based on crafting skill") Technocracy refers to (1) a political idea that we can rule social processes on the basis of technical rationality; (2) political regimes based on technocratic principles; and (3) a social group of technical experts occupying top governing positions and politically employing the ideology of technological determinism.

The concept of technocracy arose in the United States based on the work of Thorstein Veblen (1919), who proposed establishment of a rational social order by handing over political power from industrialists to technical experts. The emergence of technocratic ideas can be explained by the growing role of science and technology and of experts in the functioning of industry and the state. Technocratic ideas gained wide popularity in the 1930s, in the Western world. In the 1950s, technocratic concepts evolved within the framework of the idea that capitalism may be improved due to technical progress. Technocrats were seen then as a new class occupying leadership positions in traditional governmental structures (such as Adolf A. Berle and Alfred Frisch).

In the 1960s and 1970s, American economist and political scientist John Kenneth Galbraith asserted that in the process of development of industrial society the participation of the technostructure in political decision-making grows. The technostructure is a hierarchy of technological experts, a specific ruling elite, from ordinary engineers to professional managers and directors who at all levels ensured, by the means of technological rationality, the func-

Teaching – Transparency and Information Technology 439

tioning of the large corporate sector. Galbraith believed technocracy to be powerful even in a socialist society because all industrially developed societies share the same structural and functional characteristics, the same industrial introduction of scientific and technological achievements, technicalization, rationalization, and bureaucratization of social and economic life. This leads to the convergence of politically and socially different systems in which industry is the main institution (the theory of convergence). This theory became popular due to the necessity of a joint solution to global problems and a greater awareness of the priority of universal values.

In the same period, some concepts close to technocracy spread in the United States, such as managerialism—a doctrine of the rule of managers (James Burnham's "revolution of managers"). In the early 1980s, reacting to pressure of critics, technocratic thinking underwent reform (neotechnocratism) in an attempt to overcome the limitations of technocratic determinism and in an attempt to technocratically validate social control. Technology was still assigned a central role in social life, but its self-development and self-improvement capacities were denied. Technology could not now be considered isolated from human beings and society. All achievements of science and technology affect social and cultural life; thus, humanization of technology and public control over technical development are needed. Technological expertise should be combined with the independent evaluation of innovations from the viewpoint of personality, society, and environmental protection.

Modern neotechnocratic ideas in industrial sociology are based on the Weberian sociology of social action and theory of bureaucracy. Technology is viewed as a part of social action, as the realization of this or that development strategy that is developed and implemented by a managerial apparatus that bears responsibility for it. Sociology is to identify the degree of influence that these or those groups of technocrats exert at different levels of industrial organization.

Neotechnocratic ideas further examine technological risks that have been acquiring a global character. This requires an assessment of new technology from the viewpoint of universal ecological, moral, and humanistic values. Risk studies give rise to the idea of the responsible selection of new technologies based not only on a technical and integrated professional appraisal (economic, medical, sociological, philosophical, etc), but also on public opinion. Thus, a scientist becomes not only a professional presenting objective data, but also a scholar having an active civic and humanistic position. At the end of twentieth and beginning of the twenty-first century, technocratic ideas are being developed in theories on information society theory.

Reference: Veblen, T. *Engineers and Price System*. New York: B. W. Huebsch, 1919.

E. A. Nikitina

Technogenesis is the origination, emergence, formation, and evolution of the elements of technological reality and of the entire technogenic world. Historically, technogenesis can be presented in a specific order. Developing artifacts and relying on observations, experience, and the creativeness of human beings, it proceeded not from a single or few initial forms, but, rather, from several. It is sufficient to compare a knife, a comb, a vessel, and a needle and to examine a multitude of physical effects. Further development advanced both

440   *Global Studies Encyclopedic Dictionary*

divergently and convergently: in the course of specialization every form gives birth to a new one. A variety of distant forms generate a qualitatively distinctive one, which is the basis for specialization.

Innovations appear pseudo-randomly, embodying laws and regularities of a nature feasible for given conditions. Initially, these innovations embrace individual artifactual pseudo-species, but, under favorable conditions, they become mass-scale. A technological genotype (a set of documents) is mastered and becomes well known. For the bulk of pseudo-species further development is exercised by slow evolution. For some conquering new ecological niches, it is by leaps. The versions of the hereditary (documentary) basis of an article spread in all the directions allowed by nature. Science, creative engineering, and practice indicate such directions, suggesting methods and modalities that allow changing from a probabilistic search to an enumeration, fostering a combinative search and forming a new image. It is the technoevolution cycle that determines both the driving and the conservative aspects of informational selection, aimed at maximum utilization of the resources by the cycle. Divergent artifact pseudo-species are linked by transformations; those having appeared abruptly through mutations are distinguished qualitatively.

The process of evolution consists in the formation of new features, based on the common gene pool of scientific and engineering documentation and on the implementation of the laws of nature, including the creative capabilities appearing as intuition and insight. For a technological individual, death is a result of aging or of other physico-chemical causes, having nothing to do with survival (extinction) of the species, provided that the fact has not become significant for informational selection. Mortal aging is implemented by a physical act; for example, by disposal. The extinction of an artifact as a species results from external causes, from genotype imperfection or from internal reasons (internal imperfection). As to technogenesis, the following provisions can be considered proven (although many of them are not explained theoretically). Technoevolution replicates, on a qualitatively different level, features of bioevolution, with distinctions resulting from the separation of the document. The elementary factor governing technoevolution is informational selection, the action of which has a vector character.

Technoevolution is a creative process, the basis of which is diversification; reduction of trial and error in the assessment of innovations is achieved through cognition and utilization of the laws of natural science and inherent technological laws; technoevolution is moving toward specialization, but, while the manufacturing of an individual artifact (ontogenesis) is effected according to a pre-set document (genotype), technoevolution as a whole is a non-programmed development; succession of the technological manifests itself in the document (the fundamental property of the technoevolution). One should distinguish between microevolution, formation of pseudo-species and macroevolution. Factors of microevolution are: mutagenesis, gene flow (borrowing ideas and documents), sorting (selection), and genetic drift (organizational alterations).

Mutation frequency is continuous and differs both for different pseudo-species and for the same pseudo-species, but at different times; generally, the more sophisticated mutates more frequently. The majority of new ideas and decisions are not viable. The documentation flow is featured by migration

Teaching – Transparency and Information Technology 441

selectivity according to fields (specialties) and regions. The basic form of informational selection is univariate selection. Informational selection operates upon any stage of the cycle of article existence. Artifacts produced diversely enjoy advantages. The efficiency of the permanently active informational selection enhances, provided availability of several artifacts similar in their features and occupying the same econiche. Effective influence of the document drift occurs if the artifact pseudo-species is produced by several manufacturers; if there is a recession in demand, which requires upgrading; or if affiliated companies, aimed at separation, appear.

The cost of information selection consists in expenditures connected with the manufacturing of "no sale" artifacts and with R&D for perspective. The modifications brought in the phenotype ("grown-up" article acting in the technocenosis) are not inherited. The species formation is established by differentiation of documents. Clones and twin species represent technetics (technique, technology, material, production, and waste). The most common phenomenon is geographical speciation (company plus country). Technical isolation rests on technical distinctions stipulated in the documents. The essence of species formation consists in the creation of a document, necessary and sufficient for the article-species fabrication, to which the technical name is assigned and technical and technological documentation is ascribed. New groups of species usually originate from relatively primitive ones, the specialization of which causes mass circulation.

Single-purpose artifacts are expulsed faster. The technoevolution is directed and irreversible: the documentation system prohibits turning back to production of something taken out (because of the gene pool modifications). There exists a correlation between artifact quantity (reproduction rate) and evolutionary rate. The progressive evolution criteria are as follows: efficiency upgrading of the utilization of both energy and materials and efficiency of production due to refining and adjustment; the appearance of response to signal; reduction of environmental dependence. The essence of technoevolution is in its diversity. The technetic resemblance is the global supra-language integration of concepts, units and identification of the system of global scientific and technical documentation. There exists a similarity in the development of branches and sub branches of the economy; informational and social restrictions are also the same.

The difference between the item out of production and the one in progress grows with the time separating the events. Technoevolution is directed usually toward organization amelioration: complication, qualitative change, growth in the number of components, cephalization; but there also exist organization degradation (article simplification) and conservation in the same plane (at species replacement). The closer the econiches are occupied by two kinds of an artifact (technique, technology, material, production, and waste), the more vigorous is the struggle for existence (competition) between them. The parallels and analogies of the evolution of physico-chemical (dead), biological, technical, informational, and social worlds are not occasional, and following the implementation of the idea of global evolutionism, they will serve as the theoretical basis for making decisions on harmonization of technological reality.

B. I. Kudrin

442 *Global Studies Encyclopedic Dictionary*

Technogenic Catastrophe (Greek *techne*, "art," "mastery," "trade") A technogenic catastrophe is a grave accident with heavy consequences, caused by failure, damage, or destruction of technological systems, machines, constructions, or mechanisms, or else by the breakdown of technological processes, most distinctive for the epoch of highly-developed industrial production. Technogenic catastrophes occurred with increasing frequency and on a growing scale over the course of the Industrial Revolution (seventeenth, eighteenth, and nineteenth century) and again during the scientific and technological revolutions (1930s–1940s). Nevertheless, they happened earlier as well, for example, in Rome during the reign of Anthony Pius (138–161 CE) in the course of a gladiator fight, the top decks in the Maximus circus collapsed, injuring more than 1,100 victims.

However, the list of the largest technogenic catastrophes in the history of humankind was opened at the beginning of the twentieth century, by the wreck of the Titanic, the largest passenger liner of that time, in April 1912. Other calamities followed, brought about by objective causes—purely technological failures or defects—and subjective human factors (incompetent actions and management, and violation of safety rules). The largest among them were the demolition of the Jandzy Kjang dam in Huajan Kow during the Chinese-Japanese war in 1938, entailing thousands of victims; the A-bomb explosions in Hiroshima and Nagasaki in August 1945, which also resulted in the perishing of about two hundred thousand of people; the accident at the "Union Carbide" facility in the Indian city of Bhopal in December 1984, that killed more than 2,350 people; and the Chernobyl tragedy of April 1986, the horrible consequences of which cannot be calculated. It is also expedient to classify as technogenic catastrophes the bulk of train, ship, and aircraft accidents, accidental spills and discharges of oil into environment, and harmful chemical and radioactive contamination; e.g., the fuel oil spillage in the Mediterranean Sea not far from the coasts of Spain and France in 2002, due to the wrecking and sinking of the Prestige tanker.

I. I. Mazour, A. N. Chumakov

Technogenous Civilization is a social state, a type of development characterized by the domination of technology in various spheres of social life. The emergence of technogenous civilization is the result of the achievements of scientific and technological progress in the beginning of the twentieth century when automats and automatic modes of production were invented and massively introduced. Close interdependence between a machine and its operator was replaced with a free connection between operator and equipment. Since that time technological equipment development is not limited by human physiological capabilities. Human beings are now able to use their capabilities creatively and they are no more simply elements of the technological process. Technogenous civilization is characterized by the fact that automats and automatic systems increasingly liberate human beings from physical labor. They are introduced into all types of production: from mass production of goods to performing complex operations in the sphere of producing modern engineering constructions, machines and mechanisms. Since the end of the twentieth

Teaching – Transparency and Information Technology 443

century human beings live in the technosphere and this allows us to call this new reality "technogenous civilization."

A. N. Chumakov

Technology: (Greek tekhnologia, "systematic treatment of an art, craft, or technique"; originally, *techno*, "skill," "mastery," "ability," and *-logy*, "study," "science") Technology is an integral dynamic system of activity. It includes hardware, operations and procedures, rules, standards, models and technological activity regulations, technological process control, vital information and knowledge, energy, raw materials, and the aggregate of its economic, social, ecological, and other after-effects that influence and change its social and natural "habitat" in some way.

The word techne already had several different meanings as early as in Aristotle's time, one of which meant a tool or device for producing things and for activities going along with production. A second meaning dealt with skillfulness, activity, mastery, ability, and what now is called "know-how." The first meaning is used when one speaks, for instance, about building technology, or steel-smelting technology. The second meaning is used for the technique of piano playing, vocal performance, game technology to rather wide extent, and so on. The second component of the word "technology" (logos) signified a law, reason, the highest rational beginning or rational regulations of some kind of activity, study, or science. Here comes the word "logic" as a system of specific rational rules and substantiated deductions and conclusions. As for the science nomenclature (biology, geology, philology, and so on) the component descending from "logos" is usually understood as study and means "a study of life," "a study of the structure of the earth," "a study of language." The pattern of the term "technology" is the same. At first the word-formation "technology" signified "a study of technique," where "technique" had the first Aristotelian meaning.

Concepts of "technique" and "technology" should not be conflated, although they were hardly differentiated until the middle of the nineteenth century. Still, by the end of the nineteenth century, engineering literature took the word "technology" as a system of regulated operations, executed with the help of machinery and mechanisms within the field of industrial production. For example, this meaning was used when the technology of the industrial production of sulfur acid, loom assemblage, or railroad laying was mentioned. "Technology" was considered as a specific professional activity and did not attract any philosophical attention. Until the end of the nineteenth century, "technique" and "technology" were in essence identical.

In the second half and especially in the last quarter of the twentieth century, the terms "technique" and "technology" became noticeably different. At the same time, "technology" occurred more frequently in the literature concerning general politics, economic theory, and philosophy, and in scientific research.

A new approach to interpreting technology and understanding it as a universal phenomenon loaded with philosophical sense began to loom in the first quarter of the twentieth century, under the influence of the deep crisis of European culture, Renaissance humanism, and the fall of the modern Enlightenment values, caused by the First World War and the absolutely open dehu-

444 *Global Studies Encyclopedic Dictionary*

manization of the war technique and technology. Works by Edmund Husserl, Martin Heidegger, José Ortega y Gasset, and others marked the merging of two trends of criticism: one of machinery and technique dehumanization and the other of Enlightenment piety toward culture.

After World War II, the modern understanding of technology gradually takes place. Basically, it was for the real smashing breakthrough inventions in the high-tech and information technology fields, in nuclear power engineering, space technologies, chemical and biological sciences, agriculture, and so on that the process kept going on and growing robust. The main political institutions of superpowers and countries in the lead of scientific and educational spheres, first the United States and Soviet Union, had also changed their attitude toward technology. It became perfectly clear that national security, leadership in the globalization process, national welfare, military power, political influence, culture, and competitive ability in world markets are chiefly defined by the developmental level of science and technology. Science itself started using technologies, while technologies began to look for support in fundamental scientific achievements more and more.

During 1950–1960, the term "critical technologies" (CT) became widely used, especially in the West. CT includes technologies that play a key role in the life support of humanity, its military might, and the geopolitical and economic situation.

Being the principal factor of changes in all aspects of social life, technologies of different types, levels, and purposes gradually revealed their really general and versatile character. In the former historical epochs, this feature of technology did not show up in such a distinct way as it did at the end of the twentieth and the beginning of the twenty-first century. At the primitive stage of development and in traditional civilized systems, the basic role of technologies is hidden in the enormous variety of local and regional ethnic, religious, and cultural ways of life and other features, diversity of mythologies, social and political institutions, and all the historical sham. It makes it difficult to outline the common in the ocean of details. Racial and national features, cultural and historical traditions, mentalities, and particular ethnic life conditions are still important determining factors of the appearance, changes, and disappearance of any social-historical system. They influence the labor performance, welfare level and so on. However, we can easily see the influence of technologies, understood in the modern way, under the cover of all these details and features. Countries divided into the underdeveloped, developing and highly developed; into pre-industrial, industrial, and post-industrial societies, into "North" and "South"—all of this is in essence the consequence of scientific and technological development.

Moreover, by including different components into technological structure, such as control systems, information and knowledge, and the after-effects of technological activity of different kinds, changing the social and natural environment of technological functioning itself, we can immediately find a basic relationship between activity technologies and the status of a particular society and its situation on the historical scale. For instance, the terms "informational society," "built-on-the-knowledge society," "high-tech societies," and "post-industrial society," which are actively used in modern philo-

Teaching – Transparency and Information Technology 445

sophical, economic, and sociological literature, define only different sides and features of the development level, determined by the mix of science and technology. All of this is "high technology," based on the hardware and intellectual components of modern technologies. It is "high-tech" development that influences the solutions of ecological, public health, defense, social, cultural, and other problems, and the development of a stable society, balanced financial system, and defensive ability of any developed or developing state.

A. I. Rakitov

Technological Assessment (TA): Technological assessment concerns decision-making judgments about technology options. Such assessments evaluate impacts and consequences of technology, political and societal ways of dealing with them, potentials for contributions to societal problem-solving and innovation policy, and implementation conditions of technology. Accordingly, TA is not governed by discipline-immanent research programs but is driven by extra-scientific problems. Complex subjects of research can also be treated within individual disciplines, whereby each discipline studies its subjects from the standpoint of its own interests and methods.

That TA results are reliable and that their quality can be judged ex ante is imperative since crucial decisions rest decisively on the quality of these integratively developed results. In technology policy, furthermore, there is seldom enough time available to permit waiting for scientific consensus. Thus, despite ideal standards that have been developed for scientific research, quality assurance of rational TA as interdisciplinary research can be challenging.

The orientation of TA with respect to the formulation of social problems implies that the criteria of scientific research are constituted extra-scientifically. It has been shown that integrative research refers to socially relevant non-scientific decisions in essential pre-empirical respects. The quality of the results to be expected then depends essentially on the "quality," that is, the adequacy of these decisions as to the problem under discussion and the responsibilities for solving it. In this way, in addition to internal quality criteria for scientific work, external criteria have to be applied. Integrative research does not raise less stringent, but higher demands on quality assurance: observance of the usual disciplinary standards of quality is necessary, but not sufficient, to guarantee the quality of integrative research.

A. Grunwald

Technologies, High Human (High Tech) refers to technologies that influence social consciousness by replacing manufacturing production with increases in service, management, and education. Such technologies, utilizing miniaturization, automation, computerization and robotization, engender capabilities to manipulate individual and social consciousness. Modern mass media, the Internet, mass culture, and art permit broad changes in public consciousness. The press (via television, radio, print media, and the Internet) are the most important instruments of such an impact on public consciousness. They are able to present a large-scale tragedy as a non-significant incident to pacify public opinion and, vice versa, are able to provoke social cataclysms by presenting, for example, the discontent of several people with regard to an insignificant issue as representing broad public opinion with regard to a na-

446 *Global Studies Encyclopedic Dictionary*

tion-wide issue. High-human technologies are actively used in advertising and electoral campaigns, in providing information about military conflicts and terrorist acts, and in the course of informational and psychological wars.

A. N. Chumakov

Technological Park (Technopark): A technological park is the territorial integration of science, education, and industry in the form of a complex of research organizations, construction bureaus, educational facilities, manufacturing sites or their parts. Technoparks are organized to promote the exploration and introduction of scientific and technological achievements through concentration of highly skilled specialists and the use of fully-equipped manufacturing, experimental and informational methods. Often technoparks have fiscal preferences and exist as specialized regions of economic activity.

Since the 1970s, to enhance their competitive edge, advanced countries hurried to build technoparks. Currently, approximately 1,200 technoparks operate worldwide; e.g., Science Park in UK, Research Park in America, Sophia Antipolis in France, and Ganagawa Science Park in Japan are actively working to achieve technological innovation and nurture newly created businesses.

E. V. Mazour

Technological Progress is the interdependent, progressive development of science and technology, which stimulates a qualitative transformation of both material production and non-production spheres and affects virtually all areas of social life, being an integral part of social progress.

During the seventeenth, eighteenth, and nineteenth century, the idea of scientific and technological progress as offering the potential for endless improvement of human society and nature itself emerged on the basis of the perpetually increasing volume of scientific understanding worldwide. By the middle of twentieth century, this illusion had led to a belief in the boundless nature of human knowledge and technical action and to optimistic plans to make humanity happy with the help of the ever-increasing achievements of science and technology. In the twentieth century, scientific and technological progress became global—not only its positive consequences, but also the negative ones. Society overestimated the positive benefits of scientific and technological progress and underestimated their negative consequences.

Belief in unbounded scientific and technological progress, the absolutization of scientific research that is "free" of moral values, and the illusion of the transformability of the world on the basis of acquired knowledge resulted in the appearance of a sort of "scientific religion" based mainly on the belief in the power of scientific knowledge and the progressive nature of technological action directed by this knowledge.

One of the most important tasks of the modern philosophy of technology is to replace this obsolete dedication to unlimited technological progress in favor of "sustainable development." The goal is not just to build an efficient economy with the help of ultra-revolutionary innovations, but also to provide social and environmental sustainability, to introduce technological progress gradually and carefully into the natural environment, taking into account cultural traditions. As for sustainable development, one should, first, distinguish between evolutionary development and revolutionary development. The lat-

Teaching – Transparency and Information Technology 447

ter, being applied to society, means artificial acceleration of political, economic, and scientific technological development in all social spheres.

Two scenarios regarding the future social development have been described: one of them being called technocratic, implying the total replacement of bio-regulation by technology. This scenario cannot be practically realized, at least within a historically acceptable time frame.

Over the last two centuries, a new type of civilizational development, the technogenic one, has emerged. At the modern stage of scientific and technological development, science and technology continue to be specialized, but new productive ideas and trends emerge mainly at the junction of traditional scientific and technological disciplines; modern science and technology have a tendency toward interdisciplinarity. Therefore, the contact of modern science with the neighboring scientific areas, art, and cultural history, the constant reflection on the methodological foundations of research and development, and the tendency toward democratization and pluralism of opinions are characteristic features of contemporary scientific research and system engineering. Scientific and technological development becomes, in the industrially developed countries, a self-reflective system. This implies the parallel institutional estimation of consequences of new technology and the social ecological expertise of scientific, technological, and economic projects.

As modern technologies continue to develop, the state faces not so much the problems of compensating damages, but of preventing them. Long-term planning becomes necessary, which should include both foreseeing the new technological opportunities and estimating and removing risks.

V. G. Gorokhov

Technological Pyramid is a method of classification of countries according to the predominant level of technologies and, hence, to their competitiveness in global competition. On the top of pyramid are creators of new technological approaches, building their own markets and completely controlling these markets and sales of their product. Its efficiency is so high, that, as a rule, it actually does not enter open markets, being sold and purchased primarily within the respective transnational corporations somehow controlling the R&D. Therefore, there exist actually no markets of new technological principles as a regular event, and the turnover of such principles has an intrinsic character for major international entities. They control it not only commercially, but also, in the strictest way, organizationally.

Similar to the most effective modern technologies belonging not so much to manufacturing as to management, and especially to the development of consciousness, the said new technological principles belong not only to traditional domains of production. The efficiency of the development of these principles is associated not only with the largest part of the added value (which is steadily descending from upper to lower "levels" of the technological pyramid, reducing respectively the business efficiency) and with the highest control of distribution markets (which directly depends upon the degree of the commodity uniqueness, be it real or instilled into customers' minds through the system of trademarks, also descending from upper "levels" to lower ones), but also with the fact that it is on the basis of these principles that technical and behavioral standards are being formed, which gives an incredible competitive advantage to those who form these standards originally. This advantage is so high and yields such profits that this could be considered as a sort of rent, similar to mining, agricultural, intellectual, and other ones.

The most efficient, according to practice, is development of standards of thinking (stereotypes) and behavior, followed by those of technological activity. Hence, the main technological principles, the most vital from the viewpoint of providing national and corporate competitive ability, are those associated with the formation of perception and with management. The technologies of strategic planning and crisis management are examples of the most successful practical implementation of these principles.

However, their practical implementation to immediately embodied technologies alone brings us to the second level of the technological pyramid. Manufacturers of the goods in this group also directly control their sales, though to a considerably lesser extent than the representatives of the first level. Indeed, as distinguished from new technological principles, technologies are entering the open markets regularly in rather high volumes, though their marketing is somewhat partial and relates not to the right of property itself, but to the right of their utilization only and, sometimes, of limited reproduction.

The third, fourth, and fifth (last) levels of the technological pyramid comprise manufacturers of goods employing the "know-how" elaborated on the second level. These levels seamlessly turn into each other with simplification and reducing of the uniqueness of manufactured commodities: from unique consumer goods, sophisticated processing equipment and highly qualified services, entering the open market, but allowing the manufacturer to retain full control (on the third level) to mere complicated goods and, on the last fifth level, forming the pyramid's base, to homogeneous "trading" ones, usually raw materials or primary-processed goods, the markets for which are as a rule primarily controlled by customers and, therefore, appear unstable. A manufacturer's or a country's orientation toward them is considered traditionally and quite worthily as a serious strategic risk factor.

The classification of countries by "levels" of the described technological pyramid according to the traditional paradigm is rather established.

Today, the only countries in the world, producing new technological principles on the mass scale and in various spheres are the United States and, to a lesser degree, Great Britain. Accordingly, "one and a half" countries occupy the first level of the technological pyramid: the United States and, partly, Great Britain.

The concentration of their resources in the most efficient kinds of business, associated with the creation of new technological principles, makes every other kind of activity, potentially accessible for them, a waste of efforts and resources. It is this angle of view from which gradual spontaneous dumping of less efficient industries to other countries shall be considered.

Forms of such dumping in conditions of global competition are diverse and obscure. This is by no means only direct transfer of relatively primitive technologies and withdrawal of production departments of transnational enterprises from US territory. The quintessential, though the least visible force of structural rearrangement of the American economy is currently exactly the con-

Teaching – Transparency and Information Technology 449

centration (spontaneous and intentionally state-controlled) of all resources on developing new technological principles. This concentration leads to a relative exhausting of the resource potential (not only material and financial, but primarily organizational and intellectual) of traditional, less effective businesses, which promotes progressive lowering indices of their own development.

Global competition is the instrument of the pressing out from the American economy of relatively ineffective businesses, always wasting scarce resources and incurring a profit loss. The resulting economic restructuring is manifested not only in the stagnation of several technologically primitive branches, but also in the American representatives of the second and of the third levels of the technological pyramid lagging behind their nearest competitors in manufacturing of a variety of sophisticated technical goods and technologies. The United States, for example, is inferior to Europe in mobile communication quality, to Europe and Japan in car quality, to Japan in several computer technologies (especially, American companies have not mastered manufacturing of ultra-thin LED video screens, launched by Japanese corporations three years ago). However, it is fundamentally important that this retardation is caused not by the impossibility, but, primarily, by the uselessness of repetition or of exceeding the results already achieved. The American economy does not need to manufacture the best cars or computers in the world for reasons similar to those for which the director of a corporation or a Nobel Prize winner do not need to have the best knowledge in the world on repairing sewerage: it is not their job, they have ways of money making that are considerably more effective and, when the need arises, they simply pay for the services of respective specialists who earn much less.

Nevertheless, due to sociopolitical reasons and social inertia, the American economy is not able to get rid of industries at the lowest level of the technological pyramid that generate profit loss (some of them, for example, in the area of personal services, will have to remain forever).

From even this simplest example (concerning only two countries at the highest level of the technological pyramid) we can easily see the difference between the technological and traditional geographical division of labor among countries. The confusion increases on the lower levels of the technological pyramid: it would suffice to indicate that there is actually no country belonging exclusively to the first or to the second level, because the development of technologies appears to be inseparably associated with the manufacturing of sophisticated-technology goods.

Similarly, on the last, fifth level of the technological pyramid, the manufacturing of raw materials and primary-processing goods is bonded with the fourth one, because even export-oriented countries as a rule process a part of produced raw materials for their own needs, even when it is not profitable from a purely economic point of view.

Within each level of the technological pyramid countries compete for markets; the higher the level, the less the competition intensity, according to the above technological reasons. Between the levels of technological pyramid, the competition is for resources, since the higher efficiency of every higher level makes this competition hopeless for less developed participants. It is this kind of competition that is conducted "to annihilation." It is the most

450   *Global Studies Encyclopedic Dictionary*

significant one from the point of view of international development; it is referred to as "global" competition.

M. G. Delyagin

Technology and Responsibility: As a relational construct, "responsibility" is used in a descriptive sense; for example, someone is responsible; and as an evaluative attributional concept—someone is held (to be) responsible. This introduces the normative, even ethical, dimension of action in a stricter sense. As such, an agent is responsible for actions and their consequences with regard to some object of responsibility. Within a given framework, prescriptive, normative criteria of accountability can be identified; e.g., culture-specific social or legal norms.

Frequently, accountability questions are raised in negative cases, when one or more accountability criteria are not fulfilled. For example, a dam breaking may be the result of wrong statistical calculations, negligence, poor craftworks, or using cheap material; the level of incompetence may reach criminal levels. Negative action responsibility with regard to professional work, for example, means that professionals have a responsibility to the public to ensure high standards in their work and to avoid risk of disaster as far as possible at a reasonable cost.

Collective action (for example, by an institution) entails a responsibility of institutional or corporate actions: it may coincide with, though not be identical with, the individual responsibility of a person in a representative position. Collective (group) or joint (multiple individual participants) action entails co-responsibility, sometimes called collective or group responsibility. Other types include role and task responsibility, universal moral responsibility, and legal responsibility.

In accepting and fulfilling a role or a task, a role holder usually bears a responsibility for normally acceptable or optimal role fulfillment. These role duties might be assigned formally, informally, or even be legally ascribed. If role takers are representative of corporate or institutional role patterns, their responsibility may be connected with the associated institutional role responsibilities (as in leadership). In addition, institutions that have a special task to perform with respect to clients, the public, or members of the organization or corporation entail obligations of responsibility, which can have a legal, moral, or neutral organizational character.

In general, different levels and types of responsibilities co-exist and are interrelated—moral and legal obligations of individuals and corporations all being in effect simultaneously, for example.

H. Lenk, M. Maring

Technological Revolution is a modern stage of scientific and technological progress, where information becomes the most important strategic resource and the managerial instrument for social and natural processes and the newest technologies become the defining factor of economic development. In the early 1960s, Daniel Bell, Jean-Jaques Servan-Schreiber, Tadeo Umesao, and Fritz Machlup (who introduced the term "informational society") first mentioned this ongoing, information-based technological revolution. These studies took place against the background of the emergence and implementation of computers within specific spheres of social life. They were the result of the initial attempts to understand the changes in the economic and social sphere under the impact of the emerging telecommunications and information exchange innovations. Science formed the basis of these new technological processes, as shaped in the course of the two previous decades. As a result, computers became miniaturized and in 1975, a computer consisting of only a printed circuit was invented. The turning point in the development of scientific and technological progress was, nevertheless, in 1981, when the first mass personal computer was invented, thus beginning the informational and technological revolution. Since that, industrial and social processes in the spheres directly connected with information started to develop rapidly. The most important contributions to this new speed included the enhancement of technologies, the miniaturization of equipment, the increase in computer capacity and memory, and the cheapening of electronic devices. Under conditions of intense development of global society information and the newest technologies based on it have principally changed both interstate and interpersonal relations, which have become more mobile, more open, more transparent and less dependent on distances.

A. N. Chumakov

Technology and Science in Global Perspective: The worldwide scope of social and technological problems is vital. Science and technology, despite their origins embedded within Western political economy and cultural sources, have now become world science and world technology.

In the historically evolved division of labor, technical elites derive their power from specialized competence; they are partially insulated from others and from democratic decision making by a scientific and technological sophistication, which easily allows for esoteric secrecy (whether military or industrial). These technological threats to human societies are broadly of three sorts: political, social, and ideological.

In the political sphere, the threat due to elitism may outweigh the benefits of specialized learning and specialized practice: (1) by undermining the competence of representative democracy or by distorting the procedure of electing representatives; (2) by diverting or frustrating the development of self-management institutions (such as workers' control in the work place, the market, or other production spaces of societies); (3) by the overriding technological necessity of quick military response to security dangers with the consequent and accepted social necessity of hot-line elitism; and (4) by linking populist counter-elitism to neo-Luddism.

Scientific and technological innovations threaten to undermine the received qualities of cultural life and human consciousness: (1) by challenging the power, validity, and even the presence of the literal and the figurative icons and rites of traditional religious and aesthetic sensitivity in all their forms; (2) by promoting the psychologically symbolical fetishism of science and technology, or of anti-science and irrationalism; (3) by transforming human living relations through transforming the social relations of production, consumption, and communication; and (4) by transforming the social relations of pleasure and fulfillment, and, in the process, weakening the momentum of cultural tra-

452 *Global Studies Encyclopedic Dictionary*

ditions, leaving the individual increasingly without moorings, open prey to the immediacy and irrationality of quick-fix populist manipulation.

Ideologically, technological society poses problems, sets criteria for explanations and solutions, provides resources of people and materials, creating and distorting the cognitive culture along with daily life. Technological innovations produce their own political economy of culture along with a political economy of science; these are the objects of new work in the social sciences and, in turn, they stimulate critics who then must consider whether science and technology are themselves partial, ideological, and merely instrumental reasoning, whether seen from traditional (largely religious) premises or from humanistic and other viewpoints.

R. S. Cohen

Techno-Optimism: Cory Doctorow defines "techno-optimism" as the belief that technology could be used to make the world worse held at the same time with the hope that it can be steered to make the world better (2011). From this point of view, technical means of production (such as machines, mechanisms, computers, and innovative technologies), and scientific and technical knowledge, are the determinative factor that defines different aspects of social life and the nature and direction of social development.

Interest in technical research can be found in works as early as Aristotle, but the technocratic mood appeared for the first time much later, related to the development of engineering during the Renaissance. Then, during the seventeenth, eighteenth, and nineteenth century, as scientific and technological achievements were growing and the industrial revolution was underway, this mood grew stronger and was developed by active followers of scientific and technological advance, such as Francis Bacon, Marquis de Condorcet, Julien Mettrie, Voltaire, and Saint-Simon, who had common views on technology and saw technological advances as the most important means of solving social contradictions and achieving well being for all. At the same time, the opposite mood to techno-optimism emerged. Followers of this ideology were pessimistic about the development of scientific and technological advance.

The best-developed forms of technocratic theories appeared under the influence of impressive scientific and technological achievements in the 1920s. The author of such a theory that became widely known was an American economist and sociologist Thorstein Veblen who was one of the first to provide a philosophical ground for the leading role of industrial production and technological progress in the development of society. According to him, engineers and technical specialists should govern a modern state, because they are the only kind of people who can develop production in the interests of society (the pathos of T. Veblen's technocratic theory), and they need political power to achieve this goal.

However, by the beginning of the 1960s, the above views were obscured by a new and even more powerful wave of technocratic and progressive spirits inspired by an industrial upsurge that involved virtually all economically developed countries of the world in the post-war period. The prospects of social progress in the 1950s–1960s seemed boundless to many people both in the West and in the East. The public mind ever more acquired techno-

Teaching – Transparency and Information Technology 453

optimistic ideas that created the illusion of the possibility of solving any terrestrial and even space problem with the help of science and technology. The first achievements in familiarization of space, the new results of nuclear tests, and the discovery of several large oil and gas deposits provided a firm ground for such ideas. These views were reflected in many theories in which the technical capabilities of human beings were believed to be unlimited, and the "society of consumption" was announced as the purpose of social development. At the same time, various concepts of industrial, post-industrial, technotronic, and information society were developed.

John Kenneth Galbraith (1958) has expressed total appreciation for the scientific and technological achievements of the humankind; he paid attention to the deep transformation of economic and social structures under the influence of these achievements.

The theory of industrial society was further grounded in the works of French philosopher Raymond Aron and American political scientist Walt Whitman Rostow (1960), according to whom the traditional agrarian society has been replaced by an industrial society in which mass market production plays the most important role. The main criterion of the progressive nature of such society is the achieved level of industrial development and the degree of use of technological innovations.

Wide use of computers in all spheres of social life resulted in the emergence of new theories of post-industrial (Daniel Bell, Hermann Kahn), information (Jean Fourastier, Alan Turing), technotronic (Zbigniew Brzezinski, Jean-Jaques Servan-Shriber), super-industrial, computer (Alvin Toffler) society. In such societies, the main criteria of social progress are not merely technical achievements, but the development of science and education that are believed to play a major role. The most important criterion of progress is the use of new computer-based technologies.

For example, Bell believes that information and theoretical knowledge are strategic resources of post-industrial society and turning points of modern history. Accordingly, the first of these turning points was the change of the nature of science itself, which, being a universal knowledge became the main productive force in modern society. The second turning point is conditioned by the emergence of new technologies, which, unlike the technologies of the industrial revolution, are mobile and can be easily re-profiled.

Since the beginning of the 1970s, the interest in theories of this kind significantly dropped, especially after the first reports to the Club of Rome were delivered. In these reports, the main attention was paid to the review of the negative consequences of scientific and technological advance and to concepts of social development models oriented toward an unlimited growth of industrial production. The heated discussion of globalization processes that began in the middle of the 1990s renewed interest in techno-optimistic concepts to some degree, but the blunt actions of antiglobalists and, to an even higher degree, terrorist attacks against the United States in the autumn of 2001 significantly decreased the number of followers of such concepts.

Reference: Galbraith, J. K. *The Affluent Society*. New York: Houghton Mifflin, 1958.
Rostow, W. W. *The Stages of Economic Growth*. Cambridge, UK: University Press,

454 *Global Studies Encyclopedic Dictionary*

1960. Doctorow, C. "Techno-Optimism." Locus Online. Accessed 5 November 2013. http://www.locusmag.com/Perspectives/2011/05/cory-doctorow-techno-optimism/.

A. N. Chumakov

Techno-Pessimism is the view that scientific and technological advances are the main reason for the broken balance in the relationship between society and nature and the emergence and abrupt intensification of environmental, resource, social, and many other problems of modern social development.

The first outstanding followers of these views appeared at the time when scientific and technological advance was arising, and the industrial revolution was spreading. As early as 1750, Jean-Jacques Rousseau concluded that the revival of sciences and arts was detrimental to the refinement of morals and manners. While many associated social and moral progress with the aggregation of knowledge and the development of science, Rousseau disagreed, opining that education and erudition are bad for morals. Urging a return to nature, he said that the closer human beings are to nature, where there is no technology, nor science or art, the more virtuous they are.

The techno-pessimistic mood began to get firmly established and actively expressed at the end of the 1960s, when, in addition to environmental difficulties, other problems endangering many states and continents began to show ever more sharply: uncontrolled growth of population, non-uniform social and economical development of different countries, provision of raw materials, food, and many others. At that time, views opposite to technocratic trends actively began to develop. Sometimes, another term with a similar meaning is used—environmental pessimism—which expresses a negative attitude to human-caused action on nature and characterizes an ideology according to which an environmental disaster is inevitable. Thus, a new wave of protest was arising, a protest both against scientific and technological advance and the social progress as a whole. Many well-known scientists and philosophers, such as Herbert Marcuse, Theodore Roszak, and Paul Goodman, were vigorously protesting against scientific and technological advance, accusing their predecessors of following callous scientism, and of craving enslavement of the human being with the help of science and technology. New ideas inspired by this trend substantiated the society of anti-consumption and were aimed at persuading an "average person" to be satisfied with as little as he or she had. In the effort to find the main initiator of global problems, the main accusations were brought against "modern technology." Not only the achievements of science, but of progress as a whole was prejudiced—Rousseau-like appeals to get back to nature appeared again; there were calls to freeze, and to stop economical development at the level already achieved.

Several human-caused disasters and terrorist attacks that took place at the end of twentieth and the beginning of twenty-first century (such as the Chernobyl disaster, oil-tanker accidents, the wreck of atomic submarines, and terrorist attacks against New York and Washington) significantly increased the influence of the techno-pessimistic mood.

Teaching – Transparency and Information Technology   455

References: Rousseau, J.-J. *Si le rétablissement des sciences & des arts a contribué à épurer les moeurs* (Whether the Recovery of Science and Arts Helps Purify Morals). Geneva: Barillot and Son, 1750.

A. N. Chumakov

Technosphere: (Greek *techne*, "mastery, art, workmanship" and *sphaira*, "sphere") The technosphere is the area where technology exists; the complex of material assets for human transformative activity; the global human environment; a synthesis of the natural and the artificial resulting in the natural symbiosis of technology and humankind. The technosphere is a part of the biosphere, drastically transformed by human beings into technological and technogenic objects through the direct and indirect application of engineering tools to better meet human socioeconomic needs. It is a closed regional and global technological system of utilization and reutilization of environmental resources involved in economic turnover.

The technosphere as the "second nature" created by people in the course of the actualization of their goals, ideas, and theories constitutes an inorganic mechanical system, comprised of scientific concepts aimed at transforming the world. In the twentieth century, technology became a new artificial environment replacing the natural one. Its basic characteristics are autonomy and self-determination. Like the natural environment, technological environment is a closed system that can function without and regardless of human intervention.

In the last decades of the twentieth century, it was discovered that technology, designed with the purpose of improving life, prosperity, and world perfection, is fraught with global disasters. Natural systems split into separate components: technology pulls them out of natural relationships.

The technosphere has three stages of development: geotechnology (when construction material was timber, and water was the source of energy); paleotechnology (when iron and coal dominated); neotechnology (in the twentieth century, when metal alloys and electric power came into use). In the 1940s, Lewis Mamford foresaw the beginning of a fourth stage, biotechnology, that is, technology based on biological laws.

The technosphere integrates technological elements of various origins, contents, and functions. These elements in the course of the evolution of technology and science form relationships bearing all the characteristic traits of an epoch. In the course of the creation of a new artificial environment, the natural one should also be present to the extent necessary for a human being; humankind does not break off with nature but reorganizes it. Nature becomes, in this case, dependent and secondary. In the period of globalization, the technosphere plays a double role: it is a globalization tool and one of its causes; and it globalizes itself, absorbing and involving human beings.

A. D. Ioseliani

Technosphere, the Structure of: The technosphere is usually regarded as an integrated global system within two systemic links: "person-technosphere" and "technosphere-biosphere." Within the first link the technosphere is a natural system; within the second one it is an artificial system. The technosphere as a natural phenomenon is a continuation of the growing structural complexity of living nature, while as an artificial phenomenon it separates hu-

456 *Global Studies Encyclopedic Dictionary*

mans from nature. Technical products are the structural elements of the technosphere as a natural phenomenon. In that case we can talk about technological production methods or technological modes as a teleological principle. Territorial-industrial complexes (agro-industrial, urban-industrial, mining-production, mining-processing, energy, and recreational) are the structural elements of the technosphere as an artificial phenomenon. Transport communications connect all of these mega-objects into a common framework of the technosphere. In terms of energy, the technosphere is continuous, because electromagnetic radiation can be detected at any point on the globe.

The following classes of processes occurring therein determine the internal structure of the technosphere: (1) processes of transformation of substances; (2) processes of manufacture of items; (3) processes of use of items; and (4) processes of disposal of worn-out items.

The first group of processes creates building materials for the second group of processes and the energy necessary for the first three classes, thus performing functions similar to those of soil in the biosphere. Therefore, such a function is unavoidable. Elements extracted from the lithosphere, the hydrosphere, the atmosphere, and the biosphere are brought into technogenic circulation; some other elements become waste. Chemically, the technosphere is different from lithosphere and biosphere, which cause ecological problems. The processes of transformation of substances create network structures similar to those of trophic chains of biosphere.

Items produced by the technosphere are local, while the processes of transforming elements for producing things are global. To produce something, people set up network structures that cover significant areas. Ore can be mined at one place, metal melted out hundreds of kilometers from the mines, and metal parts manufactured in another country, these parts to be assembled to automobiles on a different continent and finally discarded at the other end of the world.

Structural elements of technotrophic chains are different levels of the transformation of substances performed at specific plants (such as mining, metallurgical, and chemical). Each plant is a natural-technical system, a structural unit of the technosphere, like an organism in the biosphere. The internal properties of the natural-technical system, which determine external behavior, can be described only through the processes of technological transformation within it. Therefore, an individual element of the technosphere is an elementary technological process. Individual processes within a plant, and technotrophic chains, are connected by material and energy flows as well as by coordination of processing technologies at various production facilities because the product of the previous unit becomes the material for subsequent technological processes.

E. B. Zolotykh

Technotronic Society refers to the qualitative characteristics of a society where electronic and computer technologies are widely used and electronic means of communication are being developed rapidly. The term was introduced in the last quarter of the twentieth century under the influence of the works of Zbigniew Brzezinski, Jean-Jacques Servan-Schreiber, Alvin Toffler,

Teaching – Transparency and Information Technology 457

and others. For them the utilization of new technologies based on electronics indicates the main criterion of social progress. They also insisted on the leading role of science and education.

A. N. Chumakov

Terror(ism): (Greek *trein* "to be afraid"; French *terreur*, "fear," "horror") refers to politically motivated violence. The term "terrorism" was introduced in the end of the eighteenth century to denote Jacobin repressive policy during the French Revolution. In modern literature, the term "terror" is usually used to describe a policy of violence and intimidation, adopted by dictatorships or totalitarian regimes and directed against their citizens; coercion on the part of "the strong," that is, of the state, of those in power. "Terrorism" is understood as coercion on the part of "the weak"-the opposition. Terrorism is a method employed by an organized group to attain its proclaimed objectives, mainly through systematic violence. Current terrorist practices include hostage-taking, detonation of socially important objects, and plane hijacking. Terrorist acts are always public and are intended to pressure a society or regime. In 1977, the member states of the European Union adopted the European Convention on the Suppression of Terrorism.

Terrorism as a sociohistorical phenomenon is úbiquitous. Many outstanding state and public figures—Julius Cesar, Albrecht Wallenstein, Henry III and Henry IV, Russian Emperor Pavel, Jean Paul Marat, Abraham Lincoln, Alexander II, Petr Stolypin, John F. Kennedy, Anwar Sadat, and Indira Ghandi—were assassinated for political reasons. Their names went down in history, as did the names of their assassins: Brutus, Jacques Clement, Francois Ravaillac, Charlotte Cordet, Booth, Grinevitsky, and Bogrov. Such events had a noticeable impact on history, but taken individually, did not pose any considerable danger to humankind, being fragmentary and apart from each other. In the second half of the twentieth century, the situation radically changed: a drastic escalation and differentiation of various forms of terrorism took place. Terrorism became an instrument of all-out warfare against current political regimes, institutions, and structures of legitimate power. The irrational side of the terrorism is manifested in the mass killing of innocent people who have nothing to do with politics. As a result, the turmoil and orgy of violence turned every person on Earth into a hostage to insane ideas and the evil will of organizers and executors of terrorist acts. By the end of the twentieth century, terrorism developed into one of the most critical global problems of modern times.

In 1960s–1970s, a wave of violence connected with the activities of extreme Left groups such as the Red Brigades in Italy and the Baader-Meinhof Group in the Federative Republic of Germany swept over Western Europe and several Latin American countries. In the beginning of the 1990s, Russia found itself in the epicenter of brutal and murderous violence generated mainly by organized crime. Subsequently, another factor was added: Islamic extremism in connection with Chechen separatism and the anti-constitutional act of the self-proclamation of the Republic of Ichkeria.

History shows that forms of terrorism are, as a rule, essentially dichotomous: revolutionary and counterrevolutionary, physical and moral, subversive and repressive, selective and blind. But purely individual forms of terror-

458   *Global Studies Encyclopedic Dictionary*

ism also exist, such as military, provocative, preventive, criminal. The most widespread, dangerous, and malign forms of terrorism currently are subversive, blind, and criminal. In reality they are all interwoven.

Methods, technology, and effective strategy for antiterrorism should rely on the experience of domestic and foreign special services: primarily on the experience of the military operations in Chechnia and operations against the Taliban and al-Qaeda in Afghanistan. Antiterrorism implies and includes a broad range of provisions and actions; the principles of war strategy are the most important of them. These principles include optimal choice of the main effort; optimal coordination between arms and weapons; creation of a superiority of forces in the direction of the main effort, and taking full advantage of the factor of surprise. An effective antiterrorism strategy implies highly professional, highly mobile and psychologically trained Special Forces having a perfect command of techniques of terrorist neutralization and elimination.

V. I. Zamkovoi

Terrorism, International, Reasons for: Though the term "international terrorism" has been widely used since the 1970s, it became universally recognized after the acts of terrorism of 11 September 2001, in New York (the World Trade Center) and Washington, DC (the Pentagon). International terrorist structures commit acts of terrorism for the achievement of global goals, but, as a rule, they do not put in any particular claims. Actions of international terrorism attract the attention of the global mass media, which provides them a considerable public resonance.

An emotional understanding that terror has no justification, and the shock and sincere solidarity with people of the United States after the acts of terrorism on 9/11, should not cover the fact that international terrorism is governed by deep reasons. Though the secret services of the developed countries manage thousands of methods of the struggle against terrorists, these bodies alone would never defeat them. The only way that leads to victory is the eradication of the reasons for terrorism.

After 11 September 2001, the basic contradiction of world history became the one between developed and developing countries, living in essentially different conditions and developing according to absolutely different models inaccessible to each other. The contradiction between developed and developing countries is most pronounced and intolerable in the essential aggravation of the living conditions of the bulk of humankind (the proportion of which is overall enlarging) that occurred in order to provide for accelerated development, growth, and prosperity for the absolute (and ever decreasing) minority of humankind. In addition, the widening gap between the two parts into which the world is being broken, actually becomes the brakes on the development of humankind.

The population of developing countries, which lag behind developed countries, is ever growing, despite all efforts. On the whole, they are becoming unfriendly toward the latter ones and their values also due to the realization (by virtue of the formation of the integrated information space that acquaints them with those values) of the absolute inaccessibility of this status for themselves and for their children. This despair grows into animosity and hostility toward developed countries, and, first, to their leader and symbol, the United States. As reasonable and civilized methods of protest against the injustice that is so obvious prove inadequate, the disposition of representatives of developing countries to barbarous, rough terrorist acts that create the illusion of efficiency will increase inevitably (both at the level of individuals and representatives of corporations, and even states). Also, terrorists will understand and effectively use this tendency.

The act of terrorism against the United States plunged American society into deep shock. The main reason for this shock became the utmost importance Americans attach not only to their domination in the world, but also to their security (both individual and collective). At the same time, the real threat to individual and collective security had a significant mobilizing influence and contributed to solving the main strategic problems of the United States.

The strategy of maintaining national and global security and domination of the United States is dictated by objective reality; hence, it is by no means subject to any alterations, either through external influence on the United States or by targeted activities of representatives of American elites. It is an objectively caused form of implementation of the basic contradiction of the modern stage of the development of humankind. Therefore, though in the short-term aspect the relative safety of the world is provided for by the vigorous struggle of the United States and other developed countries against international terrorism, in the long-term perspective global instability will probably accrue.

In developed countries (by virtue of the social activity of the population and their strained sense of justice), various protest movements appear (for example, student disturbances of the 1960s, leftist terror of the 1970s, and peace movement of the 1980s covered almost all developed countries). At the end of the 1990s, developed countries faced a new powerful and rather effective protest movement, antiglobalism. The reason for its explosive and rapid expansion (and the expansion of the protest movement as a whole) is first the fact that the development of labor productivity combined with the growth in the quality of life and reliable social protection entails that in developed countries a considerable group of people emerge who are unable to find their place in social life, but nevertheless they are protected by the society (for instance, the growth of unemployment in the developed countries of Europe in the beginning of the twenty-first century). Representatives of this "lumpen" group) uninterested in revolutionary advancement), despite rather high (relative to standards of developing countries) level of life, due to poor socialization cannot find constructive implementation of their energy and capabilities. They tend to feel unfairly offended, and so they are a fertile soil for the dissemination of all kinds of protest movements.

A mass-scale and long-term protest entails wider violence from both parts. Mutual violence will accrue inevitably; out of the amorphous and, on the whole, nonviolent protest movement, ultra-radical and terrorist groups will be forged. The terrorism born in the developed countries will very quickly begin to collaborate with the terrorism of the developing countries, receiving from it necessary financial support and essentially solving the problem of equipment and access to modern technologies, which are exceptionally important for the latter.

M. G. Delyagin

460 *Global Studies Encyclopedic Dictionary*

Terrorism, Nuclear: By "nuclear terrorism," is meant the possibility that a sub-state group might acquire the capability to produce a nuclear explosion. Contrary to what many people think, it is quite possible, indeed easy, for a small sub-national commando to manufacture a nuclear explosive device, provided such a group could get hold of a sufficient quantity of highly enriched uranium (HEU).

Keep in mind that to be detonated by terrorists, in contrast to a nuclear missile, a nuclear device need not be transportable (it could be assembled clandestinely in a rented locale in the target city, and exploded via a timer allowing ample time for getaway). It need not be reliable (its yield would be unpredictable, but reasonable to presume capable of massive destruction). It need not be equipped with any safety gadgetry. A very small group that would include at least one intelligent and well-informed conspirator, who need not have any previous knowledge of nuclear physics, could create it, without the need of any previous experimentation involving nuclear explosions. All the materials and know-how required to manufacture such a device are easily available, with the exception of HEU. A nuclear explosive device based on plutonium, however, is so much more difficult to manufacture that the possibility of a sub-national group assembling such a device is less likely (although this would be possible for many States, provided they had the necessary plutonium).

Producing such an amount of HEU sufficient to manufacture a nuclear explosive device is more than likely to be beyond the capabilities of any sub-national group; indeed few States are capable to produce HEU. On the other hand, if a sub-national group were able to obtain such a quantity of HEU, transporting it clandestinely to a target would be quite easy. The overall volume of such material could be less than *ten* liters, and its radioactive signature is quite small; indeed the material, parceled into several packets, could be safely carried without any shielding in their pockets or backpacks by smugglers.

HEU cannot be acquired legally; it might be stolen, or acquired illegally from some insider who has access to this material and who might be motivated to divert it by greed or for ideological reasons. The concern that this might eventually happen—with catastrophic consequences—is justified by the enormous quantities of HEU that have been accumulated during the Cold War era. For instance in Russia, the country that possesses the largest stock of HEU, over one million kilograms of this material are stored, in over one hundred different locations.

Another possibility is that a terrorist group will obtain an already built nuclear weapon and deliver it to their intended target, detonating it there. Because nuclear weapons are presumably more well-guarded than HEU (including the fact that each of them is individually accounted for), and are more protected by technological devices, Protective Action Links (PAL), that make their unauthorized explosion difficult if not impossible, an eventuality much less likely to occur; unless one of the eight or nine States that possess nuclear weapons deliberately transfers one such device to a terrorist group, which also seems quite unlikely. This clearly entails that the primary line of defense against nuclear terrorism is the protection of all the existing HEU, and indeed

Teaching – Transparency and Information Technology 461

the elimination of as much of it as possible as quickly as possible; and secondarily the protection and elimination of plutonium.

The complete worldwide elimination of HEU, an eventual must for the survival of our civilization, need not entail a renunciation of nuclear energy, although it will require the elimination of all peaceful utilizations of HEU, which are however confined to fuel for research reactors, for the reactors used for the propulsion of some icebreakers and especially of some submarines, all uses which can be technologically converted to nuclear fuels (including uranium much less enriched than HEU) which are not susceptible use as basic material for the easy construction of primitive nuclear explosive devices.

The risk of nuclear terrorism discussed here refers to the possible production of a nuclear explosion by a terrorist group. Another eventuality sometimes referred to as the detonation of a "dirty nuclear bomb," involves the possibility that a terrorist group may obtain some radioactive material and spread it by means of an explosive device. It is unlikely that the quantities of radioactive material involved in such a scenario would be sufficient to cause many casualties (possibly, none at all) but much panic could be produced and a very substantial economic impact.

F. Calogero

Terrorism Prevention: Attempts to defeat terrorism through military operations in rogue states may eventually defeat one foe; however the pictures of bombed hospitals and injured civilians will only create more enemies. Admittedly, military efforts have not been the only instrument in the war on terrorism, but they have been the most prominent one. To many the bombing of Afghanistan demonstrates the futility of using military means to defeat a threat that transgresses all aspects of civilized society. The belief that terrorism can be defeated by a military campaign carried out by a coalition of nation-states against rogue states demonstrates a fundamental misunderstanding of the nature of terrorism. Terrorism is transnational and feeds off the inability of sovereign nation-states to monitor their cross-border networks. The nation-state is a blunt tool, poorly equipped to engage to a long-term war against terrorism. Those who have been victimized by terrorism have a natural initial reaction to suppress the supporters of terrorism.

By criminalizing terrorism, you strike out at potential threats to peace and security. However, the history of terrorism demonstrates that simply defining terrorism as a criminal act does little to prevent it.

Northern Ireland witnessed horrific acts of inhumanity, acts that acted as a barrier, dividing communities and preventing understanding. Yet in the midst of this division and terror, some brave proponents of peace have found a way to construct a framework for reconciliation. This framework was not constructed by top-down elimination of the proponents of terror, but instead was forged through a gradual process of building trust from the bottom-up.

Instead of continually fighting the symptoms of terror, the process of building relationships that cross borders and communities is essential to addressing misconceptions that feed terror. It is only through contact that you can change the minds of those who feel they are locked in a zero-sum game

462   *Global Studies Encyclopedic Dictionary*

where the fear and alienation they feel result in support for terrorism to preserve a way of life.

Peaceful settlement of disputes in Northern Ireland provides support for the belief that the people must construct peace and not have it imposed upon them. Imposing peace may work in the short-term; however, the long-term consequences can be worse than the initial conflict. We need only look as far as the Balkans for an example of this. Is this a call to end military and legal enforcement against terrorism? No, short-term prevention of terrorism requires a degree of enforcement and the deterrence of extremist groups that are willing to perpetuate violence for the sake of violence. However, those groups gain strength from populations that feel wronged or a desire to be heard. The only way to defeat terrorism in the long term and ensure a lasting peace is to address the misconceptions of the populations that have genuine grievances. The best way to approach this is through authentic attempts at promoting contact and exchange across borders.

The events of 11 September 2001, brought a new focus to the need to respond to terrorism as part of national security and globally. Efforts at using methods such as coalition building and international law enforcement networks are laudable; however, they still fall short of the goal of providing long-term security. Peace requires a bottom-up effort. In addition to enforcement strategies a new priority should be placed on promoting understanding through bottom-up programs conducted by NGOs so that fundamentalist terrorists no longer find support from the moderate members of society.

By eroding the public support for terrorism through programs of peace-building a long-term response to ending terrorism is more likely and possible. Ireland and Northern Ireland emerged out of a history littered with poverty and terrorism to become an important member of the European Union with a strong economy and an added significance as a region that has been able to avoid all-out warfare stemming from terrorism. The lessons learned in Northern Ireland have significant implications for the rest of the world, which is beginning to struggle with the impact of terrorism.

M. Cannon

Think Tanks are research organizations engaged in the analysis of socially important local or global problems, searching for solutions and working out alternative strategies of the development of society, state, and humanity as a whole. As such, they are an essential element of post-industrial society, providing a connection between state institutions, corporations, universities and society. Think tanks develop ideology for political forces, industrial corporations, and financial institutions, which contribute to raising the efficiency of state and municipal bodies.

Think tanks are widespread in developed countries, where their services are in high demand. About 4500 think tanks have been established in the world, more than two thirds of them operating in the United States. The RAND Corporation, established in 1948, is considered to be the first think tank. The very term "think tanks"—a humorous name for brain—was first used in early 1960s, for the designation of the RAND Corporation.

Teaching – Transparency and Information Technology 463

The successful activity of think tanks increased the influence of Pentagon. Their cooperation with United States intelligence services was especially fruitful. Think tanks turned out to be indispensable for studies of various countries and regions located in the areas of United States interests. They worked out the so-called stabilization programs of fighting guerillas, methods of waging information wars, controlled crises (controlled chaos technology), psychological wars, and basic strategic political forecasting. Companies involved in scientific and technological development of new military equipment and technologies were initially as well qualified as think tanks. Later, however, this class of think tanks was transformed in the West into a separate branch: the so-called venture firms.

Among the most prominent and influential think tanks besides RAND Corporation are the Heritage Foundation, the Brookings Institution, the Institute for International Economics, and the Carnegie Fund (shaping the ideology of the United States Democratic Party). The policy of Republicans is formed by the American Enterprise Institute, Center for the Twenty-First Century, the John L. Olin Institute for Strategic Studies, the Center for Strategic and International Studies, and the American Peace Institute. The most influential think tanks outside the United States are the Club of Rome, the French Institute of International Relations, the British Institute of Public Policy Research, and the Arab Center for Arab Unity Studies.

S. L. Udovik

Third World: Three-Fourths of the World (1980) was the eighth report to the Club of Rome by Mourice Guernier, a France National Economy Ministry official, who initiated specific investigations on the "North" and "South" interrelation and the developed and developing countries' disruption problem. Having analyzed the ways of Asian, African, and Latin American countries' economical and cultural development as it was shaped in the history; the author came to the conclusion that the only way to take away poverty and backwardness would be an economic and agricultural orientation to self-sufficiency. Since humankind has got a "unified destiny," it was mentioned in the report, the selection of the social and economic development model does not play a crucial role; however, agricultural reorganization has to be accomplished, and that was emphasized, according to decentralization principles.

Reference: Guernier, M. *Tiers-monde*. Paris: Dunod, 1980.

A. N. Chumakov

Toleration: (Latin *tolerantia*, "steadfastness in patience" "ability to endure") In moral and political philosophy, "toleration" designates the attitude of acceptance of opinions that are contrary to one's own point of view, and of different individual or collective lifestyles.

The systematic discourse on toleration has its origin in the climate of spiritual effervescence that dominated the European intellectual scene at the end of the seventeenth century and the beginning of the eighteenth. It was Pierre Bayle who, in 1697, opened a sharp polemic against religious intolerance in his *Dictionnaire historique et critique* (Historical and Critical Dictionary; 1991). Initially understood as a moral principle founded on the as-

464   *Global Studies Encyclopedic Dictionary*

sumption that no one is infallible and that, consequently, everyone has the right to one's own truth, the idea of toleration only gains consistence—as Voltaire maintains in his *Treatise on Toleration* (1994)—in the sphere of politics. Voltaire forcefully denounced any tendency of imposing on others, through violence or persecution, opinions whose foundations were uncertain or at least disputable. Admittedly, no absolute truths exist in politics, but interests—mostly divergent ones.

John Locke also approached the issue of toleration from the political perspective, emphasizing the danger of reducing individuals to uniformity and setting up the principles of freedom of conscience and of respect for the citizens' private interests. Moreover, Locke asserts that a tolerant attitude will be more profitable than an intolerant one as far as political decision is concerned, since intolerance breeds intolerance, turning a frank and fair opponent into a secret, cunning and stubborn adversary 2006.

An important landmark in the theoretical discourse on toleration is the social and political thought of John Stuart Mill, who argues explicitly for a pragmatic position, declaring, "Mankind are greater gainers by suffering each other to live as seems good to themselves than by compelling each to live as seems good to the rest" (2010).

There has been a considerable amount of writing on toleration, especially beginning with the latter half of the twentieth century, and today this issue has become a constant preoccupation of philosophical and socio-political thought. This concern, which may well be considered obsessive, of theoretical thought at the beginning of our millennium for the problem of toleration is undoubtedly a reaction to the totalitarian excesses of the previous century, just as the beginnings of the discourse on toleration had their grounds in the religious persecutions suffered by Protestants after the revocation, in 1685, of the Edict of Nantes.

Now, as then, the central issue is that of the limits of toleration. Many voices, including the most authoritative one, forcefully argue that the acceptance of toleration as a universal principle, the foundation of a new humanistic attitude, and of a new ethics, would not be justified; it would even be condemnable, as it generates confusion of values, theoretical relativism and unacceptable practical situations.

In the first place, the fundamental objection to be raised is: can the intolerable be tolerated? Can terrorism, for instance, be tolerated? Moreover, excessive toleration in politics may be interpreted as a sign of indecision and weakness, with negative repercussions on the management of the social body. The rules of democracy, the only climate that permits the manifestation of the spirit of tolerance, may sometimes lead to endless debates concerning the advisability of one option or another, to sometimes extremely costly opinion polls, which delay social action, out of a desire to have an accurate estimate of the will of the majority. It is sometimes necessary that political action should prove intransigent, even intolerant, in the name of the majority's interest, in order to ensure the good working of the social whole.

We are not born tolerant. On the contrary, violence and intolerance seem to characterize our natural being and our whole history. More than that, the question may be asked whether, given the vicissitudes of history, the spirit of

Teaching – Transparency and Information Technology 465

tolerance would have been desirable at all, as it may have jeopardized our very historical existence and becoming. The communities, which could not withstand the violence and aggression of other communities, have disappeared; only the strong have resisted, imposing their own way of life and cultural model on the others. Contemporary human beings are naïve to believe that, owing to culture and civilization, they have seen the end of barbarity and violence. We have yet to learn, for the utmost enhancement of our humanity, that in the common space of humankind *individual* beings, *irreducible to each other*, exist with specific ways of life and particular choices, each of which has the absolute right to difference and to distinct expression.

The question arises whether, in the name of tolerance, we have the right to excommunicate the truth. We are not only rational but also moral beings, so that it is our profoundly human duty to promote the truth and tolerance. What if they should be at variance with each other? Is toleration still a value if its worth is measured in the detriment of truth? In the religious sphere, a tolerant attitude is reprehensible, as it indicates deficiency in faith. To accept that others are also right, that other points of view are equally justified, means acknowledging the fact that what one holds about God is incomplete and, therefore, that one's faith needs completing with elements of other faiths, which ultimately casts doubt on the authenticity of one's own belief.

It must be underlined, however, that the essence of our humanity lies in our ethical being, in our ability for the moral understanding of the world. The ethical mode of comprehending the situatedness of the ego in the world is a distinct category of knowledge, whose main task is to think the *other-in-itself*, without understanding it as an "alter ego" but as an absolute, different and irreducible other, endowed with its own identity, its own individuality, which often opposes our own way of existence.

The doubt regarding the value of tolerance in the definition of our humanity has its explanation in the fact that being tolerant is not always an indication of virtuous conduct. One should rather distinguish between *good* and *bad* toleration.

Bad toleration is conducive to confusion or absolute relativism, through the admission of the fact that all ideas and attitudes are reasonable, since they are referred to different scales of values. It is in this sense that Bossuet deemed toleration to be the poison containing the seed of the Babel-like confusion of the social whole: in the absence of firm criteria of value, of a hierarchy of truths, this kind of toleration leads to indifferentism and social passivism.

Conversely, *good toleration* is a true moral virtue founded on the understanding and acceptance of the others' right to difference and identity. As a personal attitude toward the others, good toleration is the minimal condition of humankind's existence in common—not any kind of co-existence, however, but that which is achieved in freedom and fosters an increase in humanity.

In a world developing under increasing globalization, in which the conflicts among communities and civilizations may endanger the future of humankind, the humankind's survival may depend on tolerance.

References: Bayle, P. *Historical and Critical Dictionary*. Translated by R. H. Popkin and C. Bush. Indianapolis: Hackett, 1991. Voltaire. *A Treatise on Toleration and Oth-*

466 *Global Studies Encyclopedic Dictionary*

er Essays. Edited by J. McCabe. Amherst, NY: Prometheus, 1994. Locke, J. *An Essay Concerning Toleration and Other Writing on Law and Politics, 1667–1683.* Edited by J. R. Milton and P. Milton. New York: Oxford University Press, 2006. Mill, J. S. *On Liberty and Other Writings.* New York: Classic Books International, 2010.

S. T. Maxim

Toxicants are substances or compounds capable of rendering a poisonous action on living organisms. Depending on the character of the influence and the degree of toxicity they are divided into two big groups: toxic and potentially toxic. According to their chemical origin toxicants can be inorganic (such as cadmium, mercury, arsenic, lead, nickel, boron, manganese, selenium, and chrome) and organic (such as nitrazole compounds, phenols, amines, oil products, surfactants, pesticides, formaldehyde, and benzapilene). The degree of toxicological influence of chemical ingredients on various ecosystems and living organisms is compared with the help of the concept of molar toxicity, which serves as the basis for series of toxicity reflecting augmentation of the molar amount of metal, necessary for manifestations of the toxicity effect at the minimum molar value regarding a metal with the greatest toxicity.

There is a classification of the danger of various chemicals getting into environment. Depending on the degree of toxicological influence chemicals are subdivided into three classes. The first danger class consists of arsenic, cadmium, mercury, selenium, lead, zinc, fluorine, and benzapilene; the second class consists of boron, cobalt, nickel, molybdenum, copper, antimony, and chrome; the third class includes barium, vanadium, tungsten, manganese, strontium, and acetophenone.

As a rule, the majority of toxicants (for example, heavy metals) in the character of their interaction with various ligands are considered as mediate acceptors between hard and mild acids. In the first case they are characterized by low polarizability and electronegativity, high oxidation state and formation of ionic compounds, and in the second case they are characterized by formation of mainly covalent compounds.

The behavior of toxicants in various natural mediums is stipulated by the specificity of their basic biogeochemical features: complexing ability, motility, biochemical activity, mineral and organic forms of diffusion, susceptibility to hydrolysis, solubility, and the efficacy of accumulation. For example, copper and zinc are characterized both by the greatest chemical activity, allowing them to be considered as good indicators of terrigenous drainage and sedimentation and by highly effective accumulation in seaweed and plankton that determines their special significance for biota. Nickel and cobalt are biologically active and carcinogenic toxicants. The comparatively low motility of these elements stipulates their equal distribution in natural mediums. Among the geochemical features of lead are low motility and a short lifetime in the atmosphere and the solution phase of natural waters. In surface waters this lifetime amounts to several years, and in deep waters it amounts to a hundred years. Cadmium inclines to active bioconcentrating that results in its accumulation in excessive bio-available concentrations in a rather a short time; therefore it is one of the strongest soil toxicants. Mercury is the most toxic element in the natural ecosystems. Both inorganic and organic compounds of mercury are highly dissoluble. Out of all mercury compounds mercury-organic compounds are the most toxic for humans and biota.

N. N. Roeva

Transparency and Information Technologies: Timely and accurate information is a prerequisite of any successful military operation. The importance placed on reconnaissance and communication throughout history is evidence of that premise. The need for information has been a constant; only the methods used for gathering and transmitting that information have changed. As methods have expanded quantitatively and qualitatively another factor has entered the security equation as well: Information once available only to governments and militaries is increasingly becoming available to the public at large. This phenomenon has resulted in an unprecedented degree of global transparency. Debating this occurrence as positive or negative, stabilizing or destabilizing, is an academic exercise. It has happened. Subsequently, understanding the ramifications is critical, as heretofore unavailable opportunities and challenges are both presented. Consideration of two components of Information Technology (IT), remotely sensed data and the Internet-sometimes separately and sometime combined-well illustrates the dilemma.

In 1986, the world waited nervously as rumors abounded that a nuclear accident had taken place at the Chernobyl facility near Kiev in the Soviet Union (now Ukraine). When government officials stonewalled about what was going on, the American news media purchased satellite imagery from the French SPOT, and then the United States' LANDSAT system, to verify and report on the situation. Satellite imagery has also provided convincing evidence on the disturbing rate of deforestation in the Amazon and other tropical forests and on the effects of oil pollution along coastal areas. Indeed, in August 1996, an image taken by a European satellite was used in a civil case in Singapore to decide legal liability for an oil spill in the harbor. Withholding or denying information is no longer possible in many cases.

When satellite imagery began to be available to the public in 1985, it was primarily through government-developed or owned systems, such as the LANDSAT system, the commercially owned French SPOT system which went online in 1986, and declassified imagery from Russian military satellites in the early 1990s. The Russian imagery provided resolution as sharp as two meters. That forced Washington to rethink its policy, culminating with Presidential Decision Directive (PDD) 23 in March 1994, which set specific regulatory parameters for US commercial remote sensing ventures. Through the 1990s, both government policies and technology evolved regarding what was possible, including panchromatic (black and white) and multispectral (color) imaging to 1-meter resolution.

As access to an increased number of observation satellites using various types of imaging sensors (such as radar, thermal) occurs, even more types of information will become publicly available. Mere availability, however, does not necessarily mean that the information will be used wisely, or even correctly. Indeed, part of the dilemma with global transparency is that a key aspect of the fundamental remote sensing process, trained analysis, may be omitted. The public, news reporters, and amateur analysts may or may not

468 *Global Studies Encyclopedic Dictionary*

know what they are looking at and be able to put it into an appropriate context. During the EP-3 incident, for example, when Chinese and United States planes collided and the United States plane was forced to land on Chinese territory, the media showed an image of the EP-3 plane with what appeared to be part of the wing removed. This led to speculation that the Chinese were disassembling the plane. What had happened, however, was that there had been an error in the image processing, omitting several pixels.

The potential for misuse raises the issue of whether satellite images should be made available for public viewing. One argument is that transparency is always good, as an informed public is better able to make their decisions rather than relying on a privileged few. On the other hand, there is the potential for misuse. The Federation of American Scientists had a Web site called Private Eye that generated considerable debate with its provision of imagery of missile launch sites and other sensitive facilities. The site went dormant after its primary advocate left the organization and then was abolished after 11 September 2001, but the potential for another similar site remains.

Processes more open and accessible increase accountability. With its emphasis on open architecture and global connectivity, the Internet is the ultimate example of global transparency. It should be thought of as more than a massive network of computers and data base linkages—it is perhaps more important as a tool for communication and information sharing among people. As such, the Internet provides its own set of challenges and opportunities.

The Internet's architecture, standards, and communications technologies significantly affect transparency in several arenas. The Internet changes the way governments relate to their citizens. It is responsible, in part, for a new form of warfare. Finally, the Internet amplifies the individual's ability to more clearly understand his or her world.

Increasingly, governments, from the local city to the United Nations, are using the Internet as a tool to improve the efficiency of providing services to their citizens. In Singapore, for example, citizens can go to the Web to find, fill out, and submit the various forms needed for permits and taxes. In larger nations that have less developed transportation systems, such Internet access to public forms reduces the need for citizens to spend hours or days going to public offices simply to get the paperwork they need to obtain governmental services. Reports are posted on the Web, reducing the cost of production and widening the audience that has access to the material in the reports. The Indian state of Assam is using the Internet as a tool for the Governor to communicate with his district level leaders, enabling him to better coordinate and supervise policies as they are implemented throughout the state. The increased transparency of governmental processes reduces uncertainty and fear in societies where government officials have traditionally had much latitude in how they dealt with the public.

Budgets, plans, and minutes from meetings posted on the Internet allow citizens to see what their government is doing and, where appropriate, challenge those actions and hold officials accountable for their words and deeds. As the automated filing and processing of forms spreads, it becomes more difficult for bureaucrats to take a portion of the associated fees or to demand graft for providing service.

Teaching – Transparency and Information Technology 469

While transparency of information and processes is generally seen as a positive development, the ease of obtaining and spreading government information is also making it more difficult to protect information that legitimately ought not be made public. In the United States, for example, it was discovered that posting military promotion orders on government sites allowed criminals to use the individual identification information contained in those orders to open credit card and loan accounts for fraudulent purposes. Businesses have been forced to produce electronic records of employee e-mail messages that were then used against the business in civil court cases.

Transparency is clearly a two-edged sword. Remotely sensed data has already been shown a positive factor in such instances as the Dayton Peace talks and the Ecuador-Peru peace process. In both of those instances, more and more accurate geographic information was valuable to policy-makers in identifying viable options for mitigating disputes. In the South China Sea, monitoring activities of the multiple claimants of the islands has in some cases discouraged adventurism and provided data to in some cases minimize and in some cases confirm speculation. Humanitarian relief operations is another area where transparency can be of value, in everything from allowing better estimates of refugees to assessing deforestation and drought and monitoring soil moisture levels.

The Internet is opening the world of information in previously unimaginable ways to previously unthinkable numbers of people. The potential benefit to bring education, medical information, economic data, and simply glimpses of a far-away world to multitudes of individuals makes the word "remote" increasingly inappropriate. But challenges of availability and language, and concerns regarding political and cultural pollution, remain.

Equally important, transparency is also providing information to those Thomas Friedman (1999) refers to as "Super Empowered Angry Men." Their network becomes globally linked through the same technologies that link hospitals, businesses, and international organizations. Technology is blind to the goals and intentions of its data recipients. Vigilance becomes the responsibility of the human providers.

References: Freidman, T. *The Lexus and the Olive Tree*. New York: Farrar, Straus, Giroux, 1999.

J. Johnson-Freese, H. Finley

❦ U ❧

Umma: (Ancient Arabic *lumiya*, "confederacy of related ethnic groups; or Arabic *umm*, "mother," borrowed from Hebrew or Aramaic) The term "*umma*" (sometimes spelled "*ummah*"), meaning nation or community, occurs sixty-four times in the Koran. Islamic authorities reject derivations suggested from other languages, however, since they consider the beginning of all the Arabic language to be revealed in the scripture of the Koran.

In the Koran, *umma* denotes a variety of meanings, for instance, an ethnic group (the Arabs, Franks or Slavs), a religious group (the Muslims, Christians or Zoroastrians), a moral community (good or bad people as a group), the followers of a prophet (Abraham or Muhammad), a subgroup of believers or followers, a group related to Muhammad by lineage rather than by religion, and even a period of time. However, in the *hadith* literature (reports on the Prophet's words and deeds compiled during the ninth, tenth, and eleventh century), *umma* is given its usual meaning of the single universal Islamic community embracing all the lands in which Muslim rule is established and the Islamic law prevails.

The early Islamic legal tradition gave rise to the principle of the unity of the *umma* and saw it as the ultimate source of political authority. As such this concept legitimized the institution of the caliphate. The caliph (*khalifa*) meaning the "viceregent, the deputy or successor [to the Muhammad]," was thus the sole religious and political leader of all the Muslims and their lands. In Islamic political thought the inhabited world is divided between the "land of Islam" (*dar al-Islam*) and "the land of war" (*dar al-harb*) dominated by unbelievers; the *umma* ideally coincided with the geographically defined meaning of the former term. But with the geographical extension and political fragmentation of the land of Islam, the insistence on the unity of the *umma* and the corresponding unity of the caliphate became largely symbolic. Also non-Muslims were tolerated in *dar al-Islam* if they were "people of the book" (*ahl al-kitab*). The *dhimmi* (in *sharia* 'those who are in the covenant of protection extended by the *umma*") include Jews, Christians and Sabaeans though the category was often extended to cover Zoroastrians or even Hindus. The Western colonial domination limited the extent of lands where Muslim rulers reigned and brought many immigrant Muslim minorities to non-Islamic states. Hence, the more geographically defined concept of *dar al-Islam* was decisively divorced from the increasingly universalistic and non-spatially determined concept of the *umma*.

This disconnection was sealed during the early 1920s, first, by the breakup of the Ottoman Empire (which functioned as imperfect though nevertheless political embodiment of *dar al-Islam*), and secondly in 1924, by the abolishment of the caliphate. A year later, what remained from the empire had been overhauled into the ethnic Turkish nation-state. This radical departure from the Islamic legitimization of statehood in favor of the Western national one brought an unprecedented trauma to Muslim leaders and masses. From the Islamic point of view, this was the lowest point in the entire history of *dar al-Islam*.

Umma – *Urbanization* 471

The beginning of the twentieth century was characterized by an ideological struggle over the "rightful" political ownership of the concept of *umma*. The Ottoman sultan (literally "authority" or "government," or in the Seljuk and Ottoman usage "supreme political and military head of Islam" as opposed to an increasingly religious caliph) Abdüllhamid II's (reigned 1876–1909) attempts at redeeming Islamic unity by reviving the idea of *umma* were extremely popular among Muslims from Morocco to India. Equally popular was the call by Jamal al-Din al-Afghani (1839–1897, born in Persia domiciled from London and Cairo to India) for Islamic solidarity for the reinvigoration of the *umma*. Their thought constituted the foundation of Pan-Islamism. On the other hand, the Syrian Ottoman thinker Abd al-Rahman al-Kawakibi (1854–1902) revived the Koranic understanding of the *umma* as an ethnic group. He employed the term *watan* (literally "homeland" or "place of birth" coined at the end of the eighteenth century in emulation of the French "*patri*'" or "fatherland") when he spoke of what united Muslim with non-Muslim Arabs. In 1907, the political Umma Party (*Hizb al-Ummah*) was established in Egypt. In this instance the term *umma* meant the "Arabic-speaking nation of Egyptians' irrespective of their religion. These events became the underpinning of Pan-Arabism (Arab nationalism) enshrined in the adoption of *umma* to designate the Western term "nation" in the Arabic language. (In Turkish and Persian "nation" is rendered *millet* and *milli*, respectively. In the Koran *millah* meant "religion," and in the Ottoman Empire the inhabitants were divided into millets or religio-political communities).

Paradoxically, in the twentieth century, the concept of *umma* and its variegated legal and intellectual connotations have legitimized both of the most significant movements in the Islamic world: Pan-Arabism and Pan-Islamism. Respectively, the Arab League (established in 1945) and the Organization of the Islamic Conference (established in 1971) lead these movements. Many conferences organized to discuss the political situation of the Muslim *umma* after the abolition of the caliphate failed to achieve any significant results. This led to the popularity of secular Arab nationalism during World War II. Thereafter, in the wake of decolonization, Muslim eyes perceived *dar al-Islam* as promising liberation from the unbelivers' (*kafir*) oppression. Since the 1960s, Arab nationalists have spoken in favor of complete separation of religious and national identities. But this has not brought about the unification of all the Arabs in a common ethnolinguistic nation-state; neither has it led to progress and an equally shared prosperity. These failures of Pan-Arabism 9in contrast with the achievements of the West have resulted in Pan-Islamists gaining the upper hand especially after the foundation of the Islamic Republic of Iran (1979).

The Indo-Pakistani Pan-Islamist Muhammad Iqbal (1877–1938) asserted that the Islamic *umma* is the model for human unity, and that nationalism can coexist with humanism as long as Muslims believe in *tawhid* (the unity of God). Other Pan-Islamic authorities reject this view and assert that Islam and nationalism are mutually incompatible. However, neither the Islamic Republic of Iran nor any other Islamic state, or any group of states has attempted to unify the *umma* in a single polity. In practice this means that all the Islamic states define themselves as nation-states regardless of the official rhetoric. As

472   *Global Studies Encyclopedic Dictionary*

a consequence, in the twentieth century and in the beginning of the twenty-first century the idea of the Islamic *umma* continues to conflict with the international model of the nation-state.

T. Kamusella

Underground Rivers, Energy of: Underground rivers can be regarded as a source of water ensuring the necessary water balance on the Earth and geodynamical and hydrogeological stability at the lythological level. The energy of the underground rivers can be transformed into electrical or other kinds of energy thanks to the thermodynamic characteristics and geochemical properties of the underground water flows.

E. V. Mazour

United Nations Millennium Declaration is a document adopted by the General Assembly of the United Nations Organization on 8 September 2000, with participation of heads of states and governments of the member countries of the United Nations. The world leaders expressed their deep concern for the fundamental values of freedom, equality, solidarity, tolerance, care for nature, and claimed their duty "to control the global economic and social development." The Declaration's "Millennium Development Goals" include: By 2015, the number of people whose income is less than one dollar per day should be halved, as well as the number of people suffering from hunger and of those deprived of clean drinking water. All children should be provided with primary education and with equal access to all levels of education; the maternal mortality rate should be decreased by 75 percent. The death rate of children under five years old should be reduced by two-thirds. The spread of AIDS, malaria and other dangerous diseases should be curbed. By 2020, the living standard of at least 100 million slum dwellers should be raised.

A. B. Veber

Universe: The term, "universe" refers to "everything that exists," "the all-embracing global whole," the totality of all things; the sense of these terms is polysemantic and is defined by a conceptual context. It is possible to distinguish at least three levels of the concept:

(1) The Universe as a philosophical idea has a meaning close to the concepts of "*universum*" or "world": "material world," and "created existence." This conception of the universe plays an important role in European philosophy. Images of the universe in philosophical ontologies were included in the philosophical grounds for the scientific research of the universe.

(2) The universe in the physical cosmology, or the universe as the whole, is the object of cosmological extrapolations. In the traditional sense, it is a universal, unlimited, and principally unique physical system, a material world considered from the physical-astronomical point of view. Within the framework of the non-traditional approach the universe, in cosmology, is "everything that exists" not in some absolute sense, but only from the point of view of the present cosmological theory or model, that is, a physical system of the greatest scale and order, whose existence is defined by a particular system of physical knowledge. It is the relative and transitory border of the perceived megaworld, defined by possibilities of cosmological extrapolations. In

Umma – *Urbanization* 473

many cases the universe as the whole is taken as one and the same "original." Other theories can choose as their object different "originals"; i.e., physical systems of different orders and scales in the structural hierarchy.

Classical Newtonian cosmology has created an image of the universe as infinite in space and time. Relativistic cosmology constructed the theory of the extending universe, whose features appeared completely different from the Newtonian; its object is our metagalaxy. According to this theory, our universe can be finite and infinite in space, but in time, it is definitely finite—it had the beginning. For a long time, the metagalaxy was considered the all-embracing physical whole, and no other metagalaxies existed. The most popular theory in modern cosmology, though, the theory of the inflationary ("inflated") universe, allows other universes (outer metagalactical objects) with qualitatively different features. The totality of these universes is the "Metauniverse." Inflationary cosmology often calls objects, similar to the metagalaxy, "miniuniverses." Miniuniverses originate from spontaneous fluctuations of a physical vacuum. The initial moment of a metagalaxy's expansion should not necessarily be considered as the absolute beginning of everything. It is just an initial moment of evolution and self-organization of an uncountable number of megascopic space systems. In some versions of quantum cosmology the concept of the Universe coordinates with the existence of the observer ("a principle of participation").

(3) The universe in astronomy (the observed or astronomical universe) is the area of the world opened to observations and to some space experiments as well, that is, "everything that exists" from the point of view of the means and resources of observation available to astronomy. Throughout almost the whole history of astronomy this science studied the same types of celestial bodies: planets, stars, gas-dust substance. Then, in the second half of the twentieth century principally new types of astronomical objects were discovered, including super-dense objects in galactic nuclei (that might be black holes). Many states of celestial bodies in the astronomical Universe appeared to be non-steady, unstable. The suppressing part (up to 90–95 percent) of the astronomical universe substance is supposed to be concentrated in invisible, yet not observed forms ("hidden mass").

Our universe, or metagalaxy, can be considered as an environmental niche of humankind and cosmic intellect as a whole. Such an interpretation is a little bit broad, but nevertheless it has some sense.

Metagalaxy is a place where a human being is one of the forms of cosmic intellect, opposed to other miniuniverses whose features do not allow the existence of anthropomorphic creatures. The actual natural habitat of human beings is the Earth and near space so far, but potentially this area will extend accordingly to the speed of human exploration into space up to the limits that so far cannot be specified.

Not only integral features of the metagalaxy stipulated the possibility of the beginnings of life on Earth, but also by factors of smaller scale. At early stages of evolution of the Universe the first to appear were hydrogen and helium, and heavy elements that constitute a human body were synthesized during the evolution of stars. Further on, in the process of the universal evolutionism, they were captured by the interstellar medium and became a chemical

474 *Global Studies Encyclopedic Dictionary*

basis of intelligent life. In this context a person is said to "consist of ashes of faded stars." The present location of humankind in the extending Universe is stipulated by a complex totality of space factors (distance between the Sun and Galaxy center, star type of the Sun, stability of the Solar system, distance between the Earth and the Sun).

Our universe also acts as a trophic niche for humankind since the sources for maintenance of human beings as biological creatures are space-originated (such as cosmogonic factors, and solar energy).

Social activity of humankind appears to be closely dependent on space factors. Alexander L. Chizhevsky articulated some of these factors. The essence of his interpretation is that unsteady processes on the Sun, repeating with average periodicity of 11.2 years, influence human psyches (instincts, collective unconscious) in a specific way. It is a "trigger mechanism" for many events in the history of humankind (wars, revolutions, other events of social activity). The image of space plays a huge role in all spheres of culture.

Finally, global problems of a human-caused civilization most likely can be satisfactorily solved only in the process of exploration of outer space and its resources. Scenarios of space expansion of humankind (not having a considerable sociopolitical support so far) oppose the now most fashionable "geocentric" scenarios (splitting of the world community, a new Middle Ages, or eco-catastrophe), but stay close to the scenario of a post-industrial society. They proceed from the imperative of unity of humankind.

Thus, space is the multivariate environmental niche, defining not only the beginnings of life on the Earth, but its natural habitat (actual and potentially possible) and its behavior and activity in many essential aspects.

V. V. Kazyutinsky

Urbanization: (Latin *urbs* "city"; *urbanus*, "related to city life") Urbanization refers to the growing role of cities in social development; a cultural phenomenon of consolidation and dissemination of urban lifestyle. This process is one of the most important factors of globalization, changing human beings and their habitat. Geographically, it is closely related to spatial concentration of population and industries in relatively few centers and areas of socioeconomic development.

Urbanization today is characterized by some common traits, such as the fast growth of the urban population in the developing countries and concentration of population and economy in the major cities (megacities, megalopolises). Among them one may distinguish cities with a population of over one million (Rome has been such a city since the times of Julius Cesar). Currently, over thirty "megacities" of the world have a population of five million each.

Urbanization is a comprehensive spatial process covering territorially not only a city, but increasingly more of the countryside as well. It causes rapid development of suburbs around big cities-suburbanization ("urbanization of suburbs"). At the same time, some urban features and life standards penetrate rural settlements bringing about their qualitative change ("rurbanization" or rural urbanization). A most peculiar feature of modern urbanization is a transition from a compact city to territorial groupings of urban and

Umma – *Urbanization* 475

rural settlements-city agglomerations-followed by further transformation to still greater formations-megalopolises.

Despite shared traits of urbanization as a worldwide process, it has some local peculiar features in different countries and regions manifested above all as different levels and pace of urbanization. In most of the economically advanced countries with a high urbanization level, the proportion of city population has been growing recently at a relatively slower pace, often decreasing. Former concentration of population in large cities and agglomerations was replaced by de-concentration tendencies, due to urbanization covering evergreater territories. Some scholars (e.g., American geographer Brian Berry) talk about a radical turn in the nature of urbanization, about the beginning of a new period of counter-urbanization or de-urbanization. However, urbanization continues to progress assuming new forms.

In developing countries where the level of urbanization is lower, it continues to increase. These countries account for over four-fifths of the total annual increase of the urban population, while the absolute number of city dwellers is higher in economically advanced countries. Urban population growth in the developing regions considerably exceeds the rate of their development, which causes many problems related to the so-called urbanization of slums, pointing to the spontaneous, disorderly nature of urbanization.

N. V. Logina

ೞ V ೞ

Values, Universal: Universal values are norms and principles of life reflecting the essential interests of the human community and generally valid for all people worldwide. Natural objects, the results of human material, and spiritual activity are recognized as unique and significant all over the world. Some of these values were initially formulated in the Bible as (universal) Ten Commandments. Later, persons became the initial reference point for the universal morality and the general system of values. Many authors asserted that it is possible and necessary to apply the "golden rule" to relations among people, states, and nations. In the seventeenth century, John Locke declared the inviolability of the inalienable human rights to life, liberty, and property. The slogans of the French Revolution: "freedom," "equality," "fraternity," and "justice," strengthened people's belief in universal values. These values are presented in the Universal Declaration of Human Rights accepted by the United Nations General Assembly on 10 December 1948, and approved by the Helsinki Conference in 1975.

In conditions of increasing and worsening global problems and the increasing destructive activity of people, a planetary-scale value is given to unique objects of nature and material culture such as the planetary genetic pool, non-renewable power and raw materials resources, clean atmosphere, Siberian and Amazonian woods, the Baikal Lake, the Great Chinese wall, and the Egyptian pyramids. The necessity of caring about these objects leaves humankind no other choice but to overcome disunity and disagreement and to preserve cultural originality, historical traditions, and specific features of separate cultures. A paramount task of the globalizing world is the necessity of recognizing universal values and their priority over other human interests.

A. N. Chumakov

Values, Western: There is no single set of "Western values," certainly not a uniformly accepted one. The term has only become popular in the last several decades to identify the informal institutions accepted by most countries in Western Europe, much of North America, Australia, New Zealand, and, to some degree, in Japan, South Korea, and Taiwan. "Human rights" probably enjoy the broadest consensus as being core Western values. Freedom of speech, assembly, religion, and private property are also generally included. Differences arise, however, when economic welfare is included as a state duty as opposed to the individual's responsibility.

Human rights, civil society, political pluralism, social and economic justice, and forms of art, literature, and music, things said to be Western values, would be inconceivable today without the political idea expressed in the *Magna Carta* (1215), which acknowledged that rulers are limited and that individuals have inalienable rights.

Economic liberalism is no less important to Western values than political limits on the state. Private property, a God-given right under "natural law" according to the theological reasoning of Richard Hooker, plays both economic and political roles in sustaining Western values.

If we see the liberal concept of political and economic institutions as the foundation for Western values, we can also see why they need not be confined to countries in the West. A major question in many parts of the world at the turn of the twenty-first century is the degree to which Western values can or should thrive in non-Western civilizations.

The American Declaration of Independence declares "all men," meaning people in all civilizations, have "inalienable rights." This claim implies that where governments insure such rights, a lot of things known as "Western values" will emerge, but it does not mean that all native cultural values will disappear. Only those contravening "inalienable rights" must be abandoned. Today, for example, Japan retains more of its traditional values than it has abandoned. The real struggle is not actually over "Western values." It is between individuals within a civilization and their rulers over the distribution of political and economic power. Once they move from "rule of the one" to "rule of the few" and perhaps on to "the rule of the many," they are creating grounds for some version of "Western values," but not to the exclusion of all their native values.

W. Odom

Virtualization (Virtual Technology): In simplest terms, "technology" means practical application of knowledge. Virtual technology refers to creation (including, but not limited to using computers) of a virtual (as opposed to actual) version of something, as a means of creating the perception of experiencing and operating in reality ("virtual reality") without actually doing so; e.g., using computer technology to simulate the experience of performing operations in outer space prior to actually traveling to outer space.

But before the existence of anthropogenic technology, meaning technology created by human beings, humanity had a prior technologies constitutive of its very humanity. The "higher" human technologies, human spirit and soul, are given to us by birth and enable us to immerse ourselves in spiritual and emotional realities. We also acquire "natural" technologies by birth, viz. the five senses. With their help, a human being can become immersed in physical and intellectual realities, perceive and experience them, and operate in them.

Human beings create the "artificial" (virtual) technologies in likeness of higher and natural virtual technologies. These allow immersing into artificial realities, created by human hands, intellect, emotions, and spirit, to perceive and experience them, and operate in them. To such technologies belong languages, arts, literature, social institutions, mass media, image-making technologies, and computer virtual technologies. Computer virtual technologies are believed to be the most perfected type of artificial virtual technologies, in which users are immersed as spectator and creators.

Presently, the development of computer virtual technologies is the most successful branch in the modern computer industry. The main task in this branch is to extend the scale of human capabilities. This task is being accomplished by the convergence of the individual and computer virtual technologies, bringing them together into a single conglomerate, capable of performing tasks otherwise insoluble by either human beings or computers programmed by human beings. The place to carry out these tasks is the computer

478   *Global Studies Encyclopedic Dictionary*

virtual reality. The results of such experiments are extrapolated to the real world either immediately during the course of such experiments (with the help of robotics) or after the experiment by the individual himself or herself.

Due to computer virtual technologies, people acquire extraordinary vision, hearing, smell, touch and taste capacities, new intellectual and creative resources, special abilities to move in space and time, and many others.

Internationally, because of its great effectiveness and proven potential, this field of study is being abundantly financed by state agencies and is developing at a rapid pace. Many foreign universities, laboratories, research centers, and companies are engaged in the process of creating new computer virtual technologies. They have already made about twenty computer virtual devices of the fifth and sixth generation. Among them are virtual helmets, glasses, gloves, trackers, suits; smell-forming and contact-interaction devices, systems of power feedback, entire virtual portals for groups of users-visionariums, reality centers, and many other original devices. Many spheres of life exist internationally where they have begun to apply computer virtual technologies: state governance, armed forces, education, health care, science, architecture, arts, entertainment, energy industry, oil and gas industry, banking, finances, stock exchanging, communication, research, office management, and many others. Development and application of the computer virtual technologies today bring revenues of many millions to the economies of developed countries.

Computer virtual technologies are quickly becoming ubiquitous. Possibilities of boundless interaction for large groups, united in one cyberspace, will lead us to even tighter globalization and great changes in world culture.

A. V. Yukhvid

⚙ W ⚙

War is a social political phenomenon associated with a fundamental change of the character of relations among states, peoples, nations, when confronting parties stop using nonviolent forms and methods of struggle and start to use weapons and other violent mediums directly to reach political and economic goals. Today, war is the sharpest form of direct political confrontation of enemy parties. The main instrument used to conduct war is armed forces and other paramilitary units. Affecting all aspects of life and activities of a society, war brings it into a special condition. The main specific content of war is armed struggle. At the same time, in order to achieve the set political goals, other methods and means of struggle are used in war as well: such as diplomatic, economic, and ideological.

War is closely related to politics and economics. Politics determines the goals and the social nature of war; it also constitutes the key factor determining its intensity and methods of execution, the direction of concentrating of main efforts, and the degree of mobilization of human and material resources. Economics determines the material basis of war, the instruments used to conduct it, its scale, forms, and methods of armed struggle. The course and results of hostilities to a significant degree depend on the level of development of economics. At the same time, war itself requires a fundamental reorganization of economics, mobilization of all of its resources on the basis of the needs of the armed forces.

In any war, two interrelated aspects operate; one social and political, the other military and technical. The first aspect reflects the relation of the war to basic trends of the historical development and content of this epoch. It determines the goals of the war depending on the interests of various social forces; the second aspect determines the strategic nature of the war, the means and methods of execution of hostilities, and the content and order of the solution of the main strategic military problems.

The main and decisive form of struggle in war is armed struggle, which involves organized use of armed forces and other paramilitary units in order to achieve specific political, military, and economic purposes. It consists of various combinations of offensive, defensive, support, and other activities, and regrouping of troops (forces), and maneuvering the instruments of war.

Normally goals of wars include devastation of enemy states or their coalitions and their compulsion of acceptance of specific political conditions. The final goal of a war includes achievement of its individual and intermediate goals: for example, to get confronting states out of war, to smash their armed forces or military political groups, to occupy the territory of such states or a part of it, or to deprive an enemy of allies. These goals are normally achieved by consistently performing military political tasks and other military strategic tasks originating from them in specific periods (stages) of the war, a campaign, or individual major operations.

Modern wars are normally characterized by determination of goals, a huge intensity of struggle, a destructive and ruinous character of hostilities, large scale, fast and abrupt change of forms and methods, expansion of hostil-

480 *Global Studies Encyclopedic Dictionary*

ities to all geographic spheres, sharp struggle aimed at getting and keeping the lead, and other distinguishing characteristics.

Traditional (classic) and non-traditional ways to conduct war can be described. Traditional methods of war are characterized by consistency and a continuity of typical techniques of hostilities. Non-traditional (non-typical) methods of war are fundamentally different; they are conditioned by the possibilities related to the use of qualitatively new means of armed struggle.

In war, economic struggle is used; it is the aggregate of economic measures and activities intended to undermine the military and economic potential of the enemy and to reach economic preeminence. Economic struggle is based on military economic mediums and measures. At the same time, in such a struggle, military mediums are widely used, too, first, to deliver blows against the enemy's economic centers, the most important military and industrial production facilities, and public administration bodies. The course of economic struggle is much influenced by the destruction of processing centers for strategic raw materials and power centers and by the impairment of internal transport systems and communications connecting the enemy with other states, especially the disruption of the transport of oil products.

Diplomatic struggle in war uses various types of diplomatic activities to undermine military and political positions of confronting states and to consolidate the positions of the state conducting such struggle, to disunite the enemy's coalition by initiating conflicts inside it, to bring over allies and to improve relations with them, to spread information, and to undertake other activities that help achieve political and strategic aims of the war. In wartime, diplomatic struggle is normally subordinate to armed struggle; the former is aimed at creation of the most favorable conditions for the latter. At the same time, one of its goals is to achieve the most advantageous conditions of peace.

One of new forms of struggle used in modern war is ecological struggle, which consists of a package of measures and activities intended to create unfavorable environmental conditions in the area occupied by the enemy and to pose problems for its armed forces' activities, its economics, and the life of its people. In its extreme forms, ecological struggle involves creation of a climate in which the normal life of states is impossible. The struggle is conducted using both military and non-military mediums. It includes destruction of nuclear power plants and centers of production of especially dangerous chemicals for a long-lasting contamination of large land areas, the atmosphere, and water, intentional destruction of the ozone layer, formation of catastrophic phenomena, floods, fires, and so on, and prevention and disruption of the same kind of activities performed by the enemy. The effectiveness of such struggle can significantly increase in the case of the creation of weapons based on new physical principles; for example, geophysical weapons; and in the case of the use of highly toxic substances and other long-lasting chemicals able, to disturb natural exchange and to destroy vegetation over large areas.

In modern war, ideological struggle is conducted as well: ideological and political, psychological, and informational influence on the armed forces' personnel and the population of enemy states, aimed at undermining their morale and weakening their will to win and stand up to the enemy. The most important forms of ideological struggle are propaganda and counterpropagan-

War – World Wars 481

da, agitation, spreading of purposeful information and misinformation, indoctrination of people, ideological diversions, counteraction against the military ideological measures taken by the enemy, and special ideological and psychological operations performed according to a single plan during a specific period of time in the seat of war with the purpose to support upcoming large-scale military operations and to achieve their goals.

D. O. Rogozin

Water Resources of the World are fresh and mineralized natural or processed waters (such as desalted or purified), which are used currently according to the established economic objectives and which can be used in the future. They include all kinds of water, which are on the surface of the Earth, in its atmosphere, and in its bowels—the reserves of surface and underground waters of the land. Almost all types of water in the hydrosphere that are suitable for employment are included.

As one of the biosphere's components, water has decisive importance for the provision of life on Earth and for preservation of ecological systems. It takes part in almost all spheres of production activity.

The total volume of the hydrosphere is about 1.5 billion kmi. This volume is distributed on the globe in the following way: 94 percent is in the ocean; 4 percent is underground waters, the greater part of which consists of abyssal brines (the share of fresh underground water is 4000 kmi); 1.6 percent is polar glaciers. Surface fresh waters come to about 0.25 percent, and the volume of atmospheric vapor is 0.001 percent. At the same time, humankind has always used fresh water to the utmost; this tendency will remain.

According to forecast, by 2025, the population of the world will increase up to eight billion people, and an average water supply will decrease 1.3 times in comparison with the present level.

Water supply per capita on different continents varies in a wide range: from 3.5 in Asia up to 80 thousand m^3/year in Australia and Oceania. In the regions with the deficit of water resources, desalination of mineralized waters, including seawater, is practiced for increasing the water supply. This is especially widespread in the countries of the Middle East. But total production is still rather small and it does not exceed 10 km^3 of desalinated water per year.

Brazil, Russia, Canada, the United States, China, and India are especially rich in renewable water resources. Over 40 percent of total annual outflow of rivers on our planet are formed on their territories.

Data concerning the resources of river waters are referred to as the conditions of stability; i.e., they do not take into account possible human-caused changes of the global climate. In the periods of climate warming and an increase of temperature contrast between ocean and continents, we can observe the intensification of circulation processes in the atmosphere with the increase of transfer from west to east in the Northern Hemisphere. This leads, in particular, to the growth of cloudiness and atmospheric precipitation in Europe and to the increase of river outflow.

Global climate warming due to emission of greenhouse gases causes the change of the whole complex of hydro-meteorological factors and elements of the hydrological cycle, which determine the forming of water resources.

482 *Global Studies Encyclopedic Dictionary*

Several different climatic models have been developed for prediction of changes of climate and water resources. Unfortunately, the obtained results are still rather contradictory and cannot become a basis for quantitative evaluation of water resources.

In the second half of the twentieth century, the consumption of fresh water increased almost four times (from 1,060 in the middle of the century to 4,130 km^3/year at the close of the century). The total taking of fresh water from springs constitutes not more than 10 percent of river outflow. But the long-term fickleness of river outflow, its annual fluctuations, the disparity between annual distribution and need for water, and inadmissible levels of pollution of water resources lead to their considerable deficit in several regions.

In the majority of river basins the bulk of outflow (60–70 percent) is formed during flood periods. In Europe 46 percent of annual outflow falls in the period from April to June; in Asia, 54 percent in the period from June to September; in Africa, 46 percent in the period from September to December; in North America, 49 percent in the period from May to August; in South America, 45 percent in the period from April to June; in Australia and Oceania, 46 percent in the period from January to April. In some river basins the share of low flow (3–4 months) comes to only 2–10 percent of the annual outflow.

These problems are partially solved around the world by regulation of river outflow by means of water reservoirs and by transfers between different basins. Water reservoirs are created on almost all continents. The 2,200 largest water reservoirs have a bulk of more than 100 million m^3, their total capacity exceeding five thousand km3, their effective capacity three thousand km3. The water area of these reservoirs comes to about 600 thousand km3 (0.3 percent of the land).

M. G. Khublaryan

Weapons of Mass Destruction (WMD): The term "weapons of mass destruction" typically refers to nuclear, chemical, and biological means for killing large numbers of people. The target is generally civilians or noncombatants. The intent is to strike terror into the population. Nuclear, chemical, and biological weapons have all been used in war: nuclear weapons at the close of World War II, chemical weapons in World War I, and biological weapons for over two millennia. The United Nations and many countries have called for bans against such weapons, even terming them genocidal. Several significant treaties have also been ratified that ban use or even production and stockpiling of various weapons of mass destruction.

Nuclear Weapons are the most grave among weapons of mass destruction. Each of the two atomic bombs dropped on Japan in August 1945, killed 50,000 to 100,000 people. The Soviet Union tested its first atomic weapon in 1949, and the United States tested the first hydrogen bomb in 1952. At the height of the Cold War the United States and the Soviet Union each possessed about 10,000 strategic nuclear weapons and many times more tactical nuclear weapons. Both countries could deliver strategic nuclear weapons by aircraft, from land-based intercontinental ballistic missiles (ICBMs), and by submarine-launched ballistic missiles (SLBMs). The doctrine of Mutual Assured Destruction (MAD) was based on a "balance of terror." Either super-

War – World Wars   483

power could obliterate the other, and either superpower, even after being so devastated by the other, could launch a second strike that would inflict equivalent destruction. In the post-Cold War, many analysts now regard the risks posed by continuing nuclear proliferation as the greatest threat for the next use of nuclear weapons.

Chemical weapons are composed of compounds constructed artificially, as opposed to compounds existing naturally in an inorganic or organic state. In chemical facilities around the world, disabling and deadly compounds (such as chlorine and mustard gas) are engineered for use almost exclusively against human beings. Unlike nuclear weapons and many traditional or conventional weapons, chemical weapons destroy people rather than property, but clean up can be a significant problem. Chemical weapons are heavily weather dependent; rain can dilute them or wind can disperse them. During World War I, over 100,000 deaths and 1,300,000 casualties resulted from the use of chlorine gas and other chemical agents. As a consequence, the 1925 Geneva Protocol prohibits poisonous gases. Despite this protocol, limited use of chemical weapons occurred in World War II and various countries have subsequently some use of chemical weapons.

Biological Weapons are living microscopic organisms and are largely uncontrolled once they are released. They use infectious agents, such as bacteria or viruses, to inflict physical or psychological damage or death on their victims; the diseases that they can cause include tetanus and diphtheria. They are generated by microorganisms or plants or are animal in origination. Like chemical weapons, they are heavily weather dependent; rain can dilute them and wind can disperse them. A further problem with the use of bacteria, viruses, and toxins is that these poisons are usually unstable; so, their long-term storage often presents greater challenges than the storage of chemical weapons. However, like chemical weapons, they destroy people rather than property. Among biological agents anthrax has recently received the most attention. The number of viral agents is staggering, and the prospect for the reintroduction of small pox is generating increased concern.

Preventing Nuclear War: Despite the Nonproliferation Treaty, more than a half dozen countries now possess nuclear weapons. Nuclear states include the United States, Russia, Great Britain, France, China, Israel, India, and Pakistan. South Africa had some nuclear weapons during the period of apartheid. Several more countries and terrorist groups have tried to develop or obtain nuclear weapons, including Iraq under Hussein, Libya, North Korea, and al-Qaeda. Some nuclear materials have been bought or stolen making possible at least radiological devices that could broadly disseminate radioactive contaminants. Nevertheless, no further use of nuclear devices has occurred since the US atomic bombings of Japan.

Preventing Chemical War: Given the limited use of chemical weapons in World War II and subsequently, some writers stress that humanity does not always rely on every pernicious weapon at its disposal. In addition, in 1990, the United States and the Soviet Union signed an agreement to stop producing chemical weapons and to reduce their stockpiles to 5,000 agent tons. The Chemical Weapons Convention, signed by over 165 countries, went into effect in 1997. Nevertheless, unless militarily weak and economically impover-

484 *Global Studies Encyclopedic Dictionary*

ished states and sub-national groups feel they have a voice internationally, the prospect for escalating use of chemical weapons will continue.

Preventing Biological War: Agreements that constrain production and use of biological weapons are less developed than ones pertaining to chemical weapons are. By 1997, the UN resolution "Convention on the Prohibition of the Development, Production, and Stockpiling of Bacteriological (Biological) and Toxin Weapons and on their Destruction" had been ratified by 142 countries. Nevertheless, during the twenty-first century biological weapons may increasingly become the weapons of choice of the weak against the strong.

Ethical Issues Involving WMD: Fundamentally, WMD are instruments of terror. As moral philosophers have noted, both sub-national groups and governments can resort to the use of weapons of terror. Clearly, wars generally kill far more people than do typical terrorist attacks. Principles of just war forbid the intentional killing of non-combatants. Nevertheless, especially since the obliteration bombing in Europe and against Japan at the close of World War II, cities and their civilian populations have become targets. An important ethical lesson about weapons of mass destruction is that they can be (and have been) used by individuals and by governments. In this regard, the difference is one of degree, not of kind.

Given its diverse use, the term "weapons of mass destruction" needs careful analysis. When we refer to weapons of mass destruction, we are drawing on a condemnatory connotation. Moreover, the fallacy of special pleading occurs in use of this term. For example, the United States presented its use of nuclear weapons in World War II as a means to end the war and save lives, yet the United States condemns as weapons of mass destruction ones with far less destructive capability when they are possessed by a "rogue" state or terrorist group. The time has come to realize that most violence, terrorism, and war needs condemnation, regardless whether the instruments used are termed "weapons of mass destruction."

References: Combs, C. C. *Terrorism in the Twenty-First Century*. Upper Saddle River, NJ: Prentice Hall, 1997. Gay, W. C., and M. Pearson. *The Nuclear Arms Race*. Chicago: American Library Association, 1987. Solomon, B., ed. *Chemical and Biological Warfare*. New York: H. W. Wilson, 1999. United Nations. Report of the Secretary General. *Chemical and Bacteriological (Biological) Warfare and the Effect of Their Possible Use* (A/7575/Rev. 1, 1969). New York, 1969.

W. C. Gay

Weapons and Materials of Mass Destruction, Global Partnership against: The Global Partnership against Weapons of Mass Destruction is a program fighting against proliferation of weapons and materials of mass destruction, resulting from a decision of the Great Eight (G-8) leaders to unite their efforts in fighting this threat. This program is the logical development of a Russian-American agreement better known as the Nunn-Lugar Program. The G-8 has undertaken an obligation to allocate up to $20 billion over ten years for implementation of the program of global partnership. The United States is expected to offer half this sum, with the rest to be given by European countries, Japan, and Canada. Thus, Germany has provided an aid of 1.5 billion Euros. Italy intends to provide $1 billion, Great Britain and France, 750 million Euros each, Canada, $ billion Canadian, and Russia, $200 million USD. The Netherlands, Norway, Poland, Finland, Switzerland, and Sweden have announced their intention to join the Global Partnership. Russian-Norwegian contracts for utilization of two multi-purpose nuclear submarines (NSM), an agreement with Japan and Italy on utilization of NSMs, a Russian-French project on utilization of the Lepse depot ship have been signed within the framework of the Global Partnership. Five projects related to solving nuclear environmental problems in the North-West of Russia have been coordinated with Great Britain. An agreement has been reached with Germany on implementation of a joint project aimed at creation of a complex for the preparation of one-compartment reactor units of utilized NSMs, their transportation and long-term storage.

V. N. Kuznetsov

Weathering is a complex of processes related to the erosion of minerals and rocks due to contact with the Earth's atmosphere, biota, and water. Sometimes, weathering is interpreted rather loosely, including several processes related to the mass transfer and even the formation of new mineral phases, because, in some cases, the destruction of the original minerals and the formation of new ones that are more stable in the hypergene conditions can hardly be separated both in space and time. The present review will focus on interpreting weathering only in the narrow sense.

Traditionally, three types of weathering are recognized: physical, chemical, and biological. This subdivision is relative, since none of these types of weathering occur in nature in pure form. It only makes sense to consider the physical, chemical, or biological processes as predominant in the destruction of minerals and rocks.

Physical weathering is generally related to the formation and expansion of microcracks in rock as a consequence of the repeated heating and cooling or freezing and thawing of the latter. This process, apart from causing the disintegration of the solid rock base, the transfer of derivatives, and, finally, the formation of loose sedimentary rock, is a necessary condition for chemical and biological weathering to begin. This is because chemical and biological reactions that are responsible for the weathering of soils and rocks take place on the surface of the solid phase, their rate depending directly on the specific surface area of solid particles, which is proportional to the square of linear dimensions of the latter.

Chemical and biological weathering, which begins simultaneously with the physical one and intensifies with fragmentation of mineral grains, is quite important for the biosphere's functioning. It is only due to weathering that the elements of mineral feeding are extracted from the lithosphere and are involved in biological circulation. The elements released in the process of weathering can be either utilized by the biota directly on the spot or deposited in soil and in surface waters.

Apart from the elements of mineral feeding, the weathering processes release several microelements that are toxic to the living organisms in large concentrations. For example, aluminum, the third most abundant element (after after oxygen and silicon) and the most abundant metal in the Earth's

486 *Global Studies Encyclopedic Dictionary*

crust, is needed in only negligible quantities by living organisms. But high concentrations of aluminum in surface waters and soils is harmful to some life forms and leads to the increasing acidity of soil solutions.

The release of the biophilic or toxic elements during weathering is a most important factor regulating the spatial distribution of biogeocenoses.

Weathering is a universal process on the Earth's surface. Most often, it is coupled with soil formation. This is attributable both to the polydispersity of soil systems, which provides a larger specific surface area necessary for the surface reactions, and to the high biological activity of most soils.

The most significant difference of weathering in soil from that in rock is a much greater participation of living organisms and their decomposition products in the destruction of minerals. Microorganisms are well known to affect minerals without soil environment, but, in the soil, minerals are subjected to the mass attack of microflora, fungi, higher plants, and soil mesofauna. This is why models based purely on chemical thermodynamics prove extremely difficult to apply to soil weathering. The soil biota is an active weathering agent, enabling several endothermic reactions that are impossible in the abiogenous rocks. Only a small number of exogenous chemical reactions are possible in rock. In soil, due to participation of living organisms, many more chemical reactions are possible.

Generally accepted is the climatic conditionality of weathering. Climatic conditions determine the temperature of chemical reactions, the relation between liquid and solid phases (the concentration of the reacting solution), and the participation of biota in weathering processes. Correspondingly, the extent of substrate weathering, at the same age, depends on the climatic zone, and the products of weathering also differ.

The rate of chemical reactions increases with temperature. In tropical regions, the weathering process, even on relatively young surfaces, is more advanced compared to cold regions. In the permafrost region, only exothermic chemical reactions are possible without the participation of biota.

Considerable changes can result from the changing water regime of soil. The conditions of sufficient humidification are optimal for weathering, while the solutions of mineral decomposition products should be removed from the reaction zone. Such a situation takes place in the humid regions, where soils are characterized by the flushing water regime. The weathering intensity decreases with decreasing humidity. At the same time, the composition of the new-formed products also changes. The carbonates, sulfates, and halogenous compounds begin to appear in soils, as weathering products, with the increasing aridity of the territory.

On a global scale, some equilibrium takes place: the diminished extraction of feeding elements caused by inadequate humidity is compensated by the mechanisms of fixation of extracted elements in the form of relatively easily soluble minerals. In the case of excessive humidification (swamping), the character of weathering changes considerably. In general, the rate of destruction of minerals increases due to the protective ferrous films being removed from the grain surface, and to the ferrolysis, a special type of hydrolysis, which is characteristic of overly damp soils.

War – World Wars   487

All the processes related to the biogenic destruction of minerals can be conventionally divided into three groups. Lithotrophic microorganisms exist at the expense of some reactions involved in the weathering processes. A more many group of microorganisms needs the elements of mineral feeding and purposefully affects minerals to extract the necessary substances. Almost all organisms excrete products of a various extent of corrosiveness that participate in the weathering processes.

The intensity of weathering can change to the most extent due to the change of biocenoses on the soil. The global climate changes affect both the composition of the Earth flora and the microbiological activity of the soil. These changes are complex in nature and some further studies on this subject are necessary. On the one hand, the increase of temperature should lead to increasing biological activity and, consequently, to intensified weathering. On the other hand, climate warming can lead to the coniferous forests being replaced by mixed and deciduous ones, which produce fewer acids, and the weathering could be diminished.

P. V. Krasilnikov

Wide-Area Networks (WAN) are computing systems integrating computing and information resources that are distributed in the geographical space. The origination of global information systems in the second half of the twentieth century was brought about by the trend of industries and national economies toward integration. These networks comprise telecommunications intended for data and message transmission via telephone cable, wireless and mobile communications (for example, cellular and paging communication), the Integrated Services Digital Network (ISDN) enabling data and voice message transmission through the same communication channels.

The development of Wide-Area Networks passed through several generations. By the end of the 1880s, the implementation of general-purpose information networks was at its beginning. The networks primary purpose consisted in the creation of a fast and reliable personal communications service on the basis of telephone links. Then, cellular and paging (more cost-effective) telecommunication began expanding at a rapid pace, providing additional facilities of mobile communication. Lucent Technologies/Bell Labs had projected that the number of wireless communication customers would surpass that of cable-line customers by 2010. The initial period has revealed international leaders in the domains of computer manufacturing (IBM, Hewlett-Packard, Compaq, Digital Equipment, Dell Computer), telecommunications (AT&T, Deutsche Telecom, France Telecom, Bell Atlantic, British Telecom, SBC Communication, LM Ericsson), software (Microsoft, Sun Microsystems, Apple Computer), and Internet providers, search and navigation engines, and information servers.

The leaders of equipment manufacturing and telecommunication providers are countries with highly developed economies; their business organizations are the principal users of the IT production. By 1996, telecommunication companies have obtained investments of more than 109 billion dollars, which served as the basis for the production of the following generations of global digital networks development (World Telecommunication Indicators,

488   *Global Studies Encyclopedic Dictionary*

1998). At that time, the modern directions of the global IT-industry had been established: applications for electronic document circulation, electronic billing, e-commerce (Internet shops, WEB-stalls), corporate networks of Enterprise Resources Management (ERP), Customer Resources Management (CRM), and Human Resources Management (HRM). Manufacturers of such products are BAAN, SAP, People Soft, and Navision.

In the area of environmental protection, the GRED information network has been established within the framework of the UNEP program. Building and operation of modern distributed information systems is carried out on the basis of international information protocols and standards. The architectures OMG/CORBA, Microsoft/COM, Java/RMI are universal development environments, providing cross-platform adaptability and scalability.

The new WAP generation is oriented to the process of intellectualization of wide-area networks. Such information networks have powerful mathematical platforms, providing for processing and retrieval of not only of the data but also of any kind of knowledge. Development of intellectual information networks has found a platform for the creation of integrated active information space for solving the most important tasks of global society on the basis of the activation and integration of powerful informational and computational resources, distributed in space and time, and the possibility of fast access to these resources and of exchange of information presented in various forms (numerical, textual, graphical, analytical, and cartographic). Such networks exist in the form of distributed information analysis centers, of monitoring systems, of audit and decision support, and of complicated industrial objectives and areas control systems. One of the effective methodological bases for the creation of such systems is the Bayesian approach, ensuring operation of intellectual global networks in conditions of considerable uncertainty in data and knowledge. Developers of intellectual servers and of ADE (application development environment) for such WAPs are Business Object, SAS Institute, SPSS, CINTECH, and other companies.

S. V. Prokopchina

World Dynamics is the approach in system dynamics developed by the American scientist Jay W. Forrester, which constitutes the theoretical foundation for global modeling. Relying upon the theory of information systems with feedback, decision-making research, and computer-based simulation of complex processes, Forrester developed a new type of models—simulation ones—combining the advantages of both mathematical and analog models.

An industrial enterprise, a modern city, and the world as a whole are, from Forrester's point of view, complex systems with essentially non-linear relations between their elements and are not subject to linear descriptions. The analytical apparatus of modern mathematics adapted to studies of linear dependencies in simple systems is not adequate for these systems. Forrester emphasized that processes occurring in complex systems do not admit exact unequivocal description, as they are not strictly determined and stochastic. Elements of such systems affect each other, generating a host of positive and negative feedback loops. The current state of any element at any moment is determined not only by the totality of interrelations of other elements, but also by the history of the system. Parts of complex system are, in turn, the systems of lower complexity. He emphasized that a complex system with its constituents and elements could be characterized by the state of stable equilibrium. In its behavior, a complex system is counter-intuitive. An element of a complex system does not change immediately as a result of some local influence. The element would only change when several influences within a specific time interval add up to a critical level. When the critical level is proximate, even an insignificant pressure upon the system is capable of entailing its radical restructuring.

Basic concepts of simulation modeling are "level" and "rate." The concept of level reflects the interrelation between the discontinuity of values and continuity of their accumulation. The essence of this concept is similar to the concept of "phase coordinates" used in the construction of formalized models of dynamic processes. The concept of rate deals with dynamics of complex systems, characterizing the ratio of the first derivative (more often in the context of time) to the initial function (for example, speed of acceleration). In order to define the time interval required for the change of levels to bring about the change of rates, Forrester introduced the term "delay." He believes that the description of a dynamic system requires the analysis of the rates and levels of interdependence and the construction of a cause-effect system. Simulation methods are not as refined and laconic as mathematical ones, but according to Forrester, they possess a higher heuristic power because a conceptual model is capable of penetration through deep intrinsic levels of a complex system, thus surpassing the research tasks and its developer's intention.

System dynamics is an instrument that allows experts of specific areas to build successful mathematical models of the processes in question and to comprehend better the qualitative behavior of complex systems without special training in the field of management theory and theory of complex systems. Forrester emphasizes that modeling should not be applied as a method of predicting events taking place at a specific moment or as a guarantee of decision correctness. It serves only the purposes of better understanding of the managerial process and as a support in making right decisions, not guaranteeing, however, their unconditional correctness. At the same time, Forrester overestimates modeling as a universal method of cognition, ignoring the specificity of extrapolation, analogy, or experiment and considering these methods as but special cases of modeling.

Social systems, according to Forrester, also belong to the class of complex counterintuitive nonlinear systems with many feedback loops. A human cannot comprehend the way social systems operate or clearly track every possible consequence resulting from incomplete, vague, and inaccurately formulated mental model. Forrester stresses that with civilization going out of control, any attempts of traditional theoretical comprehension and practical solution of essential problems are fruitless. Forrester invokes W. Ross Ashbey's idea of employing a cybernetic amplifier of human thinking abilities in the analysis of these processes.

Forrester (1971) offers a preliminary methodological model of the world. He considers the world as an integral complex system of various interconnected levels (six phase variables): population, industrial assets, agricultural assets, natural resources, environment pollution, and time. Population

490 *Global Studies Encyclopedic Dictionary*

and time, according to Forrester, are the absolute values. Forrester also formulates the basics of the theory of the "limits to growth." By simplifying the real economic situation, Forrester offers not a model of economic development, but a model of extensive growth. Experiments with this model allow Forrester to conclude that extensive growth cannot last forever, as it has "physical limits" and is intrinsically inconsistent. Natural resources cannot be replenished, and the amount of land suitable for cultivation is limited. Land productivity and the saving of raw materials both increase in the arithmetical progression while population, consumption, and pollution follow geometrical progression. According to Forrester, humankind has irreversibly passed the peak of the living standards of the mid-1950s. Provided key economic parameters the same exponential growth of key economic parameters, a crisis in the interrelations of society and nature is inevitable in the middle of the twenty-first century.

Forrester's contribution to scientific knowledge is the creation of the model demonstrating the possibility of translating verbal models into the language of formalized models and obtaining quantitative estimations in areas where only qualitative categories were formerly used. Forrester suggests a convenient and rational method of processing expert estimations decomposing a problem to the required detail level that helps avoid bulky data banks. All the studies of Forrester are devoted to complex poorly-formalizable problems and comprise not only an attempt of their quantitative estimation, but also a creative intensive search of alternative ways of social development within the framework of the interrelation of economic processes with the evolution of general ecological parameters.

N. V. Shulenina

World Economic Forum (WEF): The World Economic Forum (also known as the Davos forum) is an international NGO with headquarters in Geneva, Switzerland. WEF is traditionally held annually. It was created in 1971, in Switzerland as a kind of club for informal discussion among European businesspersons. Over the following years, the geography of participants and the range of matters they discussed expanded significantly. The agenda includes political and economic issues concerning both Europe and other regions, problems of world trade and economic globalization, various approaches to paying off debts of developing nations, and others. WEF declared its principal objective as improvement of the global state of the world through assistance to the world community in solving issues of economic development and social progress. WEF promotes advancement of partnership between business, political, intellectual and other leading figures of the world community for discussion and solution of most crucial problems of global development. Heads of states and governments, the world business leaders, prominent researchers, economists and political scientists are engaged in its work. In addition, WEF holds annual regional and national summits in various countries of the world. The founder and the first President of WEF is K. Schwab. Since 2003, T. Middelhoff, who used to head Bertelsmann, one of Germany's major media concerns, has chaired WEF.

V. N. Kuznetsov

World Federalism is a term used largely in the post World War II period among some idealists active in international relations theory and practice to describe the proposal to organize the various democratic nations of the world into an overall world federation. While retaining national sovereignty in most areas, the nations would pool their sovereignty into a federal power on matters of common defense, security, and foreign policy. Some founders of the United Nations and supporters of its creation shared this perspective, hoping the United Nations would eventually grow into such a body. Others who helped bring the European Union into being had their eventual sights set on a world federation, but felt that the European contintent should first organize its own federal structures on a regional basis. However, despite interest in many countries, the advocacy of world federalism has been largely neglected by leading intellectual thinkers, with the notable exception of Immanuel Kant.

T. C. Daffern

World Future Society is an organization of people dedicated to exploring the future. Since its establishment in 1966, the Society has endeavored to serve as a neutral clearinghouse of ideas on the future. Its stated mission is to enable thinkers, political personalities, scientists, and lay people to share an informed, serious dialogue on what the future will be like. The Society has representative offices in each of the United States and in more than seventy countries; it holds annual congresses and monthly forums for specialists. The Society has its own publishing house and a bookstore specialized in literature about the future; publishes *The Futurist*, a monthly magazine; *Future Research Quarterly*; and a monthly bibliographic bulletin *Futures Survey* containing regular catalogs of literature about the future (in English). The headquarters are located in Bethesda, Maryland.

I. V. Bestuzhev-Lada

World Future Studies Federation is an international NGO uniting specialists in Future Studies. In the beginning it was designed as a federation of all or most of the national and international academic societies of futurologists that appeared in the 1960s. This issue was discussed by the international future studies conferences in Oslo (1967), Kyoto (1970), and Bucharest (1972), while meetings of the leading futurologists in Paris and Moscow preceded the former. Nevertheless, the idea to unite the existing associations into a federation was not realized. A new association was created at the Paris conference (1973). The federation's main form of functioning is annual conferences held in various university centers of the world to discuss the practical and theoretical problems of future studies at their plenary sessions and panels. Also regional seminars are organized bringing together about ten participants to discuss one specific problem. The federation publishes a quarterly Bulletin. By 2002, it numbered more than 500 members. The headquarters are located where the currently elected President or Secretary General works.

I. V. Bestuzhev-Lada

World Heritage: World heritage refers to cultural heritage common to all peoples of the Earth. While simple enough to state, actually designating real

492 *Global Studies Encyclopedic Dictionary*

objects that embody such a concept is far more difficult. As such, it is closely linked to the project of global ethics as it strives to identify values that are common to all the people in the world. Accordingly, it contains a remarkable amount of diversity and dissent, and, as such, remains controversial.

The significance of values that are presupposed in the attempts to judge the value of cultural objects are partly descriptive, partly normative, and always interpretive. The concept is descriptive in a sense that an object is deemed to have an important role to play in all sorts of histories, and this could be hardly denied, for example in the case of Rome. Furthermore, denying that Auschwitz is a place where a mass of people was systematically murdered would be denying a plain fact. The concept is normative in the sense that the interpretation of our physical environment is guided by values that are considered worthy of our moral or aesthetic approval, or sometimes they serve as reminders of the tragedies caused by the negligence of these values. Most significantly, however, the concept is interpretive because it judges sites in terms of their presumed role in world heritage, and thus in world history.

World Heritage can be understood as a complex net of interactions and, to some extent, contradictions between the ideals of truth, beauty, and goodness. Clearly the historical truth of objects deemed to represent world heritage is selective, but it cannot be accused of being straightforwardly false either.

The understanding that emerges from our attempts to make sense of the traces of our past is inevitably both "scientific explanation" and "humanistic understanding." Human history cannot fully be understood except by reference to the meanings, values, and beliefs of historical agents and their relationships, but these focal points of understanding must not be separated from their natural contexts. The debate on cultural merits is a rational debate calling for the exercise of judgment.

In addition, it would be an error to dismiss all ideas about empathy playing an important role in understanding cultural heritage. To really appreciate cultural heritage demands a sophisticated way of understanding its creators as a way to reach an insight into past cultures. The puzzle created by the information concerning the artifacts, values, and members of a culture is resolved after adapting a perspective based on a laborious process of using all the relevant information.

P. Elo, A. Haapanen, M. Kabata, H. Lämsä, J. Savolainen

World Order Models Project is an international project and institution, acting in an effort to realize the challenges of humankind and to solve today's global problems, seen in the light of social and cultural preferences.

The World Order Models project included wide research connected with various images of future. Founded in 1968, with the support of the World Order Institute and United States, led by American scientist, Saul H. Mendlovitz, the project was expected to consider different variants of "preferred worlds" or global societies related to the last decade of twentieth century on the basis of their apparent desirability.

The World Order Models project was chiefly aimed at studying political events, social processes and cultured life forms. Attention was mainly paid to political unsteadiness, social injustice, and national disconnection, considered as the principal catalysts for regional conflicts and global crisis of humankind.

Some changes in the World Order Models have taken place concerning research aims. At the start of the activity, attention was focused on social and political problems of a global kind. Nevertheless, ecological and cultural problems were given attention later on as well from the point of view of local and global interconnections. As research was starting, connected with the forming and consideration of different images of "preferred worlds," the scientists involved in the project made their way from abstract pondering over the just world order and government of the world to devising some specific programs concerning the transformation of existing systems, perfecting government institutions, cultural ethnic changes, and value preferences.

The World Order Models Project publishes the *World Policy Journal*. In addition, published research, made according to the World Order Models project, attracted the attention of scientists and many other people, due to the images of the future covered there from the point of view of human life values. World Order Models developers are inclined to research the kinds of future for humanity, which are rid of political conflicts and global skirmishes, economic inequality and social injustice, racial hatred and national irreconcilability. The project developers paid increasing attention to the analysis of the situation in developing countries, specifics of religious movements, and the originality of world cultures. Their vision of the future took its place next to the discussion of concrete political strategies, and that is what brought research on global problems to a practical trend. Measures to provide peace and safety were suggested and discussed (Galtung, 1980). Mendlovitz considered ways and possibilities of reaching global stability. Works connected with saving natural resources and substantiation of human ethical attitude toward the nature (Falk, 1975), blaming the neo-colonialist development strategies and their westward orientation and providing measures to create a federation of developed and developing states (Kothari, 1974) were brought to priority and the programs of economic and social welfare.

The World Order Models project's influence upon the public opinion found itself quite modest in comparison to that of the Club of Rome, which attracted noticeable attention of researchers, public figures, and people in business. Nevertheless, the World Order Models project's publications and ideas attracted attention of those scientists who devoted their efforts to realizing the human future. Models of the world order, world government, and world culture and their images of the "most preferred world" may be considered as global studies, though they were never admired by the scientists who had a critical attitude toward the ideas of creating world government. These models and images also stimulated people who conducted research on global problems and the analysis of alternative ways of human development.

References: Kothari, R. *Footsteps into the Future*. 1974. Falk, R. A. *A Study of Future Worlds*. 1975. Mazrui, A. *A World Federation of Cultures*. 1976. Lagos Matus, G., and H. H. Godoy. *Revolution of Being*. 1976. Galtung, J. *The True Worlds*. 1980. *All New York: Free Press.

V. M. Leibin

494 *Global Studies Encyclopedic Dictionary*

World Political Forum (WPF) was created and established by Mikhail Gorbachev to foster contacts between renowned politicians, scientists, priests, writers and artists who seek to develop holistic strategic approaches to modern global problems. The founding conference was held in May 2003, in Turin, Italy. The mission of the forum is to encourage a constructive, unbiased dialog on modern world political problems among renowned and competent politicians, representing a wide multiplicity of confessions, nationalities and cultures. The forum focuses on the methods and techniques for the control of the processes associated with the globalization, an increase of the efficiency of international institutions, and a search for solutions to world political dilemmas. The headquarters of the Permanent Secretariat is in Italy (Turin and Alessandria). The forum holds annual General Assemblies and regional and local seminars.

A. S. Grachev

World Trade Organization (WTO) is an international organization with the headquarters in Geneva, Switzerland, established as a result of the Uruguayan round of negotiations in the context of the General Agreement on Tariffs and Trade (GATT); it has operated since 1 January 1995.

WTO is intended to liberalize international trade, promote economic growth, and improve the economic well being in the world. Among its tasks and functions are: (1) Control of the fulfillment of trade agreements if they were concluded between the members of the WTO and if they concern tariff reduction and elimination of other obstacles for international trade and anti-discrimination in the business field. (2) Organization of trade negotiations of the WTO members; monitoring of the WTO members' trade policy; collaboration with other international organizations. (3) Admission of new members; settlement of trade disputes among the members of the organization.

Unlike GATT, WTO is a permanent body possessing larger powers and instruments of influence enabling it to resolve international trade disputes and to use sanctions. Member countries of the WTO interact in the context of a non-discriminatory trade system where each country has the guarantees of an equitable and consistent attitude toward its export in the markets of the other countries, undertaking to provide identical conditions of import in its own market.

The non-discrimination principle is carried out by way of the mechanism of the most favored nation (MFN), when a country guarantees equal terms of trade for all member countries of WTO and the national regime when the imported goods are not discriminated against in the domestic market.

In addition to an application of the national regime and MFN, access to the market is guaranteed by the abolition of quantitative restrictions on imports in favor of customs tariffs, which are the more effective means of turnover regulation, and by the publicity and openness of the trade regimes of member countries.

In accordance with the decisions of the Uruguayan round, the rates of manufactured goods should be reduced by one third; also, subsidies and quota on the import of agricultural products, automobiles, and textiles (GATT regulations are not applied to them) should be reduced. It was also intended to widen the freedom of action in banking and other areas of services, and to take more effective measures to protect intellectual property on a global scale. The greatest disputes were touched off by questions concerning the abolition of subsidies and protective measures regarding agriculture. After the meeting in Seattle (1999) the WTO leaders reconsidered the structure of the organization.

At present, 144 countries comprise WTO, headed by a Director-General. The executive organ of WTO is the Ministerial Conference. It is convened at least once biannually, as a rule, at the level of Trade and Foreign Ministers of the WTO member countries. The Conference makes decisions concerning matters of principle. The organization's routine work is managed by the General Council consisting of ambassadors or other representatives of the member countries of WTO. Among the duties of General Council are the settlement of the trade disputes among the members of WTO and conducting periodical surveys of their trade policy.

The Goods Council supervises implementation of the General Agreement on Trade and Tariffs, the Services Council supervises implementation of the General Agreement on Trade in Services, and the Intellectual Property Council supervises implementation of the Agreement on trade aspects of intellectual property rights. These councils all report to the General Council.

Within the framework of WTO, many different working groups, commissions of experts, and specialized committees have been established. They deal with issues concerning the fulfillment of individual parts of multilateral agreements and some other questions of interest to member countries of the WTO (for example, relating to competition, investment, regional trade agreements, and trade aspects of environmental protection).

Legally, WTO is based on the General Tariff and Trade Agreement (GATT) according to the edition of 1994, the General Agreement on Trade in Services (GATS), and the Agreement on Trade Aspects of Intellectual Property Rights (TRIPS). The latter is a body of rules concerning the trade and investment in ideas and creativity; it formulates how intellectual property should be protected if trade is involved. The conception of the intellectual property includes copyrights and trademarks, and geographical names used to identify products, industrial designs, integrated circuit layout-designs, and undisclosed information; for example trade secrets.

Member countries of the WTO undertake obligations within the framework of GATS where they declare which of their business sectors and to what degree are ready to be open for the foreign competition.

There is also the so-called Dispute Settlement Agreement that provides for the making of a system, which can help countries settle their differences through consultations or by means of the adherence to a procedure providing the possibility of a ruling by a panel of experts and the right of appeal against accepted decisions.

The agreements of WTO were ratified by the parliaments of all the member countries. Within the framework of WTO decisions are made on the basis of consensus, although by the terms of its statute it is possible to carry out a voting procedure. Within the framework of WTO negotiations are carried on behind closed doors, causing criticism from the organization's opponents who consider this practice undemocratic.

A. L. Demchuk

496 *Global Studies Encyclopedic Dictionary*

World Wars: Wars referred to as "world wars" are those that have been waged in the twentieth century between the major world powers for redistribution of the world, caused by extreme escalation of conflicts in the struggle for spheres of influence, raw material sources, redistribution of colonies, outlets for investments, and enslavement of other peoples. The world wars, due to their scale (theater of operations, number of participating countries, total number of troops of both parties, their populations, volume of employed resources, and military and technical facilities, including the most advanced ones; application of all means of struggle, and scale of casualties and destruction) had exerted tremendous influence on the entire world destiny. There were two world wars in the twentieth century: World War I (1914–1918) and the World War II (1939–1945).

V. M. Smolkin

०ॐ X ৪०

Xenogamy, Cultural and Political: (*xeno*- "foreign," "strange" and -*gamy* "marriage") Xenogamy is a biological term that refers to fertilization of a flower by pollen from a flower on a genetically different plant. The common terminology for this process is cross-pollination. Borrowing this concept, in global studies, the term "cultural xenogamy" refers to cross pollination of cultures, meaning that cultures can be enriched by integrating elements of other cultures into them.

Another term for this process is "intercultural synergistics." This process obviously dates to the earliest contact of one culture with another, but has been greatly accelerated with every advance in communication and travel technologies. Students of this process trace the development of how cultural items have come into being, highlighting the different cultures that contributed to the final product, and suggesting improvements to existing technology derived through a careful study of cultural artifacts or ideas.

Such cross polinization can also occur between and among fields of study, with the process enriching both sides of the contact, applying principles of brain functioning to computer models being a classic example. However, care must be taken not to draw conclusions from over generalization or drawing comparisons that are less than precise (where the metaphor fails), in which case confusion or detrimental results may entail.

E. D. Boepple

Xenophobia (*xeno*- "foreign," "strange" and -*phobia*" fear") Xenophobia is an irrational or unreasoned fear of that which is perceived to be foreign or strange. Racism is a common form of xenophobia. For example, in 2005, United Nations Special Rapporteur on human rights Doudou Diène concluded after a nine-day investigation in Japan that the authorities were not doing enough to tackle what he called Japan's "deep and profound racism" and xenophobia, particularly against its former colonial subjects (Debito, 2008). "The phenomenon of increasingly noisy demonstrations against Koreans residing in Japan by members of right-wing groups—after festering for several years mostly on alternative news sites on the Internet—is finally attracting the notice of the mainstream print media, and Diet members have even begun to discuss the possibility of drafting a new law banning inflammatory 'hate speech'" (Schreiber, 2013).

Political xenophobia refers to the willingness to use public policy to discriminate against foreigners. An example of such from the early days of the United States can be found in the 1787 convention held to revise the Articles of Confederation. Then, the constitution read: "Representatives and direct Taxes shall be apportioned among the several States which may be included within this Union, according to their respective Numbers, which shall be determined by adding to the whole Number of free Persons, including those bound to Service for a Term of Years, and excluding Indians not taxed, three fifths of all other Persons (art.1, sec. 2, par. 3). This essentially constituted state-sponsored discrimination, in that it "(dis)counted" blacks at only 3/5 for

498 *Global Studies Encyclopedic Dictionary*

purposes of representation (and, of course, withheld suffrage from all blacks and women).

Even a high level of xenophobia may not produce political repression if the foreigners are considered to be few in number, harmless, or easily controlled. But even a small number may result in such repression if seen as a threat to culture or competition for scare social resources (jobs) (Watts, 1996). Since the 1990s, xenophobic, deeply conservative, and extreme right-wing parties have arisen across much of Europe (Mayfield, 2013). Xenophobia has been suggested as a factor that threatens full support for the European Union (De Master and Le Roy, 2000).

References: Watts, M. W. "Political Xenophobia in the Transition from Socialism," *Political Psychology* 17:1 (1996), pp. 97–126. Debito, A. "Japan's Entrenched Discrimination toward Foreigners," *The Asahi Shimbun*, 5 October 2008. Accessed 9 January 2014. http://www.japanfocus.org/-The_Asahi_Shimbun_Culture_Research_Center-/2932. De Master, S., and M. K. LeRoy. "Xeonophobia and the European Union," *Comparative Politics* 32:4 (July 2000), pp. 419–436. Mayfield, J. "Explaining the Rapid Rise of the Xenophobic Right in Contemporary Europe," *Geocurrents*. Accessed 9 January 2014. http://www.geocurrents.info/cultural-geography/the-rapid-rise-of-the-xenophobic-right-in-contemporary-europe. Schreiber, M. "Xenophobia Stretches from the Street to the Dinner Table," *The Japan Times*, 26 May 2013. Accessed 9 January 2014. http://www.japantimes.co.jp/news/2013/05/26/national/xenophobia-stretches-from-the-street-to-the-dinner-table/#.Us-itZ7iZNs.

E. D. Boepple

ଓ Y ଓ

Youth: With some culture-specific variance, "youth" generally designates a biosocial stage between childhood and adulthood. These individuals are generally experiencing physical adolescence and are have progressed past childhood cognitive and emotional developmental stages but have not yet reached intellectual or emotional maturity. Youth are the product of socialization occurring in all the spheres of the society. During this stage, individuals gradually develop productive activity and progress from social and civil dependency to the status of adult, active citizens and formed individuals. Membership of social group is constantly in the making. Each year a new cohort of adolescents (teenagers) joins the group as a new group of adults leave. According to UN data, youth make up approximately one-third of the global population.

Youths of economically developed countries participate in globalization processes to differing degrees, depending on several factors, including the class, social-group, ethnical affiliation of the youths, the system and content of education and parenting, and dominant standards and cultural rules.

Young members of the middle class, workers, farmers, and students have played an active role in the broad anti-globalist movement that developed in the late twentieth century, mostly in the Western countries,. In the process of this movement's evolution, a project of alter-globalism was created. It suggests that we concentrate on searching for and implementing alternative forms of globalization, allowing nations and people to achieve truly humane living conditions in harmony with nature.

V. V. Pavlovsky

ങ Z ഇ

Z-Generation (Gen Z; Net-Generation) is the title for the generation born between the 1990s and early 2000s, between the end of Cold War and the beginning of the Global financial crisis. The term is mainly used in sociology and philosophy of media. Z-Generation is compared with X-Generation (those born between 1965/69 and 1975/80) and Y-Generation (those born between 1975/80 and 1991). Z-Generation has grown up with the Internet and new modes of social communication (mobile, net, digital TV, etc.). This is the first generation of globalization, which largely determined its features. It is characterized by a high degree of mobility, universal values, and virtualization. Z-Generation prefers to build links by principles of network rather than hierarchy. Its employment preferences are related to information technologies, economics, and the humanities.

I. Ilin, T. Shestova

Zinovyev, Aleksandr Aleksandrovich: (1922–2006) was a prominent Russian philosopher who helped create the theory of global supra-society, and critic of the pro-Western model of globalization and global capitalism. Zinovyev was a co-founder of Moscow Logic Circle (1953–1958), which contributed much to overcoming the dogmatism of the official concept of Marxist logic after Stalin's death.

After *Yawning Heights* was released in the West (1976), Zinovyev was persecuted by official authorities and exiled from the USSR in 1978. He settled in Munich, where he lived until 1998. While in exile, he published *The Reality of Communism* (1980), *The Yellow House* (1980), *Homo Soveticus* (1982), and other works, in which he criticized anti-democratic manifestations of socialist society. Zinovyev often emphasized that he had never been anti-communist; all his criticism was directed against distortions of communist ideals by political regimes in "countries of victorious socialism."

In the 1990s, Zinovyev spoke out against the break-up of the USSR and the crash of soviet power. At that time, he started to most radically condemn the West and global expansion of Western values. The most famous work of the last decade of his life was the anti-utopian *The Global Human Hill* (1997), in which he depicted the society of "victorious Westernism," the future society of global "one party rule" following the triumph of bourgeois values. The new god of the global human hill is a computer, and human relations are substituted by the system of non-human communication through information technologies.

Zinovyev returned to Russia in 1998, and was awarded numerous awards, including the "Star of Moscow University" (2005) and the medal for "65 Years of Defeat of Fascist Army under Moscow" (2006).

I. Ilin

CONTRIBUTORS

ADROV, VALERI M., PhD, Assistant Professor, Astrakhan, Russia

AGADZHANYHAN, NIKOLAI A., MD, Professor, Moscow, Russia

AGEEV, ALEXANDER I., PhD, Doctor of Economic Science, Professor, Moscow, Russia

AHARONI, ADA, PhD, Professor and President of The International Forum for the Literature and Culture of Peace (IFLAC), Nesher, Israel

AITMATOV, CHINGIZ T,* Writer, Member of the National Academy of Sciences of the Kyrgyz Republic, Bishkek, Kyrgyzstan

AKOPYAN, KAREN Z., PhD (Philosophy), Doctor of Philosophical Science, Moscow, Russia

ALEKSEEVA, TATIANA A., PhD, Doctor of Philosophical Science, Professor, Moscow, Russia

ALESHKOVSKY, IVAN A., PhD (Economics), Assistant Professor, Moscow, Russia

APSITE, LUDMILA, PhD (Philosophy), Professor, Riga, Latvia

ASHIN, GENNADI K.,* PhD, Doctor of Philosophical Science, Professor, Moscow, Russia

ASTAFIEVA, OLGA N., PhD, Doctor of Philosophical Science, Assistant Professor, Moscow, Russia

ATTA, ZEBADA, PhD (Philosophy), Professor, Cairo, Egypt

AYATOLLAHY, HAMIDREZA, PhD (Philosophy), Professor, Tehran, Iran

BABAEV, MIKHAIL M., PhD (Law), Doctor of Juridical Science, Professor, Moscow, Russia

BAICHUN, ZAING, PhD (Philosophy), Professor, Beijing, China

BAITALA, VASILI D., Senior Research Fellow, Kiev, Ukraine

BAKSANSKY, OLEG YE, PhD, Doctor of Philosophical Science, Professor, Moscow, Russia

BANSE, GERHARD, PhD, Professor, Institute for Technology Assessment and Systems Analysis (ITAS), Karlsruhe, Germany

BARANOVSKY, SERGEI I., PhD, Doctor of Technological Science, Professor, Moscow, Russia

BARISHPOLETS, VITALYI A., Doctor of Technological Science, Professor, Moscow, Russia

BARLYBAYEV, KHALIL A., PhD (Economics), Professor, Moscow, Russia

BASAVE, AGUSTIN, PhD, JD, Professor, Monterrey, Mexico

BECHMANN, GOTTHARD, JD, Institute for Technology Assessment and Systems Analysis (ITAS), Karlsruhe, Germany

BELIKOV, ALEXANDER A, Engineer, Moscow, Russia

BESTUZHEV-LADA, IGOR V., PhD, Doctor of Historical Science, Professor, Moscow, Russia

BILALOV, MUSTAFA I., PhD, Doctor of Philosophical Science, Makhachkala, Russia

BOEPPLE, ELIZABETH D., PhD, Psychologist, Academic Editor, Gilbert, Arizona, US

BOERSEMA, DAVID, PhD (Philosophy), Professor, Pacific University, Forest Grove, Oregon, US

BOUCHILO, NINA F., PhD (Philosophy), Moscow, Russia

BRENIFIER, OSCAR, PhD (Philosophy), Paris, France

BRITKOV, VLADIMIR B., PhD (Physics and Mathematics), Moscow, Russia

BUSCH, WERNER, PhD, President of Association Internationale des Professeurs de Philosophie, Melsdorf, Germany

BUZGALIN, ALEXANDER V., PhD, Doctor of Economic Science, Professor, Moscow, Russia

502 *GLOBAL STUDIES ENCYCLOPEDIC DICTIONARY*

CADY, DUANE L., PhD (Philosophy), Professor Emeritus, Hamline University, St. Paul, Minnesota, US

CALOGERO, FRANCESCO, Laurea in Physics, Professor of Theoretical Physics, University of Rome, Chair, Pugwash Council, Rome, Italy

CANNON, MATTHEW, PhD, Irish Peace Institute, University of Limerick, Limerick, Ireland

CARPENTIER, NICO, PhD (Communication Studies), Professor, Free University of Brussels, Brussels, Belgium

CHERNYAEV, ANATILYI S., PhD (History), Assistant Professor, Moscow, Russia

CHETVERIKOV, ARTEM O., PhD (Law), Assistant Professor, Moscow, Russia

Chizhov, Alexei Ya., MD, Professor, Moscow, Russia

CHUBAIS, IGOR B., PhD, Doctor of Philosophical Science, Professor, Moscow, Russia

CHUMAKOV, ALEXANDER N., PhD, Doctor of Philosophical Science, Professor, Moscow, Russia

CHURCHILL, ROBERT PAUL, PhD (Philosophy), Professor, George Washington University, Washington, DC, US

COHEN, ROBERT S., PhD (Physics), Professor Emeritus, Boston University, Boston, Massachusetts, US

COOKE, WILLIAM, PhD, Senior Lecturer, School of Visual Arts, University of Auckland, Manukau, New Zealand

COX, SEAN M., PhD, Professor, Istanbul, Turkey

DAFFERN, THOMAS C., BA, DSc, PGCE, Director of the International Institute of Peace Studies and Global Philosophy, Powys, UK

DANILOV-DANILYAN, Viktor I., PhD, Doctor of Economic Science, Professor, Moscow, Russia

DASHKOV, GENNADYI V., JD, Doctor of Juridical Science, Professor, Moscow, Russia

DELYAGIN, MIKHAIL G., PhD, Doctor of Economic Science, Moscow, Russia

DEMCHUK, ARTUR L., PhD (Political Science), Assistant Professor, Moscow, Russia

DENISOVA, LUBOV V., PhD, Doctor of Philosophical Science, Professor, Omsk, Russia

DIURCHIK, VLADIMIR, PhD (Philosophy), Assistant Professor, Bratislava, Slovakia

DMITRIEV, ANATOLI V., PhD, Doctor of Philosophical Science, Professor, Moscow, Russia

DOBREN'KOV, VLADIMIR I., PhD, Doctor of Philosophical Science, Professor, Moscow, Russia

DRACH, GENNADI V., PhD, Doctor of Philosophical Science, Professor, Rostov-on-Don, Russia

DREYER, LUISE, PhD, Editor of *Europa Forum Philosophie*, Minden, Germany

DROBOT, GALINA A., PhD, Doctor of Philosophical Science, Professor, Moscow, Russia

DROBOT, VIKTOR N., PhD (Philosophy), Assistant Professor, Moscow, Russia

DROZDOV, NIKOLAI N., PhD, Doctor of Biological Science, Professor, Moscow, Russia

DYOGTEVA, OLGA V., Ecologist, Moscow, Russia

EGOROV, VALERI K., PhD (Philosophy), Moscow, Russia

EHRHART, HANS-GEORG, PhD, Senior Research Fellow, Institute for Peace Research and Security Policy, University of Hamburg, Hamburg, Germany

ELO, PEKKA,* Finnish National Board of Education, President of Finnish Humanist Union, Editor of *Finish Humanist Journal*, Helsinki, Finland

ENCHTUYA, SANDAG, PhD (Philosophy), Lecturer, Ulan-Bator, Mongolia

FINLEY, HERMAN, BA, MA (Information Technology), Associate Professor, Asia Pacific Center, Honolulu, Hawaii, US

FOURSE, VLADIMIR N., PhD (Philosophy), Assistant Professor, Minsk, Belarus

Contributors

FRIEDMAN, HOWARD, PhD (Philosophy), Professor Emeritus, University of Connecticut, Thomaston, Connecticut, US

FROLOV, VIKTOR V., PhD, Doctor of Philosophical Science, Professor, Mytischchi, Russia

GALICHIN, VIKTOR A., PhD (Philosophy), Assistant Professor, Moscow, Russia

GANZHA, ALEXANDER G., (History), Research Fellow, Moscow, Russia

GAPCHUKOV, VLADIMIR V., Engineer, Moscow, Russia

GAVRILOV, IGOR T., PhD, Doctor of Geographical Science, Moscow, Russia

GAY, WILLIAM C., PhD, Professor Emeritus, University of North Carolina at Charlotte, Charlotte, North Carolina, US

GELOVANI, VIKTOR A., PhD, Doctor of Technological Science, Professor, Moscow, Russia

GINZBURG, VITALI L., PhD, Doctor of Physics and Mathematical Science, Professor, Moscow, Russia

GIRUSOV, EDUARD V., PhD, Doctor of Philosophical Science, Professor, Moscow, Russia

GIVISHVILI, GIVI V., PhD, Doctor of Physics and Mathematical Science, Moscow, Russia

GLUKHOVA, ALEXANDRA V., PhD, Doctor of Political Science, Professor, Voronezh, Russia

GORBACHEV, MIKHAIL, Former President of the USSR.

GOROKHOV, VITALI G., PhD, Professor, Institute for Technology Assessment and Systems Analysis (ITAS), Senior Scientific Fellow, Institute of Philosophy, Moscow, Russia

GORSHKOV, MIKHAIL K., PhD, Sociology, Doctor of Philosophical Science, Professor, Moscow, Russia

GORTON, MATTHEW, PhD (Marketing), Newcastle University Business School, Newcastle-upon-Tyne, UK

GRACHEV, ANDREI S., Engineer, Moscow, Russia

GRAHAM, JORGE E., PhD, Mexico City, Mexico

GRININ, LEONID E., PhD, Doctor of Philosophical Science, Volgograd, Russia

GRUNWALD, ARMIN, PhD, Professor, Institute for Technology Assessment and Systems Analysis (ITAS), Karlsruhe, Germany

GU, SU, PhD (Philosophy), Professor, Nanjing, China

GUDSKOV, NIKOLAI L., PhD (Biology), Moscow, Russia

GUSEV, MIKHAIL V., PhD, Doctor of Biological Science, Professor, Moscow, Russia

GUSEYNOV, ABDUSALAM A., PhD, Doctor of Philosophical Science, Professor, Moscow, Russia

GUSOV, KANTEMIR N., JD, Doctor of Juridical Science, Professor, Moscow, Russia

HAAPANEN, ARI, Philosophical Education Coordinator, University of Vaasa, Helsinki, Finland

HIRSCHBEIN, RON, PhD (Philosophy), Professor Emeritus, California State University at Chico, Chico, California, US

HÖCHSMANN, HYUN, PhD, Professor, New Jersey City University, Jersey City, New Jersey, US

HÖSLE, VITTORIO, PhD, Professor and Director of Notre Dame Institute for Advanced Study, University of Notre Dame, Notre Dame, Indiana, US

HOWARD, MICHAEL T., MA (Philosophy), University of North Carolina at Charlotte, Charlotte, North Carolina, US

HUSSAINI (AKHLAQ), SAYYED HASSAN, PhD, Director of *Raieheye Azadi Magazine*, Kabul, Afghanistan

504 *GLOBAL STUDIES ENCYCLOPEDIC DICTIONARY*

ICHIKAWA, HIROSHI, PhD (Technology), Associate Professor, University of Hiroshima, Hiroshima, Japan

ILIINSKII, IGOR M., PhD, Doctor of Philosophical Science, Professor, Moscow, Russia

ILIN, ILYA, PhD, Doctor of Philosophical Science, Professor, Moscow, Russia

IONTSEV, VLADIMIR A., PhD, Doctor of Economic Science, Professor, Moscow, Russia

IOSELIANI, AZA D., PhD, Doctor of Philosophical Science, Assistant Professor, Moscow, Russia

IRKHIN, YURI V., PhD, Doctor of Philosophical Science, Professor, Moscow, Russia

ISHKHANOV, ALEXANDER V., PhD, Doctor of Economic Science, Professor, Krasnodar, Russia

ISHKOV, ALEXANDER G., PhD, Doctor of Chemical Science, Professor, Moscow, Russia

ISRAEL, YURI A.,* PhD, Doctor of Technological Science, Professor, Moscow, Russia

IVAKHNYUK, IRINA, PhD, Doctor of Economic Science, Professor, Moscow, Russia

JASPARRO, CHRIS, PhD, Associate Professor, US Naval War College, Newport, Rhode Island, US

JOHNSON-FREESE, JOAN, PhD, Professor, Naval War College, Newport, Rhode Island, US

KABATA, MIIKA, Philosophical Education Coordinator, University of Helsinki, Helsinki, Finland

KABO, VLADIMIR R.,* PhD (History), Canberra, Australia

KALACHEV, BORIS F., JD, Senior Research Fellow, Moscow, Russia

KALINICHENKO, PAUL A., JD, Doctor of Juridical Science, Moscow, Russia

KAMUSELLA, TOMASZ, PhD (Political Science), Formerly taught at Language Teachers' Training College, University of Opole, Opole, Poland, Currently Professor, Centre for Transnational History, University of St. Andrews, Fife, Scotland

KANZIAN, CHRISTIAN, PhD, Assistant Professor, Institute for Christian Philosophy, University of Innsbruck, President of the Austrian Wittgenstein Society, Inssbruck, Austria

KAPITZA, SERGEI P.,* PhD, Doctor of Physics and Mathematical Science, Professor, Moscow, Russia

KAPLAN, LAURA DUHAN, Rabbi, PhD (Philosophy), Professor Emerita, University of North Carolina at Charlotte, Charlotte, North Carolina, US. Now resides in Vancouver, British Columbia, Canada

KAPUR, JAGDISH C, Professor of Information Technology, Indian Institute of Public Administration, New Delhi, India

KASHKIN, SERGEI YU., JD, Doctor of Juridical Science, Professor, Moscow, Russia

KATSURA, ALEXANDER V., PhD (Philosophy), Moscow, Russia

KAZYUTINSKY, VADIM V.,* PhD, Doctor of Philosophical Science, Professor, Moscow, Russia

KELBESSA, WORKINEH, PhD (Philosophy), Associate Professor, Addis Ababa, Ethiopia

KELLE, VLADISLAV J., PhD, Doctor of Philosophical Science, Professor, Moscow, Russia

KELLER, DAVID R., PhD (Philosophy), Director of Center for the Study of Ethics, Utah Valley University, Orem, Utah, US

KELLY, THOMAS A. F.,* PhD, Professor, National University of Ireland, Maynooth, Ireland

KHALILOV, SALAKHADDIN, PhD, Doctor of Philosophical Science, Professor, Baku, Azerbaijan

KHARABET, KONSTANTIN V., JD, Assistant Professor, Moscow, Russia

KHASBULATOV, RUSLAN I., PhD, Doctor of Economic Science, Professor, Moscow, Russia

KHATAMI, SEYYED MOHAMMED, BA, Minister of Culture (1982–1992), President of the Islamic Republic of Iran (1997–2005), Tehran, Iran

KHRIBAR, SERGEI F., Sociologist, Moscow, Russia

KHUBLARYAN, MARTIN G., PhD, Doctor of Technological Science, Professor, Moscow, Russia

KIRABAEV, NOUR S., PhD, Doctor of Philosophical Science, Professor, Moscow, Russia

KISELYOV, SERGEI G., PhD, Doctor of Philosophical Science, Moscow, Russia

KISS, ENDRE, PhD (Philosophy), Doctor of Science, Professor, University Eötvös, Budapest, Hungary

KLADKOV, ALEXANDER V., JD, Assistant Professor, Moscow, Russia

KLIGE, RUDOLF K., PhD, Doctor of Geographical Science, Professor, Kursk, Russia

KLUCHAREV, GRIGORI A., PhD, Doctor of Philosophical Science, Professor, Moscow, Russia

KNUTSSON, LOTTA, President, Swedish Association of Teachers of Philosophy and Psychology, Stockholm, Sweden

KOLLONTAY, VLADIMIR M., PhD, Doctor of Economic Science, Moscow, Russia

KOROLEV, ANDREI D., PhD (Philosophy), Moscow, Russia

KOSICHENKO, ANATOLY G., PhD, Doctor of Philosophical Science, Professor, Alma-Ata, Kazakhstan

KOSTIN, ANATOLI I., PhD, Doctor of Philosophical Science, Professor, Moscow, Russia

KOZHIN, PAVEL, PhD, Doctor of Historical Science, Senior Research Fellow, Moscow, Russia

KRASILNIKOV, PAVEL V., PhD (Biology), Petrozavodsk, Russia

KRAVCHENKO, IGOR I., PhD, Doctor of Philosophical Science, Moscow, Russia

KRIVOLUTSKY, DMITRI A., PhD, Doctor of Biological Science, Professor, Moscow, Russia

KRUSHANOV, ALEXANDER A., PhD, Doctor of Philosophical Science, Professor, Moscow, Russia

KUÇURADI, IOANNA, PhD, Professor and President, Fédération Internationale des Sociétiés de Philosophie (FISP), Maltepe University, Istanbul, Turkey

KUDASHOV, VYACHESLAV I., PhD, Doctor of Philosophical Science, Professor, Krasnoyarsk, Russia

KUDRIN, BORIS A., PhD, Doctor of Technological Science, Professor, Tula, Russia

KULI-ZADE, ZUMRUD A. K., PhD (Philosophy), Baku, Azerbaijan

KUNKEL, JOSEPH C., PhD (Philosophy), Professor Emeritus, University of Dayton, Dayton, Ohio, US

KUPTSOV, VLADIMIR I., PhD, Doctor of Philosophical Science, Professor, Moscow, Russia

KURBANOV, RASHAD A., JD, Doctor of Juridical Science, Professor, Moscow, Russia

KUST, GERMAN S., PhD, Doctor of Biological Science, Professor, Moscow, Russia

KUTAFIN, OLEG E.,* JD, Doctor of Juridical Science, Professor, Moscow, Russia

KUVAKIN, VALERI A., PhD, Doctor of Philosophical Science, Professor, Moscow, Russia

KUVALDIN, VIKTOR B., PhD, Doctor of Historical Science, Professor, Moscow, Russia

KUZNETSOV, MIKHAIL M., PhD, Moscow, Russia

KUZNETSOV, VALERYI N., PhD, Doctor of Philosophical Science, Professor, Moscow, Russia

KVARATSKHELIA, VAKHTANG A., PhD (Philosophy), Zugdidi, Georgia

LÄMSÄ, HANNA, Executive Director, Association of Cultural Heritage, Helsinki, Finland

LAYUG, ALLAN S., Professor, Philippine Institute for Development Studies, Makati City, Philippines

506   *GLOBAL STUDIES ENCYCLOPEDIC DICTIONARY*

LEBEDEVA, MARINA M., PhD, Psychology, Doctor of Political Science, Professor, Moscow, Russia

LEIBIN, VALERI M., PhD, Doctor of Philosophical Science, Professor, Moscow, Russia

LENK, HANS, PhD, Professor and Vice-President, Fédération Internationale des Sociétiés de Philosophie (FISP), Karlsruhe, Germany

LESHKEVICH, TATIANA G., PhD, Doctor of Philosophical Science, Professor, Rostov-on-Don, Russia

LODEWYCKX, HERMAN, MA (Philosophy), Philosopher, Province of West Flanders (Bruges Area), Oostende, Belgium

LOGINA, NATALIA V., PhD (Philosophy), Assistant Professor, Moscow, Russia

LOPATIN, VLADIMIR N., PhD, Doctor of Technological Science, Professor and Deputy Minister, Moscow, Russia

LOS, VIKTOR A., PhD, Doctor of Philosophical Science, Professor, Moscow, Russia

LOSEV, KIM S., PhD, Doctor of Geographical Science, Professor, Moscow, Russia

LOSKUTOV, VLADISLAV K., PhD, Doctor of Technological Science, Moscow, Russia

LÜBBE, HERMANN, PhD, Professor, University of Zurich, Zurich, Switzerland

LUZHKOV, YURI M., Former Mayor of Moscow, Moscow, Russia

LYAKH, VITALY V., PhD (Philosophy), Doctor of Philosophical Science, Professor, Kiev, Ukraine

MALKOVSKAYA, IRINA A., PhD (Philosophy), Assistant Professor, Moscow, Russia

MANTATOV, VYACHESLAV, V., PhD, Doctor of Philosophical Science, Professor, Ulan-Ude, Russia

MANTATOVA, LARISA V., PhD, Doctor of Philosophical Science, Assistant Professor, Ulan-Ude, Russia

MARING, MATTHIAS, PhD (Philosophy), Karlsruhe Institute for Philosophy, Karlsruhe, Germany

MARKIN, VYACHESLAV V., PhD (Philosophy), Moscow, Russia

MAROCCO, ANGELO, PhD, Professor, University of Rome, Rome, Italy

MARTIN, GLEN T., PhD (Philosophy), Professor, Radford University, Radford, Virginia, US

MASLOBOEVA, OLGA D., PhD (Philosophy), Associate Professor, St. Petersburg, Russia

MATRONINA, LILIA F., PhD (Philosophy), Assistant Professor, Korolev, Russia

MATSKEVICH, IGOR M., JD, Doctor of Juridical Science, Professor, Moscow, Russia

MAXIM, SORIN T., PhD, Doctor of Philosophical Science, Professor, Suceava, Romania

MAZOUR, ELENA V., PhD, Doctor of Physics and Mathematical Science, London, UK

MAZOUR, IVAN I., PhD, Doctor of Technological Science, Professor, Moscow, Russia

MCBRIDE, WILLIAM, PhD, Professor, Purdue University, West Lafayette, Indiana, US

MCHEDLOVA, MARINA M., PhD, Doctor of Political Science, Assistant Professor, Moscow, Russia

MEKHRYAKOV, VLADIMIR D., PhD, Doctor of Economic Science, Professor, Moscow, Russia

MENON, E. P., PhD, India Development Foundation, Bangalore, India

MIRONOV, VLADIMIR V., PhD, Doctor of Philosophical Science, Professor, Moscow, Russia

MITINA, IRINA V., PhD (Philosophy), Assistant Professor, Rostov-on-Don, Russia

MITROFANOVA, ANASTASIA V., PhD, Doctor of Political Science, Professor, Moscow, Russia

MOROZOV, VITALI I.,* PhD, Doctor of Geological and Mineral Science, Professor, Moscow, Russia

MOSHCHELKOV, YEVGENI N., PhD, Doctor of Political Science, Professor, Moscow, Russia

Contributors

MOSHKIN, SERGEI V., PhD, Doctor of Political Science, Professor, Ekaterinburg, Russia

MOULADOUDIS, GRIGORIS,* PhD, Associate Professor, Aristotle University of Thessaloniki, Thessaloniki, Greece

NARVESON, JAN, PhD (Philosophy), Professor Emeritus, University of Waterloo, Waterloo, Ontario, Canada

NASSER, SSEHAM, PhD, Professor, Sinai University, Cairo, Egypt

NAZARETYAN, AKOP P., PhD, Doctor of Philosophical Science, Professor, Moscow, Russia

NEKLESSA, ALEXANDER I., Economist, Russia

NEKVAPILOVA, IVANA, PhD (Philosophy), Vishkov, Czech Republic

NĚMEC, VACLAV, PhD, Doctor of Natural Science, Professor, Prague, Czech Republic

NIKITIN, ALEXANDER I., PhD, Doctor of Political Science, Professor, Moscow, Russia

NIKITINA, ELENA A., PhD (Philosophy), Assistant Professor, Moscow, Russia

NIKITINA, VERA V., PhD (Philosophy), Assistant Professor, Kharkov, Ukraine

NIKOLAICHEV, BORIS O., PhD (Philosophy), Moscow, Russia

ODOM, WILLIAM, PhD, Professor, Hudson Institute, Washington, DC, US

OLSHANSKY, DMITRYI A., PhD (Philosophy), Assistant Professor, St Petersburg, Russia

OLSON, ALAN M., PhD (Philosophy), Professor, Boston University, Boston, Massachusetts, US

PANARIN, ALEXANDER S.,* PhD, Doctor of Philosophical Science, Professor, Moscow, Russia

PANTIN, VLADIMIR I., PhD, Doctor of Philosophical Science, Professor, Moscow, Russia

PAVLOV, YURII M.,* PhD, Doctor of Philosophical Science, Professor, Moscow, Russia

PAVLOV, MIKHAIL YU., PhD (Economics), Research Fellow, Moscow, Russia

PAVLOVSKY, VALERYI V., PhD, Doctor of Philosophical Science, Professor, Krasnoyarsk, Russia

PEREBEINOS, PAVEL A., Political Scientist, Moscow, Russia

PÉREZ-BORBUJO ÁLVAREZ, FERNANDO, PhD (Philosophy), Doctor of Humanities, Doctor of Political Science and Public Administration, Professor, Pompeu Fabra University, Barcelona, Spain

PLAUL, PETER, PhD, Bitterfeld, Germany

POPOVENIUC, BOGDAN, PhD (Philosophy), Suceava, Romania

PORUS, VLADIMIR N., PhD, Doctor of Philosophical Science, Professor, Moscow, Russia

POSTANOGOV, VLADIMIR K., PhD, Moscow, Russia

PREDBORSKAYA, IRINA M., PhD, Doctor of Philosophical Science, Professor, Symi, Ukraine

PRODANOV, VASIL, PhD, Doctor of Philosophical Science, Professor, Sofia, Bulgaria

PROKOFIEVA, TATIANA V., PhD (Biology), Assistant Professor, Moscow, Russia

PROKOPCHINA, SVETLANA V., PhD, Doctor of Technological Science, St. Petersburg, Russia

PUCHALA, DONALD J., PhD (Political Science), Professor, University of South Carolina, Columbia, South Carolina, US

PYNZARI, SVETLANA, N., PhD (International Relations), Chisinau, Republic of Moldova

QING-LIN, GUO, PhD (Philosophy), Professor, Xiangtan, China

QINIAN, AN, PhD (Philosophy), Professor, Beijing, China

RAJKOVICH, JUGOSLAV, PhD (Philosophy), Professor, Belgrade, Serbia

RAKHMANOV, AZAT B., PhD (Philosophy), Assistant Professor, Moscow, Russia

RAKITOV, ANATOLI I., PhD, Doctor of Philosophical Science, Professor, Moscow, Russia

RATSIBORINSKAYA, DARIA, PhD, Professor, Erasmus University, Rotterdam, The Netherlands

508   *GLOBAL STUDIES ENCYCLOPEDIC DICTIONARY*

REZHABEK, BORIS G., PhD (Biology), Moscow, Russia

ROEVA, NATALIA N., PhD, Doctor of Chemical Science, Professor, Moscow, Russia

ROGOZIN, DMITRI O., PhD, Doctor of Philosophical Science, Moscow, Russia

RUBENE, ZANDA, PhD (Philosophy), Professor, Riga, Latvia

RUSIN, IGOR I., PhD, Doctor of Economic Science, Professor, Moscow, Russia

RYBALSKY, NIKOLAI G., PhD (Biology), Professor, Moscow, Russia

SAFRONOV, IGOR A., PhD, Doctor of Philosophical Science, Professor, St. Petersburg, Russia

SANAH, MEDHI, PhD, Professor, Teheran University, Ambassador of the Islamic Republic of Iran to the Russian Federation, Teheran, Iran

SANTONI, RONALD E., PhD (Philosophy), Professor Emeritus, Denison University, Granville, Ohio, US

SAVOLAINEN, JUHA, International Philosophy Olympiad, Fédération Internationale des Sociétiés de Philosophie (FISP), Helsinki, Finland

SCHAFFER, MARVIN B., PhD, Adjunct Professor, School of Public Policy, Simon Fraser University, Santa Monica, California, US

SEMRADOVA, ILONA, PhD (Philosophy), Prague, Czech Republic

SENATSKAYA, GENRIETTA S., PhD (Geology), Senior Research Fellow, Moscow, Russia

SERGEEV, MIKHAIL YU., PhD (Philosophy and Religion), Adjunct Professor, The University of the Arts, Philadelphia, Pennsylvania, US

SHAKHRAMANYAN, MIKHAIL A., PhD, Doctor of Technological Science, Professor, Moscow, Russia

SHARIFY-FUNK, MEENA, PhD, Associate Professor, Wilfred Laurier University, Waterloo, Ontario, Canada

SHCHELKUNOV, MIKHAIL D., PhD, Doctor of Philosophical Science, Professor, Kazan, Russia

SHESTOVA, TATIANA, PhD, Doctor of Philosophical Science, Associate Professor, Moscow, Russia

SHISHKOV, YURI V., PhD, Doctor of Economic Science, Professor, Moscow, Russia

SHULENINA, NADEZHDA V., PhD (Philosophy), Assistant Professor, Moscow, Russia

SHULENINA, ZINAIDA M., PhD (Economics), Moscow, Russia

SIPINA, LARISA V., MD, Moscow, Russia

SMIRNOVA, NATALIA M., PhD, Doctor of Philosophical Science, Professor, Moscow, Russia

SMOLKIN, VLADIMIR M., PhD (Philosophy), Blaricum, The Netherlands

SNAKIN, VALERI V., PhD, Doctor of Biological Science, Professor, Moscow, Russia

SNOOKS, GRAEME, PhD (Economics), Australian National University, Canberra, Australia

STEGALL, DAVID LEE, PhD (Philosophy), Senior Lecturer, Clemson University, Clemson, South Carolina, US

STEINER, DAVID M., PhD (Education), Professor, Boston University, Boston, Massachusetts, US

STENCHIKOV, GEORGY, PhD, Professor of Physics and Mathematics, Rutgers University, New Brunswick, New Jersey, US

STEPANYANTS, MARIETTA T., PhD, Doctor of Philosophical Science, Professor, Moscow, Russia

STEPASHIN, SERGEI V., JD, Doctor of Juridical Science, Professor, Moscow, Russia

STIOPIN, VYACHESLAV S., PhD, Doctor of Philosophical Science, Professor, Moscow, Russia

STOLOVICH, LEONID N., PhD, Doctor of Philosophical Science, Professor, Tartu, Estonia

STROGANOVA, MARINA N., PhD, Doctor of Biological Science, Professor, Moscow, Russia

Contributors 509

SYTIN, ADREI G., PhD (Philosophy), Associate Professor, Moscow, Russia

TOLSTYKH, VALENTIN I., PhD, Doctor of Philosophical Science, Professor, Moscow, Russia

TOPALOGLU, AYDIN, PhD, Research Fellow, The Center for Islamic Studies, Istanbul, Turkey

TUULI, IRINA YU., PhD, Coordinator of the Paideia Project, Boston University, Boston, Massachusetts, US

UDOVIK, SERGEI, PhD (Political Science), A director of Vakler publishing house, system analyst, author, Kiev, Ukraine

URSUL, ARKADI D., PhD, Doctor of Philosophical Science, Professor, Moscow, Russia

UTKIN, ANATOLI I., PhD, Doctor of Historical Science, Professor, Moscow, Russia

VASILCHUK, YURI A., PhD, Doctor of Philosophical Science, Moscow, Russia

VASILENKO, IRINA A., PhD, Doctor of Political Science, Professor, Moscow, Russia

VASILIEV, VLADIMIR S., PhD, Doctor of Economic Science, Professor, Moscow, Russia

VEBER, ALEXANDER B., PhD, Doctor of Historical Science, Professor, Moscow, Russia

VENKATAKRISHNAN, V., PhD, Professor, Institute of Rural Management, Anand, Gujarat, India

VERMISHEV, MIKHAIL K., PhD, Yerevan, Armenia

VORONINA, OLGA A., PhD (Philosphy), Associate Professor, Moscow, Russia

VRABIC, DARINKA, PhD (Philosophy), Velenje, Slovenia

WALTERS, GREGORY J., PhD, Professor , Saint Paul University, Ottawa, Quebec, Canada

WILLIAMS, NANCY, PhD (Philosophy), Associate Professor, Wofford College, Spartanburg, South Carolina, US

YABLOKOV, IGOR N., PhD, Doctor of Philosophical Science, Professor, Moscow, Russia

YAKOVENKO, OLGA V., PhD (Philosophy), Moscow, Russia

YAKOVETS, YURI V., PhD, Doctor of Economic Science, Professor, Moscow, Russia

YANSHINA, FIDAN T., PhD, Doctor of Philosophical Science, Moscow, Russia

YELIZAROV, VALERY V., PhD (Economics), Assistant Professor, Moscow, Russia

YERIN, NIKOLAY F, ENGINEER, Moscow, Russia

YETSIN, LIN, PhD, Doctor of Economic Science, Professor, Beijing, China

YOU-JIN, LIU, Economist, Xiangtan, China

YUANZHOU, YU, Philosopher, Beijing, China

YUDIN, BORIS G., PhD, Doctor of Philosophical Science, Professor, Moscow, Russia

YUKHVID, ALEXEI V., PhD (Philosophy), Assistant Professor, Moscow, Russia

YURTAEV, VLADIMIR I., PhD (History), Assistant Professor, Moscow, Russia

ZAGLADIN, VADIM V.,* PhD, Doctor of Philosophical Science, Professor, Moscow, Russia

ZAMKOVOI, VLADIMIR, PhD, Doctor of Philosophical Science, Professor, Moscow, Russia

ZAVARZINA, YEKATERINA P., Historian, Moscow, Russia

ZELENKO, BORIS I., PhD (Philosophy), Doctor of Political Science, Moscow, Russia

ZHUCHENKO, ALEXANDER A.,* PhD, Doctor of Agricultural Science, Professor, Moscow, Russia

ZOLOTYKH, ELENA B., PhD (Geology), Moscow, Russia

ZOTOV, ANATOLI F., PhD, Doctor of Philosophical Science, Professor, Moscow, Russia

*Deceased

LIST OF ENTRIES

Acid Mine Drainage, 3
Age of Great Unity, 4
Aggression, 5
Aggression as an International Law
    Concept, 8
Agricultural Adaptive Strategies, 9
Agricultural Afforestation, 11
Agricultural Decollectivization, 12
AIDS, 13
Alarmism, 14
Alienation, 15
Alterglobalism, 16
Alternative Movements, 17
Anthropocentrism, 17
Anthropogenic Crises, 17
Anthropogenic Environmental
    Changes, 19
Antiglobalism, 20
Arendt, Hannah, 22
Asian Development Bank, 23
Atheism, 23
Atmosphere, 25
Augustine of Hippo, Perspective on
    Peace, 26
Authoritarianism, 28
Autotrophy of Humankind, 28
Axial Time, 29

Bahá'í Faith, 30
Baikal Lake, 31
Barefoot Revolution, 32
Beyond the Age of Waste, 32
Beyond the Limits to Growth, 32
Biocentrism, 33
Biodiversity, 34
Bioethics, 36
Biogeochemical Cycle, 38
Biogeography, 38
Biosphere, 40
Biosphere Resources, 42
Biotechnology, 42
Bipolar World, 43
Bologna Process, The, 44
Brain Drain, 44
Buddhism, 45

Capitalism, 47
Capital Flight, 49
Capital Punishment, 49
Capital, Transnational, 50
Center and Periphery, 51

Chernobyl Catastrophe, 51
China, Globalization Problems in, 52
China, Liberalism in Contemporary,
    55
Chinese Enterprises, Transitional
    Activity of, 57
Chinese Philosophy and Global-
    ization, 59
Christianity, 60
Christianity, Orthodox, 61
City, Creative, 63
City Development, Urban and
    Ecological Aspects of, 64
Civil Disobedience, 66
Civil Society, 68
Civilian Police, International, 69
Civilization, 70
Civilization Development, Types, 72
Civilizations, Diversity of, 74
Civilization, Evolution of, 74
Civilization, Global, 75
Civilizations, Local, 75
Climate, 75
Club of Budapest, 77
Club of Rome, 78
Club of Rome, Reports to the, 79
Cluster, 80
Coevolution, 81
Cold War, 82
Colonialism, 85
Communication, 85
Communication, Global, 86
Communism, 87
Competitiveness, 89
Composite Development Indices, 89
Computer Ethics, 90
Concerned Philosophers for Peace, 92
Conflict, Political, 92
Conflict, Social, 93
Consciousness, Global, 94
Consolation, 96
Constitutional Ideal, Global, 96
Constitutional State, 99
Consumer Advocacy, 100
Contamination with Heavy Metals,
    102
Corporations, Transnational, 103
Corruption, 104
Cosmism, 105
Crime, International, 107
Culture, 109

512 GLOBAL STUDIES ENCYCLOPEDIC DICTIONARY

Culture, Cognitive, 111
Culture Globalization, 112
Culture, Local, 112
Culture, Mass, 113
Cultural Heritage, 115
Cultural Identity and Globalization, 116
Cultural Identity and Standardization, 118
Cultural Universals, 119
Cyberculture, 120

Debt Crisis, 122
Demographic Policy, 122
Demographic Transition, 123
Depopulation, 125
Determinism, Geographic, 125
Development, Human, 127
Development, Social, 128
Dialogue, 129
Dialogue among Civilizations and the Islamic Factor, 130
Dialogue on Wealth and Welfare, 132
Dignity, Human, 132
Diplomacy, 133
Diseases of Civilization, 136
Diseases, Global Threat of Infectious, 138
Drug Addiction, 140
Dynamic-Strategy Theory, 141

East-West, 142
East-West Problem, 144
East-West as a Social Model, 144
Ecology, Deep, 146
Ecology, Engineering, 147
Ecology, Global, 149
Ecology, Industrial, 150
Ecology, Social, 150
Ecological Audit, 151
Ecological Balance, 153
Ecological Crisis, 153
Ecological Expert, 154
Ecological Information, 155
Ecological Law, International, 156
Ecological Medicine, 158
Ecological Monitoring, 158
Ecological Optimism, 160
Ecological Portrait of Human Beings (Ecoportrait), 161
Ecological Problems, Global, 161
Ecometry, 164
Economic Crisis, World, 165
Economic Relations, Int'l., 165

Economic Security, 168
Economy, World, 169
Ecotoxicology, 171
Education, 172
Educational Globalism, 173
Education, Global Philosophy of, 173
Education, Globalization of, 174
Education, Philosophy of, 174
Education in Postmodern Age, 174
Education, Transnational, 175
E-Governance, the Case of India, 175
Elite, 177
Energy, The Countdown, 179
Energy Crisis, 179
Energy Paradigm, 179
Environmental Changes, Global, 180
Environmental Liability, 180
Environmental Psychology, 183
Environmental Safety and Security: Advantages of Pebble-Bed Nuclear Reactors, 183
Environmentalism, Western, 184
Epogenesis, 186
Epometamorphosis (Epomorthosis), 187
Equality, 188
Equality of Gender (Sexual Equality), 190
Esperanto, 191
Ethics, Global, 192
Ethnic Cleansing, 192
Ethnic Group, 195
Ethos, Global, 197
Euroislam, 197
Europe, 198
European Constitution, 199
European Union, 201
European Union Charter on Fundamental Rights, 202
European Union Law, 203
Evolutionism, Global, 204

Factor Four, 207
Financial Market, Global, 207
First Global Revolution, 209
Formation Theory, 209
Freedom:, 210
Frolov, Ivan Timofeevich, 211

Gandhism, 212
Gender Oppression, 213
Geneva Conventions, 214
Genocide, 215
Geocivilization, 216

Index

Geoeconomics, 217
Geoethics, 219
Geology, 219
Geopolitics, 220
Geosystem, Natural and Technical, 221
Global Modeling, 221
Global Studies, 223
Global Studies in Philosophy, 225
Global World Outlook, 226
Globalization, 229
Globalization from Below, 231
Globalization, Fundamental, 231
Globalization, Hyperglobalist and Transformational, 232
Globalization, Limits of, 233
Globalization, Historical Stages of, 234
Globalization, Local Characteristics of (The Example of Georgia), 235
Globalization and Localization, 237
Globalization at Macro-, Meso-, and Micro- Levels, 239
Globalization, Multiaspect, 239
Globalization and the New World, 240
Globalization, Social Effects of, 241
Globalization as a Stage in the Transformation of the European Value System, 241
Globalization and Universal Values, 242
Glocalization, 242
Goals for Mankind, 242
Golden Billion, 243
Gorbachev Foundation, 244
Government, World, 245
Great Eight (G-8), 247
Green Revolution, 247
Green Taxes, 247
Greenhouse Effect, 248
Growth Limits, 248
Gumilev, Lev Nikolaevich, 250

Hegemonism, 252
History, Global, 253
Homeostasis, 254
Human Development Index, 254
Human Development Reports, 255
Human Experience, Objective and Subjective Factors in, 255
Human Potential, 256
Human Rights, 257

Human Rights from the Philosophical Point of View, 259
Human Security, 260
Humanist Manifestos, 263
Humankind as a Family, 263
Humus, 265
Hunger, 265
Hydrosphere, Global, 267
Hypothesis, 269

Imperatives to the Cooperation of North and South, 270
Industrial Revolution, 270
Inequality, Digital, 270
Inequality, Global, 270
Information, 271
Information Field of the Universe, 272
Information Revolution, 273
Information Rights and Human Rights, 274
Informatization, 275
Integrity of Human Existence, 275
Intellectual Revolution, Global, 276
Interdisciplinarity, Types of, 277
Interfaith Studies, 279
International Federation of Philosophical Societies, 279
International Framework for Collaboration in Legal Matters, 280
International Green Cross, 280
International Law and Ethics Conference Series, 281
International Law, International Legal System, and Global Studies, 281
International Organizations, 282
International Philosophers for the Prevention of Nuclear Omnicide, 285
International Relations, 285
International Relations, Democratization of, 287
International Relations, Political Philosophy (Theory) of, 287
International Relations, Realism as an Approach to, 289
International Security Governance, 290
Internationalism, 291
Internationalism and Global Studies, 291
Internationalization, 293
Islamism, 294
Islam and Globalization, 294
Islam in the Modern World, 297

514   *GLOBAL STUDIES ENCYCLOPEDIC DICTIONARY*

Islamism's Participation in Global-
ization, 299
Issyk-Kul Forum, 301
Japanese Post-War Economic
Miracle, 304
Jihad from a Nonviolent Perspective,
305
Jonas, Hans, 307
Judaism and Nonviolence, 307
Justice, Global, 308
Just War Doctrine, 310

Khozin, Grigorii Sergeevitsh, 312
Knowledge Economy, 312

Labor Market, World, 313
Land Ethics, 314
Land Resources, World, 315
Languages, 316
Law and Legal Systems,
International, 318
Learning, Constructivist Theories of,
319
Limit, Philosophy of, 320
The Limits to Growth, 321

Macroshift, 322
Malthus, Thomas Robert, 322
Management, Holistic, 323
Management, Strategic, 323
Mandela, Nelson, 325
Mankind at the Turning Point, 327
Marxism and Global Values, 327
Maximum Concentration Limit, 329
Media and Global Studies, 330
Media Participation (Participatory
Communication), 331
Mergers and Acquisitions, 332
*Microelectronics and Society: For
Better or for Worse*, 333
Migration, 333
Modernity, Global Challenges of, 335
Modernity, Industrial, 336
Moiseyev, Nikita, 338
Mondialisation, 339
Monopolar World, 339
Multiculturalism, 339
Myth, 341

Nation(alism), 342
National Security, 343
NATO Response Force, 343
Neutrality, 344
No Limits to Learning, 345

Nomadic Empires, 346
Nonaligned Movement, 346
Noosphere, 347
North-South (The Rich North and the
Poor South), 348
Nuclear Deterrence Theory, 349
Nuclear Power Engineering, 351
Nuclear Strategy, American, 352
Nuclear Warfare and Morality, 352
Nuclear WeaponsTesting, 354
Nuclear Winter, 355

Opinion, World Public, 358
Our Common Future, 359
Our Global Neighborhood, 360
Ozone Layer Depletion, 360

Pacifism, 363
Paideia (Education) and Globalism,
364
Pan-Islamism, 365
Pantheism, 368
Peace Culture, 368
Peacemaking Operations, 369
Perestroika, 371
Personalism, 372
Philosophy, 372
Policy, 373
Politics, World, 374
Political Concensus, 376
Political Culture, 377
Political Science and Global Studies,
377

Al-Qaeda, 379
Quarter, Brook No, 379
Quixote Center, 379

Racism, Environmental, 381
Radioactivity, Pollution of the
Environment by, 382
Rawls, John, 384
*Red Data Book of the Russian Feder-
ation*, 385
Refugee, 386
Regionalism/Regionalization, 387
Regional Industrial System, 389
Religion, 390
Religions and Global Studies, 391
Religious Movements, New, 392
Religion in Primeval Societies and
Modern Religious Movements, 393
Religious Politization, 394
Reshaping the International Order, 396

Index

Responsibility, Global, 396
Responsibility to Protect (2001), 397
Rio-92, 397
Risk, 400
Risk, Ecological, 401
Risk Management Technologies, 403
Risk and Postmodern Society, 403
Road Maps to the Future, 406
Roerich, Nikolai, 406
Russian Philosophical Society
   (RPhS), 406

Sartre, Jean-Paul, 408
Security Policy, Global, 408
Self-Government, Local, 409
Semiosphere, 409
Social Change, 409
Social Contract, 411
Social Doctrine of the Roman
   Catholic Church, 411
Social Entities, 413
Social Partnerships, 413
Social Theory, Critical, 414
Society, 415
Society, Closed, 417
Society, Network, 418
Society, Open, 419
Sociogenesis, 420
Soil, 420
Soil Degradation, 421
Somerville, John, 422
Sovereignty, 422
Soviet Union, Global Studies in, 424
Space Era, 425
Space Exploration and Global Devel-
   opment, 426
Spheres of the Earth, 427
Stateless Nation, 428
State System, Global, 429
Stockholm-72, 430
Superpower, 432
Sustainable Development, 433
Sustainable Life, Permanently, 434
Sustainable Development, World
   Summit on, 434

Teaching, Ethical Aspects of, 438
Technocracy, 438
Technogenesis, 439
Technogenic Catastrophe, 442
Technogenous Civilization, 442
Technology, 443
Technological Assessment, 445
Technologies, High Human, 445

Technological Park, 446
Technological Progress, 446
Technological Pyramid, 447
Technology and Responsibility, 450
Technological Revolution, 450
Technology and Science in Global
   Perspective, 451
Techno-Optimism, 452
Techno-Pessimism, 454
Technosphere, 455
Technosphere, the Structure of, 455
Technotronic Society, 456
Terror(ism), 457
Terrorism, International, Reasons for,
   458
Terrorism, Nuclear, 460
Terrorism Prevention, 461
Think Tanks, 462
*Third World: Three-Fourths of the
   World*, 463
Toleration, 463
Toxicants, 466
Transparency and Information
   Technologies, 467

Umma, 470
Underground Rivers, Energy of, 472
United Nations Millennium Declar-
   ation, 472
Universe, 472
Urbanization, 474

Values, Univeral, 476
Values, Western, 476
Virtualization (Virtual Technology),
   477

War, 479
Water Resources of the World, 481
Weapons of Mass Destruction, 482
Weapons and Materials of Mass
   Destruction, Global Partnership
   against, 484
Weathering, 485
Wide-Area Networks, 487
World Dynamics, 488
World Economic Forum, 490
World Federalism, 491
World Future Society, 491
World Future Studies Federation, 491
World Heritage, 491
World Order Models, 492
World Political Forum, 494
World Trade Organization, 494

516 *GLOBAL STUDIES ENCYCLOPEDIC DICTIONARY*

World Wars, 496
Xenogamy, Cultural and Political,
 497
Xenophobia, 497

Youth, 499

Z-Generation, 500
Zinovyev, Aleksandr Aleksandrovich,
 500

INDEX

Abrahamic faith, 297
acid mine drainage, 3
advocacy, consumer, 100
Afghanistan War, 83
Age of Great Unity, 4
Agenda 21, 52, 422, 434
aggression, 5, 67, 291, 305, 344, 371, 465
  colonial, 328
  deterrence against Russia's, 350
  genetically based vs. genocide, 216
  geopolitical, 17, 339
  as international law concept, 8
  military, 213
Agreement on Trade Aspects of Intellectual Property Rights, 495
agriculture, 5
  adaptive strategies, 9
  afforestation, 11
  decollectivization, 12
AIDS, 13, 137 174, 271, 326, 472
alarmism, 14
al-Afghani, Jamal al-Din, 365, 471
alienation, 15, 101, 118, 263, 333, 337, 409, 462
  a. zone, 52
Al-Qaeda, 291, 294, 379, 458, 483
alterglobalism, 16
alternative movements, 17
American Enterprise Institute, 463
American Peace Institute, 463
American Revolution, 230, 428
Amnesty International, 258
animism, 368, 390
anthropocentrism, 17
anthropogenic:
  crises, 18
  environmental changes, 17
antiglobalism, 20, 459
Antigone (Sophocles), 67
Antimicrobial Resistance Information Bank, 139
Apolo 11, 425
Approximate Allowable Quantity, 102
Approximate Safe Impact Level, 102
Arab Center for Arab Unity Studies, 463
aridization of land, 162
Aristotle, 22, 27, 34, 160, 189, 227, 310, 415, 443, 452

Politics, 188
Armenian Apostolic Church, 61, 63
Asian Development Bank, 23, 284
Asian-Pacific Economic Cooperation Forum, 285
Association of South-East Asia Nations, 284
Assyrian Church of the East, 61, 62
atheism, 23
atmosphere, 24
attestation, 129
Augustine of Hippo, 96, 363, 415
  Nicomachean Ethics, 26
authoritarianism, 28
autocephalous churches, 62
autonomy, 231, 300, 455
  cultural, 16
  individual, 68
  managerial, 13
  national/political, 196, 246, 309
  of system, 129
  of Ukranian Orthodox Church, 62
  university, 44
autotrophy of humankind, 28
axial time, 19, 29, 187

Bahá'í faith, 30
Baikal Lake, 31
Barefoot Revolution, 32
Beyond the Age of Waste (Gabor), 32
Beyond the Limits to Growth (Pestel), 32, 249
bio:
  centrism, 33
  diversity, 18, 20, 34, 42, 146, 163, 181
  empathy, 315
  ethics, 36, 307
  geocenosis, 40, 427
  geochemical cycle, 38
  geography, 38
  sphere, 10, 28, 33, 40, 64, 76, 81, 102, 146, 149, 154, 156, 160, 162, 187, 206, 223, 315, 350, 355, 398, 402, 415, 420, 424, 427, 433, 481, 485
  technology, 42, 80, 455
Biological Diversity Convention, 157
bog reclamation, 65
Bologna process, 44
Bolshevik Revolution, 56, 84, 88, 143

518 GLOBAL STUDIES ENCYCLOPEDIC DICTIONARY

The Book of Great Unity (Youwei), 4
brain drain, 44, 172
British Institute of Public Policy Research, 463
Brookings Institution, 463
brook no quarter, 379
Buddhism, 45, 59, 160, 391, 393

Camus, Albert, 129, 353
capital punishment, 49, 423
capital(ism), 47
    flight, 49
    transnational, 50
Carnegie Fund, 463
Case for Animal Rights (Regan), 314
Catholics Speak Out, 380
Center for Health Applications of
    Aerospace Related Technologies, 138
center and periphery, 51
Center for Strategic and International
    Studies, 463
Center for the Twenty-First Century,
    463
Centesimus annus (John Paul II), 413
Chalcedonian Christianity, 61
Chekhov, Pavel, 337
Chernobyl accident, 51, 94, 183, 383,
    442, 454, 467
China:
    Agenda 21, 52, 422, 434
    enterprises, transitional activity of,
        57
    globalization problems, 52
    liberalism, 55
    philosophy and globalization, 59
China that Can Say No (Zhang
    Zangzang et al.), 56
Chinese National Space Administration, 425
Chisholm, Roderick, 413
choice, human, 127
Christianity, 24, 31, 60, 198, 306,
    391, 393
    Chalcedonian/Orthodox, 61
    Marxism and, 341
    pacifism and, 363
    state religion in Georgia, 236
city:
    creative, 63
    development, urban and ecological
        aspects, 64
    revolution, 19

civil:
    disobedience, 66, 146, 212, 231,
        325
    society, 17, 56, 67, 68, 100, 203,
        280, 289, 343, 360, 391, 416,
        435, 476
civilian police, international, 69
civilization(s), 82
    development, types, 72
    diversity, 74
    evolution, 74
    geo-, 216
    global, 75, 94, 130, 338
    local, 70, 75, 144, 217
    techno(genous)(logical), 14, 230,
        442
    Western, 144, 146, 212, 263, 298,
        336
climate, 10, 25, 31, 42, 75, 94, 125,
    267, 316, 350, 355
    global/planetary changes, 14, 94,
        161, 362, 481, 487
    impact on economy, 217
    impact on social development, 253
    paleoclimate, 161
Climate Fluctuation Frame Convention, 157
Club of Budapest, 77
Club of Rome, 78, 243, 283, 463, 493
    reports, 79, 95
cluster, 80
    industrial, 389
coevolution, 33, 81
Cold War, 56, 82, 95, 116, 143, 235,
    245, 280, 285, 287, 289, 301,
    304, 322, 328, 339, 347, 350,
    352, 372, 376, 387, 422, 460,
    482
Colombo Plan, 284
colonialism, 85, 117, 142, 212, 365,
    367, 408
    cultural, 435
    neo-, 237, 271, 339
collectivism, 55, 83, 298, 418
Committee for Biological Education, 34
Common Foreign and Security Policy,
    204
Commonwealth of Independent
    States, 284
communism, 83, 87, 366, 387
    capitalism and, 54, 210, 346
    Soviet, 88, 353
Communist Manifesto (Marx and
    Engels), 54

Index

competi(tion)(tiveness), 5, 19, 59, 69,
80, 81, 111, 145, 169, 323,
375, 389, 441
economic, 57, 89, 218, 234
foreign/global, 447, 495
for natural resources, 179, 449, 498
political, 199
sports, 95
superpower, 376
Competitions Acts, 332
composite development indices, 89,
254
computers, 78, 86, 100, 114, 120, 172,
174, 449, 451, 452, 456, 477,
487
c.-based communication, 234, 272
c. ethics, 90
in crime, 107
globalization aided by, 334
c. models, 122, 222, 249, 327, 497
c. networks, 107, 272, 418, 468
c. surveillance systems, 140
c. viruses, 250
Concerned Philosophers for Peace,
92, 285
conflict:
political, 92, 108, 386, 493
social, 14, 93, 241, 287, 414
Confucianism, 59
Congress of Europe, 292
consciousness, global, 96, 188, 239,
258
consolation, 96
constitutional ideal, global, 96
Constitution:
European Union, 199, 201, 204
India, 201
Soviet Union, 200
United States, 101, 258
constructivism, 319
consumer(ism)(s), 47, 63, 100, 114,
118, 167, 183, 237, 264, 302,
332, 348, 435, 448
c. law in Russia, 102
contamination w. heavy metals, 102
Convention on Biological Diversity,
36, 181, 399
Convention of the Council of Europe for
Protection of Human Rights
and Dignity of the Human Being, 37
Convention on the Elimination of All
Forms of Discrimination
against Women, 258

"Convention on the Prohibition of the
Development, Production,
and Stockpiling of Bacteriological (Biological) and Toxin Weapons and on their Destruction," 484
Coptic Church, 61, 63
corporations, transnational, 44, 50, 57,
103, 135, 207, 234, 237, 293,
313, 374, 414, 434, 447
corruption, 49, 56, 104, 175, 251, 282,
318
cosmism, 105, 228, 256
Council of Europe, 37, 175, 284
Council for Mutual Economic Aid
Council of Persian Gulf Arab States
Cooperation, 284
creationism, 60
crime, international, 107, 245, 280,
292
Cro-Magnons, 18
cultur(al)(e), 109
cognitive, 111, 452
cyber-, 120
globalization, 112
heritage, 115, 298, 491
c. identity, 116, 118, 340, 367
local, 112, 237, 242, 341
mass, 87, 112, 240, 302, 378, 445
universals, 119, 372
"Cultural Liberty in Today's Diverse
World," 128
Cultural Revolution, Mao's, 56

damage, 442, 447
ecological/environmental, 19, 152,
155, 157, 159, 180, 328, 330,
400, 404
discrimination between d. to
persons/property, 379
d. inadmissibility, 349
mutagenic, 353, 361
d. prevention/liability, 182
psychological, 483
Darwin, Charles, 150, 315, 322
Origin of the Species, 34
social Darwinism, 229
Day, Dorothy, 364
debt crisis, 122
Declaration of Independence, 69, 258,
477
Declaration Prohibiting Human Cloning, 37
Declaration of the Rights of Man, 258

520   *GLOBAL STUDIES ENCYCLOPEDIC DICTIONARY*

defense, unprovocative, 289
deforestation, 19, 65, 431, 467
democracy, 16, 19, 20, 51, 53, 55, 97,
        100, 118, 131, 144, 170, 209,
        212, 231, 232, 239, 242, 245,
        255, 263, 267, 271, 287, 326,
        336, 373, 377, 385, 391, 395,
        414, 419, 430, 451, 464
   Christian, 343
   communism vs., 83
   direct/indirect, 57
   e-democracy, 312
   in Egypt, 366
   EU and, 200, 201
   Islam and, 197
De la démocratie en Amérique (de
        Tocqueville), 96
demographics, 89, 122
deontology, 310
depopulation, 125
de-Stalinization, 84
determinism:
   geographic, 125, 417
   technological, 306, 438
development, human, 125, 127, 156,
        239, 254, 261, 346, 398, 427,
        423
dialogue, 21, 129, 308
   between Catholic Church and
        others, 412
   intercivilizational/cultural, 60, 71,
        75, 86, 113, 120, 130, 135,
        263, 287, 289, 291, 293, 298,
        340, 377, 418
   between Muslims/others, 197, 297
   philosophical, 96, 373, 422, 491
   Plato's, 364
   political, 378
   pseudo-, 87
Dialogue on Wealth and Welfare
        (Giarini), 132
Dictionnaire historique et critique
        (Bayle), 463
dignity, 100, 132, 189, 202, 241, 259,
        263, 266, 274, 308, 412, 435
diplomacy, 133, 166, 308
diseases:
   of civilization, 136
   global threat of infectious, 138
displaced persons, 333
diversity, 62, 492
   bio-, 18, 34, 39, 42, 146, 156, 163,
        181, 385, 420, 431, 435
   civilizational, 74, 264

cultural, 53, 75, 98, 115, 197, 217,
        240, 303, 339, 435
   ecological, 81, 159
   local/regional, 237, 388
   of mythologies, 444
   philosophical, 174
   religious, 301, 384
   of technoevolution, 441
   unity in, 145, 201, 247, 305
dominion, 27, 257
drug addiction, 140, 172
dynamic strategy theory, 141

East-West, 142, 328, 329
   problem, 144
   Schism, 61
   as social model, 144
ecological:
   audit, 151
   balance, 42, 153, 159, 328
   crisis, 82, 153, 183, 224, 293, 358,
        398
   expert(ise), 152, 154, 183, 447
   information, 148, 155
   e. law, 156
   maintenance, 148
   medicine, 158
   monitoring, 158, 432
   optimism, 160
   portrait of human beings, 161
   problems, 81, 150, 157, 161, 186,
        228, 280, 338, 398, 430, 456
ecology:
   deep, 146
   engineering, 147, 151, 164
   global, 149
   industrial, 147, 150
   social, 150
ecometry, 149, 159, 164
econiches, 441
economics:
   geo-, 217
   hexagonal model, 218
   policy, 56, 74, 104, 135, 168, 263,
        293, 326, 375
   regional e. blocs, 170
   security, 168
   world, 209, 217, 286
Economic and Social Council, 70, 258
ecosystems, 11, 19, 34, 39, 81, 102,
        138, 148, 151, 153, 162, 180,
        248, 277, 314, 405, 430, 433,
        466
eco-techno-sociosystems, 278

Index

521

ecotoxicology, 171
Ecumenical Councils, 60
Ecumenical Movement, 391
Edict of Nantes, 464
education, 172
egalitarianism, 146
e-governance: the case of India, 175
Eisenhower, Dwight D., 351
Electronic Industrialization Infrastructure Development, 176
elite, 177, 194, 316, 336
   administrative/ruling, 230, 237, 366, 392, 395, 438
   American, 459
   communist, 83
   intellectual, 118, 229, 318, 365
   mass vs. e. culture, 115
   mobile, 241
   political, 179, 408
   pseudo-/spiritual, 178
   rural, 12
   technical, 451
emancipation, 210, 323, 337, 365
Encyclopedia of the Stateless Nations (Minahan), 317
energy:
   crisis, 179, 375
   energy paradigm, 179
Energy, The Countdown (De Montbrial), 179
engineering-ecological maintenance of production, 148
environmental(ism), 107, 146, 160
   American/European, 185
   global changes, 154, 180
   e. liability, 180
   as political force, 382
   e. safety and security, 183
   Western, 184
Environmental:
   Equal Rights Act, 382
   Health Equity Information Act, 382
   Justice Act, 382
   Protection Act, 422
   Protection Agency, 381
epogenesis, 186
epometamorphosis, 187
equality, 4, 19, 91, 133, 188, 202, 240, 263, 335, 366, 380, 409, 412, 416, 436, 472, 476, 493
   biocentric, 146
   in China, 56
   democratic, 69
   digital/global inequality, 270

economic, 45, 346
ethnic, 60
gender, 190, 392
moral, 308
of opportunity, 340
political, 189
sovereign, 99, 157, 291, 297, 422
Equality and Democracy Conference, 190
Eritrean Church, 61
Esperanto, 191
ethics:
   computer, 90
   geo-, 219
   global, 192, 307, 492
   of teaching, 438
Ethics of Humanitarian Interventions (Jokic), 281
ethnic(ity), 30, 143, 214, 242, 296, 318, 342
   cleansing, 192
   group, 75, 145, 174, 193, 195, 217, 229, 250, 317, 370, 429, 470
   polyethnicity, 253
Ethiopian Church, 60, 63
ethos, global, 77, 197, 243
Euroislam, 197
Europe(an), 29, 118, 123, 154, 163, 198, 243, 298, 355
   civilized vs. barbaric, 142
   E. climate, 481
   Cold War in, 83
   communism, 88, 348, 387
   E. Constitution, 200
   decline of, 209
   decollectivization in Central, 12
   Eastern, 19, 49, 84, 208
   economics in, 165, 169, 490
   ecosystems in, 20
   ethnic cleansing in, 193
   green taxes in, 247
   Medieval, 292
   mobile communication in, 449
   Muslim communities in, 198
   nation-states in, 428
   neutrality in, 345
   obliteration bombing in, 484
   politics in, 498
   pre-modern, 306
   Protestant ethics in, 337
   refugees in, 386
   ruling elite of, 366
   schism in, 372
   technogenic civiliation in, 72

522 GLOBAL STUDIES ENCYCLOPEDIC DICTIONARY

Europe(an), *con't.*
  unemployment in, 459
  Union, 99, 118, 167, 181, 199, 201,
     231, 247, 284, 286
  Western, 13, 17, 44, 49, 51, 75, 85,
     140, 144, 207, 236, 287, 457,
     476
European:
  Atomic Energy Community, 388
  Coal and Steal Community, 388
  Convention for Protection of Hu-
     man Rights and Basic Free-
     doms, 190, 201
  Court for Human Rights, 201
  Investment Bank, 284
  Social Charter, 190
  Union Charter on Fundamental
     Rights, 202
eutrophication, 163, 180, 431
evolution(ism):
  global, 204, 423, 441
  philetic, 35, 39

Factor Four (Weizsäcker), 207
Fair Labor Standards Act of 1938,
    414
Federal Actions to Address Environ-
    mental Justice, 382
Federal Anti-Monopoly Committee,
    102
feminism, 92, 213, 288
fetishism, 337, 390, 451
financial market, global, 171, 207,
    436
First Global Revolution, 79, 209
Flunet, 139
formation theory, 209
Forrester, Jay W., 78, 249, 488
  World Dynamics, 249
Foucault, Michel, 289, 414
Fourteen Points, 134
Frankfurt School, 414
freedom, 9, 21, 27, 87, 97, 118, 126,
       132, 157, 210, 228, 230, 231,
       241, 242, 245, 291, 297, 303,
       326, 329, 340, 353, 414, 472
  of choice, 18, 44, 219, 256, 296
  of conscience, 464
  democratic, 68, 267
  economic, 69, 494
  individual, 72, 91, 145, 189, 201,
     337, 415
  national vs. personal, 55
  political, 92
  of press, 267

  religious, 391
  of speech, 476
Freedom 7, 425
French Institute of International Rela-
    tions, 463
French Revolution, 230, 365, 428,
    457, 476
Frolov, Ivan Timofeevich, 211
Fukushima Dai-ichi plant accident,
    383

Gandhi(sm), 67, 212, 264, 364, 366
Gargarin, Yuri, 425
G-8, 247, 349, 484
gender, 30, 270, 309, 330
  equality, 190, 392
  oppression, 213
Gender Influenced Development In-
    dex, 255
General:
  Agreement on Trade in Services,
    495
  Tariff and Trade Agreement, 495
general systems theory, 278
genetic recombination, 35
genetically modified organisms, 180,
    432
Geneva Conventions, 214, 386
genocide, 14, 18, 107, 195, 308, 392
  auto-, 215
Genocide Convention, 216
Geomatics engineering, 278
geology, 39, 219, 278, 443
geo:
  civilization, 216
  economics, 217
  ethics, 219
  politics, 83, 142, 218, 220, 253
  systems, 148, 165, 221, 278
  technology, 455
glasnost, 84, 372
global(ization):
  asymmetry of, 229, 294
  from below, 20, 80, 231
  fundamental, 231
  Georgia example, 235
  historical stages of, 234
  hyperglobalist and transforma-
    tional, 232
  limits of, 233
  and localization, 237
  macro-, meso, micro- levels, 239
  g. modeling, 221, 327, 488
  multiaspect, 188, 239

Index 523

global(ization), *con't.*
  and new world, 240
  as stage in transformation of
      European value system, 241
  social effects of, 241
  and universal values, 242
Global Emerging Infections System, 139
Global Human Hill (Zinovyev), 500
Global Infectious Disease Threat (CIA), 138
Global Partnership against Weapons
    of Mass Destruction, 484
Global Public Health Information
    Network, 139
Global Reports on Human Develop-
    ment, 128
global village, 293, 315, 368
glocalization, 242
Goals for Mankind (Laszlo), 242
God, threefoldness of, 60
golden billion, 230, 243, 336, 434
Goodman, Paul, 454
Gorbachev, Mikhail, 84, 280, 303, 371, 494
  Foundation, 244
government, world, 5, 213, 245, 289, 493
Great Depression, 165, 412
green:
  revolution, 247, 266, 271
  taxes, 247
greenhouse gases, 26, 77, 162, 248, 481
Grotius, Hugo, 257
growth limits, 248
Gumilev, Lev Nikolaevich, 220, 250

Haiti invasion, 311
harmony, 4, 46, 87, 112, 114, 117, 160, 168, 264, 279, 291, 307, 434, 499
Havel, Vaclav, 68
Hayek, August, 189
Health Management Information Sys-
    tem, 139
heavy metal contamination, 102
hegemon(ism)(y), 74, 235, 238, 252, 310, 360, 367, 377
  corporate, 292
  cultural, 53, 263, 292
  economic, 295
  Euro-Atlantic, 70
  military/political, 328

Ottoman Empire over Muslim
    world, 366
Heidegger, Martin, 22, 307, 444
Helsinki Agreement, 258
Heritage Foundation, 463
high-human technologies, 445
Hinduism, 212, 391, 394
Hiroshima bombing, 227, 311, 350, 352, 354, 442
history, global, 29, 253
Hobbes, Thomas, 7, 185, 246, 257, 289, 310, 411, 416
holism, 146, 247, 314
Holocaust, 194, 215, 258, 308
Holy Scriptures, 391
homeostasis, 253
human:
  experience, objective and sub-
    jective factors in, 255
  potential, 86, 256, 412
  rights, 30, 46, 55, 61, 69, 72, 97, 100, 118, 133, 145, 157, 190, 201, 216, 245, 257, 259, 273
  security, 260
Human Condition (Arendt), 22
Human:
  Development Reports, 255
  Poverty Index, 255
  Rights Watch, 258
Humanist Manifestos, 263
Hume, David, 189, 315
humus, 11, 65, 265, 420
hunger, 164, 265, 271, 309, 322, 472
  h. eradication, 42, 266
  pandemic, 15
  in Sub-Saharan Africa, 267
Husserl, Edmund, 264, 307, 444
hydrosphere, 40, 76, 125, 220, 267, 384, 402, 420, 427, 456, 481
hypothesis, 269

immigration,
*Imperatives to the Cooperation of
    North and South* (Saint-
    Geours), 270
imperialism, 115, 118, 213, 231, 293, 366
  cultural, 300
  informational, 229, 294
  neo-, 339
  Russian, 142
Index of Physical Quality of Life, 90
Indian Remote Sensing Satellite, 138
individualism, 55, 83, 298

524   *GLOBAL STUDIES ENCYCLOPEDIC DICTIONARY*

industrial:
   contemporaneity, 337
   crisis of 1825, 165
   revolution, 19, 64, 169, 207, 227,
      254, 270, 293, 348, 442, 452
industrialization, 52, 169, 170, 227
inequality, digital/global, 270
informati(on)(zation), 86, 116, 271,
      275, 435
   age/field of universe, 272
   revolution, 19, 139, 273, 293, 411
   rights and human rights, 274
Information Kerala Mission, 176
Institute for International Economics,
      463
integrity of human existence, 275
intellect of humankind, collective,
      276
intellectual revolution, global, 276
interdisciplinarity, 277, 447
Interfaith Studies, 279
international(ism)(ization), 291, 293
   framework for collaboration in
      legal matters, 280
   law, 8, 83, 98, 109, 134, 156, 166,
      202, 215, 240, 257, 280–283,
      308, 343, 344, 370, 386, 409,
      429, 462
   organizations, 37, 97, 123, 135,
      166, 192, 261, 280, 282, 358,
      367, 376, 423, 430, 469, 494
   relations, 54, 56, 118, 128, 130,
      133, 186, 187, 218, 224, 225,
      240, 253, 260, 285, 287, 290,
      325, 335, 343, 374, 387, 391,
      396, 419, 422, 429, 431, 491
   security governance, 290
International:
   Association of Religious Freedom,
      391
   Biopolitical Organization, 34
   Court of Justice, 70, 216, 258, 282,
      284, 318
   Federation of Philosophical
      Societies, 279, 284, 407
   Green Cross, 244, 280
   Helsinki Federation for Human
      Rights, 258
   League for Human Rights, 258
   Law and Ethics Conference Series,
      281
   Monetary Fund, 21, 122, 208, 283,
      286, 376
   Peace Bureau, 292

   Peace Research Association, 292
   Philosophers for Prevention of
      Nuclear Omnicide, 285, 422
   Union of Biological Sciences, 35
Iranian Mithraism, 60
Islam(ism), 143, 216, 294, 305, 319,
      391, 393, 470. *See also* jihad;
      Koran
   cultural/scientific trends from, 118,
      131
   dialogue with civilizations, 116,
      130
   Euro-, 197
   I. extremism, 379, 457
   and globalization, 295, 299
   in modern world, 297
   Pan-, 365, 471
   political, 394
Issyk-Kul Forum, 301

Jacobyte Church, 63
Japanese post-war economic miracle,
      304
Jefferson, Thomas, 258
Jewish Enlightenment, 308
Jews, 131, 144, 193, 296, 308, 391,
      470
jihad, 305, 379
John L. Olin Institute for Strategic
      Studies, 463
Jonas, Hans, 307
   Imperative of Life, 307
   Phenomenon of Life, 307
Judaism, 300, 306, 307, 391
just war, 308, 310, 326, 363, 484
justice, 4, 12, 20, 30, 50, 55, 70, 81,
      91, 100, 119, 156, 188, 202,
      257, 271, 287, 298, 308, 328,
      335, 339, 364, 368, 379, 391,
      408, 412
   collective, 131
   for criminals, 216
   distributive, 288, 309, 384
   economic, 476
   environmental, 381
   injustice, 27, 53, 56, 66, 214, 305,
      309, 354, 382, 459, 493
Justice and Development Party (Tur-
      key), 294

Kant, Emmanuel, 22, 36, 49, 96, 126,
      132, 189, 309, 320, 364, 491
   Perpetual Peace, 246
Keynes, John Maynard, 322
Khozin, Grigorii Sergeevitsh, 312
Khrushchev, Nikita, 88

Index   525

Kibena and Attis cults, 60
King Jr., Martin Luther, 68, 364
knowledge economy, 312
Koran, 130, 296, 391, 470
Korean War, 83, 353
Kuwait invasion, 311
Kyoto Protocol, 162, 248

labor market, 241, 313, 334
Lamark, Jean Baptiste, 427
LANDSAT system, 467
languages, 316
law and legal systems, 68, 181, 263
    constitutional, 97, 101, 202
    international, 8, 83, 98, 109, 134,
        156, 166, 202, 215, 240, 257,
        280–283, 308, 318, 343, 344,
        370, 386, 409, 429, 462
    natural, 27, 73, 97, 126, 238, 257,
        288, 476
    Roman, 22
    Saxon, 204
Law of Peoples (Rawls), 384
League of Arab States, 284
League of Nations, 134, 192, 283,
    289, 292, 388
learning, constructivist theories, 319
legal matters, international framework
    for collaboration in, 280
legitimacy:
    clash of, 415
    of conclusions, 154
    of criteria, 291
    of government/power, 100, 411,
        428
Lenin(ism), 88, 287, 318, 322
Leopold, Aldo, 146, 185, 314
liability, 181
libertarianism, 48, 381
liberty, individual, 55, 189, 201, 306,
    365, 384, 466, 476
Life on the Screen (Turkle), 121
The Limits to Growth (Meadows), 32,
    122, 321
limit, philosophy of, 320
Locke, John, 246, 257, 314, 411, 464,
    476
love, 4, 26, 276
    compatible with killing, 363
Lowe, Jonathan, 413
Luna-3, 425
Lyotard, 341, 364

MacLuhan, Marshall, 120
macroeconomic factors, 169, 273, 436

in China, 59
macroshift, 322
Mafia, 95, 108, 292
Magna Carta, 257, 476
Malabar Church, 61, 63
Malankar Syriac Church, 61, 63
Malthus, Thomas Robert, 126
    *Essay on the Principle of
        Population*, 123, 249, 322
neo-Malthusianism, 123, 266, 322
management, 209, 222, 254, 273, 445
    agricultural/nature/resource, 10, 12,
        66, 102, 151, 155, 399
    coastal zone, 278
    crisis, 448
    engineering, 148
    hierarchic, 418
    holistic/strategic, 323
    incompetent, 442
    political, 178, 436
    production and, 82, 332, 414
    project, 176
    quality, 244
    risk, 219, 401, 403
    self-, 64, 331, 451
    UN, 70
Mandela, Nelson, 325
Mankind at the Turning Point (Mesar-
    ovic), 222, 327
Marcuse, Herbert, 88, 454
market anarchists, 48
Marx(ism), 7, 15, 22, 47, 53, 87, 93,
    97, 185, 189, 209, 228, 282,
    287, 289, 318, 322, 325, 327,
    341, 396, 414, 416, 422, 425,
    500
Maximum Allowable Concentration,
    102
maximum concentration limit, 329
May Fourth Movement, 56
media:
    electronic sources, 140, 445, 497
    ethical, 368
    m. exposés, 293
    and global studies, 330
    m. independence, 104
    m. information, 116
    mass, 21, 69, 74, 90, 94, 101, 114,
        117, 120, 135, 173, 313, 316,
        358, 426, 433, 445, 458, 477
    m. participation, 331
    global, 117, 196, 423
    surveillance, 139
mega-society, 293

526　　*GLOBAL STUDIES ENCYCLOPEDIC DICTIONARY*

mergers and acquisitions, 332
metagalaxy, 473
metamorphism, 187
metanarrative, 341
Microelectronics and Society, 33, 333
Migrating Species of Wild Animals
　　　　Conservation Convention,
　　　　157
migration, 15, 108, 123
　　biogenic, 41, 155
　　classification of types, 333
　　culture, 174
　　ecoportrait retained after, 161
　　forced, 173, 333
　　illegal, 334
　　labor, 313, 334; *see also* brain
　　　　drain
Mill, John Stuart, 49, 322, 464
minorities, 98, 145, 189, 196, 327,
　　　　391, 429, 470
modernity, 79, 87, 184, 229, 246, 300,
　　　　316
　　criticism of, 414
　　democratic, 230
　　focus on limit, 320
　　global challenges of, 98, 224, 335,
　　　　358, 417, 425
　　industrial, 336
　　nationhood and writing in, 317
　　religion in, 395
Moiseyev, Nikita, 338
mondialisation, 339
monopolar world, 264, 339, 348, 432
Monopolies and Mergers Acts, 332
Montesquieu, Baron de, 50, 125, 249,
　　　　258
moral behavior, 314
Moscow Logic Circle, 500
multiculturalism, 64, 196, 339, 429
multidimensionality, 410
Multi-User Domains, 121
Muslim reformers, 197
mutually assured destruction, 350
myth(ology)(omoteur), 19, 106, 160,
　　　　195, 239, 264, 341, 444
　　m. consciousness, 394
　　m. of descent, 195
　　religious, 393
　　social m. of the masses, 114
　　technocratic, 249
　　m. worldview, 187, 226, 372

Nagasaki bombing, 227, 311, 350,
　　　　354, 442

Napoleon(ic wars), 142, 282, 319
NASA, 138, 425
nation(alism), 8, 55, 59, 62, 85, 116,
　　　　134, 194, 246, 297, 310, 342,
　　　　365. *See also umma*
　　Arab, 471
　　n. building, 342, 356, 365
　　ethnic, 193, 195, 196, 197, 342,
　　　　414
　　nation-state vs., 428, 430
　　political, 414
　　state vs., 428
　　stateless, 342, 428
nation-states, 6, 83, 103, 133, 143,
　　　　174, 232, 245, 287, 318, 376,
　　　　414, 423, 428, 472
　　Antarctica division into, 430
　　Central European, 194
　　civic, 194
　　defense of, 308, 461
　　democratic, 430
　　ethnic, 193, 342, 429, 470
　　with nuclear arsenals, 354
National Electronic Disease Surveil-
　　　　lance System, 139
National Electronic Telecommunica-
　　　　tions System for Surveillance,
　　　　139
National Labor Relations Law of
　　　　1935, 414
National Notifiable Disease Surveil-
　　　　lance System, 139
NATO, 283, 291, 373, 433
　　Response Force, 343
natural selection, 34, 229
Neanderthals, 18
Nebula of Andromeda (Efremov), 88
neo-Luddism, 451
neo-Pythagoreanism, 60
neosulfates, 3
networks, 276, 290, 323, 456
　　computer, 107, 272, 418, 468
　　criminal, 241
　　cross-border, 461
　　digital, 175, 487
　　n. engineering, 175
　　global/international, 139, 239, 276,
　　　　292, 469
　　hierarchy vs., 500
　　information, 74, 139, 170, 275, 488
　　law enforcement, 462
　　nested, 139, 323
　　n. society, 418
　　trade, 103

Index 527

neutrality, 330, 344
Nicaraguan human development organizations, 380
Nicene Creed, 61
No Limits to Learning (Botkin), 345
nomadic empires, 346
Nonaligned Movement, 143, 346
nonlinearity, 120, 410
Nonproliferation Treaty, 483
noosphere, 28, 156, 160, 206, 256, 347, 409, 427
Nordic Council, 284
Norms of Nuclear Safety, 352
North American Free Trade Agreement, 284
North-South, 142, 271, 328, 348, 375, 430
Nozik, Robert, 189
nuclear:
    deterrence, 83, 349
    power engineering, 351, 431, 444
    strategy, American, 352
    warfare and morality, 352
    weapons, 44, 92, 245, 282, 285, 319, 338, 349, 352, 379, 382, 431, 482
        moral response to, 353, 422
        testing, 354
    winter, 338, 350, 353, 355

obesity, 137
October Revolution of 1917, 89
Oil Producing and Exporting Countries, 283
Old Believers, Russian, 63
ontological status, 413
opinion, world public, 358
Organization of Central-American States, 284
Organization for Security and Co-operation in Europe, 284, 374
Oriental studies, 298
Origins of Totalitarianism (Arendt), 22
Orkhuss Convention, 156
Ortega y Gasset, José, 178, 444
Osiris and Isis cults, 60
Our Common Future (World Commission on Environment and Development), 359
Our Global Neighborhood (Commission on Global Governance), 360

overpopulation, 126, 164, 245, 271, 322, 399, 419
Ozone Layer Conservation Convention, 157
ozone layer, 26, 94, 163, 360, 435, 480

pacifism, 363
paideia and globalism, 364
Paleolithic era, 18
    p. climate, 161
    p. technology, 455
Paneuropean Strategy of Biological and Landscape Diversity, 39
Pan-Islamism, 365, 471
pantheism, 368
    peace, 4, 9, 14, 20, 26, 46, 52, 83, 92, 109, 130, 157, 191, 215, 223, 245, 252, 261, 279, 282, 287, 289, 291, 301, 318, 352, 358, 363, 380, 391, 414, 416, 422, 459, 461, 480, 493
    Age of Great Peace, 5
    p. culture, 368
    between East/West, 408
    Ecuador-Peru p. process, 469
    global/world, 329, 369
    international, 351
    Jewish tradition of, 307
    jihad and, 305
    p.-making military operations, 344
    nuclear weapons to assure, 350
    positive, 307, 328
    UN peacekeeping, 371, 376
peacemaking operations, 369
Peace Treaties of Westphalia, 422
Pebble-Bed nuclear reactors, 183
perestroika, 84, 244, 371
periphery, 51, 171, 208, 230, 235, 271, 286
personalism, 227, 256, 372
person-nature system, 148
philosophy, Chinese vs. Western, 58
Piaget, Jean, 254, 319
Platonism, 60, 236
pluralism, 21, 22, 69, 114, 145, 197, 240, 301, 410, 447, 476
    cultural, 340
    political, 372
    religious, 306
Poland's relationship with USSR, 84
politic(al)(ization)(s), 374
    concensus, 376

528 GLOBAL STUDIES ENCYCLOPEDIC DICTIONARY

politic(al)(ization)(s), *con't.*
culture, 377
international p. theory, 288
religious, 394
science & global studies, 377
p. systems, 239, 283, 290, 346, 377
polytheism, Roman, 60
positivism:
logical, 227, 256
post-, 288
postmodernism, 120, 228, 256, 288, 298, 300, 340
Potsdam Protocol, 83
poverty, 51, 53, 138, 173, 177, 241, 245, 255, 266, 270, 309, 322, 326, 347, 358, 386, 399, 430, 435, 462
pre-Efesian Churches, 63
Priests for Equality, 380
Prigozhin, Illya, 410
prognostication, 14, 155, 164, 231, 269, 321, 408
Programme on Population and Development, 123
providentialism, 60
PulseNet, 139

Quadragesimo Anno, 412
Quine, W. V. O., 413
Quixote Center, 379

RABNET, 139
racism, 142, 212, 263, 300, 381, 497
anti-, 291
environmental, 381
radioactivity, 51, 382
RAND Corporation, 462
rationalism, 54, 298, 366, 410, 419
vs. reflectivism, 288
realism, 289, 413
Reality of Communism (Zinovyev), 500
Reclus, Jacques Élisée, 427
Red Crescent, 258
Red Cross, 258
Red Data Book of Russian Federation, 156, 163, 385
refugee(s), 14, 69, 333, 386, 469
regional(ism)(ization), 197, 203, 233, 286, 387, 429
regional manufacturing synthesis theory, 389
religion, 391
new movements, 392
politization, 394

in primeval societies, 393
Report on the Banality of Evil (Arendt), 22
Rerum Novarum (Leo XIII), 411
Reshaping the International Order (Tinbergen), 396
responsibility, 9, 33, 52, 100, 104, 131, 191, 228, 242, 262, 397, 405, 412, 417, 427, 435
collective, 281, 437
criminal, 107
differentiated, 399
ecological, 152
financial, 9, 182
global/international, 9, 396
individual/personal, 227, 263, 401, 431, 450, 476
government, 409
moral, 56, 450
social, 219, 353, 414
subsidiary, 181
technology and, 450
of United States, 286
Responsibility to Protect (2001) (International Commission), 397
Ribbentrop-Molotov Pact, 84
Rio Declaration, 398, 435
Rio-92, 397, 432
risk, 129, 257, 290, 324, 368, 400, 403, 448, 450
damage, 182
disease r. groups, 13, 136
ecological, 148, 152, 154, 181, 401
global, 408
hypothetical, 401
investment, 51
r. management, 219, 403
from nuclear weapons, 461, 483
r. reduction, 10, 332, 447
research costs and, 37
technological, 439
r. zones, 410
rivers, 267, 402, 481
energy of underground, 472
Road Maps to the Future (Hawrylyshin), 406
Roerich, Nikolai, 106, 406
Roman Declaration on World Food Security, 267
Roszak, Theodore, 454
Rousseau, Jean-Jacques, 96, 289, 411, 416, 454
Russian Empire, 83, 142, 198
Russian Federal Space Agency, 425

Index 529

Russian Military Doctrine, 350
Russian Philosophical Society, 406

Sartre, Jean-Paul, 353, 408
scientism, 454
Searle, John, 413
security:
  international, 134, 244, 252, 283,
    290, 344, 351
  national, 261, 343, 350, 426, 444,
    462
  policy, global, 408
Selden, John, 257
self:
  government, local, 409
  help, 289
  organization, 68, 114, 206, 253,
    473
semiosphere, 409, 427
Sen, Amartya, 266, 308
Serbian Philosophical Society, 281
Shanghai Cooperation Organization,
  285
shari'ah law, 306
Shin Fein party, 108
Smith, Adam, 315
Snooks-Panov vertical, 141
social:
  change, 17, 93, 215, 295, 307, 358,
    409, 433
  contract, 97, 211, 373, 411, 415
  critical s. theory, 414
  doctrine of the Roman Catholic
    Church, 411
  entities, 413
  partnerships, 413
  structures, 82, 128, 185, 295, 322,
    405, 413, 416, 453
society:
  Chinese, 52, 54
  civil, 17, 56, 67, 100, 203, 280, 289,
    343, 360, 391, 416, 435, 476
  closed/open, 229, 235, 372, 384,
    417, 419
  coevolution of nature and, 81
  communist, 87, 210
  democratic, 68
  individual as basic unit, 55
  industrial/postindustrial, 46, 73, 79,
    101, 224, 233, 271, 321, 337,
    347, 438, 444, 453, 462, 474
  informational, 63, 250, 312, 444,
    450

market, 44, 48, 372
  network, 418
  socialist, 47, 87, 439, 500
  stability of, 53
  transitional, 71
  Western, 16, 400
sociogenesis, 420
soil, 420
  degradation, 180, 421,
  xerotization of, 162
Sollicitudo (John Paul II), 411
Solzhenitsyn, Alexander, 98
Somerville, John, 285, 422
Sorokin, Pitirim, 410
South African Freedom Charter, 325
sovereignty, 8, 100, 131, 168, 202,
    224, 245, 257, 346
  individual, 373
  national, 55, 230, 262, 294, 367,
    371, 374, 422, 491
  s. of relativity, 300
  over seas, 430
  state, 52, 237, 397, 423
Soviet Union, 143, 200, 371, 424
  Afghan War, 379
  bipolar world, 43
  breakup of, 195, 218, 237, 339,
    342, 348, 352, 354, 411, 425,
    428
  capital flight after collapse, 49
  chemical/nuclear weapons/bans in,
    350, 351, 383, 432, 482
  communism in, 83
  elite in, 179
  ethnonational principle in, 194
  glasnost and perestroika, 84, 244,
    371
  Marxism-Leninism in, 318, 327
  ozone layer protection, 362
  Poland's relationship with, 84
  radiation status after Chernobyl
    accident, 51
  republics of former, 208
  socialism in, 88, 328
  space exploration, 425
space era, 425
SpaceShipOne, 425
species formation, 35, 39, 441
state system, 170, 196, 200, 425, 429
stateless nation, 196, 317, 342, 428
stochasticity, 410
Stockholm-72, 397, 430

530 GLOBAL STUDIES ENCYCLOPEDIC DICTIONARY

Suess, Eduard, 40, 427
suffering, 4, 45, 56, 68, 90, 96, 240,
244, 263, 265, 271, 340, 363,
397, 464, 472
superpowers, 44, 83, 235, 347, 376,
424, 432, 444, 483
supersociety, 293
sustainable:
development, 29, 31, 45, 52, 82,
219, 244, 347, 398, 409, 425,
426, 433, 446
permanently s. life, 434
Syriac Church, 61, 63

Taoism, 23, 59, 160
teaching, ethical aspects of, 438
technocracy, 438
technogenesis, 147, 439
technogenic catastrophe, 442
technological park (technopark), 446
techno-:
evolution cycle, 440
genesis, 147, 439
optimism/pessimism, 452, 454
sphere, 277, 428, 443, 455
structure, 438
trophic chains, 456
technolog(ical)(y), 32
t. assessment, 445
bio-, 42, 80, 455
computer, 91, 477
t. in global perspective, 451
information, 117, 138, 175, 300,
444, 467
t.progress, 33, 72, 113, 130, 136,
187, 222, 408, 442, 446, 450
t. pyramid, 447
and responsibility, 450
t. revolution, 14, 43, 136, 234, 239,
243, 358, 442, 450
technotronic society, 456
Teilhard de Chardin, Pierre, 160, 264,
347
telecommunications, 100, 138, 235, 250,
273, 284, 332, 418, 451, 487
territorial expansion, 126, 203, 217
terror(ism) , 75, 95, 97, 108, 131, 169,
233, 241, 264, 280, 290, 294,
299, 321, 326, 343, 352, 358,
464
anti-, 458
nuclear, 460
Theory of Justice (Rawls), 288, 339,
384

thermodynamics, 205, 324, 410, 486
think tanks, 462
3rd International Conference of the UN
on Problems of Women, 190
Third World (Guernier), 270, 463
third world countries, 15, 32, 43, 83,
116, 123, 143, 173, 213, 229,
233, 309, 347, 432
Thirty Years War, 193, 374, 422
Thoreau, Henry David, 67, 185, 212
Three-Mile Island accident, 183, 383
toleration, 463
topos, 160, 400
totemism, 390
toxicants, 102, 171, 466
Treatise on Toleration (Voltaire), 464
Twelfth Report on Carcinogens, 137
Twenty-First Century Agenda, 398

umma (nation), 131, 470. *See also*
nation(alism)
UNDP Human Development Index,
90, 247, 254
United Nations:
Convention on Aridization, 162
Convention on Biodiversity, 422
Convention on Struggle against
Desertification, 422
Convention on the Law and the Sea,
430
Convention on the Prevention and
Punishment of the Crime of
Genocide, 258
Declaration of Human Rights, 190
Declaration on the Right to
Development, 258
Development Programme, 260
Framework Convention on Climate
Change, 248, 399
International Agreement, 258
International Covenant on Civil
and Political Rights, 258
International Covenant on Eco-
nomic, Social and Cultural
Rights, 258
Millennium Declaration, 411, 472
Relief and Works Agency for Pal-
estine Refugees, 386
Security Council, 9, 70, 97, 262,
290, 371, 397, 430
specialized agencies, 284
World Trade Organization, 21, 57,
235, 283, 286, 349, 374, 494
underdevelopment, 266, 335, 358,
365

Index

UNESCO Universal Declaration of Human Genome and Human Rights, 37
UNESCO World Heritage Committee, 31
Uniate churches, 63
Universal Declaration of Human Rights, 260, 317, 318, 331, 476
urbanization, 52, 64, 123, 138, 227, 474
USSR collapse, 49, 116, 143, 347, 387

values:
  anti-, 378
  communal, 274
  cultural, 44, 109, 111, 131, 242, 251, 340, 378, 406, 477
  ethical, 91, 306
  fundamental, 101, 308, 376, 472
  global, 295, 327
  moral/spiritual, 115, 219, 240, 264, 301, 340, 358, 420, 446
  shared, 15
  tradtional, 174, 336, 395, 435, 477
  universal, 71, 99, 225, 242, 360, 439, 476, 500
  Western, 300, 340, 476, 500
Vatican II, 61, 412
Veblen, Thorstein, 438, 452
Vernadsky, Vladimir, 28, 40, 106, 150, 160, 347, 424
  Proliferation of Organisms, 427
Vienna Convention on Protection of the Ozone Layer, 361
Vietnam War, 83, 304
Virtual Community (Rheingold), 121
virtualization, 114, 415, 477, 500
virus-genetic theory, 137
Vygotsky, Lev, 320

Wall Street Crash, 207, 412
Wallace, Alfred, 322
war, 479
War Crimes and Collective Wrongdoing (Jokic), 281
water resources, 153, 481
weapons:
  of mass destruction, 9, 21, 83, 95, 108, 245, 282, 319, 244, 350, 354, 482
    global partnership against, 484
weathering, 485
Westphalian world system, 374

What Is to Be Done? (Chernyshevsky), 89
wide-area networks, 487
Wilson, E. O., 315
Wilson, Woodrow, 134
wisdom, 5, 59 132, 160, 198, 239, 260, 302, 373, 393
Workmen Act of 1947, 414
world:
  Bank, 21, 152, 166, 175, 267, 284, 286, 289, 396, 436
  civilization, 21, 56, 63, 70, 75, 79, 108, 144, 209, 212, 233, 335, 378
  dynamics, 249, 488
  federalism, 245, 491
  heritage, 491
  order models, 492
  political forum, 494
  trade organization, 494
  wars, 496
World:
  Business League, 436
  Citizens Movement, 245
  Climate Program, 432
  Constitution and Parliament Association, 245
  Council of Churches, 391
  Drugs Report (2012), 141
  Economic Forum, 490
  Federalist Movement, 245
  Food Summit, 267
  Future Society, 491
  Future Studies Federation, 491
  Mundialist Movement, 245
  Order Models Project, 223, 492
  Summit on Sustainable Development, 399, 434
  Union, 245
  War I, 123, 143, 165, 194, 232, 245, 289, 296, 428, 482, 496
  War II, 43, 83, 97, 109, 134, 143, 192, 208, 245, 259, 263, 283, 304, 308, 375, 379, 386, 388, 444, 471, 482, 491, 496

xenogamy, cultural/political, 119, 497
xenophobia, 391, 497

Yawning Heights (Zinovyev), 500
youth, 499

Z-Generation, 500
Zinovyev, Aleksandr A.

VIBS

The **Value Inquiry Book Series** is co-sponsored by:

Adler School of Professional Psychology
American Indian Philosophy Association
American Maritain Association
American Society for Value Inquiry
Association for Process Philosophy of Education
Canadian Society for Philosophical Practice
Center for Bioethics, University of Turku
Center for Professional and Applied Ethics, University of North Carolina at Charlotte
Central European Pragmatist Forum
Centre for Applied Ethics, Hong Kong Baptist University
Centre for Cultural Research, Aarhus University
Centre for Professional Ethics, University of Central Lancashire
Centre for the Study of Philosophy and Religion, University College of Cape Breton
Centro de Estudos em Filosofia Americana, Brazil
College of Education and Allied Professions, Bowling Green State University
College of Liberal Arts, Rochester Institute of Technology
Concerned Philosophers for Peace
Conference of Philosophical Societies
Department of Moral and Social Philosophy, University of Helsinki
Gannon University
Gilson Society
Haitian Studies Association
Ikeda University
Institute of Philosophy of the High Council of Scientific Research, Spain
International Academy of Philosophy of the Principality of Liechtenstein
International Association of Bioethics
International Center for the Arts, Humanities, and Value Inquiry
International Society for Universal Dialogue
Natural Law Society
Philosophical Society of Finland
Philosophy Born of Struggle Association
Philosophy Seminar, University of Mainz
Pragmatism Archive at The Oklahoma State University
R.S. Hartman Institute for Formal and Applied Axiology
Research Institute, Lakeridge Health Corporation
Russian Philosophical Society
Society for Existential Analysis
Society for Iberian and Latin-American Thought
Society for the Philosophic Study of Genocide and the Holocaust
Unit for Research in Cognitive Neuroscience, Autonomous University of Barcelona
Whitehead Research Project
Yves R. Simon Institute

Titles Published

Volumes 1 - 240 see www.rodopi.nl

241.   William Sweet and Hendrik Hart, *Responses to the Enlightenment: An Exchange on Foundations, Faith, and Community.* A volume in **Philosophy and Religion**

242.   Leonidas Donskis and J.D. Mininger, Editors, *Politics Otherwise: Shakespeare as Social and Political Critique.* A volume in **Philosophy, Literature, and Politics**

243.   Hugh P. McDonald, *Speculative Evaluations: Essays on a Pluralistic Universe.* A volume in **Studies in Pragmatism and Values.**

244.   Dorota Koczanowicz and Wojciech Małecki, Editors, *Shusterman's Pragmatism: Between Literature and Somaesthetics.* A volume in **Central European Value Studies**

245.   Harry Lesser, Editor, *Justice for Older People,* A volume in **Values in Bioethics**

246.   John G. McGraw, *Personality Disorders and States of Aloneness (Intimacy and Aloneness: A Multi-Volume Study in Philosophical Psychology, Volume Two),* A volume in **Philosophy and Psychology**

247.   André Mineau, *SS Thinking and the Holocaust.* A volume in **Holocaust and Genocide Studies**

248.   Yuval Lurie, *Wittgenstein on the Human Spirit.* A volume in **Philosophy, Literature, and Politics**

249.   Andrew Fitz-Gibbon, *Love as a Guide to Morals*. A volume in **Ethical Theory and Practice**

250.   Ronny Miron, *Karl Jaspers: From Selfhood to Being*. A volume in **Studies in Existentialism**

251.   Necip Fikri Alican, *Rethinking Plato: A Cartesian Quest for the Real Plato*. A volume in **Philosophy, Literature, and Politics**

252.   Leonidas Donskis, Editor, *Yet Another Europe after 1984: Rethinking Milan Kundera and the Idea of Central Europe.* A volume in **Philosophy, Literature, and Politics**

253.   Michael Candelaria, *The Revolt of Unreason: Miguel de Unamuno and Antonio Caso on the Crisis of Modernity.* A volume in **Philosophy in Spain**

254.   Paul Richard Blum, *Giordano Bruno: An Introduction.* A volume in **Values in Italian Philosophy**

255.   Raja Halwani, Carol V. A. Quinn, and Andy Wible, Editors, *Queer Philosophy: Presentations of the Society for Lesbian and Gay Philosophy, 1998-2008.* A volume in **Histories and Addresses of Philosophical Societies**

256.   Raymond Angelo Belliotti, *Shakespeare and Philosophy: Lust, Love, and Law.* A volume in **Philosophy, Literature, and Politics**

257.   Jim Kanaris, Editor, *Polyphonic Thinking and the Divine.* A volume in **Philosophy and Religion**

258.   Michael Krausz, *Oneness and the Displacement of Self: Dialogues on Self-Realization.* A volume in **Interpretation and Translation**

259.   Raymond Angelo Belliotti, *Jesus or Nietzsche: How Should We Live Our Lives?* A volume in **Ethical Theory and Practice**

260.   Giorgio A. Pinton, *The Conspiracy of the Prince of Macchia & G. B. Vico.* A volume in **Philosophy, Literature, and Politics**

261.   Mechthild E. Nagel and Anthony J. Nocella II, Editors, *The End of Prisons: Reflections from the Decarceration Movement.* A volume in **Social Philosophy**

262.   Dorota Koczanowicz, Leszek Koczanowicz, and David Schauffler, Editors, *Discussing Modernity: A Dialogue with Martin Jay.* A volume in **Central European Value Studies**

263.   Pekka Mäkelä and Cynthia Townley, Editors, *Trust: Analytic and Applied Perspectives.* A volume in **Nordic Value Studies**

264.   Krzysztof Piotr Skowroński, *Beyond Aesthetics and Politics: Philosophical and Axiological Studies on the Avant-Garde, Pragmatism, and Postmodernism.* A volume in **Central European Value Studies**

265. David C. Bellusci, *Amor Dei in the Sixteenth and Seventeenth Centuries*. A volume in **Philosophy and Religion**

266. Vasil Gluchman, Editor, *Morality: Reasoning on Different Approaches*. A volume in **Ethical Theory and Practice**

267. Jakob Lothe and Jeremy Hawthorn, Editors, *Narrative Ethics*. A volume in **Philosophy, Literature, and Politics**

268. Greg Moses and Gail Presbey, Editors, *Peace Philosophy and Public Life: Commitments, Crises, and Concepts for Engaged Thinking*. A volume in **Philosophy of Peace**

269. Bartholomew Ryan, *Kierkegaard's Indirect Politics: Interludes with Lukács, Schmitt, Benjamin and Adorno*. A volume in **Philosophy, Literature, and Politics**

270. Patricia Hanna, Editor, *Reality and Culture: Essays on the Philosophy of Bernard Harrison*. A volume in **Interpretation and Translation**

271. Piotr Nowak, *The Ancients and Shakespeare on Time: Some Remarks on the War of Generations*. A volume in **Philosophy, Literature, and Politics**

272. Brian G. Henning and David Kovacs, Editors, *Being in America: Sixty Years of the Metaphysical Society*. A volume in **Histories and Addresses of Philosophical Societies**

273. Hugh P. McDonald, *Environmental Philosophy: A Revaluation of Cosmopolitan Ethics from an Ecocentric Standpoint*. A volume in **Studies in Applied Ethics**

274. Elena Namli, Jayne Svenungsson and Alana M. Vincent, Editors, *Jewish Thought, Utopia, and Revolution*. A volume in **Philosophy and Religion**

275. Wojciech Małecki, Editor, *Practicing Pragmatist Aesthetics: Critical Perspectives on the Arts*. A volume in **Central European Value Studies**

276. Alexander N. Chumakov Ivan I. Mazour and William C. Gay, Editors, *Global Studies Encyclopedic Dictionary*. A volume in **Contemporary Russian Philosophy**

Printed in the United States
By Bookmasters